Richard Pococke's Letters from the East (1737-1740)

The History of Oriental Studies

VOLUME 9

The titles published in this series are listed at *brill.com/hos*

Richard Pococke's Letters from the East (1737-1740)

By

Rachel Finnegan

BRILL

LEIDEN | BOSTON

Cover illustration: "Portrait de l'Archéologue et Théologien Richard Pococke", miniature red-and-black chalk drawing by Jean-Étienne Liotard. Courtesy of Louvre, Paris: Fonds des Dessins et Miniatures (RF 1379, Recto).

Library of Congress Cataloging-in-Publication Data

Names: Finnegan, Rachel.
Title: Richard Pococke's letters from the East (1737-1740) / by Rachel Finnegan.
Description: Leiden ; Boston : Brill, 2021. | Series: History of Oriental Studies, 2405-4488 ; volume 9 | Includes bibliographical references and index. | Summary: "In Richard Pococke's Letters from the East (1737-1740), Rachel Finnegan provides edited transcripts of the full run of correspondence from Richard Pococke's famous Eastern voyage from 1737-41.
In this new volume, Finnegan combines updated biographical accounts of the traveller and his correspondents (his mother, Elizabeth Pococke and his uncle and patron, Bishop Thomas Milles) from vol. 1 of the original edition of Letters from Abroad (2011) with transcriptions of the letters from vol. 3 of the series (2013), together with new material that has hitherto been unpublished. Thus, in a single volume, she sets the context of the life and times of the traveller and his family against the background of this voluminous corpus of fascinating correspondence, which can be read in conjunction with Pococke's own published account of his travels, A Description of the East and Some Other Countries (1743-45)"-- Provided by publisher.
Identifiers: LCCN 2020037610 (print) | LCCN 2020037611 (ebook) | ISBN 9789004440043 (Hardback : acid-free paper) | ISBN 9789004440050 (eBook)
Subjects: LCSH: Pococke, Richard, 1704-1765--Travel. | Pococke, Richard, 1704-1765--Correspondence. | Voyages and travels--Early works to 1800. | Middle East--Description and travel--Early works to 1800.
Classification: LCC G460 .R54 2020 (print) | LCC G460 (ebook) | DDC 915.604/15092--dc23
LC record available at https://lccn.loc.gov/2020037610
LC ebook record available at https://lccn.loc.gov/2020037611

Typeface for the Latin, Greek, and Cyrillic scripts: "Brill". See and download: brill.com/brill-typeface.

ISSN 2405-4488
ISBN 978-90-04-44004-3 (hardback)
ISBN 978-90-04-44005-0 (e-book)

Contents

Preface

By the time Richard Pococke travelled in the Eastern Mediterranean, Westerners were no longer an uncommon sight for many Ottomans. Initially he wore Oriental clothes over his European clothing, but soon after he had arrived in Egypt he switched to Egyptian garb altogether. Even then Pococke will have remained recognizable as a foreigner, but that was not necessarily problematic. Travellers from Europe seldom went out alone, being accompanied either by a longer-term Western resident (often his or her host or someone from his household) or by one of the Janissaries (Ottoman infantrymen) who invariably guarded the 'consular house'. It should also be remembered that Pococke travelled with "an excellent servant turned of 50, very faithfull & diligent, he lived 12 years in Constantinople, with the Emperors Ambassadors as a steward, talks Turkish, Italian, German, French, Latin". It is typical of the period that Pococke informed his mother that the man was from Friuli, but does not tell her his name. The correspondence shows that the two men often went on trips together. He was evidently a good shot, because in Samos he "killed 3 larks for my supper" (many Westerners hunted for sport in the Ottoman Empire), and when they were in quarantine in Messina together, it was this man who would prepare Pococke's food. More importantly, which routes Pococke took to the sites he visited and, when it came to encounters with Ottomans, what he understood of what was said, will largely have depended on this anonymous intermediary. His importance for the Letters should therefore not be underestimated.

Pococke's servant from Friuli will also have been an indispensable guide to the diplomatic infrastructure in 'the Levant' as the Eastern Mediterranean was often called in the West in the premodern period. Because a network of Western consuls and vice-consuls had already existed in the region for several centuries. By Pococke's time British 'diplomatic representatives' were found across the Ottoman Empire, and his correspondence shows that he invariably visited them first whenever arriving in a new location. This was exactly what was expected of new arrivals; visitors were often few and far between, so everyone wanted to hear the 'latest' news from home – or at least Christendom. In return the traveller would invariably be invited to lodge with the consul in the consular house. Most consulates were only institutionalized in the nineteenth century, so in Pococke's time the consulate was wherever the consul chose to live. In many cases it was only practical to take over one's predecessor's home, so in practice there was usually some degree of continuity, even if it was not the result of any kind of formal policy. Travellers staying with 'the consul' also tended to eat at his table – the were no female consuls yet in this period – and enjoy other benefits of being part of his household, like borrowing a Janissary guard when going on trips.

Pococke's letters show that this system worked very well in the 1730s. Although a great deal of research has been done on these consular networks in the pre-modern period, Pococke also provides new information about who was acting as consul where. This is particularly true for Egypt, which was not very popular among European merchants because the relative independence of the local rulers from the sultan in Istanbul often meant that the commercial privileges Western traders enjoyed in the Ottoman Empire – known as *adhnames* in Turkish and 'capitulations' in the West – were not honoured as consistently in Egypt as elsewhere in the Levant.

There was a second mechanism that also ensured that few European travellers arrived in any Ottoman centre of trade without an introduction: the delivery of mail. Wherever a traveller would arrive and tell people about his intended itinerary he would be entrusted with letters to be handed over personally to family members, friends, or business contacts. So delivering letters was as much a part of everyday life for a Western traveller in the Ottoman Empire as writing them. In Akhmin, in Upper Egypt, Pococke even delivered a letter from a high-ranking Ottoman official – not actually the Governor-General of Egypt, that was Abu Bekir Paşa in 1737, but evidently someone very powerful behind the scenes – to the Amir of the Bedouin in the area, explaining to his mother how the mechanism of delivering post worked:

> I went this evening to wait on the Emir or Prince of Akmim. [...] I delivered my letter to him, from Osman Bey, who is in reality the Governor of all Egypt: – He received me with great civility, said he had a regard for the Franks, told me without the letter, he would have done anything for me, in his power, much more with it; told me he would send his own people with me whenever I pleased, & bid me be afraid of nothing, he would do me what service he could, both now & on my return, & offered me a letter to another person, & received very kindly the little present I made of Venice glass & 2 or 3 boxes of sweetmeats.

The Prince's assurances of safety were not hollow phrases. In the early-modern period travelling was hazardous by definition. Contrary to today, it was unusual for individuals to move around without a clear purpose, so travellers raised suspicions. This was the case in Europe as much as in the Ottoman Empire, where a travel permit from the central Ottoman authorities in Istanbul was needed. Pococke had applied to the British embassy in Istanbul to arrange this travel permit (T. *yol emri*) for him and send it to Venice, where it was indeed waiting for him when he arrived there.

This 'passport' was a personal document that was valid for Pococke and his retinue (in his case, just one servant) for a specific number of places. The exact route was not mentioned in the Ottoman document, which allowed the traveller to wander about unhindered by the local authorities. The travel permit also meant that Pococke could make use of the Ottoman roads and the caravanserais along them, where lodgings and refreshments were provided free of charge. He only seems to have used them once or twice (e.g. en route from Alexandria to Cairo), however, usually finding shelter elsewhere.

The travel permit guaranteed that Pococke would not be bothered by the local representatives of the Ottoman government, but in many locations this was not enough. In Akhmim, as we have seen, Bedouins effectively controlled the area and their protection good not be taken for granted simply because Pococke was allowed to travel. The Bedouins had interests of their own, which did not always coincide with those of the central authorities. If any local power faction – be it the Bedouins in Egypt, or the Kurds in the vicinity of Iskenderun (Alexandretta) – decided to rob an individual European merchant or traveller, then obtaining satisfaction was practically impossible. Pococke not only made himself useful to local notables like the Emir of Akhmim, he was also wise enough to have small presents for them, which also tended to work well. Later Pococke also delivered letters for other Ottoman and Egyptian dignitaries, while doing the same for the many Catholic missionaries he met on his journey. The missionaries formed an additional network in the Levant that a traveller like Pococke relied upon; they tended to have convents and houses of their own, where guests were welcomed irrespective of their religious background.

The traveller bringing letters was also often offered coffee (mentioned no less than 190 times in the *Letters from the East*), and sometimes he would be invited for a meal. Pococke is exceptionally informative about such occasions. Where other travellers tended to focus on the sites – which in many cases had already been described extensively by previous visitors from the West – Pococke also includes copious notes about what he ate and drank. For example, in Cairo,

> I drank Caffelet at Mr Congreves, made in this manner; – first you make Coffee of what strength most agreeable, you may try, let it settle & pour it off, & then with that Coffee, make Chocolate, about half the quantity you would boyl in so much water, well sweetened; – it is esteemed more wholesome than either Coffee or Chocolate.[1]

1 Francis Congreves, presumably a merchant, was a recent arrival in Egypt himself, having settled there earlier in 1739 and staying on until 1741 The family archives of the Congreves are

This almost reads like a recipe. Did Pococke imagine his mother making the drink for herself at home, or having it made for her in accordance with his instructions? In any case, Pococke tended to have three meals a day; breakfast, dinner – the principal hot meal around what we now call lunch time – and supper in the evening. From Alexandria he informed his mother that:

> this country is very plentifull & a table well served & well dressed – fish in abundance, good shrimps & oysters, small but sweet. Egyptian onions, with the inside scooped out & filled with forced meat, an excellent dish, & cucumbers in the same manner: – at supper an Egyptian custard, which is milk turned so as to be like sowre milk, 'tis turned with a [...], made of a seed they feed birds with, & is a litle sowr but not disagreeable; – as to the rest after the French, & English way. A water melon, Pomegranates after every meal, & sometimes those rich figs growing about Rosetto [Rosetta], cultivated from the more Eastern parts of the leaves of which they say, our first parents made their aprons, in shape like a date, four inches long, an inch diameter being eat alone, 'tis in the mouth like Almond flummery [pudding], & tastes most like the richest pears, & eat with white wine & sugar, something like that flummery: we have also Dates, rough & not agreeable to me, the fruit of the Date or Palm tree but different from the Palm of the West Indies, of which they make the wine. The Tree here is most useful, of the branches they make strong baskets of the leaves slight ones, of the bark ropes, which & the wood are not susceptible of the plague: the wood is good, the Dates men eat, the kernel they grind & give to camels & in some places make bread of it, & also a paste which is made into balls being well dried.

Pococke's *Letters* constitute a surprisingly human mix of perceptive observations of mundane elements of life in the Levant like these –on foodstuffs in Egypt, and the local use of fig leaves to make baskets – and reassurances from a far-away son to his mother at home that he is taking good care of himself. For that reason this edition of Pococke's correspondence by Rachel Finnegan is of great value for the modern reader who is interested not only in reports about the exact measurements of the pyramids, the state of certain pharaonic temples, and the availability of ancient coins and seals from local antiquities' salesmen, but also in the every-day lives of Europeans in foreign parts in the early modern period.

Maurits van den Boogert

---

kept in the Staffordshire and Stoke-on-Trent Archive Service: Staffordshire County Record Office.

Acknowledgements

I wish to thank the following organisations and individuals for their valuable assistance in the production of this book. First, I thank the British Library Board for permission to transcribe Pococke's eastern travel manuscripts in full (Add MS 22998, Add MS 15779, Add MS 22995 and Add MS 15774), and to cite and quote from other manuscripts both from the Pococke collection and from other related collections (Add MS 22995, Add MS 4824, Add MS 14263, Add MS 52362).

I am also grateful to the boards of the following institutions for allowing me to cite and quote from records in their collections, as listed in the Bibliography: the National Maritime Museum, London, St Canice's Cathedral, Kilkenny, the National Archives of the UK, the National Library of Ireland, Hampshire Archives and Local Studies, Derbyshire Record Office, Waterford Museum of Treasures, Warwickshire County Record Office, the Bodleian Library, the Royal Irish Academy, Waterford City Archives, the National Archives of Ireland, the Fonds des Dessins et Miniatures, Louvre, the Cotton Library, St Carthage's Cathedral, Lismore, and the Royal Society.

I am also very grateful to the Very Reverend Katharine Poulton, former Dean of St Canice's Cathedral, Kilkenny and the Very Reverend Paul Draper, Dean of the Cathedral and Rector of Lismore Union of Parishes, for permission to consult the respective cathedral archives.

Regarding the illustrations, I thank the Fonds des Dessins et Miniatures, Louvre Museum, Paris for permission to reproduce Jean-Etiènne Liotard's miniature red-and-black chalk drawing of Richard Pococke on the cover of this volume; David Rymill and the Earl of Carnarvon for permission to reproduce the portrait of Revd Isaac Milles, from the collection in Highclere Castle; Donnchadh Ó Ceallacháin and Waterford Treasures for permission to reproduce the painting of Waterford Cathedral; the owner of the early photograph of Newtown House, Newbury (Mrs Pococke's house), who wishes to remain anonymous; Dr Susan Hood, Librarian & Archivist at the Church of Ireland Representative Church Body Library, for permission to reproduce the portrait of Bishop Pococke from the Episcopal Portrait Collection; and Tony Hand for permission to reproduce his photo of the Montgomery Monument, in Ardbraccan (Pococke's grave).

I also thank Canon Adrian Empey, for advice on matters of Irish church history and Professor Luigi Lehnus of the University of Milan, for help in deciphering Italian proverbs in the manuscripts.

I am indebted to Dr David Wilson for producing and making available to me a document entitled "More Notes on Richard Pococke's Letters from Abroad, vol. 3", in which he identified some of the hitherto unidentified consular officials and merchants in the Levant, from his reading of my original book (2013).

My greatest debt of gratitude is to Maurits H. van den Boogert, who continued in this quest and, through dedicated research into many obscure manuscripts and published works, managed to identify numerous individuals who featured in the letters. I am also grateful to him for writing the Preface to this volume.

I wish to thank my husband, David Kane, for his photographs of the portrait of Richard Pococke and of all the illustrations from *A Description of the East and Some Other Countries.* I also thank David and our sons Sammy, Jacob and Ben for their enduring patience.

This volume is dedicated to Professor George Leonard Huxley, MRIA, who encouraged and advised me throughout the whole of the original undertaking (the series of three volumes of Pococke's travel correspondence) and has continued to do so in my subsequent academic and editorial work.

Rachel Finnegan

Illustrations

CHAPTER 1

Introduction

1 General Background

This is a second, revised edition of the eastern travel correspondence of the Revd Dr Richard Pococke, originally published in 2013 as the third volume in a series entitled *Letters from Abroad: The Grand Tour Correspondence of Richard Pococke & Jeremiah Milles*.[1] Whereas the first edition of volume 3 reproduced only the eastern letters from Pococke to his mother (excluding five, which were summarised), the revised volume contains the whole collection of eastern letters from the author to his mother, Mrs Elizabeth Pococke, née Milles, who resided in the parish of Newtown (near Newbury), in Berkshire, England. Also added was the entire corpus of letters or "accounts" sent by Pococke to his uncle and patron, Bishop Thomas Milles, who occupied the Episcopal House in Waterford, Ireland.[2] This revised edition also contains an updated introduction and biographical accounts of the three individuals concerned (Pococke and the two recipients of his letters).[3] These biographies were included in volume 1 of *Letters from Abroad*,[4] but have been substantially revised here. The original edition also contained a biographical account of Jeremiah Milles, who was Pococke's travel companion on his earlier travels. However, as Milles turned back before Pococke journeyed to the east, it was not thought necessary to include his biography here, though a brief note has been included below.[5]

1 R. Finnegan, *Letters from Abroad: The Grand Tour Correspondence of Richard Pococke & Jeremiah Milles, Volume 3, Letters from the East, 1737-41*, Pococke Press, Piltown, 2013, hereafter Finnegan, *Letters 3*.

2 Between 2014 and 2017, some of the eastern letters excluded from Finnegan, *Letters 3* were edited and posted on the Pococke Press website and on the author's Academia.edu site.

3 At the time of his eastern voyage, Pococke (1704-65) was Precentor of Lismore Cathedral and later became Archdeacon of Dublin (1745-55), Bishop of Ossory (1756-65) and Bishop of Meath (1765); Dr Thomas Milles (1671-1740) was Bishop of Waterford & Lismore from 1708-40; and Mrs Elizabeth Pococke (?-1760), who was Bishop Milles's widowed sister and Richard Pococke's mother, lived in England with her unmarried daughter.

4 See Chapter 1 of R. Finnegan, *Letters from Abroad: The Grand Tour Correspondence of Richard Pococke & Jeremiah Milles, Volume 1, Letters from the Continent, 1733-34*, Pococke Press, Piltown, 2011, hereafter Finnegan, *Letters 1*.

5 Jeremiah Milles (1714-84) was Pocococke's younger cousin (his father, Revd Isaac Milles, was a brother of Pococke's mother) and was educated at Eton and Corpus Christi College, Oxford. The cousins journeyed together to France and Italy (1733-34) and then to France, the Low Countries, Germany, Austria, Poland, Hungary and the Balkans (1735-37). While the intention

 | DOI:10.1163/9789004440050_002

Pococke became a notable egyptologist, antiquarian and collector, sealing his fame with his pioneering book *A Description of the East and Some Other Countries*, published in two lavishly illustrated folio volumes in 1743 and 1745.[6] Why it took more than 270 years for Pococke's grand tour correspondence to be published is not clear, particularly when his fame rests largely on his eastern voyage, and given that the assorted manuscripts have been in the British Library since the 19th century. For those who are familiar with *A Description of the East*, it may be of interest to compare the letters with the printed version of this voyage. For those who do not have access to this rare set (though it is now available online), the collection of letters reproduced in the present volume will give a complete account of the travels, observations and experiences of this enthusiastic and slightly eccentric clergyman, who was dubbed by one of his Victorian biographers, "an Englishman by birth, and an Irishman by adoption".[7] As outlined below, the conversational letters addressed to his mother are in a diary format and tell of personal matters such as the friendships he formed while abroad, the nature and significance of his ever-changing attire, and the progress of his beard and "whiskers", together with descriptions of his journeys, accommodation arrangements, the climate, the flora and fauna of the east and, in particular, its food and wine. They also contain valuable information about his collecting methods and his growing collections of antiquities, natural curiosities, botanical specimens and coins. In contrast, the letters to his uncle, Bishop Milles, constitute more formal and detailed accounts of the history, geography and architecture of the places he visited, with minimal ornamentation or personal detail. To differentiate the two collections in this volume, individual letters to Mrs Pococke are referred to as "Letter 1", "Letter 2", etc., whereas those to Bishop Milles are referred to as "Account 1", "Account 2",

was for Jeremiah Milles to continue with Pococke on his voyage to the east, he was summoned back to Ireland to tend to his ailing uncle Bishop Milles and did so because he was his intended heir. At the time of his second grand tour, Jeremiah was Treasurer of Lismore Cathedral. He later became Dean of Exeter (1762-84) and was President of the Society of Antiquaries of London from 1769 until his death. For further biographical details, see Finnegan, *Letters 1*, pp. 56-66. See also R. Finnegan, *Letters from Abroad: The Grand Tour Correspondence of Richard Pococke & Jeremiah Milles, Volume 2, Letters from the Continent, 1736-37*, Pococke Press, Piltown, 2012, hereafter Finnegan, *Letters 2*.

6 The full titles of these books are given in the Bibliography and henceforth in the footnotes are abbreviated to *Description 1* and *Description 2*, with reference to the appropriate volume and section. However, when the book is mentioned in general, the title is abbreviated to *A Description of the East*.

7 See Daniel William Kemp (ed.), *Tours in Scotland 1747, 1750, 1760 by Richard Pococke Bishop of Meath*, Publications of the Scottish History Society, vol. 1, University Press, Edinburgh, 1887; Reprinted Maryland, 2003, p. xxxi.

etc., "accounts" being the word used by Pococke when updating his mother on his correspondence with his uncle.

2 Pococke's Eastern Travels in Context

Most English and Irish grand tourists of the 18th century were content to visit the classical and renaissance sites of Italy, and it did not become fashionable to "wish" for "something more than just the common course of what is called the tour of Europe" until the period 1763-97.[8] While there was of course much to be seen in the Eastern Mediterranean, a voyage to such distant and unfamiliar regions (including Greece, which had been under Ottoman rule since 1453) was not only considered arduous and costly, but was also perceived as hazardous and difficult.[9] The few existing eastern travel books did little to encourage such ventures, since they frequently contained accounts "in which dangers [were] made or magnified."[10] For a detailed account of the historical and literary contexts of the eastern voyage of the period, the reader is referred to the introductory chapters of a recent publication on the travels and travel books of three contemporary writers: the Reverend Dr Thomas Shaw, Dr Charles Perry, and Pococke.[11]

For those few who were intrepid and curious enough to make such a journey (many of whom were members of the clergy), travel to the east generally involved the academic pursuits of antiquarianism (whether classical, Egyptian or biblical) and linguistics. Pococke, who has been described as the "first of the new kind of traveller" to visit such parts,[12] was interested in all these pursuits. In particular, he is recognised as a pioneer of Homeric topography (a field considered by Lord Byron to be "antiquarian twaddle"),[13] and it is thought that his identification of the Plain of Troy inspired Robert Wood, 15 years later, to make his more penetrating examination of the Troad, in his book, *An Essay on the Original Genius and Writings of Homer*.[14]

---

8 See Edward Chaney, who cites Boswell, in *The Evolution of the Grand Tour*, Frank Cass, London, London, 1998, p. 125.

9 For an interesting account of the dangers and discomforts of travel in Greece in Ottoman times, see George Huxley, *Homer and the Travellers*, Athens, 1988, pp. 5-7.

10 See Letter 10, where Pococke warns his mother not to read such travels.

11 R. Finnegan, *English Explorers in the East (1738-1745). The Travels of Thomas Shaw, Charles Perry and Richard Pococke*, Leiden, 2019.

12 See W.B. Stanford, *Ireland and the Classical Tradition*, Dublin, 1976, p.133.

13 See Huxley, *Homer and the Travellers*, p. 11.

14 Published in London, 1775. See J.V. Luce, "Robert Wood and Homer", in J.V. Luce, C. Morris and C. Souyoudzoglou-Haywood (eds), *The Lure of Greece: Irish Involvement in Greek*

However, it should not be supposed that the only English and Irish visitors to the "Sultan's dominions" were grand tourists and/or scholars, or that these were the first Westerners to the east. On the contrary, the initial interest was kindled during the Crusades, whose pilgrims travelled on routes and laid down communications that were to be used for generations. With the rise of British sea power during the war of the Spanish Succession, the Mediterranean was opened up to commerce, principally involving the sale of English cloth. The Levant Company, formed in 1581, had factories in Smyrna (Izmir), Constantinople (Istanbul) and Aleppo, the latter centre serving as the company's headquarters in the Middle East. So important was this company that it led to the establishment of an English ambassador at Constantinople, which in turn made travel to the Ottoman Empire less difficult. Other (and often related) reasons why the British spent time in the east included taking part in naval campaigns in the Mediterranean; serving as captains of ships trading in these parts; working as "factors" or agents operating the "factories" (or communities of merchants) connected with the textiles trade; or serving as consuls, physicians or chaplains to the English factories.

Many of these individuals (particularly the chaplains and physicians), together with family members who visited them, were engaged in scholarly pursuits and examples of all these types feature in Pococke's letters, since they were the people with whom he mixed when staying in cities with English trading connections, and from whom he received "great civilities". Not only was he wined and dined in true English style, and invited to play cards and attend concerts,[15] but he was looked after in other ways: being accompanied around the antiquities; being escorted to places where he could copy inscriptions or acquire medals; being given letters of recommendation to British officials and other residents elsewhere in the Levant; and on his departure from certain places, being ceremoniously accompanied by his compatriots for some considerable distance along the road.

He particularly favoured Aleppo, a city from which he sent three letters to his mother, and soon after his arrival was able to assure his uncle: "The short time I shall stay [here] I find will be very agreeable, as I receive all manner of

---

Culture Literature History and Politics, Hinds Publishing, Dublin, 2007, p. 85; and J.M. Cook, *The Troad: An Archaeological and Topographical Study*, Oxford University Press, Oxford, 1973, Special Edition, 1999, p. 20.

15 As noted by Wood: "The English merchants carried with them into the Levant as much as was possible of the social life and recreations of their native land; and society in the factories was a faithful, though necessarily imperfect, reflection of that in England.... In good living and conviviality they surpassed those of any other nation." See Alfred C. Wood, *A History of the Levant Company*, Oxford, 1935, p. 240.

civilities from the Gentlemen of our nation, who in every respect answer the character of their Preceptors, and make one of the most agreeable and easy societies in the world."[16] Though he rarely refers to their forenames, it has been possible to identify most of the merchants, agents, clergymen and consular officials mentioned in the letters, by consulting modern works of scholarship on the British in the Levant and by seeking the help of experts.[17]

The hospitality Pococke experienced in the east was not confined to the English: he was treated with cordiality both by locals (sheiks, emirs and villagers alike) and by other resident foreigners, in particular the dignitaries of the various religious missionary orders he encountered during his travels, especially the Franciscans. He spent much time, for example, travelling and socialising with Father Jacomo (a Bohemian) and Father Francisco (a Piedmontese), the former being the "Procurator, Prefect or Superior" of the the Fransciscan missionaries in Egypt, stationed in Cairo, Father Francisco having previously held that position. Father Jacomo, described as "a great traveller all the world over" (Letter 5), had written a journal of a voyage made in 1722 from Cairo to Mount Sinai, which is mentioned in Pococke's book.[18] The manuscript was subsequently acquired and translated by Revd Robert Clayton, Bishop of Clogher, and published in 1753.[19] It is not clear how Bishop Clayton obtained this manuscript, but Pococke (who is acknowledged in the first paragraph of the work) may well have had some involvement, since he was at the time Archdeacon of Dublin and is believed to have lived in the vicinity of Clayton House, on St. Stephen's Green (see Chapter 2).[20]

16 Pococke to the bishop, Aleppo, 31 July/12 August 1738 (BL, Add MS 15779, reproduced below as Account 19). Pococke is cited as one of the visitors to Aleppo who painted an idyllic picture of life there, even though "every factor whose letters survive was longing for the day when his fortune would be adequate for him to return home." See Ralph Davis, *Aleppo and Devonshire Square: English Traders in the Levant in the Eighteenth Century*, Macmillan, London, 1967, p. 80 note 1.

17 David Wilson's *List of British Consular Officials in the Ottoman Empire and Its Former Territories, from the Sixteenth Century to about 1860* (available at: <http://www.levantine heritage.com/>) has been an invaluable resource in this regard, as has Davis, *Aleppo and Devonshire Square*. Above all, I am indebted to Maurits H. van den Boogert for the generous assistance he has given me in identifying so many of the individuals in the letters who went unidentified in the first edition of this correspondence.

18 *Description 1*, p. 147.

19 *A Journal from Grand Cairo to Mount Sinai and Back Again ...*, London, 1753. A copy of this quarto volume is listed as Lot 55 in Pococke's library sale catalogue. See L. Flin, *A Catalogue of the Library of the Late Right Revd. Dr Richard Pococke, Lord Bishop of Meath Deceased*, Dublin, 1766, p. 2.

20 A similar manuscript by another of his associates in the east (Arthur Pullinger, a merchant in Aleppo) was part of a collection of documents bequeathed by Pococke to the British

It is clear from contemporary accounts that those travellers courageous enough to venture into eastern parts in the first half of the 18th century inspired great awe in others. During this period, visitors were so rare that their return to Britain merited a mention in the newspapers;[21] and the inveterate letter writer Revd Joseph Spence, who met John Montagu, 4th Earl of Sandwich, at Turin shortly after his return from the east (1740) remarked:

> A man that has been all over Greece, at Constantinople, Troy, the Pyramids of Egypt, and the deserts of Arabia, talks and looks with a greater air than we little people can do that have only crawled about France and Italy.[22]

As can be seen from the surviving accounts of such journeys, the more scholarly of these Levantine voyagers engaged themselves in investigating, observing and recording the most important aspects of eastern culture. This was particularly true of its monuments and architecture, on which very little, by the middle of the 18th century, had been published in the Western world. While some who braved this lengthy and hazardous journey kept diaries or travelogues for their own personal use, others set out with the express purpose of producing a book on some aspect of Greek or eastern culture. Pococke did both, which is why his testimony is considered such an important source of information for this subject.

Museum on his death: "Geographical and Antiquarian Notes Relating to Greece and Asia Minor", BL, Add MS 4824. See Chapter 4, note 66, below.

21 See Jeremy Black, *The British Abroad: the Grand Tour in the Eighteenth Century*, Sutton Publishing, Stroud, 2003, pp. 66-67.

22 See Slava Klima, ed., *Joseph Spence: Letters from the Grand Tour*, McGill-Queen's University Press, Montreal and London, 1975, pp. 248-49. The earl's journey to Constantinople, via Malta, Sicily, Greece and the Greek Islands, is described in his memoirs, edited by his chaplain, John Cooke, and published posthumously under the title *A Voyage Performed by the Late Earl of Sandwich round the Mediterranean in the Years 1738 and 1739, Written by Himself*, London, 1799. The original manuscript version exists in the National Maritime Museum, entitled "A Voyage Round the Mediterranean in the years 1738 & 1739" (Ref. SAN/F/50). On his return to London, Lord Sandwich and Pococke met and established both the Egyptian Society and the Divan Club, Sandwich proving to be of great help and an inspiration to Pococke in the writing up both volumes of *Description*. Incidentally, Spence cites Pococke's book in a related letter concerning Lord Sandwich, also from Turin (Klima, *Joseph Spence*, p. 268), as well as noting that he had met him in Rome during his (Spence's) third grand tour in 1739. Pococke was also highly aware of the presence of the Earl of Sandwich in the east and mentioned this several times to his mother (see Letters 18, 22 and 43, below).

Whether a trip to the Levant was part of the traveller's original scheme (which was clearly the case with Pococke, see below) or whether (as with the Earls of Sandwich and Bessborough)[23] it evolved in the course of his European sojourn, a considerable amount of planning was necessary to put the idea into action. Particularly important was the organisation of passes, permissions and letters of introduction, together with the associated problems of arranging finances. With the right connections, much of this could be done from home prior to the trip, though such preparations were often made in Italy in the months leading up to the voyage.

Pococke, having spent a year performing the more traditional grand tour of Italy (1733-34), must have begun to make preparations for his eastern voyage soon after his return, since he was to set out again in the summer of 1736 on a tour of the Low Countries and the Austrian Empire, en route to Egypt. As noted in a letter from Venice, "But when I came here I met a Firman[24] or Passport from the Ottoman Porte sent to us by Sr Everard Falkener [Fawkener] the English Ambassador at Constantinople,[25] which encouraged me to undertake the Eastern voyage being well prepared for it in pursuance of our first design."[26]

Although these wealthy travellers could afford to engage the services of a dragoman (a locally recruited interpreter) to see them through many of the practical difficulties encountered in the east, connoisseurs and scholars such as Pococke found it beneficial to gain, beforehand, some knowledge of the

---

23 See R. Finnegan, "The Classical Taste of William Ponsonby, 2nd Earl of Bessborough", in *Irish Architectural and Decorative Arts. The Journal of the Irish Georgian Society* 8 (2005), pp. 12-43.

24 An official Ottoman document permitting the bearer to travel.

25 Sir Everard Fawkener (1684-1758) had early connections with the Levant through his family trading company and in 1716 was sent to Aleppo, where he stayed for nine years. He arrived in Constantinople in 1735, shortly after the Ottoman Empire had declared war on Russia. However, after the Treaty of Nemirov, in 1737, the French dominated the negotiations, leaving the British and Dutch embassies very little to do, and in 1742 Fawkener asked to be recalled home for a time but did not manage to return. His appointment was formally terminated in 1746. Having returned to London, he became a member of the Divan Club, an oriental society founded by several returning eastern voyagers, including Pococke (see below). However, he attended very few sessions, as his new post as Secretary to the Duke of Cumberland required his presence outside London. See R. Finnegan, "The Divan Club: An 18th-Century Club for Travellers to the Ottoman Empire", in *Electronic Journal of Oriental Studies*, 9, 2006, pp. 47-48. Available at: <http://www.pocockepress.com/DivanClub.pdf>.

26 Pococke to Bishop Milles, Venice, 19/30 August 1737, BL, Add MS 15774. See *Finnegan, Letters* 2, p. 257. This letter from Pococke mysteriously found its way into the volume of correspondence from Jeremiah Milles to Bishop Milles.

Turkish language.[27] Others, like Lord Sandwich, preferred to take lessons when abroad. Furthermore, most foreign visitors would have been advised to familiarise themselves with the particular etiquette required when dealing with foreign officials. As is evident from both his book and his letters, Pococke kept to hand a ready supply of gifts for these officials, including money, food, drink, soap, shoes and even pistols, in order to secure his personal protection and to gain access to antiquities and places of local interest.

An examination of extant libraries and library records (such as library catalogues or the auction catalogues of library sales) shows that travellers to the east had often acquired a considerable amount of relevant literature on Greece and Turkey before they embarked upon their travels.[28] John Bartlett discusses the classical sources and the works of scholarly predecessors available to Pococke for his eastern voyage.[29] His analysis and accompanying tables (based on references from *A Description of the East* and certain books listed in the sale catalogue of the late traveller's library) is extremely useful and in particular illustrates the literature available to Pococke during the writing of his book. However, it is his travel correspondence that indicates precisely which books he consulted (or in some cases tried unsuccessfully to obtain) during his actual voyage. For example, his letters from the Continent, while travelling to Livorno, en route to Egypt, refer to a box containing "some Eastern books", which he wishes his mother to send to him. He also asks her to send on various other volumes, "but not the Spon, Pausanias or Tournefort",[30] which suggests he may

27 Though there is only one reference (Letter 7) to Pococke learning "Arabick" when travelling in the East (and a further reference to his getting "some litle knowledge" of modern Greek, Letter 31), evidence from his library sale catalogue shows that he owned several relevant dictionaries and grammars whose publication pre-dated his eastern voyage. The following titles indicate his interest in oriental languages, even if he may not have put them into practice: Johann Christian Clodius, *Compendium Lexicon Latino-Turcico-Germanicum* (Leipzig, 1730); Simon Ockley, *Introductio ad Linguas Orientales* (London, 1705); J.C. Clodius, *Grammatica Turcica* (Deer, 1729); Martinus Trostius, *Lexicon Syriacum* (Köthen Anhalt, 1623); Petrus Victorius Caietanus Palma, *Paradigmata de Quator Linguis Orientalibus Praecipuis, Arabica, Armenica, Syra, Æthiopica* (Paris, 1596); Johann Joachim Schröderus, *Thesaurus Linguæ Armenicæ, Antiquæet Hoderniæ, cum Varia Praxios Materia* (Amsterdam, 1711) [Pococke's notes say 1660, but Schröder was only born in 1680]; Edward Pococke, *Liturgiæ Ecclesiæ Anglicæ* [...] *in Linguam Arabicam Traductæ* (Oxford, 1674). See Flin, *A Catalogue of the Library*.

28 See, for example, R. Finnegan, "The Classical Library of William Ponsonby, 2nd Earl of Bessborough, 1704-93", in *Hermathena, A Trinity College Dublin Review* (In Honour of George Huxley) 181 (Winter, 2006), pp. 149-87.

29 See John. R. Bartlett, "Richard Pococke in Lebanon, 1738", in *Archaeology & History in Lebanon* 16 (Autumn, 2002), pp. 29ff.

30 Clearly, the famous book by Spon and Wheler arrived, since Lot 485 of the sale catalogue of Pococke's library is described as, "Voyages per Spon & Wheler, 2 Toms, Hay 1724" (see

have acquired these abroad. Instead, he requests Paul Lucas' *Voyage* and George Guillet's book on ancient and modern Athens.[31]

While there is no reference in Pococke's posthumous library sale catalogue to Henry Maundrell's two famous travel books on the east,[32] he clearly had access to this work, since he includes the following entry in his letter from Larnaca (dated 10/21 November 1738): "I set out & viewed the Antiquities of a fossee, with three stairs on each side for seats. I take it to have been an Hippodrome or some place of sports; – then went to an open temple cut in the rock, with a throne in the middle; 4 sepulchral towers, & the grotts under them all mentioned by Mr Maundrel."[33] He also refers to this source several times in his letters or "accounts" to his uncle, Bishop Milles (see below).

While some visitors to the east hired the services of a draughtsman or an artist to take "views" of the many fascinating sites visited,[34] Pococke did not, and though he suggested, in the preface to *Description 1*, that the drawings and plans of Egyptian buildings were his own,[35] there are no surviving sketch books

---

Flin, *A Catalogue of the Library*, p. 13). However, there is no entry in the catalogue for the Greek geographer Pausanias's *Description of Greece*, nor for Joseph Pitton de Tournefort's *A Voyage into the Levant Perform'd by Command of the Late French King, Translated by J. Ozell*, London, 1718.

31 Paul Lucas, *Voyage du Sieur Paul Lucas, Fait en* MDCCXIV, *&c. Par Ordre de Louis* XIV, *Dans la Turquie, l'Asie, Sourie, Palestine, Haute & Basse Egypte, &c., Rouen, 1724; and Guillet de Saint-Georges, Athenes Ancienne et Nouvelle, et l'Estat Present de l'Empire des Turcs, Contenant la Vie du Sultan Mahomet* IV..., Paris, 1675. See letters from Pococke to his mother, Buda, 7/18 May 1737; Pettaw [Ptuj] in Stiria, 23 May/3 June 1737, Letter 20; and St Veit in Carinthia, 21/27 June 1737, Letter 21, BL, Add MS 19939. These are reproduced in Finnegan, *Letters 2*, pp. 193, 202 and 212 respectively.

32 Henry Maundrell, *A Journey from Aleppo to Jerusalem at Easter A.D. 1697*, Oxford, 1703 and *A Journey to the Banks of the Euphrates at Beer, and to the Country of Mesopotamia*, Oxford, 1699. See Bartlett, "Richard Pococke in Lebanon, 1738", p. 29. Maundrell (1665-1701) was the chaplain to the Levant Company at Aleppo from 1696 until his death there in 1701, aged only 35.

33 See Letter 22.

34 Lord Sandwich's party, which was travelling in the east at the same time, included the then unknown artist, Jean-Étienne Liotard. He departed from his travelling companions in Constantinople and set up in his own studio where he made his name taking the portraits of wealthy grand tourists (including Pococke) and other foreign residents. Though becoming a very successful portrait artist, he was, according to Revd John Cooke, originally engaged by Lord Sandwich and his company "to draw the dresses of every country they should go into, to take prospects of all the remarkable places which had made a figure in history; and to preserve in their memories, by the help of painting, those noble remains of antiquity which they went in quest of." See Cooke, *A Voyage Performed by the Late Earl of Sandwich*, p. iii.

35 His words are: "The publisher of these observations at first had it only in his thoughts, to give the world the plans he had taken of the Egyptian buildings, together with some

from his travels, nor are any drawings included in his original letters and travel journals.[36] This has naturally led to speculation concerning the authenticity of his artwork and the possibility that Dr Charles Perry, a rival author of eastern travels, and one-time fellow traveller, may have been alluding to Pococke in his tyrade concerning the temple buildings of Egypt:

> We have given the geometrical Plans of some ancient Temples, which are of our own drawing. But for the Designs of intire Edifices, and other whole Pieces, we have borrowed them ... from other Travellers; having first compared them with their respective Originals, to see that they were just and right. This, we presume is what our Betters have sometimes done before us; and that, perhaps, with less Candour and Ingenuousness, not acknowledging that they borrowed them, but given them to the World as the genuine Fruits of their own Pains and Dexterity.[37]

Though clearly Perry refers to the past, and Pococke's eastern book came out at approximately the same time as his own (the spring of 1743), Perry will undoubtedly have been aware of his rival's method of passing off other people's artwork as his own (see below). We do not know where Pococke obtained his more elaborate and professional Egyptian drawings but part of the answer may lie in Lot 66 of his library sale catalogue (in the section devoted to "Prints, Maps, &c"), described as "Twenty Egyptian Antiquities".[38] Antiquities was a word used during the period to describe both buildings and objects, and it is possible that Pococke may have acquired sketches or prints of famous Egyptian monuments while travelling in Egypt and then adapted them for use in his book, under the pretence of his own authorship.

---

drawings of them, and to add an account, and designs of all the different orders of Egyptian architecture." *Description 1*, preface, p. iii.

36 By contrast, his travelling companion Jeremiah Milles kept a "Sketch Book of Antiquities in Germany", now in the BL (Add MS 14263) containing drawings of buildings and statutues from throughout his travels with Pococke; and in addition, his grand tour letters to the bishop contain some very proficient (and controversial) drawings of Roman monuments, four of which are reproduced as Plates 1, 3, 5 and 7, in Finnegan, *Letters 2*. Three plates from *Description 1* bear the signature of the engravers George Child and Charles Mosley, the latter of whom (1720-c. 1770) engraved portraits and genre scenes and made a number of political caricatures after his own drawings. See Emmanuel Benezit, *Dictionary of British Graphic Artists and Illustrators*, vol. 1, Oxford, 2012, p. 123.

37 Charles Perry, *A View of the Levant: Particularly of Syria, Egypt, and Greece. In Which Their Antiquities, Government, Politics, Maxims, Manners, and Customs, (with many Other Circumstances and Contingencies) are Attempted to be Described and Treated on*, London, 1743, preface, p. xv.

38 Flin, *A Catalogue of the Library*, p. 23.

There may have been additional personal reasons for Dr Perry's attack on Pococke (assuming that it was indeed an attack on him), apart from the obvious one of regarding him as a rival travel writer. While Pococke informed his mother in his letters that he found Dr Perry to be a pleasant travelling companion, it appears that he later made a nuisance of himself in the Egyptian Society, the abrupt end to which has been attributed to the dissension caused by his insistence on moving the motion "that a Law might be made in order to oblige every member of the society to purchase one copy of every book which either has, or shall hereafter be published by any member of this society."[39] Such a proposal was introduced at the meeting of 28 May 1742 (when it was agreed to postpone consideration of it until the winter) and was raised again at a further five almost consecutive meetings, until it was finally rejected.[40] Though voting was done in secret, and we therefore do not know the identities of those who objected to the proposal, it is clear from the minute book that Pococke (who was Secretary to the society and therefore recorded the minutes) was present at every meeting and may have voiced particular objections.[41] One reason why he and several other members might have objected to the idea was because they had already paid a subscription towards Perry's book, whose publication date was either imminent or had already passed at the time of the final rejection of his motion,[42] and would therefore have had their own copy or copies.[43]

39 Egyptian Society "Minute Book" or "Journal", BL, Add MS 52362, folio 28, verso.

40 On 3 December 1742 the motion was rejected, nine votes to two; on 21 January 1743 it was rejected six to one; on 18 February 1743 it was "BURNED"; on 18 March 1743 it was again rejected; and on 31 March 1743, having been presented "viva voce", the motion was not received, a member insisting on Dr Perry delivering it in writing. The final meeting of the society was 16 April and the matter was not raised, after which it disbanded and replaced with the Divan Club, from which Dr Perry was excluded.

41 In fact, the records were kept by Pococke in more than one sense, as after its demise he held on to them for the duration of his lifetime (see Chapter 2). For further details of the society and of the rift, see M. Annis, "The First Egyptian Society in London (1741-1743)", in *Bulletin de l'Institut Français d'Archéologie Orientale* 50 (1952), pp. 99-105 and Finnegan, "The Divan Club", pp. 15ff.

42 Proof of the date of publication can be found in an announcement in a periodical, where it appears in the "Register of Books for March 1743". See *The Gentleman's Magazine and Historical Chronicle. Volume XIII. For the Year M.DCC.XLIII*, London, 1743, p. 168.

43 Dr Perry dedicated his book to Lord Sandwich, Vizier of the Egyptian Society, and of its other officers and members, the following subscribed to it: Revd Mr Cutts Barton, Revd Thomas Dampier, Lord Duncannon (William Ponsonby), Smart Lethieullier, John Mackey, Jeremiah Milles, Richard Pococke, the Duke of Richmond, the Earl of Sandwich (who took 10 copies), his wife and mother (who took a further four), Charles Stanhope and the Revd Dr William Stukeley. The following entries are included in Pococke's library sale catalogue (original wording): "Perry on Diseases 2 Vols (Lond 1741), "Perry's View of the Levant"

A more serious and sustained attack was made against Pococke in 1746, when Revd Thomas Shaw (1647-1751), former chaplain to the English Factory at Algiers, and author of the previously published *Travels or Observations Relating to Several Parts of Barbary or the Levant* (1738), published a folio volume (110 pages in length) as a *Supplement* to his original work, the purpose of which was to: "vindicate it … from some Objections, that have been lately raised by the Author of The Description of the East, and some other Countries".[44] Shaw accused Pococke of contraverting a number of theories proposed in his own book, largely concerning the geography of Egypt, and in his preface made the following statement:

> Before I perused the Description of the East, I flattered myself, as well on Account of the great Intimacy and Friendship, that had, for many Years, subsisted betwixt the Author and myself, as of our seeming Unanimity and Agreement, whenever the Conversation turned, as it often had done, upon the Subjects in Dispute; that, without giving me some previous Notice and Advice, He would not, in so unexpected a Manner, have drawn me into a Controversy; which, in Justice to my own Observations, how contrary soever it may be to my Inclinations, he might well imagine, I could not possibly avoid. Neither has this Author been more friendly to me, in other Respects. For, as it will sufficiently appear, on comparing such Parts of our Descriptions, as correspond to each other, that, in several Points, hitherto overlooked or disregarded or mistaken by former Travellers, I had carried the Torch and marked out the Way before him; yet he has not vouchsafed, so much as once, in this Voluminous Work, to acknowledge the Assistance, or the many useful Hints, at least, which he had received from my Book of Travels. In like Manner, whenever these Observations of mine were not agreeable to his Sentiment; and, instead of being adopted by Him, were to be controverted, as so many supposed Errors and Mistakes; there the Author of them still continues to be equally concealed.[45]

(Lond 1743) and "Perry on Spaw Waters" (London, 1734). Flin, *A Catalogue of the Library*, Lots 37, 77 and 128 (pp. 2 and 4) respectively.

44 Thomas Shaw, *A Supplement to a Book Entituled Travels, or Observations, &c. Wherein Some Objections, Lately Made Against it, Are Fully Considered and Answered: with Several Additional Remarks and Dissertations*, Oxford, 1746, preface, p. i, hereafter Shaw, *Supplement*.

45 Shaw, *Supplement*. preface, p. xv.

Though Shaw was elected a member of the Egyptian Society, there is no record of his attendance, which suggests he did not take up the offer.[46] Nor was he a member of the Divan Club.[47] However, since both he and Pococke were Fellows of the Royal Society,[48] it is likely that they would have had contact with each other during the period when Pococke was writing up *Description 1*. Furthermore, the two complainants (Perry and Shaw) were no doubt irritated by the fact that their rival had dedicated many of the 75 plates in his book to men of great influence (including members of the English and Irish aristocracy, church dignitaries, and academics) but not to them, when they had been so helpful to him.[49] Whatever the case (and to a certain extent the travel correspondence itself tells the story) we know that Pococke was capable of plagiarising the artwork of even his closest travel companion. His act of passing off several of his cousin's drawings as his own, by having engravings made of Roman antiquities in Pola and Lucca and including them, unacknowledged, as plates in *Description 2*, annoyed Jeremiah Milles, who when editing his own letters from these sites made it very clear that he, and not Richard Pococke, was the original artist.[50]

It is interesting that, while Pococke failed to respond to any of the printed accusations (or indeed to the criticisms of others, such as Revd William Warburton),[51] he made no further claims in *Description 2* to the illustrations being his own. Nevertheless, with the exception of the decorative frontispiece, vignettes, and some of the tailpieces (designed variously by C. Frederick and Hubert-François Gravelot), together with the maps (by Thomas Jefferys) and the botanical prints (by Georg Dionysius Ehret), this larger and more extravagantly illustrated volume contains no mention in the preface or elsewhere of an artist or draughtsman; nor do any of the actual engraved plates (as opposed

46 He was proposed by the "Sheik", Lord Sandwich, at the first meeting, held on 11 December 1741. See the "Journal" or "Minute Book" of the society, BL, Add MS 52362, folio 15.

47 These London clubs are discussed in Chapter 2, below.

48 Shaw was elected on 13 June 1734 and Pococke was elected on 4 February 1742.

49 The table of illustrations lists in *Description 1* lists 75 plates (including two maps and numerous plans and views). While Plate 33 is not listed and was never issued, the figure of 75 is retained because the final plate, representing an Egyptian statue in the collection of the Duke of Richmond, was acquired and inserted at the last minute and so failed to appear in the table.

50 See Finnegan, *Letters 2*, pp. 179 and 237, note 12.

51 In his *Remarks on Several Occasional Reflections in Answer to the Rev. Dr. Middleton, Dr Pococke... and Others...*, London, 1744, Warburton challenged Pococke for controverting some of the views he had proposed in his *Divine Legation* (1737-42). In particular, he ridiculed Pococke for what he considered to be his naïve and ignorant views on hieroglyphics, objecting strenuously to his theory that Egytpian hieroglyphs represented objects rather than words. See especially *Remarks*, pp. 24-25.

to the decorative elements mentioned above) bear the signature of an artist. However, the careful observer may note the minute inscription "S W sc" or "Sam Wale sc",[52] at the foot of 12 engraved architectural drawings, including those depicting the monuments at Baalbek, but the Latin "*sculpt*" denotes the name of the engraver, rather than the artist. This suggests that Pococke may have wished the reader to assume he was the original artist of certain drawings, particularly when he frequently included in the foreground the characteristic "cartoon" of himself engaged in viewing, measuring and drawing the antiquities, or at times simply posing in front of them.[53]

There are also doubts over the authorship of some of the equally elaborate plates (in *Description* 2, Part II) of Athenian monuments, especially as Pococke was only in Athens for six days. Letter 43, dated Corinth, 6/17 October 1740, outlines how Pococke had spent his time in this city, and nowhere does he mention having actually *drawn* any of the monuments. Furthermore, in the book he presents several other plates depicting Athenian monuments that are not even mentioned in the correspondence. Regardless of how or where he obtained them (which has always been a mystery), he is recognised by Michael McCarthy as having provided the earliest scaled drawings of monuments in Athens.[54] The mystery remains unsolved and though the matter has been raised in a recent book,[55] a full examination of Pococke's drawings from the east, in the context of those appearing in the later and more influencial

52 Samuel Wale was an English painter and book illustrator. It is thought that he was apprenticed to a London goldsmith in 1735 and then went on to study drawing at St. Martin's Lane Academy, in London. Aged only about 25 at the time, this was possibly one of his first serious commissions for this type of work. See M. G. Sullivan, "Wale, Samuel (1721?–1786)", in *Oxford Dictionary of National Biography*, Oxford University Press, 2004. Available at: <https://doi.org/10.1093/ref:odnb/28453>.

53 He is easily identifiable by the descriptions he so frequently gives of himself in the correspondence (particularly relating to his facial hair and headgear) and from the famous portrait of him in eastern dress by Liotard, the preliminary sketch for which is featured on the cover of the present volume. For a detailed discussion of his eastern costume, hairstyle and beard, see Michael McCarthy, "'The Dullest Man That Ever Travelled?' – A Re-assessment of Richard Pococke and of his Portrait by J.-E. Liotard", *Apollo* 143 (1996), pp. 1996.

54 The original signed typescript of an unpublished lecture in which this point is made, was kindly sent to the editor by Professor McCarthy shortly before he died in 2010. The same point is made in a posthumously published article, where he notes, "Pococke travelled alone to Athens, recorded the major monuments there and produced illustrations far more comprehensive and correct than those published by that other Irish artist Richard Dalton in 1752." See McCarthy, "Extending the Grand Tour: Dr Pococke's Travels, 1737-41", in P. Ní Chatháin and S. Fitzpatrick (eds), *Pathfinders to the Past: The Antiquarian Road to Irish Historical Writing, 1640-1960*, Dublin, 2012, p. 33.

55 See Finnegan, *English Explorers in the East*, Chapter 8 and pp. 280-83.

publications of Robert Wood, James Stuart and Nicholas Revett, would prove a fascinating topic for a future study in art and architectural history. What may be of significance is the fact that Lot 374 in the sale catalogue Pococke's library is listed as, "Antiquities of the Eastern Countries, on a great number of Plates."[56] This is a folio volume (unpublished) that may well have comprised a collection of drawings acquired in the east and subsequently included as plates in *Description 2*. Nevertheless, it is clear from the correspondence that some of the drawings from the earlier part of his eastern travels were indeed his own. In Letter 13, from Jaffa, for example, Pococke states: "When anything arrives for me from these parts, I desire you not to let any of my sketches in a pencil out of your hands, on no account, for fear any of the figures or lines should be defac'd; – & so likewise what is writ with a pencil, & if Mr Milles should be with you & desire to see anything, pray turn them turn them [sic] over for him – but it is better to keep them unseen."

Though the eastern voyage was undoubtedly the highlight of Pococke's foreign travels, and involved a considerable degree of planning, he did not mention it to his anxious mother right up until two days before he boarded the ship from Livorno to Alexandria, though it is clear from the context that she had already found out by some other means (presumably from her brother, Bishop Milles, or her nephew, Jeremiah Milles). The letter in which Pococke revealed his plan (dated 6/17 September 1737) was worded very carefully, in order to put her at ease about the management of his diocesan duties during his absence, his finances and his laudible motives for extending his travels to the East:

> I think I can answer in every respect my undertaking this voyage. As to the cures [ecclesiastical duties] that were upon me, the Bishop called me from them, to a more extensive office, as Chancellor of his Diocese & substituted others in my room [place]; – & now his Lordship has taken immediately on himself, the care of every thing that relates to that office. I have 300 pr An. [& spend nearly two]: – & I hope to gain such knowledge & experience as to return better qualified to do service to my fellow creatures in the station in which it has pleased God to place me. And whatever is the motive, if seeing those places which were the scenes of our Saviours Life & passion, may touch my heart with a greater sense of my duty & make greater impressions as to what he did for us; I hope this journey on that account alone, will not be thought a fruitless one.[57]

---

56 See Flin, *A Catalogue of the Library*, p. 10.

57 See Letter 0. This letter is not in the manuscript volume of eastern letters but is the last in the volume of his travels on the Continent (prior to his passage to Egypt). BL, Add MS 22997.

Once he had dispatched this letter and boarded the ship, he kept in constant touch with his mother, sending her letters in a daily journal format approximately every three to four weeks and sometimes more frequently. From the correspondence it appears that there were several reasons for this (apart from a strong sense of filial devotion), as follows: Mrs Pococke "insisted" on his keeping a journal and telling her "the most trifling particulars" of how he spent his time and what he saw (Letters 5 and 6); by his own admission (Letter 10) he felt guilty for being the cause of her anxiety ("If I had thought you would have taken my voyage so very much to heart, I would not have gone…"); he was totally reliant on her to manage his travel manuscripts and the boxes of objects (antiquities, "trumperies", seeds, specimens of natural history, etc.) regularly sent back home to her home; and in addition he expected her to continually forward letters and relay messages to her brother (Bishop Milles) and her nephew (Jeremiah Milles), the latter of whom he had charged with organising his "License of Absence" from the diocese of Waterford & Lismore.

During the first half of his eastern voyage, Pococke composed and sent a series of letters which he describes as "accounts", comprising detailed observations on the history, geography and antiquities of the east, and often accompanied by references to and quotations from the works of classical authors such as Herodotus, Strabo, Virgil, Vitruvius and Pliny. Though only 34 of these letters have survived, we know from correspondence with his mother (see below) that Pococke continued to write to his uncle even after his death in 1740, unaware that he had finally succumbed to gallstones, from which he had suffered for many years.

3 Sources and Transmission of the Texts

We are fortunate both in the extent and variety of the primary sources for Pococke and Milles's foreign travels, which took place from 1733-34 and 1736-41. The documentary evidence spans an extensive collection of correspondence and numerous illustrated manuscript journals assiduously kept by each traveller. All of these manuscripts are housed in the British Library.[58]The manuscript sources for the correspondence for the the third (eastern) phase of Pococke's foreign travels (late September, 1737 to the end of August 1741) are the focus of this volume and are as follows: a collection of 66 letters to his mother, Elizabeth Pococke (copies), covering the period 30 September/11 October 1737 (from

58 With the exception of three original letters from the first grand tour (Jeremiah Milles to the bishop), which are located in the Gloucestershire Archives (Ref. D2663/Z8).

Livorno) to 29 August 1741 (from Dover);[59] a collection of 34 letters or accounts to his uncle, Bishop Milles (copies), covering the period 6/17 September 1737 to 7/18 March 1739;[60] and a note book or journal entitled "A Voyage From Leghorn to [Alexandria]" (original), almost half of which is taken up with an account of Africa.[61]

We are again fortunate in the condition in which the correspondence has survived for this phase of Pococke's travels, since fair copies were made of both collections (sadly the original letters are not extant). The copies of letters addressed to Mrs Pococke were neatly transcribed by her into a quarto volume and cover her son's daily activities for a substantial period of time, often up to three or four weeks. Internal evidence in the form of a note at the beginning and end of the manuscript volume shows that she began her transcript on 3 September 1746 and completed it on 17 October of that year. This was precisely five years after her son's return from the east a year and a half after the publication of *Description 2*.[62]

The copy letters in the collection addressed to Bishop Milles are composed in a large neat hand, presumably that of a secretary or clerk, and preserved in the form of large folio volumes.[63] Both collections of copy letters bear evidence of later editing, though in two different hands. This includes the addition of page numbers, marginal notes, brackets, minor amendments, and explanatory comments written on the backs of pages. Furthermore, numerous words, lines and in some cases whole passages from Mrs Pococke's copy letters have later been scored out with heavy black wavy lines, making the original wording impossible to decipher. An analysis of the handwriting shows that the letters addressed to the bishop have been edited in his own hand (including notes regarding the order in which letters should be placed), while those addressed to Mrs Pococke have been edited in her son's hand. This suggests that both Bishop Milles and Pococke had intended to publish these memoirs, based on the copy letters, rather than on the originals. Clearly, it was ill health that prevented the bishop from carrying out this intention. As for Pococke, he may

59 BL, Add MS 22998. This volume contains 255 quarto pages.

60 BL, Add MS 15779. This volume contains 61 folios, verso and recto.

61 BL, Add MS 22995. This volume, the title of which is incomplete, contains 60 quarto pages.

62 According to the printer's records, the date of printing was 4 February 1745. See Paul W. Nash, Nicholas Savage, Gerald Beasley, John Meriton and Alison Shell (eds), *Early Printed Books, 1478-1840: Catalogue of the British Architectural Library Early Imprints Collection*, K.G. Saur, Munich, 1994-2003, vol. 5, p. 2745.

63 In the British Library catalogue, they are described as "Registers of Letters", which gives the false impression that they are merely lists, rather than actual copies of the correspondence.

either have been too busy in later life to return to this, when appointed as an archdeacon and then a bishop in Ireland; or perhaps the scathing attack from his rival travel writer Thomas Shaw may have deterred him from publishing anything further on the subject.

The transmission of Pococke's travel manuscripts can be summarised as follows: Jeremiah Milles deposited these items in the British Museum on 7 June 1767, according to the terms of his late cousin's Will, but seems to have later appropriated them for his own use.[64] They were then absorbed into his own library collection, which was inherited by his heirs on his death, in 1784. According to a two-page write-up in *The Gentleman's Magazine* 60 years later (1843), this important library collection had three months earlier been "brought under the hammer of Mr Leigh Sotheby on the 10th of April and four following days".[65] There were actually three parts to the sale. The first auction took place on 10 April, in which his "valuable topographical, theological, classical and historical library" was sold in 1,211 lots; the second took place on 25 April, in which his "valuable and interesting assemblage of autograph letters" was sold in 233 lots; and the third took place on 14 August, in which "the remaining portion of the books and manuscripts of a distinguished collector deceased" were sold in 276 lots.[66] The same issue of the *Gentleman's Magazine* confirms that while portions of the collection had previously been offered for sale, the library was still reasonably complete at the time of the Sotheby's auction, having "received from time to time considerable accessions".[67] It also confirms that the main bidder on Pococke's travel manuscripts was Thomas Thorpe, a leading London bookseller celebrated for his extensive dealings in old books and manuscripts. While *The Gentleman's Magazine* for 1843 does not specify the two collections of eastern letters in question (they form part of combined lots), it does indicate the prices paid by Mr Thorpe for the "many volumes of the journals of Dr.

64 This was perhaps on his appointment as President of the Society of Antiquaries in 1769, at which point he automatically became an official trustee of the British Museum.

65 See "Library of Dean Milles", *The Gentleman's Magazine* XX, New Series, July to December Inclusive, London, 1843, p. 72.

66 The titles of the catalogues for these three sales are: *S. L. Sotheby, Catalogue of the Valuable Topographical, Theological, Classical and Historical Library: with Some Highly Interesting Manuscripts*, London, 1843; S. L. *Sotheby, Catalogue of a Valuable and Interesting Assemblage of Autograph Letters ...*, London, 1843; and S. L. Sotheby, *Catalogue of the Remaining Portion of the Books and Manuscripts of a Distinguished Collector Deceased*, London, 1843. The section relating to Milles in this last catalogue is titled "Imperfect Books from the Sale of the Late Dr. Jeremiah Milles, Dean of Exeter" and runs from pp. 57-70 (lots 1,150-1,426).

67 *The Gentleman's Magazine*, 1843, p. 72.

Pocoke, the Oriental traveller".[68] His total outlay for these and other related items (including the original manuscript version of *A Description of the East* and the Minute Book of the Egyptian Society)[69] was £103 and 13 shillings, though he also acquired many other lots from the collection.

Mr Thorpe immediately incorporated very detailed descriptions of the acquired manuscripts into his 1843 catalogue of recent purchases,[70] making unambigous declarations regarding their national importance and claims for them to be "preserved in some of the public archives in this country".[71] This was clearly an invitation to the British Museum, whose officers, though present at the Sotheby's auction (and who had even bought several of the related items),[72] had either failed to bid on the rest of the collection or had been outbid by Mr Thorpe. It is likely that the British Museum acquired the remainder of the collection direct from Mr Thorpe, though at double the price, as the descriptions given in the current British Library catalogue for the Pococke Collection are very similar to Thorpe's detailed listings.

4 Scope and Format of this Revised Edition

The combined manuscript source material for the Milles/Pococke grand tours is extensive, and a full edition of the entire collection (including the journals of both travellers for their continental voyages), though a huge undertaking, would be a most welcome addition to this important area of scholarship.[73] The scope of this revised editon of *Letters 3* is less ambitious than that but more ambitious than than the first edition, since it is based on *all* the eastern letters from *both* collections of Pococke's correspondence, rather than reproducing selected letters only from his mother's collection. This revised edition does not constitute a comparative analysis between Pococke's eastern travels and those

68 Ibidem, p. 73.

69 The book manuscript (Lot 1199) was sold for £3, 3 shillings and offered for sale by Mr Thorpe for double the price, £6, 6 shillings; the manuscript of the Egyptian Society (lot number unspecified) was purchased for £2 and put up for resale at £5, 5 shillings.

70 Thomas Thorpe, *Manuscripts, upon Papyrus, Vellum and Paper, ...* London, 1843, pp. 125-27. For the full title of this volume, see Bibliography.

71 Thorpe, *Manuscripts, upon Papyrus*, p. 127.

72 For example, Lot 1202, a small collection of Pococke's travel manuscripts from his tours of the British Isles, costing £33. Jeremiah Milles's grandson bought at this auction a quarto volume of observations made by Milles on the antiquities, churches and palaces at Rome. This was from his first grand tour with Pococke in 1733-34 (Lot 1197 £2, 2s).

73 However, such a study would inevitably involve considerable levels of repetition and overlap, which might prove tedious for the general reader.

of other grand tourists of the period, or between Pococke's travels in Ireland, Scotland and England.[74] Its purpose is simply to present a series of wonderfully detailed and engaging memoirs of a highly influencial 18th-century figure, who was passionate about his foreign experiences, both when abroad (as is evident from his correspondence) and for many years after his return (as is clear from his subsequent antiquarian activities and publications).

Pococke's letters to his mother are of general interest, since his intention in writing them is to inform, amuse and placate her, as well as to instruct her on how to manage his business and personal affairs during his absence. The style of writing is relaxed, chatty and often humorous, as might be expected in correspondence between two exceptionally close and devoted relatives. While his mother and sister were the intended recipients of the original letters, other members of the extended Milles/Pococke family are likely to have received edited versions, as requested by the traveller.[75] The letters cover the whole of Pococke's journey in the east and only are couple are missing. While they are one-sided, it is often possible to work out, from the content, the type of questions Mrs Pococke had asked or how she had reacted to her son's news. For example, he frequently corrected her on certain preconceptions she had about the dangers of travelling in the east, as well as constantly advising her on how to manage her anxiety and her general health, of which she was presumably complaining.

Mr Thorpe advertises the importance of the Lot 472 (the full collection of 21 volumes, of which the last is the corpus of eastern letters from Pococke to his mother), as "forming a most valuable body of Historical Materials for future Writers, for which purpose they ought unquestionably to be preserved in some Public Library …". He further states:

> The importance and interest of these travels can only be appreciated by inspection … The whole is detailed in a collection of letters from Dr. Richard Pococke to his honoured mother, and continued until [his] return, with increased spirit and indefatigable research, written in an easy style with reference to classic authors, with which [his] acquaintance rendered [him] so pre eminently qualified for so arduous and praiseworthy a task. The names alone of the places visited by [him] would fill a large

74 Such considerations would be a worthy subject for a PhD thesis.

75 Because his time was spent in writing so many different versions of his travels (his notebooks, the daily epistolary journal written for his mother, and the lengthy "accounts" compiled for the bishop), Pococke frequently asked his mother to abridge his personal letters to her and send them on to her brother in Waterford. However, these abridged versions do not appear to have survived.

> volume; it is, therefore, hoped the following brief notice will create sufficient interest for inspection.[76]

While the collection of 19 manuscripts volumes of grand tour letters was acquired by Mr Thorpe (Lot 1203) for £20, he offered for sale 21 volumes (the last of which comprised the eastern letters) three months later for more than double the price, at £42.

In contrast, Pococke's letters to Bishop Milles are more scholarly and at times pedantic, since their purpose is presumably to show his uncle that he is a keen scholar and antiquarian, and a worthy recipient of his continued and very indulgent patronage. The style of writing in these letters is studied and self-conscious, as though anticipating a wider readership, and in many ways these manuscripts can be seen as a precursor of his published account. As already noted, though in letter format, the extant letters are described by the author as "accounts" of particular places and contain very little in the way of personal information. The manuscript volume starts in Livorno, before the author sets off for Egypt (September 1737) and continues for a year and a half, incorporating formal descriptions of his travels in Egypt, the Holy Land, Lebanon and Syria, and Cyprus, and ending with his return visit to Egypt (March 1739). Though they have not survived, these epistolary accounts continued to be written and sent right up until Pococke received the "melancholy news" of his uncle's death in September 1740. We know this because he regularly updates his mother on his literary progress, the last mention of this being on 12/23 August 1740 (Letter 44), when he notes: "Writ all day to my Lord Bishop of Waterford", who by then had been dead for three months.

As the manuscript survives only as far as March 1739,[77] we do not know if the missing letters were ever received by the bishop or if they "miscarried". It is possible that they may have arrived in Waterford when he was too ill to deal with them, but they were certainly not among the manuscripts acquired by Mr Thorpe at the 1843 Sotheby's, since Lot 473 of his catalogue describes the manuscript as a single volume: "Pococke's (Dr. Richard's) Letters to Dr. Thomas Milles, Bishop of Waterford, during his Travels in Egypt, Syria, Cyprus, &c. neatly copied, with Remarks' in the Bishop's Autograph, with a view to publication, folio, bound in vellum, 5*l.* 5s. 1737-9"[78] In a further description of this lot, the catalogue says:

---

76 Thorpe, *Manuscripts, upon Papyrus*, p. 125.

77 A cover note from Pococke accompanying the final batch from Egypt mentions that he is sending 13 letters, whereas only eight (and nine if the cover note itself is included) appear in the manuscript volume.

78 Thorpe, *Manuscripts, upon Papyrus*, p. 127.

> These interesting letters describe the places through which this learned traveller passed, and their principal objects of curiosity, in a manner that unquestionably claims to be preserved in some of the public archives of this country. *They* commence at Leghorn, and others are dated from Alexandria, Grand Cairo, Damatia, Jerusalem, Bethlehem, Joppa, St. Jean d'Acre, Nazareth, Tiberias, Tyre, Sidon, Beyrout, Tripoly, Cannobine, Balbeck, Aleppo, Famagusta, Nicosia, Baffa, Colosse, Limesole, Larnica, Mehulla, Faiume, Saccara, &c."[79]

Clearly, from this description, this was the only volume of copy letters from Pococke to the bishop from his eastern voyage.

Letters 1 (volume 1 of the original series of the Milles/Pococke travels) reproduced extracts from 46 out of the 53 extant letters from the cousins' first tour of the Continent (1733-34), which was cut short after 10 months, when Milles, straight from Oxford and not yet determined on a church career, was persuaded to hasten to Ireland to take up a preferment in his uncle's diocese. Also reproduced in that volume were: a transcript of the bishop's Will, a letter from the bishop to his sister outlining his arrival in the Diocese of Waterford & Lismore, and a further letter from Mrs Pococke to her son on the occasion of his ordination. *Letters 2* reproduced the entire collection of original letters (plus two copy letters) from Pococke to his mother from the cousins' second tour of the Low Countries, Central Europe and Italy (1736-37), up to the point at which he departed alone from Livorno to Alexandria. Also reproduced in that volume were: a letter from Mrs Pococke to her brother the bishop suggesting that he should recall the travellers from abroad, a note to her London banker Mr Hoare, and selected letters/excerpts from both Pococke and Milles to the bishop. *Letters 3* reproduced an extract from the final letter of the 1736-37 tour from Livorno, 43 of the 48 letters sent to Mrs Pococke from the east (with summaries of the remaining five) and the first letter in a series of four from Messina (Sicily), where Pococke was was quarantined before making his return journey through Europe. In addition, one letter from Pococke to Bishop Milles was reproduced, the only example in that collection of a personal nature rather than an "Account".

The present new, revised edition of *Letters 3* reproduces all the letters/accounts from the two manuscript volumes (addressed to Mrs Pococke and Bishop Milles), together with two additional letters written before he set sail for Alexandria, which are included for context: one to Bishop Milles from Venice (Account 0) and the other Mrs Pococke from Livorno (Letter 0).

---

79 Ibidem.

Chapter 2 of the present volume comprises biographical accounts of Pococke, his mother and his uncle; Chapter 3 provides a summary of the itinerary for Pococke's eastern voyage; Chapter 4 reproduces all the letters to Mrs Pococke from Livorno up to the end of her son's first tour of Egypt; Chapter 5 reproduces all the corresponding letters (or accounts) sent to Bishop Milles, as well as the earlier letter from Venice;[80] Chapter 6 reproduces all the the letters to Mrs Pococke from the Holy Land, Lebanon and Syria; Chapter 7 reproduces all the corresponding letters (or accounts) to Bishop Milles; Chapter 8 reproduces all the letters to Mrs Pococke from her son's second tour of Egypt, and from Turkey, Asia Minor and Greece (including the first letter from Messina, which relates to details of his journey from Greece); Chapter 9 reproduces all the letters (or accounts) to Bishop Milles from Cyprus and the second tour of Egypt; and Chapter 10, the Conclusion, gives an outline of Pococke's return journey and comments on his reputation as a traveller and scholar. The Appendix lists the letters and accounts transcribed in this volume, giving the date and place of dispatch; and following the Bibliography of manuscripts and printed books and articles is an Index of Selected People and Places.[81]

5 Editing of the Texts

As already pointed out, no original letters have survived from this phase of Pococke's travels, and the reader should note that the extant copy letters have undergone significant editing by Mrs Pococke who, as can be judged from the difference between the original and copy manuscripts surviving from the 1733-34 and 1736-37 tours (reproduced in volumes 1 and 2 of the original series), took liberties with her son's script, tidying up his grammar and syntax, changing words to numerals, altering his vocabulary and introducing a whole new mode of punctuation. In addition, she frequently omitted words and sometimes lines, presumably in error, and occasionally skipped over certain sections which she considered to be repetitive or otherwise unnecessary. Missing words, or those that are in some way obscured, are indicated thus [...]. Square brackets are also used to indicate the correct spelling or the modern version of placenames,

80 To avoid confusion, the letters and accounts are dealt with separately, with the letters addressed to Mrs Pococke placed first, in a block, and followed by the accounts addressed to Bishop Milles, in a block.

81 This relates primarily to the countries, towns and villages mentioned in the text and to Pococke's family members and associates. It does not include buildings, monuments, biblical, classical or historical figures.

where they can be ascertained.[82] Any ordinary brackets appearing in the text are Pococke's. It should also be noted that Mrs Pococke sometimes used two or more different spellings of placenames or other words unfamiliar to her, even within the same letter. The editor has remained faithful to her text, retaining all abbreviations and peculiarities of spelling as well as her curious punctuation and the errors she made when transcribing Greek words.[83]

The same method of editing has been applied to the the accounts sent to Bishop Milles. However, if we are to judge from the level of editing made by the transcriber of the correspondence from earlier phases of Pococke and Milles's travels, for which original copies exist, there appears to have been less intervention. Likewise, the bishop made far fewer later amendments to the text than Pococke made to his mother's transcripts. As is evident from volume 1 of the first edition, the deleted or scored passages from Mrs Pococke's copy letters (which can be reconstructed from the originals) generally relate to personal matters such as her son's financial problems, the inadequacies of Mr Bagwell, his Irish banker, his uncle's failing health, an errant relative or his own sartorial taste. While some deletions from her copy letters from the second tour are relatively innocent, concerning for instance her son's consumption of lavish meals and wines, others contain allusions to more titillating matters or to his increasingly disdainful comments about his cousin, and it is understandable why a man of his standing took the decision in later life to remove such sensitive or embarrassing passages from the pages of these manuscripts. Sadly, though, as there are no originals for the collection of letters from the third phase of Pococke's travels (his eastern voyage), the excised lines in his mother's copies must remain a mystery.

Finally, brief notes are provided where it is thought the reader may benefit from an explanation of historical, geographical and topographical points arising from the letters, or on the identity of people encountered during the course

82 In general, this is done only once, the first time the place is mentioned in the text, after which the original spelling or placename is retained. However, it should be noted that in some instances the modern placename could not be identified and that Pococke is the only source for the given name.

83 Examples of words consistently misspelled and/or written in a style that might be unfamiliar to the modern reader are: "chearfull", "dutifull", "lye" ("lie"), "strait", "shew" ("show"), "heighth" ("height"), and the frequent use of "writ" or "wrote" instead of "written" and "rid" for "rode". Common abbreviations, found in the original letters and which are retained in the present transcripts, include, "'tis" ("it is"), "tho'" ("though"), "Bp." ("Bishop"), "Ld" ("Lord"), "Xtmas" or "Xtian" ("Christmas" or "Christian"), "yt" (that).

of Pococke's travels.[84] These are more plentiful in the letters to his mother than to his uncle.

---

84 The sources used for identifying such individuals include various peerages, especially Peerage.com and the *Oxford Dictionary of National Biography*, Oxford University Press, 2004; John Ingamells (ed.), *A Dictionary of British Travellers in Italy, 1701-1800*, New Haven and London, 1977; *The Clergy of the Church of England Database* 1540-1835. Available at: <https://theclergydatabase.org.uk>. and Wilson, *List of British Consular Officials*. Since explanations of biblical references are not included in the notes, it is recommended that the reader, where applicable, should consult a copy of the Bible. The reader who wishes to follow *A Description of the East* is directed to the appropriate volumes and sections of this work.

CHAPTER 2

Biographical Accounts of the Pococke and Milles Families

1 Life of Mrs Elizabeth Pococke (?-1760)

We begin the biographical accounts with a consideration of the life of Mrs Elizabeth Pococke and her family. She is a central figure in this story, since she was not only the mother of Pococke, but also the sister of Bishop Milles and therefore Jeremiah (the traveller's) aunt. She was, as we shall see, a true matriarch, always involved in the affairs of the Milles/Pococke family and retaining this role by correspondence even when separated from its members by long distances. At the time of the cousins' first trip to the Continent, she had been widowed for 33 years and was living with her unmarried daughter Elizabeth ("Besse") in the parish of Newtown, in the extreme north-west corner of Hampshire, bordering with Berkshire.

The year of her birth is not known but she was certainly an only daughter of Revd Isaac Milles and Elizabeth Luckin (who were married in 1670) and the youngest in the family. The last of her brothers (Isaac) was born in 1676/77, which places her date of birth somewhere between 1679 and 1682, making her between the age of 16 and 19 at her own marriage, in 1698. In 1680 her father was nominated by the lord of the manor, Robert Sawyer, as rector to the Highclere Parish,[1] a position which he retained until his death in 1720. In addition, Revd Milles (depicted in Plate 1) ran, with his youngest son Isaac, a small private boarding school in the rectory, whose pupils included not only his own sons, but those of Thomas Herbert, 8th Earl of Pembroke and 5th Earl of Montgomery.[2]

Sadly, we know virtually nothing about Mrs Pococke's early life, but on 26 April 1698, presumably in her late teens, she married Revd Richard Pococke (1660-1710), the eldest son of yet another Revd Richard Pococke (Rector of Colmer, in Hampshire) and his wife Constance (*née*) Newlin. Revd Pococke (Elizabeth's husband) was educated at Magdalen College, Oxford, and ordained as

1 The position became vacant on the death of Revd Christopher Massey, who was buried on 9 November of that year. See "Extracts from Parish Registgers of Highclere", in J. Gough Nichols, *The Topographer and Genealogist,* vol. 1, London, 1846.

2 One of these sons, the 9th Earl, was to be an important patron of the family, and Pococke dedicated *Description 1* to him.

 | DOI:10.1163/9789004440050_003

deacon in 1689 and priest in 1690. This was his second marriage, his first wife Anne Clutterbuck (daughter of the Archdeacon of Winchester) having lasted only a year (1695-96), owing to her premature death. Since 1690, Mrs Pococke's husband had held the position of Headmaster of King Edward VI School (depicted in Plate 2), a free grammar school in Southampton, and in 1694 he was appointed Curate Sequestrator of All Saints' Church, also in Southampton, a curacy he retained until 24 August 1708. In his Will, he describes himself as "Clerk & Master of the Free-school in Southampton".[3] However, it is not clear whether the word "Clerk" refers to another position in that school or to the the clerkship at Magdalen College, Oxford, to which he was elected in 1688-89.[4]

The town of Highclere, where the young Elizabeth had grown up, was only about 30 miles from Southampton, and it is likely that she and her future husband had met through her father, since both were clergymen in the same diocese (Winchester) and both were educationalists. Clerical circles within such close proximity would have encouraged such marital opportunities, and judging from the amount of clergymen in both families (together with Revd Pococke's previous choice of matrimony), theirs would have been considered an ideal match.[5]

In addition to the social implications of such a marriage were the financial benefits, at least for Revd Pococke, since Elizabeth Milles brought with her a fortune of almost £1,000.[6] It is thought that such an amount could not have come entirely from her father, Revd Milles, with his relatively small income, but was possibly a joint contribution from his brothers, one of whom was a successful tradesman.[7] This dowry would have greatly enhanced Revd Pococke's prospects and reputation, since he is described in an anonymous biography of his father-in-law, Revd Isaac Milles as: "a Man of great Worth and Integrity, but who was very unsuitably provided for".[8] The authorship of this extensive book,[9] referred to hereafter as *Life of Isaac Milles*, is problematic.

3 Will of Richard Pococke (Senior), Hampshire Archives and Local Studies, Ref. 1710B/42.
4 See C. F. Russell, *A History of King Edward VI School Southampton*, Cambridge, 1940, p. 201.
5 In the same way Revd Sam Billingsley, who had briefly been "under-master" at King Edward's in Revd Pococke's time (c. 1696/98), subsequently married the latter's sister, Constance, after taking up a position as under-master of Christ's Hospital, in December 1698.
6 Russell, *A History of King Edward VI School Southampton*, p. 201.
7 See Kemp, *Tours in Scotland*, pp. xxxi-ii.
8 *An Account of the Life and Conversation of the Reverend and Worthy Mr. Isaac Milles, late Rector of High Cleer in Hampshire*, London, 1721, p. 147.
9 The main text is 166 pages to which are appended a three-page transcript of the memorial erected by his son, Bishop Milles, and the text of a 20-page sermon delivered by Revd Isaac Milles on "The Duty of Serving Our Generation".

PLATE 1
Portrait of Revd Isaac Milles, Rector of Highclere (1638-1720), engraving by George Vertue. Courtesy of the Earl of Carnarvon: Highclere Castle Archives Box 53 no. 3, f.73
PHOTO DAVID RYMILL

PLATE 2
"Grammar School at Southampton", *Sixty Views of Endowed Schools: From Original Drawings by J. C. Buckler: With Letter-Press Descriptions*, London, 1827

Though traditionally attributed to the subject's eldest son, Bishop Thomas Milles,[10] a copy in the Thorold and Lyttelton Library at Winchester has a handwritten note on the title page: "Said to have been written by Mrs Pococke, daughter of Isaac Milles."[11] While some sources admit that she had some hand in its authorship, but under the direction of her brother the bishop, the latest edition of the *Oxford Dictionary of National Biography* proposes that this pamphlet was in fact the work of a neighbouring clergyman, Joseph Wood, who had also composed the accompanying funeral sermon for the late Revd Milles.[12] However, given that Bishop Milles had been living in Waterford for 13 years at his father's death, while Mrs Pococke had nursed him through his last illness; and given that the pamphlet is based partly on his "Diary or Journal, now before me, which he then kept of his life",[13] and towards the end gives a very detailed and full account of his malady, there is no reason to doubt that Mrs Pococke was the author. After all, she had been continually by her father's side in his last years, had access to his papers, was very pious, respected this virtue in members of her immediate family, and was not only highly literate but well versed in matters of the church.[14] Furthermore, her brother Bishop Milles was not given to modesty and would surely have made it known that he was responsible for this publication, had he written it.

Quite apart from the question of its authorship, *Life of Isaac Milles* is a fascinating publication which, among other matters of personal and family interest, gives details of the parsonage in Highclere and the renovations carried out on this house by Mrs Pococke's father, "as to render it more commodious for the Reception of Boarders, and the teaching of a larger Number of Scholars". These improvements paid off, since, as mentioned earlier, the "illustrious" Earl of Pembroke "intrusted Mr. *Milles* with the Care of the Education of his sons".[15]

Details of the Pococke family's life are relatively sketchy, though we do know that their two children (Richard, the traveller, and his sister Elizabeth) had

10 Possibly because of its mildly disparagingly references to Mr Henry Dodwell (pp. 52-53), whom Thomas Milles, before becoming bishop, had attacked in print (1707). Other references are made to Dodwell. See, for example, pp. 56ff

11 See Russell, *A History of King Edward VI School Southampton*, p. 203, n.1.

12 See D. A. Brunton, "Isaac Milles (1638-1720)", *Oxford Dictionary of National Biography*, Oxford University Press, 2004. Available at: <https://doi.org/10.1093/ref:odnb/18749>. The latter source does not mention the possibility that Elizabeth Pococke may have written her father's biography, and in fact its only reference to this woman (whose name is omitted) is to note that during Revd Milles's last days he was "tended by his daughter".

13 *Life of Isaac Milles*, 23. There are other references to his Journal, see for example p. 43.

14 For evidence of these virtues and qualities, see her letter to her young son, reproduced in Finnegan, *Letters 1*, Appendix 2, Letter 2, p. 305.

15 *Life of Isaac Milles*, pp. 75-76

been born at West Hall, Bugle Street (see Plate 2), where they had spent the first few years of their life. This was the address of King Edward's, which suggests that the family had living quarters in the school. Some sources state that Pococke (the traveller) began his education at King Edward's, though his time there (if at all) would have been very brief, on account of his extreme youth. On 4 January 1708, Mrs Pococke's mother (Elizabeth Milles), about whom we know virtually nothing, apart from a few lines recording her female virtues,[16] was buried in Highclere Church. Dying from smallpox ("a Distemper, which she had been always afraid of"),[17] she left Revd Milles a widow at the age of 70. Two years later, in 1710, Revd Pococke also died of smallpox, leaving Mrs Pococke "a very *disconsolate* Widow".[18] Furthermore, since they lived in tied accommodation, he left her young family without a home.[19] Nevertheless, her late husband bequeathed to her the right and title of some land in neighbouring Totton, together with an investment fund worth £100 and the rest of his "goods, chattels, household stuff, apparel, mony, credits, and all [his] estate whatsoever".[20]

On her husband's death, Mrs Pococke left Southampton and returned to Highclere Rectory, where she was able to look after her elderly father and her brother Isaac, and where her young son could take up his studies at his grandfather's private school. This clearly was a mutually convenient arrangement, which continued to exist for the next 10 years, until Revd Milles himself died in 1720. His death (caused by a series of strokes, which began on 25 November and whose symptoms and course are vividly described in his *Life*)[21] is sentimentally recorded by his eldest son, Bishop Milles, on a monument erected by him in Highclere Church.[22]

16 She is described in *Life of Isaac Milles*, p. 97, as "a most frugal, faithful, and diligent Manager of all his Concerns. For Industry, and a prudent Foresight and Care in the providing for, and governing, her Family, she had few Equals. Her Children found the Benefit of her Conduct, and, as I have heard, have ascrib'd the prosperous Condition, in which they now are, chiefly to her Management".

17 Ibidem.

18 Ibidem, p. 147.

19 At this point, Pococke the traveller was six and his sister Elizabeth, who lived until 1774, possibly younger.

20 Will of Richard Pococke (Senior).

21 *Life of Isaac Milles*, p. 151ff.

22 "He was continually resident, and carefully diligent in the cure of this parish of Highclere, for thirty-nine years, two months, and seven days; when, after having contracted a great feebleness by the labours of his life, he sweetly fell asleep in Christ, without struggle, groan, or sigh, on Wednesday, the fifth day of July, 1720, and of his age the 82d year." Bishop Milles was not actually present at his father's death. Though Mrs Pococke had recalled him from England, "seeing he could no longer be serviceable to him ... towards the latter

On her father's death, Mrs Pococke would again have found herself and her family homeless, with the rectory subsequently being transferred to the next incumbent, Revd Hastings Lloyd, on 17 April 1723.[23] However, one child was immediately taken off her hands, since within a week Pococke had matriculated at Corpus Christi College, Oxford, at the age of 15. It is possible that Bishop Milles contributed towards his university education, which extended over the next 13 years.

Apart from the few lines in her father's *Life*, describing her as dutiful and diligent in nursing her dying father, there are virtually no records which might help us to form an impression of Mrs Pococke's character or the relationship she had with members of her family. We therefore have to rely on the limited sources available to us which, broadly speaking, are the contents of various family Wills and the Milles/Pococke correspondence. Testimony to her superior organisational skills can be seen, for examples, in the sentiments expressed in her husband's Will, made seven years after their marriage and two months after the birth of their son (the traveller) where he makes her his sole executrix and says: "I give unto my said wife Elizabeth Pococke, in an entire confidence that she will manage and provide for my children as well as if I had given particular directions therein."[24]

As is evident from the travel correspondence reproduced in the present edition, Mrs Pococke and her son were devoted to each other. This mutual devotion took various forms: while Pococke spent long hours abroad composing detailed accounts of his trips, for the purpose both of entertaining his mother and reassuring her that he was safe and well, she was of great assistance to him in the management of his literary output from abroad, taking delivery of the journals which he sent home for safe-keeping and later painstakingly transcribing his letters into neat volumes. Also, she played a vital role in managing his finances while he was away from home, especially during his first grand tour of 1733-34, when he experienced problems with accessing his money.[25]

End of this Month of June 1720, took his last leave of him and retired to the *Bath*." *Life of Isaac Milles*, pp. 155-56. It is not known what exactly it means that Bishop Milles retired to "the Bath" but it is perhaps a typographical error in the original, meaning that he retired to the city of Bath.

23 The reason for the lengthy delay in appointing Lloyd was that he had not actually taken Holy Orders. He was duly ordained as deacon on 23 December 1722 and as priest, 10 March 1723. We do not know if Highclere Rectory remained vacant during the intervening three years, or whether it was occupied by a curate, since the records are silent on the matter.

24 The Will of Richard Pococke (Senior) is dated 28 January 1705.

25 It should be noted that Pococke's travel correspondence to his mother was not confined to his two grand tours, but was applied just as assiduously to his tours of Scotland and Ireland, right up until her death in 1760, at which point the recipient of his letters became

Regrettably, though, the correspondence is one-sided, and we do not see the letters she sent in return to each of her son's missives. However, letters between Mrs Pococke and the bishop (including one to her son) are sufficient to illustrate her literary style, the absolute strength of her religious convictions, her determination to see her son do well in his clerical profession, and her complete devotion to her brother, the Bishop of Waterford & Lismore.[26] In addition, the content of certain letters indidcate that she was highly educated, well versed in the classics and modern travels, and had an impressive library that she must have added to after her husband's death. For example, passages in Letters 35, 40 and 44 suggest that she owned a copy of Edmund Chishull's scholarly books of inscriptions.[27] She could not have inherited this from her late husband (d. 1710) or father (d. 1720), as both these relatives predeceased the publication dates of these books (1721 and 1728). If she had acquired them for herself, which looks likely, she was clearly more of a scholar than has hitherto been acknowledged.

Thirty-seven years after his death, Mrs Pococke appears to have still felt a great sense of duty and affection towards her late husband. In her Will, she gives the following instructions to her daughter Elizabeth concerning her burial: "I remmit .... my Body to the ground to be decently and Christianly buried as to my Executrix hereafter to be named shall think fit in or as near my Dear Husbands Grave in Holy Rhood [sic] Church in Southampton as may be."[28] Further details of her character (and of course her financial situation) may be gleaned from the various legacies she received throughout her life, as well as

---

his sister, Elizabeth. His letters from Scotland, for example, address Mrs Pococke as "Honoured Madam" or "Dear Madam" up until 1760, at which point they are suddenly addressed to "Dear Sister"; and the same is true of the letters sent from his Irish tours, where he again addresses his mother as "Honoured Madam" or "Dear Madam". However, none in the latter collection are addressed to his sister, as the tours end in mid-May 1760, presumably just before his mother's death. See John McVeagh, *Richard Pococke's Irish Tours*, Irish Academic Press, Dublin, 1995. On the other hand, Pococke's letters from his tours of England are not addressed to any particular person and it is possible that he simply used the epistolary account as a literary convention. See James Joel Cartwright (ed.), *The Travels through England of Dr Richard Pococke, Successively Bishop of Meath and of Ossory during 1750, 1751, and Later Years*, Camden Society, London, 1888-89.

26 Two of these are reproduced in Finnegan, *Letters 1*, Appendix 2, Letters 1 and 2, pp. 303-5.

27 It is not clear which of these are referred to in the letters (perhaps both: *Inscriptio Sigea Antiquissima Boustrophedon Exarata/Commentario eam Historico, Grammatico, Critico Illustravi*, London, 1721 and/or *Antiquitates Asiaticae Christianam Aeram Antecedentes: Ex Primaris Monumentis Graecis...*, London, 1728.

28 See Elizabeth Pococke's Will, TNA, Prob.11/858.

from the contents of her own Will, drawn up in 1747, 13 years before her death.[29] We know from these sources that by the year 1721, she possessed her husband's entire estate plus £100 from her father; in 1740, on the death of her brother Bishop Milles, she inherited a further £500, and at some date between then and 1747, she had received another legacy of £150, together with a house and all its contents. Though the combined amount of £750 may not seem a great fortune, in today's terms it works out at about £105,000 when equated to retail values, or almost £1.25 million, when equated to average wages of the period. Also, this is the minimum we can assume she possessed at that time, since there is the unspecified sum of "all my money" left to her daughter, who at the time of her own death (1774), could afford to be extremely generous to her own relatives and family servants.

Mrs Pococke died in 1760 and though we do not know her precise age at death, we can assume that she was in her early eighties. Curiously, given his dependence on her while she lived, there is no reference in Pococke's travel correspondence to his mother's decease. At the time of her death he was on a tour of Scotland, and as mentioned in an earlier note, simply switched the addressee of his travel correspondence from his mother to his sister. Mrs Pococke was presumably buried close to the grave of her husband, as stipulated in her Will, and was joined 14 years later by her affectionate daughter Elizabeth (by then probably in her seventies), who requested the following very modest funeral arrangements in her own Will: "I resign... my Body to the Ground to be decently buried in Southampton in Holy Roods Church in my most Dear Mothers Grave so as not to disturb her Ashes. And if not already done, to have an Ash turnd just over her Coffin and my Corps laid on it and an Ash turnd just over my Coffin. I desire to have a Coffin rooved in White and everything given in White and all at the least expence consistent with common Decency and no Pall Bearers".[30]

2 Life of Revd Dr Thomas Milles (1671-1740)

We continue the biographical accounts with a consideration of the life of Bishop Milles,[31] the recipient of one collection of Pococke's eastern letters and the

29 See Mrs Pococke's own Will, TNA, Prob 11/858. See also the bequests made to her in the will of her father, Revd Isaac Milles, TNA, Prob 11/578, and likewise the bequests made to her in the will of her brother, Bishop Thomas Milles, TNA, Prob 11/717.

30 Will of Elizabeth Pococke (Spinster), TNA, Prob 11/999.

31 As in the earlier sections, in order to avoid confusion, Thomas Milles will henceforth be referred to as "Bishop Milles" or "the bishop", though he did not actually succeed to this

patron of both his nephews.[32] The bishop was, as we have seen, Mrs Pococke's eldest brother, born on 19 June 1671, and it is likely that he attended his father's school in Highclere Rectory. He matriculated at Wadham College, Oxford, on 12 March 1688-89, aged 17 and was admitted as a Goodridge Exhibitioner 1691-92. He obtained a BA on 5 February 1692 and was ordained a deacon by Bishop Hough, on 6 June 1694, in Magdalen Chapel, Oxford. Later that year, on 7 December, he was admitted Chaplain of Christ Church. The following year, 1695, he was awarded an MA from St Edmund Hall on 28 May and was appointed vice-principal of the same college. On 16 March 1701, he was ordained a priest by the Bishop of Oxford, William Talbot, in Christ Church Cathedral; on 9 May 1704 he obtained a BD from Christ Church; and on 15 July 1707 he gained a DD from Trinity College, Dublin.[33]

There is no record of his having undertaken a grand tour,[34] though the antiquary and diarist Thomas Hearne (see below), notes in a diary entry of 14 September 1705, "Mr. Milles of Xtchurch tells me he has some design of Travelling into Forreign Countries".[35] No further mention is made of this, so we do not know if his design was ever carried out. Nevertheless, the sources suggest that he did visit France. In one of his earliest letters to his uncle, for example, Jeremiah Milles, when describing a Benedictine monastery in St. Dennis', notes: "It is impossible to pretend to give your Ldship any account of this it is so large, and I do not in the least doubt but your Ldship has seen it & viewed it with more accuracy, than I could possibly do."[36]

Bishop Milles succeeded Revd Humphrey Hody[37] as Regius Professor of Greek in 1706, an appointment which incurred the wrath of his most strenu-

---

position until 1708. For a more detailed biographical account of Bishop Milles, see R. Finnegan, "Thomas Milles Bishop of Waterford and Lismore", *Decies, Journal of the Waterford Archaeological & Historical Society* 69 (2013), pp. 53-80.

32 Not only did the bishop finance the education and travels of Pococke and Jeremiah Milles, but he took a particular interest in his latter nephew, who was the son of his younger brother Jeremiah, and appointed him as his sole heir.

33 For the various dates, see R.B. Gardiner, *Registers of Wadham College, Oxford*, London, 1889-95, vol. 1, p. 357, J. Foster, *Alumni Oxonienses*, Oxford, Oxford University Press, 1887-92, vol. 3, p. 1015 and the Ordination Records from *The Clergy of the Church of England Database*, under the entries for both Thomas Milles and the bishops who ordained him.

34 No entry exists in Ingamells, *A Dictionary of British Travellers*, which only begins at 1700.

35 C.E. Doble and D.W. Rannie (eds), *Remarks and Collections of Thomas Hearne. Nine Volumes, covering the Years 1705-1714*, Oxford Historical Society, 1885-98, vol. 1, p. 44.

36 BL, Add MS 15773: Milles to the bishop, Paris, 6/17 September 1733.

37 Hody (1659-1707) was a member of the non-juror movement (whereby six bishops and about 400 lower clergy refused to swear an oath of allegiance to William and Mary, and their successors under the Protestant Succession Act of 1689) and was friendly both with Thomas Hearne and the scholar and theologian Henry Dodwell.

ous critic (or "special aversion"), the same Thomas Hearne,[38] whose diary entries contain frequent and often unwarranted attacks on his character and his scholarship. The same is true of the bishop's subsequent appointment as Greek Professor in the University of Oxford, on 17 February 1707.[39] However, his colleagues in that university were soon to be relieved of his presence, when, a year later (March 1708) he secured from his patron, Thomas Herbert, 8th Earl of Pembroke,[40] a recommendation to the See of Waterford & Lismore. This honour resulted, in turn, from the earl's own appointment, the previous year, as Lord Lieutenant of Ireland. From the church records, it appears that Bishop Milles severed all ecclesiastical links with England and took up full duties (though clearly not full residence) in Waterford.

According to a letter he wrote to his sister (Mrs Pococke), on 12 May 1708, he was "received with great ceremony, solemnity & kindness both by the Clergy & People" of Waterford; he found the town "very agreeable"; and claimed: "I shall be very happy if it pleases God to afford me Life and Health."[41] He also referred in this letter to his domestic arrangements, noting: "I lodge in my own Palace, but eat generally abroad the Clergy & People of the Town inviting me almost every Day."[42] However, it is not known how long this hospitality was to last, or the extent to which he continued to mix with his flock. His appointment as Bishop of Waterford & Lismore was unpopular in Ireland, especially for his high-church tendencies (he allegedly wore a crucifix as his pectoral cross) and his inclination to give all his important preferments to members of his own family. Though it is well known that nepotism among English-born bishops during this period was rife,[43] it must be said that the number of his younger brothers and nephews appearing in the succession lists for this diocese is quite astounding.[44]

---

38 See Joseph Wells, *Wadham College*, London, 1898, p. 123

39 For a full discussion of Hearne's catalogue of criticisms against Bishop Milles, see Finnegan, "Thomas Milles Bishop of Waterford and Lismore", pp. 56-58.

40 The family patron mentioned earlier, who, according to the anonymous life of Revd Isaac Milles, had such a regard for the integrity and piety of the latter, "as... inclined him to permit his eldest son to attend him into *Ireland* ... and caused him to recommend him to the late Queen for the vacant See of *Waterford* and *Ireland*, which he now enjoys" (*Life of Isaac Milles*, pp. 89-90)

41 Reproduced in Finnegan, *Letters 1*, Appendix 2, Letter 1, pp. 303-4.

42 Ibidem.

43 See Toby Barnard, *A New Anatomy of Ireland: The Irish Protestants, 1649-1770*, Yale University Press, New Haven, 2003, pp. 88 and 98-99

44 See the entries for Pococke & various members of the Milles family in W. Rennison, *Succession List of the Bishops, Cathedral and Parochial Clergy of the Dioceses of Waterford and Lismore, from the Earliest Times, Together with Some Hitherto Unpublished Records,*

Furthermore, despite the encouraging welcome afforded him on his arrival, much of the bishop's 32-year period of office in Waterford was marred by his unhappy relations with Waterford Corporation, with whose officers he fought for many years (from 1711-32) over their alleged mismanagement of charitable institutions and other matters. In retaliation, they accused him of impropriety, and in particular of entertaining papist servants in his house, which they considered as offensive. While details of these complaints, counter-complaints and endless court cases have already been discussed in another publication, based on the Minutes of Waterford Corporation for the period,[45] the editor recently came across an account of "*five curious documents to the civic authorities of Waterford* [original italics]", namely a series antagonistic letters between Bishop Milles, the Attorney General and the Mayor and Sherrifs of Waterford, dated 1722-35. This fascinating material, which is described in Lot 314 of Thomas Thorpe's sale catalogue and is therefore part of the library inherited by Jeremiah Milles on the death of his uncle,[46] relates to a further longstanding dispute which the bishop had with the Crown, particularly with building works that affected his living arrangements. It appears, from the summary of these letters, that he was being subjected to continued "annoyance" and wished "to be quieted, and live unmolested in his Epispcopal House". However, he was twice rebuffed: once, in 1724, when he was informed that his Excellency could not "interfere or prevent the City from making use of their Ground, &c.," and 11 years later, in 1735, when he was told that they could not "interefere without new Matter".[47]

This issue probably relates to the Corporation's complaint about the fact that the bishop did not spend enough time in Ireland. Family correspondence corroborates this fact, as he was in England very frequently and for long periods. Indeed, in the letter to his sister cited above,[48] he mentions the possibility of returning at "Whitsuntide", only weeks after his first arrival in Waterford. As though to make this point, as soon as he died (an event which passed without mention in the Corporation Minutes),[49] a more spacious palace was built for

---

Ardmore, 1921. These nepotistic preferments are also discussed in Finnegan, "Thomas Milles, Bishop of Waterford and Lismore", pp. 57-59.

45 Finnegan, "Thomas Milles, Bishop of Waterford and Lismore", pp. 63-69.

46 See Thorpe, *Manuscripts, upon Papyrus*, pp. 77-78.

47 Ibidem, p. 78.

48 See Finnegan, *Letters 1*, Appendix 2, Letter 1, pp. 303-4.

49 This was despite the fact that the board met frequently during the year of his decease (1740), when he was sufficiently ill to warrant his nephew Jeremiah Milles's return from abroad.

PLATE 3 Exterior View of Waterford Cathedral, oil on canvas, artist unknown (c. 1730). Courtesy of Waterford Museum of Treasures
PHOTO DONNCHADH Ó CEALLACHÁIN

his successor, Bishop Este, in the hope that he "may be incouraged to dwell here".[50]

When Pococke and Milles set out on their first grand tour, Bishop Milles was over 60 years of age and without an heir. We know very little of his relationship with his immediate family, though as we have seen, he showed great admiration and respect for his father on his death in 1720, attributed by some as having written his anonymous biography, and setting up a handsome memorial to him in Highclere Church. Perhaps his most generous and worthy act was that of providing an education for his nephew Jeremiah Milles, putting him through Eton and Oxford, and then, while on his first grand tour (clearly financed by him) offering him a preferment in Waterford.[51] As is shown in the correspondence, the young Milles was astounded at the offer (he had not actually de-

50 See the Minutes of Waterford Corporation, for 28 March 1741, Waterford City Archives, Ref. LA1/1/A/10, 105. For further details of these plans, see Finnegan, "Thomas Milles, Bishop of Waterford and Lismore", p. 65.

51 For the story supposedly behind this act of generosity, see under Life of Jeremiah Milles, in Finnegan, *Letters 1*, pp. 56ff.

PLATE 4
View of the South Prospect of Lismore Cathedral. *The Whole Works of Sir James Ware Concerning Ireland, Revised and Improved by Walter Harris*, Dublin, 1764, vol. 1

cided on a church career) which he considered to be a great honour that he could hardly live up to, and immediately started home from his travels, making straight for Ireland.

Why Milles felt compelled to write to his uncle every few days is understandable: since the bishop was paying for his travels, he felt obliged to keep him informed of his movements, and this also provided an opportunity to prove to him that he was a serious traveller and scholar, and worthy of becoming his heir. However, from the tone and contents of several letters, Milles seems to have had a genuine affection and respect for his uncle, who had sponsored his eduction and taken a real interest in his career.[52] Pococke, who had already benefitted so much from the bishop's patronage in terms of preferments in his diocese, also appears to have felt the same esteem and obligation towards him, and, at least in his later travels, wrote equally lengthy and regular letters from abroad. It is clear that the bishop had copies made of the voluminous letters from his nephews on their travels because he was genuinely interested in their content, recognised their literary potential, and intended to some day (health permitting) have them published.

Bishop Milles died of "gravel" (gallstones) on 13 May 1740, from which he had been suffering for at least the previous three years. It is possible that Jeremiah Milles was not actually there at his death, since the British Library holds a manuscript in the Milles collection entitled, "Notes in a Journey from London

52 In 1724, when Jeremiah was only 10 years old, the bishop had noted to his sister, Mrs Pococke: "Nothing shall be wanting that is reasonable which I can furnish for his Education. I design him for the Law, if he will take to it, & will not rather desire some other Calling". See Letter from Bishop Milles to Mrs Pococke, Waterford, 9 January 1724, in the Waterford Museum of Treasures.

to Holyhead, 6-13 May [1740?]".[53] It is likely, therefore, that he was in England at the time and hurried back to Waterford on hearing of his uncle's deteriorating health. According to some superficial biographies,[54] the bishop was buried in Christ Church Cathedral, Waterford. However, given that he specifically requested, in his Will that "[his] Body may be conveyed to High Clare [sic] near Newbury in Hampshire to be interred in the Chancell of the Parish Church there as near as conveniently may be to the Bodys of [his] ffather and Mother swiftly but with as little ffuneral Pomp as possible and over [his] Grave to be layd a Marble Stone with the following Inscription upon it: subtus jacet Thomas Milles S.J.P. Waterfordiensis & Lismorensis in Hibernia Episcopus...";[55] and since a monument of this description is still to be found in Highclere churchyard, this is highly unlikely. Also, given that Waterford Corporation had been on such bad terms with him for most of his tenure as bishop, and even failed to acknowledge his decease in any of their records; and since there is no plaque to his memory in the present cathedral in Waterford (while a handsome monument to his immediate predecessor, Bishop Foy, takes pride of place in the western corner of the building), it is probably safe to say that his executor and heir, Jeremiah Milles, carried out his funeral arrangements according to his express wishes.[56]

Though the bishop printed a number of scholarly works in his early years,[57] and had been engaged in others,[58] it appears that he published nothing after taking up residence in Waterford. However, it is clear that he never lost his thirst for scholarship, since he refers, in a letter to his sister, to his progress with his magnum opus, a History of England, which must have involved him in a great deal of painstaking research: "I am come now quite to the Conquest in the Civil History of England. The Battle of Hastings between the Conqueror & King Harold was fought on the 14 day of October in the year 1066, wherein King

---

53 BL, Add MS 15776.

54 See, for example, G. Le Grys Norgate, Revised by Julian C. Walton "Thomas Milles (1671-1740)", *Oxford Dictionary of National Biography*, Oxford University Press, 2004. Available at: <https://doi.org/10.1093/ref:odnb/18754> .

55 A transcript of the bishop's Will is given in Finnegan, *Letters 1*, Appendix 1, pp. 298-302.

56 This view is corroborated in an article by Commander A.W.B. Messenger, who notes that, "his body was brought to England and buried in his father's church at Highclere, where his nephew the Dean of Exeter erected a tablet to his memory." See Messenger, "An Eighteenth Century Dean of Exeter and his Family", in *Report and Transactions of the Devonshire Association for the Advancement of Science*, vol. 83, 1951, pp. 29-30.

57 For example, *The Happiness of Those That Suffer for Righteousness Sake ...*, Oxford, 1701; *S. Patris Nostri Cyrilli Hierosolymorum Archiepiscopi Opera Quae Supersunt, Omnia ...*, Oxford, 1703 and *The Natural Immortality of the Soul Asserted: ...*, Oxford, 1707 (see Bibliography for full titles of these works).

58 Such as an edition of Aristophanes' comedies, which he is said, by Hearne, to have abandoned. See Doble and Rannie, *Remarks and Collections of Thomas Hearne*, vol. 12, p. 327.

Harold, his two Brothers Gyrth & Leofwine, & most of the English baron nobility were slain, & an end put to the Monarchy of the English baron Kings. I have now onely the State of the church from the year 828 to the abovementioned year 1066 to set forth. But after this I design three Introductory Discourses; one concerning the state of England or rather Britain in the British times before Julius Cesar, the other concerning the state of Britain during the Government of the Romans, & the third concerning the state of Britain during the times of the barons; & this together with fitting of Maps to these three periods will take up some time." As a rather sad afterthought, he admits: "This Amusement is the onely satisfaction I have in this wretched country & place."[59]

The letter cited above was dated 1724 and therefore 16 years after his arrival in Waterford. Another 14 years further on, this major work was still not published, as is clear from clause 10 of his Will: "As for the History of England which I have drawn up as far as the Reign of King Stephen with the Greatest Care ffidelity and Exactness and if God grants me health and Life may sometime hereafter and of which I have two Copys one writ all with mine own hand in which are diverse Margins and Interlineations the other transcribed by several hands and corrected tho but imperfectly by me My Will is that it if be thought proper in time to come to print and publish it that it be printed according to the Copy of my own hand writing and adding the Subjects of the Paragraphs which are to be found in the Margin of the Transcript of this my History but are not in the Original".[60]

Clearly, the manuscript of his history was inherited by Jeremiah Milles as part of his uncle's estate, since it is listed in *The Gentleman's Magazine* as having been sold at the 1843 Sotheby's auction of the Milles library and described as follows: "Lot 1,195 consisted of the historical researches of Dr. Thomas Milles, Bishop of Waterford, thus described – 1. Concerning the state of Britain before its conquest by the Romans; 2. Concerning the state of Britain under the Romans, &c.; 3. Concerning the state of Britain under the Saxons; 4. History of England from William the Conqueror to the end of the reign of Stephen: the original autograph MS. in 7 volumes, folio, with a fair copy, in 2 very large and 2 smaller folio volumes, with notes, &c. prepared for the press but never published."[61] As with many of the Milles/Pococke travel manuscripts sold at the Sotheby's auction (see Introduction), this lot was acquired by the book dealer Thomas Thorpe, for the sum of £9.

---

59 Letter from Bishop Milles to Mrs Pococke, Waterford, 9 January 1724, in the Waterford Museum of Treasures.

60 See article 10 of the bishop's Will, in Finnegan, *Letters 1*, Appendix 1, pp. 298-302.

61 *The Gentleman's Magazine*, 1843, p. 73.

In turn, Thorpe offered the manuscripts for sale in three separate batches, listed in his catalogue as Lots 383-385. Lot 383 comprises the history from the first invasion of Julius Caesar to the conquest of 1066, together with an account of the state of the church, the succession of bishops, etc. It is described in the catalogue as: "The Original Autograph Manuscript, *neatly written, accompanied with a fair copy of the whole, with Marginal Notes in the Author's autograph, prepared for publication, but never printed, in two very large volumes, atlas folio, all in five vols, Folio, bound vellum*" (£21).[62]

Lot 384 comprises the three introductory mentioned in the bishop's letter to his sister (above). It is noted as "*a most interesting* MS. *pp. 215, neatly written, apparently prepared for the press, but never published, folio, vellum*" (£6, 6 shillings) and further described as being "a most valuable and interesting manuscript to the historian, antiquary, or county collector".[63] Though Jeremiah Milles did not have the work printed, it is possible that either he or his heirs had it bound in vellum, as this detail is not mentioned by the bishop in his Will.

The final lot in the series (Lot 385) comprises the lives of the kings of England from William the Conqueror up to King Stephen, together with an account of the state of the church. It is further described as "the original Autograph Manuscript, in 3 large folio volumes; and a fair copy of the above, *the whole prepared for publication, with Marginal Notes, in the Autograph of the Rev. Prelate, but never printed, in all 4 vols, folio*" (£10, 10 shillings).[64] It is assumed that Mr Thorpe eventually sold the three lots, for the combined sum of £37, 16 shillings, but the whereabouts of the manuscripts is unknown and there is no record of their ever having made it into print.

Regrettably, there are few flattering or positive portraits of Bishop Milles in the records, though it is well documented (particularly by him) that he spent a large amount of his own money on restoring churches in the Waterford area.[65] Also, unusually for a man in his elevated position, there is no known portrait of him, though it has been suggested (but not proved) that the large portrait of Bishop Foy formerly in Christ Church Cathedral, Waterford, but now hanging on the staircase landing of the Bishop's Palace building of Waterford Museum of Treasures, may in fact represent Bishop Milles. One argument against this is that Bishop Milles is reputed to have "possessed a large Roman nose",[66] which

62 Thorpe, *Manuscripts, upon Papyrus*, p. 95.

63 Ibidem.

64 Ibidem, p. 95-96.

65 For further details of his renovations to certain churches in the diocese, see Finnegan, "Thomas Milles, Bishop of Waterford and Lismore", pp. 71-76.

66 See Messenger, "An Eighteenth Century Dean of Exeter", p. 29, who cites an amusing anecdote where Swift humiliated Milles in public on account of his classical learning and his lengthy nose.

is not particularly apparent in this portrait. The bishop is commemorated by a monument in Highclere Church by Louis Francois Roubiliac, thought to have been moved to the present church in 1870, prior to the demolition of the old church beside Highclere Castle.

3 Life of Revd Dr Richard Pococke (1704-65)

Accounts of Pococke's life appear in the editions of his British and Irish travels, in varying degrees of detail. Of his three Victorian editors, Cartwright virtually dismisses the man, with his uncharitable claim: "Apart from his journeying, the life of Pococke presents little to interest the reader, and may be told in very few words",[67] and while George Thomas Stokes devotes several pages to his life,[68] Kemp provides the reader with a comprehensive "Biographical Sketch".[69] More recently, John McVeagh gives a brief account of his life, admitting, "Richard Pococke was a man obsessed with travel [and] no less compulsive was his habit of writing every journey down";[70] and John Bartlett gives details of his life in his article on Pococke's visit to Lebanon.[71] Biographical accounts of Pococke also appear in the histories of the various institutions with which he was associated, such as the school his father ran in Southampton,[72] and the Pococke School in Kilkenny,[73] and while deservedly the traveller has his place in various editions of the *Oxford Dictionary of National Biography*, to date he has never been the subject of a full biography. The present account, in addition to outlining the main points of Pococke's life, also considers matters not previously considered by his biographers, including his domestic arrangements, personal interests, and contribution to learned clubs and societies.[74]

Born in Southampton in 1704, the only son of Revd Richard Pococke and Elizabeth Pococke (née Milles), Pococke, as we have seen, was educated at his grandfather's school in Highclere Rectory and matriculated at Corpus Christi College, Oxford in 1720, aged only 15. He remained a student of this college for

67 The Travels through England, preface, p. i.

68 *Pococke's Tour in Ireland in 1753*, Dublin, 1891, Introduction, pp. 8-11.

69 *Tours in Scotland*, pp. xxxi-1xx.

70 McVeagh, *Richard Pococke's Irish Tours*, p. 11.

71 Bartlett, "Richard Pococke in Lebanon, 1738", pp. 17-18.

72 See Russell, *A History of King Edward VI School Southampton*, pp. 204-9.

73 See Michael Quane, "Pococke School, Kilkenny", *The Journal of the Royal Society of Antiquaries of Ireland* 80, 1950, pp. 36-43, and Lesley Whiteside with Andrew Whiteside, *Where Swift and Berkley Learnt: A History of Kilkenny College* (Dublin, 2009) 130-33.

74 Versions of this biographical account can also be found in the related works of Finnegan, especially *English Explorers in the East*, pp. 71-86.

13 years, receiving a BA in 1725, a BCL in 1731 and a DCL (Doctor of Laws) in 1733.[75] Though there is no reference in the English records to his actual ordination,[76] on graduating with a BA, he received his first preferment from Bishop Milles with the appointment of Precentor (or Chantor) of Lismore Cathedral (1725-56).[77] By the time he and Jeremiah Milles embarked on their first grand tour in 1733, Pococke had already held this position for eight years, as well as occupying a number of lucrative appointments in the Diocese of Waterford & Lismore.[78] Furthermore, while on his first voyage (or perhaps just after his return) he was appointed to the more senior office of Vicar General of Waterford and Lismore (1734-45),[79] and during his second trip became Vicar of Tullaghorton (1736). It is not known where he lived (or predominantly lived) from 1725 until the time of his first grand tour (1733), but since he is recorded as attending most of the biannual meetings of the Lismore Chapter from 1726 to 1730 and again in 1735,[80] it can be assumed that he divided his time between Oxford, his family home in Hampshire, and Waterford.

Pococke returned to England after his second grand tour in 1741, which took him to the east, by which time his uncle and patron, Bishop Milles, from whom he had recently received a bequest of £600, had been dead for a year. On his arrival in England, he was obliged to "wait on" the new Bishop of Waterford in London, presumably to discuss matters relating to the diocese from which he had been absent for the previous five years.[81] On 24 December 1741, he was elected Fellow of the Society of Antiquaries of London and a few weeks later, on 11 February 1742, became Fellow of the Royal Society, on the grounds of being "a Gentleman of Universal Learning, great Curiosity, every way well

75 Foster, *Alumni Oxonienses 1715-1886* (L-Z) (1888), p. 1124.

76 It is likely that he was ordained in Ireland, by his uncle, as was the case with his cousin, Jeremiah Milles

77 A Precentor was the member of the cathedral body responsible for the direction of the choral services, and the first dignitary in the cathedral after the dean.

78 Vicar of Kiltegan (1729), Rector of Outeragh (1730), Vicar of Cahir (1730), Curate of Lisnakill (1731), Curate of Grangemockler (1731), Rector of Fenoagh (1732), Vicar of Dysert and Kilmoleran (1732), Vicar of Mothel (1732) and Vicar of Stradbally (1732). Dates recorded in Rennison's *Succession List.*

79 The chief representative of the bishop in the administration of the diocese.

80 According to "The Chapter Book of Lismore, 1113-1829", which is in the Cotton Library, St Carthage's Cathedral, Lismore, he was present at the meetings of 12 May 1726, 6 July and 3 November 1727, 16 July 1728, 22 July 1730; and again, on 16 April and 14 May 1735. His name appears only three times after this, on 5 April 1738, 5 April 1739 and 17 April 1740, where he is recorded as absent. I am grateful to Dean Paul Draper for giving me access to these records.

81 This was Bishop Este. See Pococke's letter to his mother from Dover, dated 29 August 1741, the last in the collection but not reproduced in this volume. BL, MS Add 22998, Letter 66.

quallified and likely to be a very usefull and valuable member of the Same".[82] He was also a founder member and officer of the Egyptian Society and the Divan Club (see below).

For the next few months he worked on *Description 1*, based on his journals and the notes and sketches he had taken while travelling in the Egypt. This impressive folio volume, containing 75 etched (and occasionally engraved) plates, was printed for the author by William Bower, London, and the ledgers for this company show that a total of 1,250 copies were printed, of which 1,190 were on demy paper and 60 on royal.[83] Printing of this volume was completed on 31 March 1743, and it was dedicated to the author's old family patron Henry Herbert, 9th Earl of Pembroke and Montgomery, the so-called "architect earl". The immediate success of this work encouraged Pococke to proceed with the second volume, which was ready for print on 4 February 1745, and had the same print run as the earlier work. The two parts of this volume contain a total of 104 etched (and occasionally engraved) plates.

Ten London booksellers were appointed to sell the book on Pococke's behalf,[84] and it is fortuitous that a receipt issued to him by one of these, John Nourse, "Bookseller to His Majesty",[85] survived and was printed almost a century later in a magazine. It shows that Pococke supplied books on a sale or return basis, and though we do not know how much commission the bookseller charged, or how much the customer paid, the author received £1, 7 shillings per copy: "April the 7th 1743 Received of the Rev. Dr. Pococke Seven Copies of his first Volume of The Description of the east for which I promise to pay him Nine Guineas When Sold or in proportion for any Number I shall use to return the Remainder. John Nourse".[86]

Sale catalogues, especially those of Thomas Payne of London, tell us the prices commanded by new and second-hand copies throughout the 18th century, though most of these sources relate to the complete set (volumes 1 and 2). In 1754, for example, only a decade after their publication, Payne was selling a

82 RS citation, Ref: EC/1741/15. Among his proposers was Martin Folkes, Inspector of Medals and Coins in the Egyptian Society, of which Pococke was also a founder member (see below). Pococke and Milles met with Folkes on their first grand tour.

83 See Nash et al. (eds), *Early Printed Books*, 1478-1840, p. 2745.

84 These are printed on the title page of the first volume, as: J. and P. Knapton, W. Innys, W. Meadows, G. Hawkins, S. Birt, T. Longman, C. Hitch, R. Dodsely, J. Nourse, and J. Rivington.

85 He is thus described on the title page of Johann Heinrich Cohausen, Hermippus Redivivus, or, *The Sage's Triumph over Old Age and the Grave*, London, 1742.

86 *The Aldine Magazine of Biography*, Criticism and the Arts, "From the Collection of a Lady", vol. 1, 1839, p. 177. A note from Pococke corresponding to the Nourse receipt is held in a private archive in Suffolk, UK, and I am grateful to the owners of this archive for bringing it to my attention. A guinea was £1 and 5 pence.

copy of the set, described as "new and elegantly bound in Russia leather", for £4, 4 shillings.[87] In 1761, however, the price had dropped, so that a "new and elegantly bound" copy of the set was being offered by the same bookseller for £2, 15 shillings.[88] This was still for sale at the same price the following year, along with a copy of the first volume, at £1, 1 shilling.[89] Two years later Payne was selling the set "neatly bound in 3 vol." for £2, 10 shillings.[90] There are two copies of the set, described as "new", in the 1769 catalogue printed by the bookseller Benjamin White, who was selling the library of the famous letter writer Joseph Spence, whom Pococke had met in Rome, but unfortunately no price is given.[91] From 1770, however, the value of the book began to increase, since Payne was now offering the set, bound in Russia and with marble leaves, for £3, 3 shillings.[92] In 1779, a set was for sale at £5, 5 shillings, and, interestingly, for £10, 10 shillings he was advertising the same set "bound in 3 vol. large paper, with a portrait of Pococke inserted, proof impression".[93] It is possible that this proof copy had become available after the death of William Bowyer, in 1777, as it was obviously not part of Pococke's estate, or it would probably have been part of the 1843 Sotheby's book sale. However, its present whereabouts is unknown.

Two years later, a "very neat" copy of the first volume was being sold for £3.[94] Kemp gives a similar account of sale prices in the 19th century, by which time they had risen further, when he notes: "If auction sale prices may be taken as indicating their value, it may be mentioned that the 2 vols. brought £21 at the Marquis of Townshend's sale. [1812],[95] £16, 10s. at Heaths' [1810],[96] £14 at

---

87 T. Payne, *A Catalogue of a Very Large and Curious Collection of Books, in All Branches of Learning...*, London, 1754, p. 180.

88 T. Payne, *A Catalogue of a Large Collection of the Best Books, in All Parts of Learning, etc.*, London, 1761, p. 9.

89 T. Payne, *A Catalogue of a Large and Curious Collection of Books..., London, 1762*, p. 9.

90 T. Payne, *A Catalogue of Twenty Thousand Volumes..., London, 1764*, p. 8.

91 B. White, *A Catalogue of the Entire Libraries of the Rev. Mr. Joseph Spence...*, London, 1769, p. 19.

92 T. Payne, A *Catalogue of the Library of the Late John Grey, Esq...*, London, 1770, p. 9.

93 T. Payne, *A Catalogue of Valuable Books, in Various Languages, and in Every Branch of Science and Literature*, London, 1796, p. 22.

94 T. Payne, *A Catalogue of a Large and Curious Collection of Books...*, London, 1781, p. 7.

95 This figure is verified in a note beneath Pococke's *Description 1* and *Description 2*, in the sale catalogue which included the library of Jeremiah Milles: "A copy of these interesting Travels sold in the Marquis of Townshend's sale for Twenty Guineas". See Thorpe, *Manuscripts, upon Papyrus*, p. 84.

96 This was a three-volume set, as noted by William Clarke, author of *Repertorium Bibliographicum; or Some Account of the Most Celebrated British Libraries,...*, London, 1819, p. 468.

Townby's, and forty years ago [c. 1847] for large paper copies £10 was a common price…"[97] However, the prices of these copies could have been inflated because of their provenance.

While the various figures cited above can be verified by William Thomas Lowndes, who, in 1834, gives details of prices fetched at other library sales,[98] Revd Thomas Frognall Dibdin advises his reader a decade earlier to be "at no pains to secure a good copy" of these volumes.[99] Nevertheless, while he notes that they have been reprinted in Pinkerton's collection for "three-fourths of that price" (the price fetched at Dr Heath's sale), he asks: "But who would not prefer the primitive and embellished folio? These are noble tomes; and the author rises in estimation more and more every day. He is a facile prince – in his department."[100]

Pococke's pride in his book is evident from the fact that he was in the habit of bestowing bound copies on friends and associates, as well as institutions, years after its publication. For example, in a letter to the Revd Dr William Stukeley, dated London, 4 March 1745, he wrote: "I have orderd to Mr Manby a copy of my second volume in sheets, it is so thick that it binds best in [?] being near as thick again as the first, you will do me the honour to accept of it; & he is to receive your orders in relation to it."[101] This may have been a reciprocal arrangement, as Stukeley's own book "Stonehenge and Abury [Avebury]" in two volumes (London, 1740) was in his library.[102] Similarly, in a letter from Grosvenor Street, London, dated 19 February 1757, Pococke wrote to the architect Sanderson Miller, whom he had commissioned to carry out some work on the cathedral church in Kilkenny: "I have orderd my book binder to prepare a work of one who has a very great regard for you & yours, & shall request the favour of you to let it have the honour of a place in your Library."[103]

Likewise, a copy of the work, "in fine condition", is in the library of King Edward's School, Southampton, with the inscription "Donum Authoris"

97 Kemp, *Tours in Scotland*, p. xxxviii.

98 See *The Bibliographer's Manual of English Literature* … by William Thomas Lowndes, vol. 2, London, 1834, p. 1471.

99 T. F. Dibdin, *The Library Companion, or the Young Man's Guide, and the Old Man's Comfort, in the Choice of a Library*, London, 1824, p. 441.

100 Ibidem, p. 421.

101 Oxford, Bodleian Library, MS. Eng. misc. c. 114, fol. 147r. I am grateful to Oliver House, Superintendent, Special Collections Reading Rooms at the Bodleian for his kind help and for supplying scans of the letter.

102 See Flin, *A Catalogue of the Library*, Lot 181, p. 5.

103 Pococke to Sanderson Miller, 19 February 1757, Warwickshire County Record Office, CR125/B802.

(compliments of the author) on the title page of each volume.[104] Pococke's father, as already noted, had been headmaster of this school until his death in 1710 and had established a library there. Judging from both the inscriptions in the books and the catalogue, the collection was based on gifts from his personal friends, as well as donations from old boys and masters of the school, Oxford dons, and prominent men in the municipal life of the town. His son's donation was therefore in keeping with the school's tradition.[105]

Pococke's book was never reissued in its original form, so even though it was included, either in full or in part, in various collections of famous travels, the 1743/1745 version remains the first edition and complete sets (bound usually in two volumes though occasionally three) are very valuable, particularly when all the plates and maps are intact.

In 1744, Pococke was appointed as Domestic Chaplain to Philip Dormer Stanhope, 4th Earl of Chesterfield (1694-1773). The latter had been appointed Lord Lieutenant of Ireland in November 1744, but on account of other diplomatic duties,[106] was unable to take up the post immediately. In a letter from sent to his friend Dr Hartley, on 7 June 1745, Pococke explained: "I am going again to Ireland, in the month of August, having the honour to wait on the Lord Lieutenant as his Domestic Chaplain."[107] He and the Earl of Chesterfield eventually arrived in Dublin on 31 August of that year.

It would seem that this was the first occasion on which Pococke actually took up residence in Ireland, even though he had occupied a number of important positions in the Church of Ireland for the previous two decades. He dedicated *Description* 2 to Lord Chesterfield, who reciprocated this public honour by appointing him the same year to the Archdeaconry of Dublin. Chesterfield then returned to England on 11 April 1746, leaving Pococke in this senior position in Dublin, where he was to remain for the next 10 years, and from which he was to be translated to his first bishopric.

His pioneering book, illustrated with 179 plates, many of which are dedicated to friends and associates of the author, immediately became a standard work on the Levant, and was translated at the time into French, German and Dutch, as well as being incorporated into various compendiums of 18th-century

104 See Russell, *A History of King Edward VI School Southampton*, p. 206.

105 Ibidem, pp. 213-25.

106 He was sent to The Hague to persuade the Dutch to enter the war against France.

107 Quoted in J. Nichols, *Illustrations of the Literary History of the Eighteenth Century, Consisting of Authentic Memoirs and Original Letters of Eminent Persons; and Intended as a Sequel to The Literary Anecdotes*, vol. 2, London, 1814, p. 806.

travels.[108] Although attracting extensive criticism from contemporary literary and theological rivals such as Shaw and Warburton,[109] it gained the attention and admiration of academics and architects alike, earning him a place among the literati and architectural circles of the day. Its impact lived on well into the next century, when it continued to be held up as authority on the east, and was universally described as a learned, scholarly and desirable work;[110] and of course it is even more collectable today, though surprisingly has never been reprinted in English in its original form. In addition to the book's intellectual and visual appeal was of course its practical value, and it is a good example of a grand tourist's views on the benefits of foreign travel – a stock subject with travel writers, particularly as a conclusion to their work.[111]

It is difficult to ascertain precisely where and when Pococke resided throughout the period from his university days until his arrival in Dublin, in 1745, though of course for almost half that time (1733-34 and 1736-41) he was abroad. Also, it is clear from the earlier grand tour correspondence that he spent time at his mother's house in Newtown, since this is where Jeremiah Milles collected him when setting out on their second trip. It is unlikely that he owned or even leased a house in London, but clearly he had access to accommodation in the capital at least after his return from his travels, since he is recorded as regularly attending the various clubs and societies to which he belonged.

Pococke was a founder member of the Egyptian Society,[112] whose only surviving record is the Minute Book (or "Journal"), which was kept from 11 December 1741 to 16 April 1743.[113] This important manuscript contains a list of the members and their signatures, the Laws of the society, and reports of papers delivered by members, together with several illustrations of items which they discussed and exhibited. The manuscript has had a chequered career. Though Pococke was Reis Effendi to the society immediately before its demise and is likely therefore to have held on to the "Journal" throughout his life, it was

108 For a full list of the various editions and translations, see Finnegan, *English Explorers in the East*, Appendix, pp. 293-94.

109 See the Introduction to this volume.

110 For futher details on the reception of this book, see Finnegan, *English Explorers in the East*, pp. 273-75.

111 See Finnegan, *English Explorers in the East*, Chapter 2.

112 The other members are listed as follows: Sandwich, Charles Perry, Frederick Ludewig Norden, Thomas Lisle, Tho. Anson, Ed. Vernon, Wm Lethieullier, C. Stanhope, M. Folkes, Wm Stukeley, T. Dampier, Jeremiah Milles, Montagu, Richmond, Lenox, & Aubigny, Andr Mitchell, Cutts Barton, Smart Lethieullier, Daniel Wray, Roum [Truleman], Solomon Bagrolle, Jo. McKye, H. Legge, Lovell, Thos Hill, Wm Fauquiert, Lincoln, Duncannon, W. Bristow and E. Legge.

113 BL, Add MS 52362.

among the many Milles/Pococke manuscripts sold at the 1843 Sotheby's auction, so must have been acquired at some point (perhaps after Pococke's death) by Jeremiah Milles and absorbed into his library.

The manuscript is briefly described in *The Gentleman's Magazine*, which states that it was sold to Mr Thorpe for the sum of £2.[114] The detailed description in Thorpe's catalogue (item 187) confirms that it was sold along with "a curious brass symbol, or regalia of the Society, in a case",[115] which must refer to the sistrum (ancient Egyptian music percussion instrument) whose acquisition had occupied so much of the members' time and consideration.[116] This item was sold for £5, 5 shillings and though it may have had other owners in between, it was eventually acquired by Walter Prideaux, a lawyer and clerk to the Goldmith's Company, who wrote to the editor of *Notes and Queries* in 1885,[117] asking if any of his correspondents could give him information on the Egyptian Society.[118] Four years later, in 1889, the manuscript was sold at Christie's, after which it became the property of the Egyptian Exploration Society of London (founded in 1882), where it still belonged in 1952 (the date of Annis's article). In 1964, it was given to the British Library where it has become part of the Pococke Collection.

One of the more spectacular events of this society, which like others of its type was essentially a dining club, was the occasion on which Pococke performed the opening up and dissection of a mummy in the Duke of Richmond's house.[119] The observations were recorded by Mr Dampier at the following meeting (5 March), after which Pococke communicated to the society some "observations on the manner of embalming bodys amongst the Egyptians". Though the latter account is based largely on Herodotus, it also includes interesting points concerning what he had personally observed in the Catacombs.[120]

We can see from this source that Pococke attended all 20 meetings held between its establishment on 11 December 1741 and its last recorded meeting on 16 April 1743. As already noted, it was during the second year of of this term (from 3 December 1742 until the society's demise) that he acted as its Secretary. It was stated in the "Laws", drawn up during the first session, that meetings

114 See *The Gentleman's Magazine*, 1843, p. 72.

115 Thorpe, *Manuscripts, upon Papyrus*, p. 46.

116 This is depicted on the cover of Finnegan, *English Explorers in the East*.

117 *Notes and Queries: A Medium of Intercommunication for Literary Men, General Readers, Etc.*, vol. 10, London, 1885, p. 427.

118 See M. Annis, "The First Egyptian Society in London (1741-1743)", *Bulletin de L'Institut Français d'Archéologie Orientale* 50 (1952), p. 100, n.9

119 This took place on Monday 24 February 1742.

120 See BL, Add MS 52362, pp. 21ff.

should take place every fortnight from the first Friday in November to the first Friday in May. Furthermore (and thus establishing the fact that he spent at least one summer in London), at the meeting of 14 May 1742, it was resolved that "the further consideration of the Egyptian medals be deferred till next winter, & that in the mean time Dr Pococke, Dr Perry, & Mr Lethieullier, & any other members that shall stay in town this summer be desired to enquire after proper persons to draw & engrave the said medals…".[121]

Pococke was also a founder member of the Divan Club (originally named the Turkish Club), the other nine individuals, who qualified on account of proving "that they have been in the Sultan's Dominions", being Sir Francis Dashwood, Lord Duncannon (future 2nd Earl of Bessborough), Mr Edgcumbe, Mr Fanshawe, Sir Everard Fawkener, Mr Frolich (tutor to Lord Sandwich), Lord Granby, Mr McKye and Lord Sandwich.[122] In a similar vein to the Egyptian Society, the rules of the Divan stipulated that "this Society meet once every fortnight: viz, every other Friday, from the first Friday in December, to the first Friday in May", though this was amended later to every Sunday. Pococke was present at the club's first meeting on 8 January 1744 and acted as Reis Effendi for its fourth meeting. This was the title given to one of the chief Ottoman officers of state, but here, as a joke, refers to the secretary to the society. He then attended the next 11 consecutive sessions, during which he stood Reis Effendi for the second time, as well as Vizir – the term for a high-ranking political advisor or minister, in the Ottoman system but, in the context of the club, referring to the chairman. However, he only attended once more, at the last session before the summer recess, on 24 May 1745, being absent from the last 10 recorded meetings. His absence can of course be explained by the fact that on 31 August, as noted above, he had arrived in Dublin with his new patron, the Earl of Chesterfield.

Equally, Pococke must have retained an address in London after taking up residence in Ireland, since he regularly visited the capital, and as can be seen from his grand tour correspondence, liked to spend his winters at his family home in Newtown. Two letters in a collection of correspondence between himself and the gothic-revival architect Sanderson Miller (1716-80) regarding some alterations to St Canice's Cathdral, Kilkenny, give Grosvenor Street as Pococke's address.[123] However, the address of the sender given in such correspondence is not necessarily an indication of a permanent or even regular

121 BL, Add MS 52362, p. 28.

122 See "Al-Koran, the Minute Book of the Divan Club", National Maritime Museum, London, Ref. SAN/V/113. Short biographies of all these members, together with some who joined later, are given in Finnegan, "The Divan Club".

123 See letters from Pococke to Sanderson Miller, 8 and 17 February 1757, Warwick County Record Office, 125B/801 and 802.

arrangement. In the letter sent to Dr Hartley in June 1745 (cited above), for instance, Pococke gives his address as Rawthmell's Coffee House, Henrietta Street, Covent Garden. Since this was one of the more informal meeting places of members of the Royal Society (and was later to be the place of origin of the Society of Arts), we must not assume that he had lodgings there, but rather found it a convenient and convivial place in which to conduct his daily business.[124] Likewise, a letter dating from January 1748 gives his address as Adams Court in Broad Street, Behind the Royal Exchange, London.[125]

The matter of Pococke's accommodation arrangements for the first few years of his residence in Ireland was unclear to one of his nineteenth-century biographers, George Stokes, and has remained so until very recently.[126] However, records now show that in March 1746, seven months after his arrival in the country, he leased a property on the vast Fitzwilliam Estate (later renamed the Pembroke Estate), which was being developed on the south side of the city.[127] Though the precise location of the leased house (which had recently been vacated by Lord Prendergrasse) is not known, it is likely that it is the same property that he was occupying a year later, in 1747, in Leeson Street W.[128] This was a central and desirable area, the location of which ties in with the account of Mrs Delany (wife of Dean Patrick Delany), who says, in a letter dated 30 April 1752: "I ... was yesterday at Dr. Pocock's Drum. The Dean [Jonathan Delany] and I dined at the Bishop of Clogher's, and went from thence, it being in the neighbourhood."

She further describes Pococke's house as being: "very pleasant; he has a large room on the ground-floor of thirty-one feet long, filled with his curiosities. It opens into a very pretty garden, gaily filled with flowers."[129]

124 For a discussion of the connection between coffee-houses and gentlemen's clubs, see A. E. Musson and E. Robinson, *Science and Technology in the Industrial Revolution*, Manchester, 1969, p. 58.

125 Letter to Emanuel M. Da Costa, dated 1 January 1748, quoted in Kemp, *Tours in Scotland*, p. xlii.

126 Stokes, the editor of Pococke's Irish tour (*Pococke's Tour in Ireland in 1753*) admits, in his Introduction: "I have tried to find out where Pococke lived when Archdeacon of Dublin and Rector of St. Peter's, but have failed. Perhaps some reader may be able to throw light on this local problem" (p. 11).

127 National Archives of Ireland, Pembroke Estate Papers, 2/5/32.

128 See C. Lennon, *Irish Historic Towns Atlas (IHTA), no. 19, Dublin, Part II, 1610 to 1756*, Royal Irish Academy, Dublin, 2008, p. 37. I am grateful to Colm Lennon for verifying this and confirming the original source: Mason MSS, ii (3), 722).

129 A. Hall (ed.), *Autobiography and Correspondence of Mary Granville, Mrs. Delany*, vol. III, London, 1861; digitally printed by Cambridge University Press, 2003, p. 113. Mrs Delany moved to Dublin after her marriage to the Dean in 1743, and much of her correspondence was published in 1861-62 (five vols).

The See of Ossory became vacant in February 1756, after the death of Bishop Edward Maurice, whereupon Pococke, upon the recommendation of Duke of Devonshire, Lord Lieutenant of Ireland, was nominated as his successor in King's Letter of 5 March 1756.[130] He was consecrated in St Patrick's Cathedral, Dublin, on 21 March, by the Archbishop of Dublin, assisted by the Bishops of Kildare and Leighlin.[131] This tendency for the duke to promote only Englishmen to bishoprics did not go unnoticed, as can be seen from a draft note from Sir Robert Wilmott (secretary to the viceroy of Ireland) to Henry Fox seven years later, on 25 February 1763: "Sir R[o]b[er]t pres[ent]s his Comp[liment]s to Mr Fox & acq[ain]ts him that there are not many precedents. Appar[en]tly Ric[hard Pococke B[ishop] of Ossory is an Englishman & if so Dr Leslie and Dr Pocoke are the only Bishops made by the D[uke] of Devonshire."[132]

Two months later, on 24 March, he was enthroned by proxy, though the exact date at which he actually took up residence in the Bishop's Palace is not known. By late June, however, he described his new circumstances to Revd Charles Lyttelton (Dean of Exeter) in very glowing terms, claiming, "every one as well as myself thinks I have got the sweetest situation in Ireland, a most exceeding good house, a pleasant garden, wt. we call a noble cathedral, & good neighbourhood, in a fine country, with a good river running through it".[133] It has been held that the Bishop's Palace, erected on the foundations of the medieval palace was actually built by Charles Este, Bishop of Ossory from 1736 to 1745.[134] However, Bishop Este (who as we have seen was Bishop Milles's successor in Waterford) merely enlarged and improved this residence, the error perhaps being attributed to Chetwood's brief description, in 1748, of the palace as being "new built".[135] It is thought that, among his other alterations and addi-

---

130 The request was made in a letter from the Duke of Devonshire to Henry Fox, dated 14 February 1756, desiring him to "lay before the King [his] humble Request, that His Majesty will be graciously pleased to grant the … Bishoprick of Ossory unto the Reverend Richard Pococke, Doctor of Law, now Archdeacon of Dublin". TNA, 5P 63/414.

131 See J. B. Leslie, *Ossory Clergy and Parishes: Being an Account of the Clergy of the Church of Ireland in the Diocese of Ossory, from the Earliest Period, with Historical Notices of the Several Parishes, Churches, &c., Enniskillen*, 1933, pp. 30-31.

132 Derbyshire Record Office: D3155/C3207.

133 Letter from Pococke to Dean Charles Lyttelton, 25 June 1756, Warwickshire County Record Office, 125B/798.

134 See, for example, Mark Bence-Jones, *A Guide to Irish Country Houses*, London, 2nd revised edition, 1990, p. 167.

135 W. R. Chetwood, *A Tour through Ireland in Several Entertaining Letters Wherein the Present State of that Kingdom is Consider'd and the Most Noted Cities, Towns, Seats, Rivers, Buildings &c. are Described…*, London, 1748, p. 178. The most comprehensive modern account of this building is that of The Integrated Conservation Group, "The Bishop's Palace, Kilkenny", *Old Kilkenny Review, Journal of Kilkenny Archaeological Society* (2003), pp. 30-53.

tions to this property, Pococke created the palace library, with its characteristic "Pococke pilasters",[136] and other classical ornamentation in oak. This room is now the library of the Heritage Council, and these features have been carefully preserved.

Pococke was to remain as Bishop of Ossory for a decade, during which time he carried out many improvements on other churches in the diocese, as well as establishing and supporting a linen-weaving factory at Lintown, just outside Kilkenny.[137] Clearly, he was already a man of independent means, but his promotion to this bishopric greatly increased his income, and, being of a generous disposition, he invested a great deal of his own money in supporting such charitable works.

His transfer to Meath, after 10 years in Kilkenny, took place as follows: "He was translated, by the King's letter, from Ossory to Elphin, in June 1765, Bishop Gore of Elphin being then promoted to Meath; but Bishop Gore, finding a great sum to be paid to his predecessor's executors for the house at Ardbraccan,[138] declined taking out his patent; and therefore Bishop Pococke, in July, was translated by the Duke of Northumberland, directly to the see of Meath…".[139]

According to an Edwardian history of the diocese, it was Bishop John Evans (1716-24) who "had bequeathed certain funds for the erection of an episcopal palace at Ardbraccan and had obtained plans for the building".[140] However, Evans died before executing the work, and his three immediate successors (Bishops Downes, Lambert and Ellis) allowed the project to remain in abeyance. In 1734, the work was finally taken in hand, when Bishop Arthur Price constructed the two wings of the palace to the design of Richard Castle. However, he was then transferred to Cashel, whereupon the work once again ceased, and it was in this state that Pococke found the palace, with only the wings and no central block. Ardbraccan House was eventually completed by the non-resident Bishop Henry Maxwell between 1772 and 1775, to the combined designs of James Wyatt, Revd Daniel Augustus Beaufort, and Thomas Cooley. The same Revd Beaufort some years later praised this episcopal residence, which he described as being "a large and convenient mansion, erected

136 See R. Finnegan, "Bishop Pococke's Improvements to St. Canice's Cathedral, Kilkenny", *Irish Architectural and Decorative Studies: The Journal of the Irish Georgian Society* 11 (2008), pp. 12-55.

137 For detailed accounts of this institution, see Quane, "Pococke School, Kilkenny", and Whiteside, *Where Swift and Berkley Learnt*, especially Chapters 13 and 14.

138 In much the same was as with Bishop Milles, who has to pay £800 to Bishop Foy's heirs

139 J. Nichols and S. Bentley, *Literary Anecdotes of the Eighteenth Century*.… vol. 2, London, 1812, p. 158.

140 See J. Healy, *History of the Diocese of Meath*, vol. 2, Dublin, 1908, pp. 92-93.

by the present bishop, in a style of superior elegance, and yet with such simplicity as does equal honour to his Lordship's taste and liberality".[141]

While in his short-lived role as Bishop of Meath, Pococke acquired number 13 Henrietta Street, a large and stately town-house in Dublin's fashionable quarter. This cost him the considerable sum of £3,800, but it was obviously an episcopal purchase, since not only is there no reference to it in his Will, but the next recorded owner for the following 32 years (from 1766-98) was Henry Maxwell, his successor as Bishop of Meath.[142] Many of the other residents of this street from the time it was built until the 19th century were members of the clergy, and in particular bishops, who would have needed a respectable residence in the capital city in which to entertain important guests, and where they could stay in relative comfort while attending to their many church and social duties.[143]

With regard to his private residence, as has already been noted, Pococke's mother moved from Southampton to Newtown, near Newbury in Hampshire, after her father's death in 1720, though no further details of her address are given. However, we know from later correspondence that the family subsequently lived in "Newtown House" (depicted in Plate 5). Whether this house, built in 1740 (the year in which Bishop Milles died), belonged to Pococke or his mother is unclear, though one source maintains that the property was owned by this family for two generations until Bishop Pococke's death.[144] It is likely, however, that it was built by Pococke, on the proceeds of his uncle's recent bequest, and that he allowed his mother and sister to reside there, since in his own Will (drawn up on 10 July 1763), he states: "First I give and bequeath to my dearly beloved Sister Elizabeth Pococke, Spiner [sic] of Newtown in Hampshire my house and land in Newtown and Hampshire in which she now lives."[145]

141 Daniel Augustus Beaufort, *Memoir of a Map of Ireland; Illustrating the Topography of that Kingdom, and Containing a Short Account of its Present State, Civil and Ecclesiastical, with a Complete Index to the Map*, London, 1792, p. 117

142 Though not his immediate successor, as Arthur Smyth was bishop for a few months before being transferred to Dublin in 1766.

143 For an account of the residents of several houses in Henrietta Street, see Eoin O'Mahony, "Some Henrietta Street Residents, 1730-1849", *Irish Georgian Society Quarterly Bulletin* 2/2 (April-June 1959), pp. 9-19.

144 See D. Ellis, *Newtown at the Millenium,* Newtown Parish Council, May 2000.

145 Last Will and Testament of Richard Pococke. National Library of Ireland, D. 27, 161. This manuscript, which is a photocopy of a 20th-century copy of Pococke's Will and codicil, has minor typographical differences from that reproduced in Kemp's *Tours in Scotland,* pp. lxiii-lxvi. In the codicil to his Will (National Library of Ireland, D. 27, 161) he states, "I do leave to my dear sister Elizabeth Pocoke of Newtown in Hampshire spinster my house and land in said parish of Newtown."

PLATE 5
Early Photograph (pre 1863) of the Pococke family home in Newtown, near Newbury, Berkshire/ Hampshire.
PRIVATE COLLECTION

Had Mrs Pococke owned the property, she presumably would have bequeathed it to her son in her own Will, but she makes no reference to it.

As for his personal interests, Pococke continued his membership of clubs and associations after moving to Dublin. He belonged, for example, to the Physio-Historical Society (1744-55), whose purpose was to promote enquiries into the ancient and present state of the counties in Ireland,[146] and the Dublin Florists' Club (1747-66), established for the promotion of ornamental horticulture in Ireland.[147] Pococke was one of the founder members and an officer of the latter organisation, who was charged, at the meeting of 6 March 1747, with the task of causing "proper receptacles to be made of for the flowers..."[148] Active involvement in such associations, botany, collecting books, antiquities, coins and "natural curiosities" (including a large amount of basalt columns from the Giant's Causeway, in Antrim, which he had assembled as an ornament in his garden in Dublin and afterwards in Kilkenny),[149] travelling and travel writing seem to have been Pococke's major interests outside his actual clerical work. Though obviously close to his immediate family, he did not marry and does not appear to have formed any romantic attachments, unless we are to believe the depiction of him as a lecher in the satirical pamphlet that appeared shortly before his death (see Conclusion, below).

Pococke was very particular about what should happen to his various collections after his death, and in his Will, desires "that all my antiquities and every

146 See Royal Irish Academy, MS 24 E 28.

147 The Minute Book for this club is also in the Royal Irish Academy, MS 24 E 37.

148 For an interesting analysis of the club's minute book, see E. Charles Nelson, "The Dublin Florists' Club in the Mid Eighteenth Century", *Garden History* 10/2 (1982), pp. 142-48

149 For an account of his collecting and collections, see R. Finnegan, "The Travels and Curious Collections of Richard Pococke, Bishop of Meath", *Journal of the History of Collections* 27/1 (2015), pp. 33-48.

thing relating to natural history and all my coins and medals be sent to England to London to be sold by Publick auction as likewise all my books which will not sell here According to a just value and I desire that the Revd Mr Mervyn Archdall be requested to pack up my natural curiosities and label them for which I desire that a proper present be made to him Tho he is assigning witness to this Will."[150] In the codicil to his Will, he further adds: "I do desire that my chaplain the Mr revd Mervyen [sic] Archdall do pack up carefully my natural collection and direct the packing up of all my antiquities all to be sent to London to be sold in proper lotts at publick auction for which trouble I do bequeath him the sum of Twenty five pounds and it is my will that they be sent by long sea to London as my Executors shall direct."[151] This was duly done and (with the exception of his library, which, as already noted, was sold in Dublin rather than in London), his collections were sold by the London auctioneers Abraham and Son at two sales, each of two days' duration, in May and June 1766.[152] The sale catalogues not only provide detailed information on the nature and extent of his connoisseurship, but also contain references to objects presented for discussion at his London clubs, such as ancient medals and coins, Greek and Egyptian statuettes, and mummy cases.[153]

Though his British and Irish travels were edited and published long after his death, Pococke did produce several other minor publications, namely two sermons (1761 and 762)[154] and a piece that appeared posthumously in 1773, in his cousin's newly founded journal, *Archaeologia*. This beautifully illustrated article concerns ornaments and other antiquities found in prehistoric Irish graves and demonstrates Pococke's wide-ranging interests in ancient history and archaeology, as well as his continued and varied collecting record.[155] More im-

150 National Library of Ireland, D. 27, 161. Mervyn Archdall (1723-91) was a fellow antiquarian and Pococke's trusted chaplain. He retired periodically to his spacious country vicarage in Attanagh, just outside Durrow, County Laois (then Queen's County) for relaxation and an area in the ground is still known as the Bishop's Garden.

151 National Library of Ireland, D. 27, 161.

152 See A. Langford, *A Catalogue of a Curious Collection of Greek, Roman, and English Coins and Medals, of the Right Reverend Dr. Pococke, Lord Bishop of Meath, Collected by his Lordship, during his Travels*, London, May, 1766 and A. Langford, *A Catalogue of a Large and Curious Collection of Ancient Statues, Urns, Mummies, Fossils, Shells, and Other Curiosities, of the Right Reverend Dr Pococke, Lord Bishop of Meath, Deceased; Collected by His Lordship, during his Travels*, London, June 1766.

153 See Finnegan, "The Travels and Curious Collections".

154 For the full and very lengthy titles of these works, see the Bibliography

155 See "An Account of some Antiquities found in Ireland; communicated by the Right Reverend Richard Pococke, late Lord Bishop of Meath", in *Archaeologia: or Miscellaneous Tracts Relating to Antiquity* 2 (1773), 1773, pp. 32-41. For modern commentaries on this article, see John Waddell, *Foundation Myths: The Beginnings of Irish Archaeology*,

portant, however, was the large folio volume of Greek and Latin inscriptions collected during his travels, which he had privately printed in 1752 under the title *Inscriptionum Antiquarum Grœcae. et Latinae Liber.*[156] This book, which he must have composed in his time as Archdeacon of Dublin, is 100 pages long, including an appendix, and contains a 27-page supplement of antique inscriptions, which he co-edited with Jeremiah Milles. According to the auction records, this volume was often sold with his *Description*, though Lowndes, writing in 1834, mentions it being sold separately at the library sales of Combe, Gough, and the Bishop of Ely for the relatively low sums of 13 shillings, 10 shillings and sixpence, and £1 respectively.[157]

Pococke's practice of publishing separate volumes on his travels and inscriptions was followed later by the English antiquary Richard Chandler (1737-1810), whose expeditions to the east were financed by the Society of Dilettanti and who published a volume on the antiquities of Ionia, the journals from his travels in Ionia and Greece, and an account of the ancient inscriptions from both places.[158] Very little has been published on Pococke's volume of inscriptions, but as observed by Robert Pitt, who is currently working on a related subject, it really speaks of the learning of the age that these authors thought inscriptions were worth a volume on their own, albeit in both cases with significantly shorter print runs than the travel volumes. However, as also noted by Dr Pitt, Pococke seems to have had no idea what he was recording, and the texts make little sense. He writes only in capital letters, as he sees them on the stone, with no attempt at resolving them into minuscule Greek text, let alone with a translation; and although the book is handsome, it seems like the work of someone copying a totally unknown language. As a result, his legacy among epigraphists is of a frustrating nature: if Pococke had been the only person to see a particular inscription, then there would have been scholarly articles desperately trying to work out what he might have been looking at that could have caused him to copy down what he did. August Boeckh, in his magisterial *Corpus Inscriptionum Graecarum*, the first corpus of all known Greek inscriptions, spends much time grappling with Pococke, to obvious frustration.[159]

Wordwell Ltd., Dublin, 2005, pp. 70-71, Aideen Ireland, "A Gentle Luxury: Collectors and Collecting in Eighteenth Century Ireland", in Luce et al. (eds), *The Lure of Greece*, p. 57 and Aideen Ireland, "Richard Pococke (1704-65), Antiquarian", *Peritia* 20 (2008), pp. 253-78.

156 *Inscriptionum Antiquarum Grœcae. et Latinae Liber. Accedit, Numismatum Ptolemœorum, Imperatorum, Augustarum, et Cœsarum, in Ægypto Cusorum, e Scriniis Britannicis, Catalogus...*, London, 1752.

157 *The Bibliographer's Manual of English Literature*, p. 1471.

158 For the titles of these works, see the Bibliography.

159 I am grateful to Dr Pitt for generously providing me with this information and with his views on Pococke's book, his method, and his reputation as an epigrapher.

Interestingly, given that Pococke was chastised by Dr Shaw for his failure to acknowledge the help of his colleagues in *Description 1*, he thanks two individuals in the preface to his *Inscriptionum Antiquarum Grœcae. et Latinae Liber*, both of whom appear frequently in his letters from Smyrna (see Letters 33-34). The first is Revd Charles Burdett, a fellow Oxford scholar, who was English consul at Smyrna during the time of Pococke's visit and who acted as an agent to English and Irish collectors of antiquities; and the second is Baron de Hochepied (consistently misspelled in the letters), the Dutch consul at Smyrna, who was equally helpful to him in his quest for antiquities and inscriptions.[160]

As for other insights into Pococke's character, it is the unfortunate remark of the celebrated letter writer, Mrs Delaney (wife of Dean Patrick Delany) that seems to have made the most lasting impression on the world, since it is quoted in virtually every account of Pococke's life. After meeting him while staying in Bulstrode with the Duke of Portland, she wrote, in a letter dated 2 January 1761: "The Bishop of Ossory has been here ever since Monday .... We lose not much entertainment, for he is the dullest man that ever travelled: but he is a good man."[161] While this view has rightly been disputed by critics such as Maureen Hegarty and Michael McCarthy,[162] the present author has examined the context in which this statement was made and it appears that Mrs Delany was annoyed and frustrated by his refusal to part with some of his collection of basalt columns from the Giant's Causeway, which were coveted by her great friend and hostess, the Dutchess of Portland, and which she had promised to accomplish on her behalf.[163]

As with his chatty and amusing letters, which are far from dull, the life-size portrait of Pococke (oil on canvas) by the the Swiss artist Jean-Etiènne Liotard (the preliminary sketch for which is depicted on the cover of the present volume) shows him in a rather different light, portrays him in a distinctly bohemian manner, sporting a distinguished oriental costume, a turban and an impressive beard. It has to be said, too, that he is extremely handsome. For a full account of this portrait, including details of Pococke's meeting with the artist, the original sketch, the transmission of the work, and its reputation, the

160 *Inscriptionum Antiquarum Grœcae. et Latinae Liber*, preface [second page].

161 The comment quoted here is stated in a letter dated 2 January 1761, and is reproduced in many works, including Quane, "Pococke School, Kilkenny", p. 41.

162 See M. Hegarty, "Dr. Richard Pococke's Travels in Ireland, England and Wales", *Old Kilkenny Review: Journal of Kilkenny Archaeological Society* (1987), p. 398, who states: "Mrs. Delaney, doyen of the Dublin Blue-Stocking Society, dubbed Dr. Pococke a very dull man. Pococke? DULL? NEVER!!"; and Michael McCarthy, "'The Dullest Man That Ever Travelled?'", *passim.*

163 See Finnegan, "The Travels and Curious Collections of Richard Pococke, Bishop of Meath", pp. 44-45.

reader is directed to another related study.[164] The portrait was painted in Constantinople, 1740, and is one of many pictures by this artist of grand tourists in local costume. This mode of dress seems to have been a popular, if not necessary, part of the experience of visitors to the Levant, though the custom (at least while abroad) was often practised for considerations quite unrelated to fashion. One such reason was that of disguise to enable easy access into places of interest.[165] A further reason for adopting such dress was for self-protection by blending in with the locals, since it was considered unwise, and even dangerous, for foreign travellers (who were usually quite wealthy) to draw unnecessary and unwelcome attention upon themselves. However, whether such masquerade was an everyday mode of dress, rather than one reserved for special occasions, is not so clear and would have varied according to the individual, and both the literary and artistic evidence indicates that this type of costume for visitors was almost the norm.[166]

The only other traceable portrait of Pococke (apart from the many miniature self-portrait sketches or "cartoons" in his *Description* and his Scottish journal) is that depicted in Plate 6, below. This is doubtless a copy of the larger portrait described by Revd Leslie, in 1933, as "a fine portrait in oils of the Bishop – representing him three-quarter length seated – in the Board Room of the Incorporated Society in Kildare Street, Dublin," who further notes that there were two small engravings in the Bishop's Palace (Kilkenny), probably taken from this picture.[167] An engraving of a photograph of that painting (i.e. the one described by Leslie) appears as the frontispiece to Kemp's *Tours in Scotland*, and shows the sitter in his full episcopal roles, with a grey periwig and the same expression. A similar version of this portrait is reproduced in the history of the school in Southampton run by Pococoke's father.[168]

Sadly, the whereabouts of the Leslie/Kemp/Russell portrait is not known, and the copy portrait reproduced here as Plate 6 is on a much smaller scale, depicting only the sitter's head and shoulders, with no background detail. An intriguing reference to one or other of the portraits is made by his sister in a codicil to her Will, dated 10 April 1767,[169] in which she says: "I desire that my

164 See Finnegan, *English Explorers in the East*, pp. 81-84.

165 See W. B. Stanford, W.B. and E. J. Finopoulos (eds), *The Travels of Lord Charlemont in Greece & Turkey 1749*, Trigraph, London, 1984, pp. 163ff.

166 For an opposite view, see McCarthy, "'The Dullest Man That Ever Travelled?'", p. 26, who asserts that he "flouted convention by adopting the Turkish dress and wearing hair on his face".

167 See Leslie, *Ossory Clergy and Parishes*, p. 31.

168 Russell, *A History of King Edward VI School Southampton*, Plate 15, opposite p. 212.

169 Though the terms of Miss Pococke's Will were simple, the seven codicils, added between 1767 and a month before her death were problematic, since they were written on loose

Brothers Picture which was taken the last time he was in England in Parliament Robes be given to the School that is directed to be founded by his last Will at Kilkenny [the later Pococke School] and that it may be hung up there."[170] The fact that one of these portraits was in Elizabeth's possession suggests it must have been part of her brother's estate, which she had inherited from him two years earlier. Regrettably, we do not know who the artist was, but since Pococke was in London in March 1765 and died in Ireland six months later (see below), and given that he is depicted here in his robes, we can date the original with some certainty as having been executed around the first half of that year, when he was 65 years of age. In the June he was translated to the See of Elphin and then in July to the See of Meath, at which point he must have been back in Ireland and this date corresponds with the depiction of the sitter in late middle age. Though we cannot be sure whether it is the original or a copy, it is fitting that the only surviving version now hangs in the new Bishop's Palace, in the grounds of the episcopal residence in which Pococke had lived for nine years as Bishop of Ossory (1756-65).

Exactly two months after his transfer to the see of Meath, Pococke died of "apoplexy", following a visit to Charleville Castle, near Tullamore, Co. Offaly, where he met a relative of his host, Col. George Stoney, of Greyford [or Greyfort] House, Borrisokane.[171] Fortuitously, his last hours are recorded by Col. Stoney, thus: "1765, 15 Sept. – He this day confirmed at Tullamore; returned indisposed, went to his chamber, took a Puke, went to bed about 5 o'clock, seemed to rest quietly, but was found dead about 12. He complained of a pain in his stomach, which he could impute to no other cause than a few mushrooms eaten the day before at Ballyboy."[172] This account is verified by Thomas Johnston, who only that year (1765) had become the owner of Charleville Castle and was to remain there until 1785. Two hours after Pococke's death, in a letter dated "Charleville two o' Clock Monday Morning 16th Septr: 1765", Mr Johnston wrote the following letter to the Dublin banker David La Touche and his sons: "As the earliest Intelligence of the Bishop of Meath (Dr Pococks) Death may be of use to you, or some of your Friends, I send you an Express to

---

scraps of paper and were not signed by witnesses The additions to the original Will were relatively minor, concerning family portraits, a counterpane and chair-covers that she and her mother had worked, two more legacies worth £200 to be given to her female cousins, and a sum of £40 to be paid to her servant Mary [Stokes]. See Will of Elizabeth Pococke (Spinster).

170 Will of Elizabeth Pococke (Spinster).

171 At the time, Charleville Castle was let to a Captain Thomas Johnston, the Earl of Charleville having died the previous year and his widow having moved to Dublin.

172 From a diary published in *Some old Annals of the Stoney Family ... Compiled by Major S. ... Illustrated by C. Stoney of Portland,* London, 1879, p. 13.

PLATE 6
Portrait of Bishop Pococke, artist unknown (1765). Courtesy of the Episcopal Portrait Collection, Representative Church Body, Ireland.
PHOTO DAVID KANE

acquaint you of it. He died suddenly about two Hours ago at my House, where he has been these two Nights, if it can be of Use to you it will give very great Pleasure to Dr [Dear] Sir."[173]

This may seem a somewhat cold communication, without a hint of regret; but it probably was related to business: David La Touche was not only a banker (and possibly Pococke's banker), but he was also a member of the Incorporated Society for Promoting Protestant Schools in Ireland – the society to which Pococke had bequeathed his fortune, to come into effect after the death of his sister. La Touche is listed as an annual subscriber to the society's fund (on 29 September 1761, he donated £3.00) and he and his sons (clearly as a company) are also named as "Persons to receive Benefactions in Ireland". Clearly, Mr Johnston was aware of Pococke's bequest and the purpose of his letter was to inform La Touche in his capacity as an official recipient of bequests to the society.[174]

173 Derbyshire Record Office, D3155/C3725. I am grateful to Michael Byrne, Manager of Offaly History, for his kind help and for providing me with this information.

174 See R. Pococke, *A Sermon Preached at Christ Church, Dublin* ..., Dublin, 1762, pp. 55 and 60, respectively.

PLATE 7
The Montgomery Monument, Ardbraccan, Burial Place of Richard Pococke
PHOTO TONY HAND

Pococke was buried in the churchyard of the parish church of Ardbraccan, in a large vault constructed for his family by an earlier Bishop of Meath, George Montgomery (1569-1621). This vault, known as the Montgomery Monument, has a recessed central panel which contains three half-length figures of Montgomery, his wife and daughter, framed with a border of alternating leaves and flowers.[175]

On the south side of the monument (see Plate 7) is a simple framed plaque, which reads: "HERE LIES INTERED THE BODY OF DOCTER RICHARD POCOCKE OF MEATH WHO DIED September 15th 1765 in the 63rd year of his age". Considering the extent of Pococke's literary and scholarly reputation, it is not surprising that James Norris Brewer was to refer somewhat disparagingly to this "small slab", which he described as "a humble and ill-executed memorial".[176]

On his death, as we have seen, Pococke left his house and estate in Newtown to his sister, of the same address. The remainder of his estate was left in trust to the Incorporated Society in Dublin for Promoting English Protestant Schools in Ireland, with the interest being used to support Elizabeth and other members of his family during their lifetime, after which the fortune was to be used for "founding a school for papist boys from twelve to sixteen years old who

175 For a full description of the structure, see C. Casey and A. Rowan, *The Buildings of Ireland: North Leinster*, Yale University Press, London, 1993, p. 115.

176 *The Beauties of Ireland Being Original Deliniations, Topographical, Historical, and Biographical, of Each County*, vol. 2, 1826, London, p. 191

shall become protestants and to be bred to linnen weaving and instructed in the principles of the protestant religion…"[177] This weaving school, which later became known as the Pococke School, was eventually closed and the endowments were amalgamated with the present Kilkenny College.

177 Pococke's Will, National Library of Ireland, D. 27, 161.

CHAPTER 3

Itinerary of the Eastern Voyage

1 Explanation of the Dating

This chapter outlines the route taken by Pococke during his eastern voyage and where possible indicates the length of time spent at each place. This is based on an analysis of the letters themselves and divided into the main regions visited. Firstly, though, it may be useful to look briefly at some of the conventions he used in documenting this information. He was very methodical in his approach to dating his letters, and there are very few examples of letters omitting the date or place of issue. One aspect of 18th-century letter writing that can be confusing to the modern reader is the dual dating system used in correspondence sent from abroad, whereby letters are dated according to both the Gregorian and the Julian calendars, with a difference of eleven days.[1] For example, all letters sent from abroad are dated as follows, "Alexandria in Egypt, Septr 30. O.S.[Old Style] Oct. 11 [New Style] 1737", or, more usually, "Damiata [Damietta] in Egypt, March 8th-19th, 1737-8". With the rare exception when the writer (or the scribe) forgets, the only time this convention does not appear in our correspondence is, of course, when the letter is sent from England, just before the sender embarks from Dover to Calais, for instance, or after returning to English soil. Therefore, the last letter in Mrs Pococke's collection is dated: "Dover, August 29th, 1741".

When intending to travel to a new place, the writer typically gives a forwarding address to his correspondent, usually that of a banker, an English merchant or a diplomat.

1 The new Gregorian calendar was devised in 1582 and was immediately adopted by a number of countries including Spain, Portugal, Italy, France and the Low Countries. This was eleven days ahead of the old Julian calendar still in use in Britain until 1752. Therefore, when British or Irish travellers before this date wrote back home from the Continent, they included both dates to avoid confusion.

 DOI:10.1163/9789004440050_004

2 Itinerary of the Third Tour, from 30 September 1737 to 29 August 1741

2.1 *First Tour of Egypt* (*1737-38*)

On the evening of the 7/18 September 1737, Pococke went aboard the *Mary Galley* at Livorno and the next day set sail .for Egypt, arriving in Alexandria after a journey of 23 days. He remained there for over three weeks, after which he travelled by chaise to Rosetta (October 24/November 4) where he stayed for 10 days, next embarking on board a galley to travel up the Nile to Cairo. He arrived at Egypt's capital on 11/22 November and stayed almost a month, departing on 6/17 December, on a boat to Akhmim. He reached this place on 17/28 December, spent Christmas at a convent three miles outside town and returned to Akhmim on 23 December/3 January. After making a short excursion to Menouf, he set out to Farshut and on 3/24 January 1738 arrived at Karnak, where he viewed the antiquities there and at Luxor for a period of four days, until continuing on his way to Aswan, which he reached on 21 January/1 February.

From there he proceeded to Adfou, where he stayed the night (31 January/11 February) and two days later arrived at Etne. On 5/16 February he went to Sheikh Abd el-Qurna, where he visited the famous tombs of the kings and from here went to Quena, Dardana and Girga on his return journey to Akhmim, which he reached on 14/25 February. Two days later he departed for Cairo, arriving back at that city on 27 February/10 March, "after 3 months absence, without speaking to an Englishman. It was like coming home...."[2] Pococke spent the next few days fraternising with the Fathers of Propaganda Fide and Terra Santa and collecting from them letters of introduction for Jaffa and Jerusalem. Having packed up, he set out, on 4/15 March, in a boat with "black slaves, ten or twelve of them boys & girls for Constantinople".[3] He arrived at Damietta three days later, the plan being to depart from here immediately for Jaffa. However, contrary winds prevented this, and he eventually had to travel on a pilgrim boat, which set sail on 10/21 March 1738.

2.2 *Travels in the Holy Land, Lebanon and Syria* (*1738*)

Pococke arrived in Jaffa after a journey of four days, on 14/25 March 1738, and spent a further four days at the Terra Santa Convent in Ramla. On 18/29 March he set out for Jerusalem, arriving on the same day, and remained there throughout the Easter period. On 27 March/7 April, he took a three-day excursion to the River Jordan, where he ceremoniously bathed, returning to Jerusalem two

2 See Letter 10.

3 See Letter 10.

days later. On 2/13 April he took another excursion, this time to Bethlehem. On the way he visited (and bathed in) the Dead Sea and arrived at that city the following night. He remained here for several days, also visiting Acre and arrived back at Jerusalem ("came home") on 11/22 April, where he remained for a further 11 days. On 22 April/3 May, he set out for Jaffa, arriving there the next day, and over a week later (2/13 May) he set sail for Acre on a journey that took three days and three nights. After three days in that city, he set out for Nazareth, arriving the following day (9/20 May); and over the next few days he took various day-trips and excursions until his return to Acre on 16/27 May. Four days later he set out on another short excursion, this time to Mount Carmel Caesura and Castle Pelegrini, arriving back at Acre two days later (22 May/2 June). He remained there for just under a week, before departing for Sidon (on 28 May/8 June) which he reached in two days. On 31 May/11 June, he continued via Beirut on his journey to Tripoli, arriving at this city five days later (5/16 June). After a week in Tripoli (and having exchanged his turban for a hat & wig) he set out on 12/22 June for Mount Lebanon and from there went to Baalbek, where he spent several days viewing the antiquities.

On 21 June/2 July 1738, he set out with the caravan for Damascus, which he reached in two hours, and from this city spent over three weeks making short trips to places of interest, eventually departing for Aleppo on 16/27 July, which he reached almost two weeks later, on 29 July/9 August. He remained in this city for a month and on 15/26 September embarked on an excursion north of Aleppo for a fortnight, travelling, among other places, to Antab, Urfa and Birecik, and returning on 27 July/7 August. He remained in Aleppo for a further three weeks, socialising with the English residents, and on 19/30 August set forth on his long journey to the port of Latakia, in Syria, which he reached on 9/20 October. A week later he departed from Latakia for Tripoli (only two hours away), where he stayed a further week, and then departed for Cyprus on 24 October/4 November. He arrived at Limassol four days later and made his way to Larnaca, which he reached on 30 October/10 November. After spending 10 days here, on 10/21 November, he set out on a Tour round the island lasting 26 days and arrived back at Larnaca on 5/16 December. From there he sailed to Limassol, and then to Damietta, in Egypt, where he arrived on 24 December/4 January 1738/39.

2.3 *Second Tour of Egypt (1738-39)*

Two days later, Pococke set sail from Damietta, with the English physician Dr Charles Perry,[4] in a hired boat costing 30 shillings, and arrived at Cairo on 31 December/12 January. Here he spent time with Dr Perry and also resumed his friendship with the English residents he had met there on his previous visit. During his sojourn in the capital he visited the pyramids at Giza (8/19 February) and at Saqqara (24 February/7 March). After a prolonged stay in Cairo, he set out, on 27 March/7 April, to Mount Sinai with the caravan, reaching this place after a fortnight's journey via Suez, Wadi Wardan and El Tor. He remained there over the Easter period and departed on 23 April/4 May on his return to journey to Cairo, which he reached (after spending a few days in Suez) on 10/21 May 1739. On 4/15 June he set out for Rosetta, via Bulaq, a journey of five days. On 23 June/3 July, he took a chaise to the ferry and sailed to Alexandria, which he reached late that night, and where he remained for a week before embarking on a ship to Crete on 2/13 July.

2.4 *Travels in the Greek Islands (1739-40)*

Pococke's journey from Rosetta to Crete was long and difficult and involved a detour to Rhodes. However, he eventually landed in Sfakia on 10/21 August 1739, setting out for Chania by horse two days later and arriving at that place on the following day. Unfortunately, a letter is missing at this point, so we do not know his precise movements between 16/27 August and 3/14 September 1739. However, after travelling around the island for some time, he eventually departed from Crete on 30 October/11 November on his voyage to Scio (Chios), which he reached four days later. Having explored this island for almost a week, he visited Psara, and Lesbos, staying at the latter island from 20/31 November to 1/12 December, after which he set sail for Izmir, which he reached the following day. He spent five weeks in this city staying put for most of the time and socialising with the English residents. On 2/13 December, he travelled through Vourla, Seferihisar, Sığacık and Kuşadası, arriving at Ephesus on 12/23 December. Here he viewed the antiquities for four days until departing for Samos on 16/27 December. Having explored this island for several weeks, which at the latter end included a two-day trip to Patmos (4/15 to 6/17 February 1740), he set sail for the Aydın Province of Turkey (then called Scala Nuova) on 12/23 February, which he reached the same day.

4 The author of *A View of the Levant*, which was published just before Pococke's *Description 1*, in 1743.

2.5 *Travels in Turkey and Mainland Greece (1740-41)*

Pococke stayed in this area for a fortnight, taking day-trips to its many sites of antiquity, in company with a Dr Mamachi, and from here set out on horse-back for Ankara on 29 February/11 March 1740, on a journey of over a month, which took him through Geyre, Denizli, Hierapolis-Pamukkale, Sandıklı, Karahisar and Sivrihisar. On his arrival at Ankara, on 3/14 March, he was "met by all the English nation & most of the French".[5] He remained in this city for over three weeks, visiting the local attractions and socialising with the English, and on 29/10 April set out on a fortnight's journey to Constantinople. En route to the capital he travelled through Ghende [Gerede], Borla [Bolu?], Izmit and Pendik and stayed for three weeks at Üsküdar, home to many of the Franks (Western residents) at Constantinople. He then took a trip to the village of Belgrad Forest (T. Belgrad Ormani), 15 km north-west of Constantinople, where the English Ambassador had his country residence. Arriving on 1/12 June, he stayed there 11 days and returned to the capital on 12/23 June, where he remained for another three and a half weeks. By the time he left Constantinople (7/18 July) he had been there for more than three months, making only very short excursions such as the two-day trip he took to the Princes' Islands with a small group of companions, including the artist Liotard, which is possibly where the preparatory drawing for his famous portrait in oriental costume was taken (see the cover of this volume).[6]

Pococke's next destination was Andrianople (Edirne), which he reached on 11/22 July 1740 and where he stayed six days until he resumed his travels to the Dardanelles (17/29 July). He spent several days exploring the antiquities of this area, famously investigating (and attempting to indentify) ancient Homeric sites such as Sigeion and Troy. On 1/12 August he travelled to the islands around the Sea of Marmara, visiting Çatal, Marmora, Alonia and Erdek, later (13/24 August) setting out with a caravan for Bursa, visiting the Turkish Mount Olympus (Uludağ).

His journey from Bursa to Salonica is not recorded, as a letter dated 18/29 August 1740 is missing. Having learned of the death of Bishop Milles, Pococke felt under pressure to return home as quickly as possible, though not without visiting some of the more famous sites in mainland Greece. At midnight on 19/30 September 1740, he set sail from Thessaloniki to Karitsa and from there

---

5 See Letter 39.

6 This miniature red-and-black chalk drawing (height 210 mm/width 130 mm) is one of a series of Liotard portraits in the Louvre depicting locals and foreign residents at Constantinople, Izmir and the Greek islands during this period. It is described in the inventory as, "Portrait de l'Archéologue et Théologien Richard Pococke" (Louvre, Paris: Fonds des Dessins et Miniatures, RF 1379, Recto) and is incorrectly dated as 1738-39.

travelled by mule to Thebes, via Larissa, Catadia and Lamia. He left Thebes on 28 September/9 October, arriving that evening at Athens, where he met Charles Perry and viewed the antiquities for six days before setting out, on 5/16 October for Corinth, which he reached that afternoon. From Corinth, he travelled to Patras (where he rejoined Perry), from which he departed, on 20/31 October, on a Genoese ship for Messina, in Sicily, where he was quarantined. He arrived there, after a difficult passage, on 13/24 November 1740, and thus ended his eastern voyage.

2.6 *Travels in Europe* (*1740-41*)

Pococke left Messina on 16/27 December 1740, and the route he took back to England was via Italy, Germany, Switzerland and Belgium, with letters being sent to from Naples, Rome, Florence, Milan, Innsbruck, Munich, Kassels, Geneva, Basel and Liège.

CHAPTER 4

Letters from the First Tour of Egypt

This chapter starts with an excerpt from a letter from Leghorn (Livorno), which was sent shortly before Pococke embarked on a boat to Alexandria. Although it comes from a different manuscript,[1] is included here to set the scene. It was in this letter that Pococke informed his mother of his intentions to travel to the east. The chapter continues with 10 letters from his six-month tour of Egypt, from the end of September, 1737 to the beginning of March 1738, as follows: three letters from Alexandria, three from Cairo, one from Akhmim, one from Farshut, a further one from Cairo, and one from Damietta.

The first of the letters from Egypt outlines Pococke's journey from Livorno to Alexandria, a journey also described in an epistolary account sent to his uncle in the same post (see Account 1), whose scholarly references to Strabo, Pliny and other classical authors are in the style of his unpublished journal ("A Voyage from Leghorn to…").[2] For further details of Pococke's experiences in Egypt, and for an abundance of plates depicting the maps, buildings and antiquities of this country, the reader is directed to *Description 1*. Three of these engravings are reproduced here, as Plates 9, 10 and 11.

Pococke visited Egypt again the following year, sending a further seven letters from Cairo, Mount Sinai, Suez and Alexandria. These are reproduced in Chapter 8.

1 A volume of letters from Pococke to his mother from Dover to Leghorn (1736-37), BL, Add MS 121997.

2 BL, Add MS 22995. This manuscript is the only original document from the voyage, all others being copy letters.

 DOI:10.1163/9789004440050_005

LEGHORN
6/17 September 1737
Pococke to his mother
Letter 0[3]

Honoured Madam....

4th-15th Arrived at Leghorn [Livorno] about 8 – went to the church rather late; introduced myself to Mr Horton the Chaplain,[4] who introduced me to Mr Blackwall [Blackwell] an English merchant;[5] – walked with the chaplain; – dined with Mr Blackwall, went to prayers; – went with an English Gentleman a mile into the country to visit Mr Week an English gentleman, sat for some time with him. What follows is for yourself alone, unless as far as you think it necessary in General to speak of what I propose.... And now I pray you not to make yourself very uneasy on account of the journey I am about to take; – which some circumstances since has led me to, as being well prepared for it, in every respect, being in Venice where I could embark, but after was determined by some reasons, to come here to embark on an English bottom [ship] – & above all meeting at Venice a Firman or Passport from the Ottoman porte, which will make me entire Master of the East. As for the voyage it is much more difficult than dangerous, & neither in the degree it is represented; – It is but a fortnights voyage to Alexandria in a fine season of the year, where I hope to be a fortnight before you receive this.The plague is not there at all, to speak of; – will be over at Smyrna [Izmir] before I get there, lasting seldom above 4 or 5 months, & Europeans are not apt to catch it, 'tis not in the air, but the contagion is by a near conversation with the person. I go now aboard a large English ship, *The Mary*, a good Gentleman like Captain Parkins,[6] & two Franciscans from the Propaganda Fide go with us to Alexandria.[7] I have with me an excellent servant

3 BL, Add MS 22997. Reproduced as Letter 27 in Finnegan, *Letters 2*.

4 Revd Thomas Horton, the British Chaplain at Livorno from *c.* 1739-42. See Ingamells, *A Dictionary of British Travellers*, p. 524. The cousins had met his predecessor, Revd John Swinton (1703-77), during their earlier stay in this city. See Finnegan, *Letters 1*, pp. 115-17 and 152.

5 Either John Blackwell senior or Thomas Blackwell (d.1747). See Ingamells, *A Dictionary of British Travellers*, p. 96.

6 Pococke gives several variations of this name in manuscripts, the most complete being that in his "Journal", where he refers to "an English ship The Mary Galley Captain Thomas Parkins". See "A Voyage From Leghorn to [Alexandria]", BL, Add MS 22995, folio 1.

7 The Sacra Congregazione De Propaganda Fide was a missionary congregation founded in Rome in the 17th century to spread catholicism to countries that had recently been discovered; his two travelling companions were Father Jacomo (further described in Letter 5) and Father Francesco.

turned of 50, very faithfull & diligent, he lived 12 years in Constantinople, with the Emperors Ambassadors as a steward, talks Turkish, Italian, German, French, Latin. So I propose to be at Jerusalem next Easter, by the way of Grand Cairo; – & so make the circuit of the East; – & pray be not over concerned for I am fully perswaded we shall all meet at Newtown. It is true you must not expect to hear often from me, & probably after this you will receive no letter till next midsummer: – but you may continue writing to me once a month or 6 weeks with your Diary & direct it to me, to be left with Cutts Maydwell Esquire Turkey Merchant London;[8] – who will forward all letters to me by the best opportunity, & you may write one to me in a fortnight, which may possibly come to me before I leave Grand Cairo, & since my journey is fixed, I pray you not write anything to me that will make me uneasy; but make your self chearfull & easy; use exercise, take a dish of Coffee & divert your self by company; & do not show any great dissatisfaction before common company, at least concerning my Tour: – but you ought rather to think it an happiness, that I can see the world in this manner; – as I don't doubt but that you will when I am returned. I think I can answer in every respect my undertaking this voyage. As to the cures [ecclesiastical duties] that were upon me, the Bishop called me from them, to a more extensive office, as Chancellor of his Diocese & substituted others in my room [place];[9] – & now his Lordship has taken immediately on himself, the care of every thing that relates to that office. I have 300 pr An. [& spend nearly two]:[10] – & I hope to gain such knowledge & experience as to return better qualified to do service to my fellow creatures in the station in which it has pleased God to place me. And whatever is the motive, if seeing those places which were the scenes of our Saviours Life & passion, may touch my heart with a greater sense of my duty & make greater impressions as to what he did for us; I hope this journey on that account alone, will not be thought a fruitless one. Mr Maydwell is my friend in London, who recommends me abroad, especially to Mr John Lisle in Constantinople [Istanbul].[11] You are wrongly

8 Cutts Maydwell was a Levant Company merchant in the firm of Lisle & Maydwell, based in Constantinople, and Treasurer of the Company from 1730-32. See Finnegan, *Letters 2*, pp. 258 and 270-72, where Mr Maydwell is given as a forwarding address for Pococke's correspondence in the east.

9 According to Rennison, *Succession List*, p. 41, the position of Chancellor of Waterford & Lismore had been occupied by the Very Revd Alexander Alcock (1665-1747) for the previous 38 years, which means that the arrangement described in the letter was unofficial. On his return to Ireland Pococke found that the position had been given to John Alcock Junior in 1740.

10 These words have been scored out by Mrs Pococke but are still discernable.

11 John Lisle was Cutts Maydwell's partner and Treasurer of the Levant Company from 1736-39. See Wilson, *List of British Consular Officials*, p. 11.

informed concerning the plague, & no merchant had ever an house under ground to defend him from the infection; & those that live well hardly ever catch it, those being most liable to it who live poorly & have not proper nourishment. It is now a very fine season for the voyage, the bad weather not beginning here till November: the ship is ready to sail this day.

5th-16. We shall probably only touch at a litle Isle at the south west corner of Candia [Crete] to take in Provisions. I go at the best time of the year against winter, when the heat will not be troubling me & next summer it will advance by degrees.

Pray keep up your spirits & have a full trust in God Almightys Providence, who will dispose all things for the best. And I am fully perswaded Providence has for some nice ends lead me to this voyage, to which I have been determined by many extraordinary accidents.

One of the Fathers viz. Father Jacomo of Bohemia, has been [on] the tour of the East, & probably I shall be in his company as far as Jerusalem; & my servant is of Friuli, that country where I found such good sort of people, & where we received so many civilities; & we have the finest promising season in the world. English ships are at peace with all the world, & cannot be taken by Corsairs [French privateers], having always a pass that secures them from them....

Mr Milles is to open my parcel sealed, on which my Titles are writ, to take out my License of absence, & if he finds it not there, he may open the others & seal them up again in your presence....[12]

ALEXANDRIA IN EGYPT
30 September/11 October 1737
Pococke to his mother
Letter 1[13]

Honoured Madam,

On tuesday Septr 7th-17th [6th-17th] in the afternoon I sent my last to you from Leghorn: & that evening I took leave of Mr Blackwall [Blackwell], went to my Banker; & Mr Horton the Chaplain went with me to the Boat, & I went aboard the *Mary* Galley Captn. Parkens, bound to Alexandria in Egypt; the next day we

---

12 This relates to Pococke's License of Absence from his church duties, which he asked his cousin to arrange for him on his return to Waterford. The matter is clearly of great importance, as he reminds his mother of this again in Letters 18 and 30.

13 BL, Add MS 22998.

went in to the road of Leghorn, proceeded on our voyage at ten, but litle wind as likewise in the 8th, we made but litle way.

9th-20th. We passed by the Isles of Gorgona, Capraia, Elba, Planosa, Monte Christi & that day &

10th-21st. We had a pretty good wind, passing near Corsica & seeing the town of Bastia – this day I had a litle sickness but well again immediately.

11th, 12th, 13th & 14th. But very litle wind & that not fair. I find myself perfectly well at sea; – on the 12th in the evening I saw Sardinia. 13th in the afternoon litle wind & fair; rise take Coffee, dress myself, dine at one, drink tea sometimes in the afternoon, make a litle supper, play at Ombre [card game] with the Captain, & our good travelling Franciscan [Father Jacomo] who has been in the two Indies as well as this way before; & at leisure hours look into books so that the voyage is most agreeable. We had a grampus or small whale passed by us one day.

15th. We had wind but not fair, & in the evening saw Sardinia plainly being about the latitude of 40, 39. The carpenter aboard our ship is son of Mr Troutons brother of Sowley [next line scored out] he was born at Liverpool & is a very honest man.[14]

16th-27. At noon in the degree 39d 58. We live very well here, two dishes at dinner, bacon & fowl & sometimes roast beef, pancakes, plum-pudding, Pilaw of rice made into soup, – Ham, besides the Ship-fare, two of these things generally.

17th-28th. A litle more wind & contrary, so that I found it proper to lye only on bed, but the wind sprung up fair towards of the 18th 29th & on

19th –30th. In the morn we were very near Sicily: I saw the land which makes the Port of Palermo, passed by the port of Trepani [Trapani], the old Promontorium Drepano & near it Trepani, & near that the famous Mount Eryx [Monte San Giuliano] & where Venus was worshipped – then we saw Pharsala the old Libybeum, the old port to go to Carthage & after Mazra [Mazaara] & Xiacca [Sciacca].

20th of Septr 1st Oct. fine winds all yesterday, & this morn gentle fair winds. Saw Girgenti [Grigenti] the old Agrigentum, so famous for the residence of the Tyrant Phalaris, saw Lucatta [Licata] Terra Nuovo [Gela] & Mount Aetna very

14 A Miles or Myles Troughton is recorded as having refurbished the Sowley ironworks at Beaulieu, Hampshire, in 1722, so the carpenter is possibly his nephew. The fact that Pococke later scored out part of a line suggests that he had made some unflattering comment about a member of the Troughton family. However, since no original letter exists, we do not know what this was. The personal comments on individuals from his mother's locality is a typical feature of all his grand tour correspondence.

plainly & a fine smoak at ye top of it.[15] Yesterday even we saw the Pundem towards Africa & I don't know but we saw Cape Bona near which Carthage stood, not not being above [..]ty leagues from it.[16] We are now making the voyage which Aneas took, & at Drepanum he [..] his father Anchises. We had a fine gentle wind all day & in the eve came up with Cape Scalanius [Scaramia].

21st-2d Oct. We were opposite to the South east Promontery of Sicily called Cape Passaro [Passero], we saw the High land about Syracuse & Catania & the old Port of Ulysses where they bored out Poliphemus [the Cyclops] eye, & where Aeneas came soon after. We saw many sail, & over the low land that makes Cape Passaro we saw the high land about Syracuse & this even we lost sight of Sicily.

22d-3rd. Fair wind & high, so was somewhat sick as were some of the mariners, so kept [to] my bed. We were in Catiluce, about 36 2.

23d-4th. Fine winds but rather high, so did not get up till towards evening.

24th-5th. Fine gentle winds, just saw the tops of the high mountains cover'd with snow of the Isle of Candia, old Crete, famous for the reign of Jupiter & his wise son Minos; 'tis a great run we have made from Sicily to Candia in two days & 3 nights.

25th-6th. Fine gentle winds & in the degree of 33.3c.

26th-7th. Fine day & good wind.

27th-8th. Got up early to see the land of Egypt, which we saw very plainly about Point Solyman just on the borders of Egypt & the Kingdom of Barca.[17]

28th-9th . We came up with Cape Burazzo & with the tower of Arabia.

29th-10th . We came in sight of Alexandria, & of the famous pillar [Pompey's Pillar]:[18]& approaching to it, saw the Palm trees growing like high Pines, stript up to top, which look beautifull: & approaching the Harbour we saw the Palm trees appearing over the houses. We entered the Port, God be praised just at Sun set. Some English Captains came aboard, & I heard Mr Dobson the Vice

---

15 He further describes this sight in his "Journal": "but as it was fine weather it did not flame out, which it does most frequently in windy & wet weather, we thought notwithstanding we saw something of a light smoak or air at night where we judg'd the top of it to be." See "A Voyage From Leghorn to [Alexandria]", BL, Add MS 22995, folio 30.

16 Mrs Pococke has omitted this figure, presumably being unable to read her son's writing. The word "Pundem" is problematic and cannot be verified in the "Journal", as folios 30-31 are missing from this manuscript. It probably refers to the small island of Pantelleria in the Strait of Sicily.

17 This is the ancient Greek Cyrenaica or Pentapolis, which comprised five cities (Euhesperides, Taucheira, Barca, Ptolemais, Apollonia, and its capital Cyrene) in the northeastern part of modern Libya.

18 Though Pococke frequently mentions this monument (see Letters 2 and 3) he gives no account of it. However, he devotes a whole page to it in *Description 1* (p. 8) and includes a magnificent drawing of it as Plate IV.

Consul was dead, to whom I had letters; but that Consul Barton was in town from Cairo,[19] who I think is related to Mr Maydwell my friend in London, & I have letters to his house & company in Cairo; & you see we have been 23 days on our voyage – 13 to Sicily & 3 on this coast, which 15 days might have been performed in 5. I did not go ashore this evening because it was late, but go tomorrow morning & in two days shall be in the Turkish habit, a Turban & all as they go in these parts. I shall be at the Consuls house both here & at Cairo where I suppose I shall go in a fortnight or 3 weeks time, & probably in company with the Consul, when he has settled affairs here for which I can wait, having time enough to be at Jerusalem by Easter. The way to Cairo is by land to Rosetta & then up the Nile to Cairo; – from Cairo down the Nile to Damietta & thence on board an English ship to Joppa [Jaffa] which always against Easter go with passengers.[20] There is no plague in these parts, nor anywhere but at Smyrna, & all the road is as safe as possible. This alteration & the Consul of Cairo happening to be here will make everything much more agreeable & easy. You may look into the latter end of the last book of Caesar *De bello gallico* the beginning of that de Bello Alexandrino, where you will have some account of these parts.[21]

Pray my very kind love to my sister & my compliments to Mr Milles, who I suppose is at Augsburg or thereabouts seeing the world as fast as he can in his return home.

Pray also remember my due respects to friends at White-oak & to all your neighbours in Europe. The coast all along appears sandy & desert without either house or tree appearing; for Barca which is under Tripoli & Tripoli itself is a very poor kingdom, tho' large & they do very litle by sea or otherwise. A ship going out early this morning before I go ashore, I send this by it to Leghorn.[22] You must not expect to hear often from me & not be surpriz'd if you hear not in half a year. I find the Consul Barton has heard before of my coming & I send this the 30th Sept. [....] before I go ashore. I am to go to Cairo with the Consul & his Lady &c in great state in about 3 weeks.

---

19 Robert Barton was the English Consul at Cairo c. 1731-50. See Wilson, *List of British Consular Officials*, p. 43. Pococke was to meet his brother, George Barton, Consul at Larnaca, later; see Letter 22.

20 As it happened, he travelled on a French ship that was taking Egyptian pilgrims to the Holy Land. See *Description* 2, Part II, 2.

21 This is one of the many examples in which Pococke's betrays a confidence in his mother's appreciation of the classics and other areas of scholarship.

22 By the same post he sent two letters to his uncle, Bishop of Waterford & Lismore, both of the same date (reproduced in this volume as Accounts 1 and 2).

ALEXANDRIA IN EGYPT
5/16 October 1737
Pococke to his mother
Letter 2[23]

Honoured Madam,

On the 30th Septr. 11th October, the day after I came into the port of Alexandria, I dispatched a Letter to you & two to the Bishop of Waterford by a ship in Leghorn; & about 11 went ashore to the Consuls [Mr Barton], who received me with much civility; he is a young man, not much older than your Son, a Gentleman of great politeness, good sense & of a most easy behaviour; he is much related to the Clergy, – his Brother of Peter-house is minister of St Andrews London,[24] the Duke of Montagu being Patron of the family,[25] he is also cousin german [first cousin] to Mr Maydwell my friend in London, who recommends me to Mr Lisle at Constantinople.[26] His Lady [Mrs Barton] who dresses after the Turkish manner is here, I suppose the daughter of a merchant of our nation at Constantinople, & seems to have been bred abroad, widow I think of the Consul of Ancyra [Ankara], a Lady of great modesty & civility, & niece of Mr Coke at Cairo, I had a letter to, in company with the consul & Mr Upton.[27] After dinner we all went out a mile to the Canal that brings the water to this place from the Nile; East of the Old Town which is East of the new, for the walls only of the old Town remain, being inhabited only by a few people. This is the canal of Canopus that runs along it; – we had a carpet & cushions spread on the banks under a tree, & walking about we saw a caravan [convoy] of Moors from Tetuan [Tétouan] returning from Mecca, – we advanced & met them giving them a salamalek, the shake of [the] Head [...][28] cloathed something like a Greek Bishop with two or 3 guards with Pikes & a standard, about 30 horsemen, besides many spare horses & camels with provisions, & a Tartavan

23 BL, MS Add 22998.

24 This brother is Revd Cutts Barton, who was ordained Deacon in 1728 by Bishop Richard Reynolds, and whose various appointments included Dean of the Cathedral Church of Bristol from 1763 until his death in 1781.

25 Probably John Montagu, 2nd Duke of Montagu (1690-1749).

26 Pococke informs his uncle of these important connections, all of which "contributes to make this country very agreable" (see Account 4).

27 Further details of his family are given in Letters 4 and 5 below. This was probably John Upton, who was licensed to trade in Cairo in 1742 and who was also a subscriber to Charles Perry's eastern travels.

28 Mrs Pococke appears to have had difficulty transcribing this passage, which relates to the Arab style of greeting, and has omitted one or two words.

[*tahtırevan*, a litter] with the Harem & 2 women in it, something like a sedan, but in this shape [small sketch] carried by two camels: – one woman on a camel, in a sort of wooden frame, reaching almost up to her shoulders, & a woman or two on foot, for they reckon it not right if they have not some women with them. The Consul shot a while & we caught a sort of Lizard, like a crocodile on the trees: we after reposed on the carpet, & took a glass of wine, & thus agreeably entertained, we returned home. I went in the chaise with the Consul, the Ladies in another, & the rest after.

Octr. 1st-12th. You see how we spend our day, breakfast at nine on Tea & Coffee, with the Ladies, oysters before dinner, & small bowls of punch & white wine, if anyone pleases to drink; dined; that morn settled my things, at 4 drink Coffee: went with the Consul &c: to see the ruins near, of what they call Cleopatra's palace; – very great Kings; – an obelisk of red granite with Hieroglyphicks, as that at Rome,[29] & another near lying down; – a most magnificent round tower; – great quantities of red granite. We sup & as at dinner, drink [to] the King [George II] & our friends in a most agreeable sober manner: – this country is very plentifull & a table well served & well dressed – fish in abundance, good shrimps & oysters, small but sweet. Egyptian onions, with the inside scooped out & filled with forced meat, an excellent dish, & cucumbers in the same manner: – at supper an Egyptian custard, which is milk turned so as to be like sowre milk, 'tis turned with a [...], made of a seed they feed birds with, & is a litle sowr but not disagreeable; – as to the rest after the French, & English way. A water melon, Pomegranates after every meal, & sometimes those rich figs growing about Rosetto [Rosetta], cultivated from the more Eastern parts of the leaves of which they say, our first parents made their aprons, in shape like a date, four inches long, an inch diameter being eat alone, 'tis in the mouth like Almond flummery [pudding], & tastes most like the richest pears, & eat with white wine & sugar, something like that flummery: we have also Dates, rough & not agreeable to me, the fruit of the Date or Palm tree but different from the Palm of the West Indies, of which they make the wine. The Tree here is most useful, of the branches they make strong baskets of the leaves slight ones, of

29 This is the antiquity mentioned in the Minutes of the Egyptian Society, at the meeting of 5 March 1742. As recorded in Pococke's hand: "Mr Letheilleur [Lethieullier] shewd at the same time the drawings of four sides of a fragment of Egyptian Granite, which now serves as a Post before the church door of St Bartolomeo nel Isola in Rome. This drawing as taken in 1736. He thinks it not improbable that this may be a piece of the obelisk, which we are told once stood there...." (BL, Add MS 52362, folio 23, verso). The drawing is included on the following page (folio 24) and then completely out of context, and with no discussion, it is embellished, enlarged and reproduced as an engraving (Plate XCI) in *Description 2*, Part II.

the bark ropes, which & the wood are not susceptible of the plague: the wood is good, the Dates men eat, the kernel they grind & give to camels & in some places make bread of it, & also a paste which is made into balls being well dried.

2d-13th. Read: In the even [evening] went out attended by the Consuls Janizary [Janissary] & another servant or two out of officiousness, being Sunday; & viewed part of the walls, the fine towers in them with apartments, a mosque not to be entered, in which they say are 1001 pillars. I saw some of Granite, there being a [...] within it, with a portico round it, with three or four rows of pillars; – mounted a litle hill, from which had a fine view of the Town & old walls; – found much company at home, spent the evening with Mr Orbas [Orbus][30] a Dutchman, a factor [agent] here for the other company at Cairo, there being but two houses or companies of English houses here; the Consul being of that I am recommended to.

3d-14th. Rise early, got out at nine, began to pace the walls & take an account of the towers, & the bearings by my compass – saw the Jews burial place; & near it a small encampment of Arabs, went about a quarter round the city; in the afternoon the Fathers my fellow passengers came to see me.

4th-15th. Went out early to continue my circuit round the walls; – I am now in the pleasantest time, the Spring now beginning, & continues till April, when all the harvest is over, this being sowing time; – in upper Egypt the harvest is ever in January: – these two days have had a hot almost East & north east wind so that it wets the very beards, & is reckoned the most disagreeable [...]. Staid within, played at Commerce [card game] before supper, forced meat boyl'd as a pudding is a dish & sea-eggs [sea urchins].[31] Most delicate, they are flattish & round & being divided, send in like oysters, they have long prickles like hedgehogs, which move as well as the meal within, which is yellow & have an agreeable salt, about 2 inches diameter; – the crabs are delicate tho' small, & have prickly claws. We are 12 a clock about the time that you are 10, that you may see exactly for 11 degrees East makes an hours difference.

5th-16th. Being a fine & calm day, I went on the water to view the shore of the [...] & examine things by old Authors; – went aboard our ship & took Captn. Parkins with me; after dinner the Father [Jacomo] came to see me & we took Coffee with the Consul. I walked out on a hill in the old City, which commands a fine view, the Arabs had some tents under it; – these people do no harm now

30 Mr. Orbus/Orbas was Jan Herman Alberts, the Dutch consul in Cairo (in office 1733-1742). He died in Alexandria on 26 or 27 March 1742 from stab wounds inflicted by a "Moor". O. Schutte, *Repertorium der Nederlandse vertegenwoordigers residerende in het buitenland 1584-1810*, The Hague, 1876, p. 362.

31 "The Echini" has been added as a footnote in Pococke's script.

being at peace with the Alexandrians &c: – but I never go out without a Janizary, either my own or the Consuls, & another servant. I went out to the famous pillar [Pompey's Pillar & thence went to some catacombs near. The Arabs were encamped round; after dinner one of the fathers came to see me. I thank God I enjoy my health very well, not the least plague in these parts, – no danger from the Arabs, who are every where in perfect peace with the Turks. – I find this opportunity to write by way of Marseilles. I believe I shall stay here a fortnight or 3 weeks & shall be fully employed, – but you must not expect to hear often from me, it is a great chance I met with these opportunities, & when I leave Cairo it will be a considerable time before I meet with any; & miscarriages will happen, therefore pray make your self easy: – & please God, as I am fully persuaded that we shall meet together in Newtown – & trust no accident will prevent it, & wish ye all as secure as I am. The Janizaries are brave honest fellows & you must tell Mr Milles De Turken seint brave Lighters.[32] Pray my most kind love to my sister & kindest respects at Whiteoak &c:[33] – you may write to me once in six weeks, directed to be left with Cutts Maydwell Esqre in London; – & if you have not writ already & have anything particular to say, write immediately, & so always on any case of importance; otherwise once in six weeks.

I am Dear Madam your most obedient & dutiful Son, Richard Pococke[34]

The face of the country at present is all sandy & in some parts Palm trees, & the trees that make wine in Guinea, I think or the West Indies – If you can find it in books let me know & put all queries to me in your letters.

---

32 This obscure passage – a mixture of Dutch, German, and English – is perhaps a joke intended for the amusement of his cousin, Jeremiah Milles, who was still travelling through Germany.

33 Whiteoak is mentioned several times at the end of letters and presumably relates to the family that lived in White Oak House in Newbury.

34 This edition transcribes the valediction (whose wording rarely changes) so as to run across the line, rather than spreading the words over several lines as sometimes done by Mrs Pococke.

ALEXANDRIA IN EGYPT
23 October/3 November 1737
Pococke to his mother
Letter 3[35]

Honoured Madam,

On the 6th-17th. I sent you a letter by way of Marseilles & a week before, another by Leghorn, giving an account of my arrival at Alexandria, after a long but pleasant voyage of twenty three days: – that afternoon I walked out with Vice Consul Butler,[36] & saw a very fine castle in the town walls, three stories, curious Arch-work.

7th-18th. We went out all but the Consuls & Ladies, to see the Catacombs to the west on the sea.[37] First saw a fine double well, with a descent by about twenty steps, & then we came to the water which comes from the Lake Maris [Mariout] now near its being filled with the overflow of the Nile. Came to the sea & went into a [...] & after into some apartments cut in the rock, into which the sea now comes, & this whole rock where the sea has gain'd, appears to have been cut out into catacombs & other apartments for embalming the dead, & for the use of those that did it, this being called Necropolis, inhabited by those people, as well as serving for the mansions of the dead. We went into an ordinary Catacomb as described in my last: & then went into what they call the magazines, twenty foot high, cut out of the rock, it is a Gallery or passage, 30 paces long, 6 wide & about 20 broad, there being two passages across between divided by thick walls cut in the rock, it might be to preserve things for curing the dead. We then went into a Catacomb which I believe is the finest in the world, it consists of a saloon, all cut out of the rock, with five rooms on each side, each having three niches almost as large as the room for each for ye burial

35 BL, Add MS 22038.

36 This is one of the Butler brothers, Samuel. It has not been possible to find any biographical details of this gentleman, who could either have been English or Irish (Butler being a common Anglo-Irish name of the period), but he is mentioned in three letters (in French) from, the 1750s. Two are of the same date (1 December 1756) and are from Elbert De Hochepied (289) and Mrs De Hochepied-Boelema (290) [Constantinople] to Karme & Nicola Mahfud [in Cairo], respectively. The third, dated 1 June 1757, is also from Elbert De Hochepied to to Karme & Nicola Mahfud (298). See J.G. Nanninga, *Bronnen tot de Geschiedenis van den Levantschen Handel Vierde Deel: 1765-1826 – Eerste Stuk,* The Hague, 1964, pp. 354, 356, 362. Pococke also met him on his return visit to Egypt (see Letter 30, below).

37 These catacombs are described in *Description 1*, p. 9, though in less detail than in the letter, and plans and sections are represented in Plate v.

of one body,[38] I suppose, all adorned with Dorick pilasters; – we then went into a square room with handsome pediments over the four doors supported by Dorick pilasters, then into a round room, with an arch'd vault & three such niches with dorick pilasters; returning to the square room, saw two beyond it plain, & went into a passage from oak filled up, returning went to the right from the square room into two other large ones, & then into a large square one, with four square pillars in it at equal distances; – the rooms were some ten & some eight foot high; saw many bones brought in by wolves, & some say they saw a wolf run out, but no sign of bodies buried. We dined under a rock by the sea, the Fathers & French with us, took a walk to a hill over the Lake Maris, now near dry, return'd home.

8th-19th. I walked out to see the New Town & Old Port, where none but the Turkish ships lye & then to the walls, & went on paceing [pacing] them. After dinner went on board Captn. Boyds (Boyle's?) ship of Portsmouth, with the Vice Consul & others; – had a cold entertainment & firing guns at drinking health; returned before it was dark. We play at Commerce every evening before supper.

9th-20th. Saw a Tortoise cut up 4ft long, 2 f ½ wide, live on grass, 'tis one of those of the West Indies, & not those they make Tortoise Shell of, will live two months on its back without eating, no legs but 4 fins, it appears to have three hearts, one like a beast, the other like a birds & the third like a fish, & that the flesh towards the back eats [tastes] like fish, that on the belly like fowl, & the next to it, towards the guts tastes like beef; I found the last to be so being dress'd [prepared]& thought it eat like Rabbit. I went to the Fransiscan Convent, dependant on that of the Holy Land, but my fellow passenger not at home. Went & took the heigth of the Obelisk & examined the Hieroglyphicks. Walked in the even with the Consul & Lady &c: on the strand, & spent the evening at Mr Orbus's.

10th-21st. I went out early, about three miles west to see a canal, that was out from the Lake Maris to the sea, returned by the edge of the Lake to the canal, & din'd on the side of it, under a tree; – was pleased to see the people swim backward & forward, with their cloathes in one hand, as Caesar swam with his papers. We met a set of Arabs moving with their tents; & another sett mounted on good horses, these live in villages, – the other are the poorer sort: – went to Pompeys Pillar as it is called, & took the heigth & measures of it; paid a visit

38 The word "ye" is a later addition in Pococke's hand.

with the Consul to Mr Cassin [Chassin] the French Droggerman [Dragoman],[39] plaid at [ ][40] came home to supper.

11th-22d. I walked out to the Obelisk & to the walls, & then to the Coptics church of St Mark, where they say was the Old Cathedral; – the Coptics are of the Church of Alexandria, they show here a wooden chair of St Mark, & the place where they say, his body & head are, especially the latter; – it is a miserable place & not above 8 Coptics in the city, the Patriarch reside at Cairo. I went at 4 to a billiard table with Mr Scot, & then walked, play'd at cards, sup'd.

12th-23d. I went to see some ruins & pillars, – thence went to the Greek church & convent of St Catherine, who was beheaded on that spot, & they show the spot on which she was executed; – the Bishop Suffragan of the Patriarch was very civil, & they have a better convent & church than the Coptics, tho' but very few Greeks here. In the afternoon went with the family to see a castle in the town walls, inhabited in the summer by the Aga [lord] of the castle, & also his gardens, where I saw great ruins, & some very fine Granite pillars, & we had a collation [picnic] on the top of the castle. The French Drogerman & another one of their nation plaid at cards till supper, & then went away; at dinner we had a skuttle fish [scuttlefish], very delicate, like boyl'd oyster, & at supper an eight leg'd fish somewhat like it, having an exceeding small body. We have excellent grapes from the Greek Islands, as solid as a pear.

13th-24th. Walked about the new city, – went to a Turkish Coffee house, & in the afternoon [went] with the Vice-Consul Butler & others, on a fine Arabian horse for 3 miles along the shore towards the East, where I saw some signs of the city Nicopolis, & considerable ruins of a very great building; returned home by the sea side, being very pleasant. Spent the even at Mr Orbus's; the weather is not quite so hot. N.B. here one may ride a horse, & at Acre, Tripoli & Aleppo, – but at Cairo only the Consul.[41]

14th-25th. I walked about the new town; – in the afternoon rid two or three miles up the canal, crossed away to the sea side, where we were yesterday & came home: before two Arabs mounted went with us from the gate, under pretence of defending us, but to get money, which we gave them; the offere'd the same today, but we told 'em, we were not going far. I had an exceeding pleasant ride.

---

39 Chassin, "second dragoman in Cairo" is mentioned in Henri Dehérain, "L'Orientalisme Français en Égypte au XVIIIe Siècle", *Journal des Savants* 6 (1931), pp. 261-272, esp. 264-5. A dragoman was an official interpreter and advisor. The spelling of this title varies throughout the manuscript, but is consistent in Pococke's printed book.

40 Mrs Pococke has left this word blank.

41 In *Description 1*, p. 18, Pococke confirms that in Cairo consuls were the only Christians who were allowed to ride a horse.

15th-26th. I went out to the Pillar, dined, went to see Mr Dou, joint agent with Mr Scot for the other house in Cairo; saw his medals, a mummy &c:[42] walk'd out with him on the sea shore, at the place, where was the palace of the Kings, & saw a great many pieces of Porphiry & the finest marbles, probably used in mosaic works, now worn & perished by the sea. We have such quantities of fine quails, that we knock them down as we are walking among the ruins [being ... & fatigued].[43]

16th-27th. I went out to the Obelisk. The French Consul[44] &c: came in the even. I walked with some English, went to Mr Scots, came home & sup'd.

17th-28th. Set out at 6 with Mr Dou & went 9 miles East to Beckier [Abu Qir],[45] where is a garrisoned castle, a poor old Jew that serves there as Frank Consul went with us,[46] & we went to his house carrying provision with us. We had a note from the Governor of the castle here, to the Governour there, which we sent; – who sent his Lieutenant to us, to know if we wanted anything, & in what he could serve us, & the Lieutentant himself brought us a pot of Coffee & set down with us with much complaisance & drank Coffee. We went in a boat two leagues to an Isle, where we saw a small work under ground & some signs of foundations; – we took a litle refreshment on the shoar, pick'd up shells & returned to Beckier, after resting ourselves took a walk on the [...].

18th-19th. We set out at 7, went 2 miles by the sea shore; saw the ruins of a foundation of a temple & near 30 pieces of columns of red granite, mostly channel'd, & on the sea side near, saw ruins probably of another Temple, – all red granite, a sphinx broke to pieces, many Lyons & a statue of an Egyptian Goddess almost entire: In the Old City called at the Coptics, saw their church again & the stone on which they say St Marks head was cut off. Came home, dined, walk'd out in the Old Town.

19th-30th. Went out early 4 miles on the sea side, on the East to see some ruins, & the situation of the ground; walked out in the evening with Mr Dou & Mr Scot: there area on the sea-shore amphibians crabs, which have holes in the strand, into which they run very quick, supporting their bodies 2 inches above

42 Pococke's drawing of a mummy he sent from Egypt to England is reproduced in the present volume as Plate 16.

43 Added later in Pococke's script but indistinct.

44 The French consul in Alexandria at the time was Jean Baume; the French consul in Cairo was Pièrre-Dauphin de Martin d'Amirat. It is not clear to whom Pococke refers here. Anne Mézin, *Les Consuls de France, au Siècle des Lumières* (1715-1792), Paris, 1998, pp. 656, 675, respectively.

45 Misspelled in the manuscript, Pococke informs the reader in his printed book that the placename is called Aboukir by the locals and Bikiere by the Europeans. *Description 1*, p. 1.

46 The term "Frank" refers to Europeans. It was used throughout the eastern Mediterranean.

the ground, their eyes they can lift up, which are like a bead, & they seem to be endued [endowed] with sight all round.

20th-31st. I went out in the morn to Pompeys Pillar to finish the measures &c:, went to the bridge on the canal near, & under the shade of it dined; – two Arab boys that had accompanied us, sharing with the servants in the repast; – they are moletto A[rabs] & live in a very simple manner, after an improved nature. I rid up the canal two miles & came home.

Oct. 21st-Novr 1st. I rid westward to the seaside, to see some Catacombs, partly worn away by the sea, on this side of the fine ones we saw, & returned home riding up by the canal above a mile. Dined; – visited Mr Dou, & went with him & some other English to a Billiard table; – visited Monsr. Chasin [Chassin] the French Drogerman, who came & spent the evening at the Consuls.

22d-2d. I rid out to see & find out the situation of some old places; & in the afternoon was busy in packing up, being to set out by land for Rosetto, 30 miles, in order to embark on the Nile for Cairo. I have the pleasure to wait on the Consul & his Lady, & we expect to have a very pleasant voyage; – we shall stay two or three days at Rosetto, which is a few miles above the western mouth of the Nile; it will be probably ten days before we arrive in Cairo, – & they say nothing can be more pleasant than the country on each side of the Nile: We have had no rain since I came, but ten days of hot damp air, which is very disagreeable, tho' the heat is reckoned as far as it causes perspiration; the air here is so damp, that everything mists & spoils, & animals breed in all woollen things, not often exposed to the air. There are in winter great rains here, but hardly ever any at Grand Cairo, only for half an hour at the change of the seasons. This country here is all sandy, about the town for a mile they pay a small rent to the G. [Grand] Seignor; – but in other parts all is common, but they pay a heavy tax for everything brought into the town. I shall spend about three months in & about G. Cairo. I thank God I have my health exceeding well, & find that constant exercise agrees with my constitution. The people of this town are mostly Turks, a pretty many Jews, a few Greeks, & not above 4 or 5 families of Coptics of the church of Alexandria under their Patriarch, who resides at Grand Cairo. Tho' you see I have been here above 3 weeks, you may observe I have been fully employed; & indeed there are great difficulties in finding out how the city was of old, because the situation is so much altered by the sea, loosing in some places & gaining in others, that few cities have undergone such a change, but the old walls fix it pretty much otherwise everything would be very obscure. The city you know was built by Alexander, the residence of the Kings who descended from Lagus Ptolomy, one of Alexanders Generals, Lagus's son being the first King, & Cleopatra the last Queen. The city flourished much after it was under the Romans, & again after the middle age, when the East India trade was carried on there, by the Venetians. And when that was lost by the passage by sea, the Old Town sunk & the litle trade carried on by sea, occasioned the building of the new town on the strand to the west of it; & the trade that was carried on by the land being stop[p]ed by the hostilities of the Arabs, the old town was quite forsaken which was near the canal, & all removed to increase the new Town, about 120 or 130 years ago. This letter I leave at Alexandria to be sent by the first opportunity, but the conveyance is very uncertain, because the letters are all opened at Leghorn & tumbled about. I habit myself tomorrow in the outward garment of the Turks, call'd Benishe & is of a sort of crimson cloth, a Turban; over my European dress: – at Cairo I shall be habited entirely after the Turkish manner. Let Mr Milles know the Venetian ship is not come in. I am under very great obligations to Consul Barton, cousin german to Mr Maydwell; when you write to Mr Wood [clerk in London], pray desire him to make my complimts & acknowldgmts to Mr Maydwell on this account; & for other favours; for Mr Wood is well acquainted with him.[47] I do assure you, there is no plague anywhere, & this country agrees well with me, & therefore pray be perfectly easy & leave everything to God Almighty's providence, & don't doubt but in no very long time we will meet at Newtown.

23d -3d. Pack'd. Walked out to a hill in old town to a castle, & to see some cisterns under ground. I have been so employed, I have not been able to give the Bishop any account, so when you write, you may mention the time of the date of the letter, tho' probably letters I shall write to him may arrive as soon as this. Tomorrow morning we set out for Cairo. Pray make my kind love acceptable to my sister & due respects to all friends. I went out with Mr Dout to see 2 stones with Hieroglyphics on 'em & then went to see some medals & antiquities at Mr Orbus's, & finished packing & sent away things to go by water, we going by land. I am now habited a la Turque, & thank God I am very fortunate in everything, & have a prospect of a very agreeable journey throughout.

I am Dear Madam your most obedient & dutiful Son, Richard Pococke

47 Mr Wood (the "Reverend Montague Wood of the City of London Clerk") was a business associate of the Pococke/Milles family who regularly received instructions and deliveries from Pococke during his travels and who met the cousins on their return from their first grand tour. See Finnegan, *Letters 1* and *Letters 2*. He is not to be confused with Robert Wood of antiquarian fame (see Finnegan, *English Explorers in the East*, pp. 281-2.

GRAND CAIRO
11/22 November 1737
Pococke to his mother
Letter 4[48]

Honoured Madam,

On October the 24th Novr 4th we left Alexandria at 4 in the morning, having sent you a letter the day before, as I suppose, by Marseilles; in which as in two before from Alexandria, I acquainted you that we were three weeks in our voyage from Leghorn to Alexandria; & staid there three weeks & three days. I went in a chaise with Mrs Mackenzee a Scotch Lady, wife of a Captn. of a ship; – we travelled mostly in the same way. I went to Beckier, & after about 9 miles came to the sea side, by which we travelled 5 or 6 more to Medea Lake made since the time of the ancients by the irruption of the sea; – we crossed over the channel by which the sea enters, in a boat, & staid at a cane [inn],[49] & dined on what we brought with us, & on young mullets, each wrap'd up in straw, which being set on fire, they are so broiled, & eat very rich, almost like Eel. Vice Consul Butler, & others who came so far, returned. We were a good company, 3 chaises, & towards twenty mules & Asses. We went again after about a mile to the sea side, along which we travelled about 6 miles, & then began to cross towards a large sandy Plain, & found the sand troublesome to our eyes, & I perceived the want of my Goglers, which for this short journey I did not take; – they are glasses in cases to defend the eyes; the sand here blowing over the tracks of carriages: – there are about ten round pillars built of brick along the plain, to direct the way to Rosetto, & about the middle is a well. This plain is about five miles over, & from it to Rosetto about a mile & half over sandy hillocks; – a litle before we came to them, the English Vice Consul & some French came out to meet the Consuls family mounted on horses, for the Franks here ride horses, but not at Cairo. When we came within the hills a very beautiful Tent appeared, pitched to receive the Consul; in it we were entertained with Coffee, biskets & wine; – & then all mounted on fine horses, with a slave attending each one, sent by the chief Governor of the Town; & so made our entry

48 BL, Add MS 22998.

49 A cane, kane, khan, or (from Letter 39, in the Turkish spelling) han is a roadside inn. They were found along the main trade routes. Pococke stayed in numerous kanes during his travels in the east. He found that they varied in quality, some being very comfortable and others being sordid. The khan he stayed at in Lamia, in central Greece, was destroyed by an earthquake, apparently while he was there (see Letter 46). For a description of a khan in Turkey, see Letter 35.

into Rosetto to Mr Fabri the English Vice Consuls house, an Italian:[50] after we alighted, guns were fired, & having sup'd – we went to take our repose. Rosetto is much larger than Alexandria, finely situated on the Nile, where the most western branch divides, turning a league from the sea & yet the water is good, unless, when the wind sets in, or the Nile is very low; but the Nile here rises not above 3 feet: The country in the East is a mixture of water by the overflow, of verdure, & Palm or Date trees as they are commonly called, & appears most beautifully. The trade between Alexandria & Cairo passes by this place, & if there were a port here where ships could ride, Alexandria would be forsaken. There is otherwise a great manufacture here for linnen, but none very fine; 'tis mostly a striped sort most proper for furniture, & in all other course linnen for table & other uses. This place abounds in Pomegranates, figs, Adams figs before mentioned, oranges, Lemons, some grapes, & all sorts of fruit that they please to plant; dates are yellow & red, the last best, both blackish when ripe & then delicate like a sweetmeat, & before ripe eat well roasted. They are here in their Rice-harvest, & one sees 'em threshing it out like other corn in the grain, & the husk is taken off as from bales to make oatmeal. It is sown in May, the Nile overflows the ground, & it grows in the water, & reaping it in the water, people stand to take it as it is out & lay it where it is dry; much flax grows also in this country of Delta, as it requires water.

25th-5th. I staid within the morn; – in the afternoon went with the Consul to see the French Vice-Consul, & he & all the French came & play'd at cards at our house, at a new game call'd the Lottery.

26th-6th. I took a walk half a mile out of the Town, northward near the Nile, the country is most pleasant; & here one sees in their gardens all those rare plants which we preserved in Green:houses. The Chiaus [*çavuş*; emissary] sent a present to the house of two sheep & a dozen or two of fowls. I have shirts making [being made] for me of mugra thred [thread] made of flax, but looks like coarse thick muslin, they cost me about two shillings apiece, without gusset or gathering at neck or sleeves, & are good where people sweat much in hot countries. I took a walk in the evening the length of the town, mostly along the short Quays, which are broken by buildings going to the water. The French Consul & nation came & plaid at cards for two hours.

27th-7th. I walked out, round most part of the town. In the evening walk'd out with the Consul, met the Saida of the Janizaries on horse back, who had paid so many civilities to the Consul, exchanged a few words after the Turkish

---

50 Pococke was to meet him on his return visit to Rosetta (see Letter 30, below); he had married during the traveller's absence. Fabri is not listed in Wilson, *List of British Consular Officials*.

manner by the Droggerman & went on; – walked in a garden planted all over with Lemon & orange trees, Pomegranates &c: Basil growing two feet high to a Shrub, Cassia, a wild mulberry the fruit of which drops & comes to no perfection, as all the true mulberry trees were killed you know, in the land of Egypt: we took a walk round into some date gardens & returned by the Nile. We met a man stark naked, a Sheik or saint: on Fridays, when the women go to the places of sepulture [burial], he lies on the road, & they go & kiss his hand & then something else, & this they do out of Religion, & they think it makes them fruitful: these men are either fools or mad, or pretend to be so; & when dead they build monuments to them & often mosques over them; & at Cairo one rides about in this manner & has some respects paid to him.[51] The French plaid at cards, – & I spent two hours in my room. We saw the manner also of skinning Rice.

28th-8th. I walked out round another part of the Town, & met a young naked saint, a boy of about 18, & I saw the people kissing his hand. Our things came by sea; in the evening went in a boat on the water with the Consul, & walked a while on a fine Island opposite to Rosetto – most pleasant.

29th-9th. Went with the Consul the water into the Delta of Egypt, a shooting with some French, tis part pools of water, salt & bitter, by the soil & overflowing of the Nile. We traversed groves of Date trees, on which the people live three months in the year; – had a fine view of Rosetto, which appears like a skirt of London, with near 20 minarets, not unlike our high steeples at a distance. We sat down under a tree by the Nile, & took a litle collation, & came home at 10; the French Consul & French took a litle draught of small punch with us, as we had done with them yesterday before dinner, which is the custom here; I walked in the evening, the French came & played: – they have a dish, it is like clotted cream, which they eat with bread & sugar. Sellery [celery] green in sallad [lettuce] but not good.

30th-10th. I walked out with the Consul &c: to a garden, – saw the cream tree with its fruit like a pine-apple; were entertained in a summer house with Coffee, the people here being very civil: – dined with Monsr. Teissin [Tassin][52] the French Vice-Consul, two courses, an elegant desert, & Coffee after dinner: – he showed me his medals & presented me with a little unsett seal of onyx, yt

51 These men, whom he further describes as "natural fools" (*Description 1*, p. 14), were probably sufis (Islamic mystics).

52 Antoine Tassin, trained as an interpreter by the Capuchins in Pera, from 1711 to 1717, held posts in Tripoli and Crete before being appointed chancellor of the French consulate in Cairo in 1731. He was then appointed vice-consul in Rosetta in 1736, and vice-consul in Alexandria in 1743. Mézin, *Les Consuls de France*, p. 559.

[that] you may see on this [letter] if not opened, for which I shall present him with a medal.

31st -11th. Windy yesterday – rain today – in showers. I walked – in the afternoon went out with the Consul on the Nile in a fine three mast galley, he has hired to get to Cairo. Went up two miles & went on shore to a hill, whence a fine view like Greenwich Park, saw in a summer house the Cadilesker [*kadiasker*, military judge] of Egypt &c: on my return, the same as Mufti [chief jurisconsult of the Ottoman Empire] at Constantinople, sent once a year; French came & play'd.

Novr. 1st-12th. The French dined with us & we went a pleasuring in the Gallery a league down the river, as far almost as the Castle, & the French came & staid till 8 at cards: – but I commonly retire to my room till supper.

2d.-13th. I went in a boat to the mouth of the Nile, it divides into two branches two miles before the mouths of it, & makes an Isle; – I went on, then went to the last, not so low as the division, about 3 miles from the town: wind being contrary, rid home on asses, that met me, a most pleasant day.

3rd.-14th. In the afternoon – went with the Consuls Droggerman & Janizary to pay a visit to the Greek Patriarch of Alexandria, on his visitation coming from Rosetto, his name Cosmas, a litle old black man, with a long beard, a long lock on each side hanging on his shoulders from his hair, under garment coarse black linnen, a short one over of finer purple linnen lined, & a longer over that of some stuff [fabric], a lineing, a black stiff high Cap, with a broad band of stiff black stuff round it; – talk'd Italian a litle: – we had first Coffee, then Shirbet, & after one came with Rose-water & poured it towards the beard, which they usually receive in their hands & stroke it on their beards; – & then one comes in with a pot to incense the beard, & then they hold their hands over the nose & beard, to receive the grateful oder [sic], when this is done the company goes away: – the manner of calling a servant here is by clapping of the hands. Walked out with the Consul & French.

4th-15th. Got everything ready to embark on board the galley to go up the Nile to Grand Cairo. – Wind contrary, set not out in the morning. If my things are open that were sent from Holland & are in Waterford; – pray tell Mr Milles I desire he would take out four Indian handkerchiefs, of different sorts, if there are so many sorts, & send them by a convenient opportunity to Newtown; two of which I desire you to accept of & two to my sister. After dinner we embark'd & went up the river about two miles, but being calm we returned at night in a small boat, & spent the evening till eight with the French; I in the room of the Drogermann or interpreter of the French looking on some books.

5th-16th. We set out again, & went on with a good wind, Mr Fabri & the French Drogerman going with us till two a clock & then returned; – we got a quarter of the way this day, but lay by, good part of the night.

6th-17th. We went on a litle way, but being calmed, we fixed to land near a village, they had a market of Dates & pulse on the river side & a litle Coffee-house made up with hurdles; – we went to see the Kaimachan [*kaimmakam*, deputy governor] or Governor of the village, who ordered Coffee but we came away, & he sent a present of 50 Eggs after us: we walked out again in the evening; saw the herb nil [al-nil], of which they make the Indigo blue, has a leaf like wood sorrel or sage, & they make it as we saw, by bruising it, putting in water & heating it over a fire & stirring it, & then pouring out on thick clothes, & the water evaporates & leaves the blue in cakes, with which they here mostly dye coarse Linen, being so good, as that we use which comes from the Indies. This country [Delta is the richest spot in the world, from which they have two crops of corn, besides many crops of smaller things] – cotton, Flax , one harvest now of what [was] sown three months ago, & now they sow for a harvest in February, & never dung the ground.[53] We put on &

7th-18th. We got above half way, beyond Terrane & on the

8th-19th. In the morn we came to Ouardan,[54] we went a shoar as always, shooting we saw an ostrich in the fields belonging to the Governor; but on the neighbouring hills, they are in large flocks, & on them also are Antelopes. We walked in some gardens of this large village full of orange trees, standard Apricots & Peaches, &c:; then visited the Kaimacham, who would have given us a letter to speak in the Phrase of this country; – we invited him aboard, & he sent us a present of 100 eggs & a Lamb; he came to us on horse back, was treated with sweetmeats & Coffee, & presented us with a vessel of wine to entertain Strangers it is to be supposed, tho' the Mussulmen [Muslims] many of them drink privately that cheerful liquor. We went on in the evening, got but very litle further & on the

9th-20th. We walked out in the west side a league [& the galley went on, but we ... before], fastening to the Delta [...].[55] We walked out in the evening & shot at some large birds, called Pharaohs Hens [Egyptian vultures] – about 8 a clock a galley came up with us from Cairo, in it was Mr Upton, partner with the Consul, his mother lives at Bath & he is a very honest, plain working Gentleman

53 This sentence is problematic as Pococke has inserted a sentence between the original lines of text and it is not clear where it is meant to begin and end. The additions are represented here in square brackets.

54 Spelled "Ouarden" in *Description 1*, p. 17 and "Vardan" in Revd Clayton, *A Journal from Grand Cairo to Mount Sinai and Back Again*, p. 9, i.e. Wardan to the west of Cairo.

55 This sentence is problematic, with various words being obscured.

about 40, knows Mrs Hunter well. There lives with him as a Boarder Mr Coke, an old Gentleman, who has had a litle Paralytic Stroke, & is uncle of Mrs Barton & brother of Mr Cokes in London, he broke [was an agent] at Cyprus. In the galley was Mr [Francis] Congreve, the head partner of the other English house, Mr Vernon a very young Gentleman, much beloved by those that know him, being his [Mr Congreve's] partner,[56] there were besides some others of this country: – on their arrival we fired our canon often & sat up till 12.

10th-21st. We all walked out, & about 11 set sail, & about 8 we arrived God be praised, at Hele, two miles from Grand Cairo, lay aboard.

11th-22nd. Great numbers came out to pay their respects to the Consul & Ladies coming to wait on Mrs Barton, we retired to the other Galley, they are veiled & must ride on Asses; – men do not go into the same room ordinarily; but it is permitted sometimes to Franks, so we went in & had a view of the veiled Ladies; if any are handsome they let fall their veils, by accident, & catch it as in great confusion. Leaving the Ladies behind, the cavalcade began of our entrance into Cairo. The Consul [is] in Turkish habit, except Wig & hat mounted on a horse finely accoutred – preceded by half a dozen Janizaries; – & I observed a ceremony at first a man throwing water on the road before him, to lay the dust: – the cavalcade beside consisted of about 40 or 50 not horses – but Asses, for Christians & Jews may not mount horses here, except the Consuls, who are regarded as Ambassadors. We went two small miles through the pleasant country & entered Grand Cairo, the streets narrow to keep off the sun from the houses, which are built round courts: – We arrived at the Consuls house. I find I am like to have an occasion to exercise my office, on account of an aged Dr, who is thought to be on his death bed. I must tell you, that I meet with very great civilities here, which makes this country very agreeable, & I thank God I have my health exceeding well, rather better than in Europe. I am in hast [haste] to send this without reading, that it may go with a ship that is now ready. You will let Mr Milles know when you hear from me, or the Bishop, if he is not with him; for I have not time, as yet, to write any account to him, & I know not when I shall. I shall not leave this place till the beginning of March,

---

56 Francis Congreve (1702-42) was from Stretton Hall, Staffordshire. He became a member of the Divan Club on his return to England. See Finnegan, "The Divan Club", pp. 58-59 and his Will, The National Archives of the UK, Catalogue Ref. Prob 11/727/314, 30 June 1743). John Vernon was born in 1720, third son of Henry Vernon (1663-1732) of Hilton Park, and died in Cairo in 1747. He was related to the famous Admiral Edward Vernon (1684-1757) who was a member of the Divan Club. See Finnegan, "The Divan Club", pp. 53-54. Congreve and Vernon resumed their friendship with Pococke on his return visit to Cairo (see Letters 24-28). I am grateful to David Wilson for confirming the identity of Vernon, in his unpublished document "More Notes on Richard Pococke's Letters from Abroad, vol. 3", 2014, p. 1.

having many things to see near it; & all that I aim at, is, to be at Jerusalem at Easter, which is as much as I shall be able to compass [accomplish]. I have desired you to direct your letters to me to be left with Mr Maydwell &c: who will forward them to me, – & I desire you to write in 6 weeks. I was at Leghorn two days, 3 weeks on my voyage to Alexandria, & 3 weeks there, from which place I writ you three letters: one by Leghorn, another by France, & I believe also the 3d by France. I am like to have a Franciscan Superior of the convent here to accompany me to Jerusalem, & I believe shall never want good company, if I am so fortunate, as I have been. Pray present my very kind love to my sister & due respects to all friends. I desire you to be in no apprehensions on my account; – We shall meet at Newtown, & there are not the dangers you apprehend, the country is healthy, & here sometimes they have not the plague in 5 or 6 years; they had it this time two years, brought from other parts, & this year they had a good[...], so that the country will be healthy. You must not be unsure if you hear not from me, [..... Accidents...].[57]

GRAND CAIRO
15/26 November 1737
Pococke to his mother
Letter 5[58]

Honoured Madam,

On November 11th-22d. the day I arrived at Cairo, I forwarded a letter to you before noon; dined; – Mr Upton went with me to meet sick Dr Aram, a surgeon & great traveller on board ships as a surgeon; – I prayed by him twice; came home. The French were with the Consul in a visit. I gave Dr Aram the sacrament, he lives in the other English house of Congreve & Vernon: – The English house we are in is a large building round a court, two or three large saloons, one very handsomely furnished of the Consuls. I have a litle convenient lodging room.

12th-23d. Dr Aram died this morning. I breakfasted with Mr Upton in his Apartment. Joseph Euchy[59] of Tunis came to see the Consul, he was Ambassader

57 The end of this letter is indecipherable, since Mrs Pococke crammed the final lines into a tiny space and some editing has taken place. However, the general idea is that letters can sometimes be "miscarried" (lost) and that his mother should not worry, etc.

58 BL, Add MS 22998.

59 In his article "Islam in Britain, 1689-1750", *Journal of British Studies* 47.2, 2008, 284-300, esp. 294, Nabil Matar mentions a "Yusouf Hogia", an envoy of the Bey of Tunis, in London

from the Bey of Tunis for three years in England, & returned about a year ago; talks Italian & was much in our Queens [Queen Anne's] favour , a very clever man. I went to see the Greek priest perform a small ceremony over the Doctor in the house; putting on his stole, & saying Kyrie Eleyson & the Lords prayer &c: in Greek, – they wrapt [him] up in a sheet, covering the face, – & a small cross is laid on the breast, & put into a coffin in this shape [sketch] the top being raised thus [sketch of triangular shape]; dined, went to the greek Abbey of St George, to see the Dr buried there; – it is at the further end of old Cairo, which is a mile South west of new Cairo, & the convent is 3 miles from our house & the attendants have a wax candle in their hands lighted, & after incensing the body, which had also oyl poured on it, the people were incens'd. This was done in the church of our lady, & after we went to the convent of St George adjoining, one part for men, the other for women, mostly old, & it is more like an Hospital; – we went on the top & had a fine prospect of all the places about; – saw the Pyramids four miles off, which we saw first at Ouardan 40 miles off; – we saw also the Pyramids of Sicara [Saqqara], about ten miles. Passed by Joseph's Granaries,[60] only enclosures open for corn. We took a glass of wine we brought with us at top of the house, & went down & took Coffee & Biskit with the Abbess, a good old woman, the old ones show themselves, but the young ones do not; found Monsr. Damara [d'Amirat] the French Consul at our house at cards. The air of Alexandria only is damp, but it is not so, in other places of Egypt, unless close on the sea, but sometimes great dews; we had a shower of rain this afternoon.

13th-24th. Went with Mr Upton & Mr Vernon to the castle, which is on a hill; here the Bashaw [Pasha] lives, saw the Divan [Council of State] just meet, but

in 1733. Allowing either for Pococke (or his mother's) incorrect transcription of the name, together with the possibility that the envoy used his family name "Euchy" ("Oukhi?") in Egypt but not in London, this must be Yusuf Hoca, who indeed arrived in London in 1733. Furthermore, an official document dated 21 October 1733 records that Jassuf Cogia, an envoy from the Bey of Tunis, had arrived in England on 11 May 1733 and that he was to be given an allowance of 15 shillings (back-dated to his arrival and to continue until his departure) for food. See "Treasury Board Papers", CLLXXXIV, No. 22, *Calendar of Treasury Books and Papers*, vol. 2, 1733-34, Her Majesty's State Paper Office, London, 1898, p. 410. Available at: <https://www.british-history.ac.uk/cal-treasury-books-papers/vol2/pp406-411>. The following June (1734), he was still in London, as the same source records presents being given to "Jusuph Cogia" (see "Warrants for the Payment of Money: 1734, April-June", p. 304) and two other references to complaints about the amount of presents being given to Cogia – a "paltry" story involving a second envoy, Sidi Mahomed, who was petitioning for gratuities outside the norm (see ibidem, records dated July 17 and July 25 respectively). The Joseph Euchy who visited the consul in the present letter is undoubtedly the same envoy mentioned in Letter 6 (below), and with whom Pococke dined.

60 As described in *Genesis* 41:49.

the Bashaw not there, being on business; – it consists of the 24 Beys or Governors of so many Provinces of Egypt; went to the Coffee room of the Bashaw, where they brought us sweetmeats & Coffee; – had a fine view towards Old Cairo, from a Balcony: of Cairo, had the river run through it at the Act of the hill of the castle, it would be something of the situation of Prague – court of the castle full of fine horses of theg reat men, finely accouter'd, no wheel carriages here. The Cadeliskar mentioned in my last, is not as the Mufti at Constantinople, but is the head person in the Law; – & Sheke Beken[61] is like the Mufti of Constantinople greatly respected here, this place being much bigoted to the Mahometan Religion, even more than Constantinople. Mr Coke mention'd as Partner here in my last, is not so, but lived 25 years in Cyprus, had bad partners & met with misfortunes, & retired here to his Brothers 11 years ago, who now live in London & are at the head of this house & company. The whole family dined with Messrs. Congreve & Vernon. Pomegranate taken out of the shells a constant desert, eat well with wine & sugar, we had also Pistacho nuts which come from Aleppo, & nutmeg & its husk a sweetmeat, the outer husk or pele is like the walnut & has a tast [taste] of Ginger, under this is the mace & then the nutmeg, which appear'd to me, to have a thin shell over it & so to be the kernel of the fruit; – but its said that this shell is part of the nutmeg. We drank tea & came away toward the Evening: – They came & sup'd at our house. I staid up with Mr Upton talking about Mr Christmas & the two Masons, Harry Alcock & Billy, all whom he know[s] well at London.[62] If you see Mr Milles you may bid him tell them, we drank their healths here. Mr Uptons father was a Physician, his Grandfather Canon of Xtchurch, his sister married to Col. Upton who lives near Antrim & I think is member of Parliament,[63] & his niece is married at Bath to Dr Parrot brother to Golden Parrot, Knight of shire for Oxfordshire, who was fellow of St Johns.

14th-25th. Staid within this morning, yesterday a shower in the morn, today raw cold; dined. Visited Father Jacomo, the Procurator, Prefect or Superior of the Franciscans in upper Egypt; – they have a small convent in a house & a litle Refectory & chapel only three fixed there, besides two or three travellers. Took

61 'Beker', i.e. [Ebu/Abu] Bekr/Bakr?

62 These are connections from Waterford, whom Mrs Pococke may have met during her visits to her brother the bishop. The two Masons are John and Henry, sons of Sir John Mason of Waterford, all three of whom had served as Mayors of Waterford. Harry and Billy are Henry Alcock (1716-1811) and Colonal William Alcock (d. 1779), the sons of William Alcock (1681-1739) and his first wife Henrietta Mason (daughter of the same Sir John Mason). They were nephews of the Revd Alexander Alcock, Dean of Waterford, whose wife Susan Mason was the sister of Henrietta Mason.

63 Dr Francis Upton (1659-1711) was the only son of Revd Ambrose Upton (d. 1686) and Mary Charleton, and his sister was Mary Upton (d. 1720) who in 1691 married Henry Howard.

Coffee with the Consuls company at home, & play'd at Quadrille [card game] in the Ladies apartment. The streets of Cairo are very narrow to keep off the sun; – but there is a street or two that are long but winding & appear tolerable, but the others but mean: – no houses of the great appear well without; you see only [a] great arch'd gate, & they are built round a court. But the Great men have magnificent saloons within.

15th-26th. Father Francesco came to see me, the Predecessor of Father Jacomo, who came over with me, in the Procuratorship of the Franciscans of the Propaganda Fide in upper Egypt; – he is a Piedmontese, a great traveller all the world over; – he has received many favours of the English & is very affectionate towards them, & I find myself very fortunate thro' but my whole voyage, & shall always have a sense of God Almighty's goodness towards me, in protecting me, as he has hitherto done. An Aga [military officer] is come from Constantinople, to demand 3000 men fat & rich, which will soon be raised, & as an Embargo will be laid on all the ships of Alexandria, to transport them to Europe, so it will probably, be a very considerable time before you will hear from me again. But if my letters arrive, I have writ four within these six weeks, & I believe some of them will arrive to Cologne; know that I left Leghorn Septr. 6th-17th & arrived at Alexandria on the 29th O.S. of Septr, that I had staid there above three weeks & in less than three weeks after arrived here, staying near a fornight in Rosetto, & that I enjoy my health very much & was very happy in in meeting Consul Barton at Alexandria, with whom I came here, & am in his house & receive very great civilities from him & Mr Upton, & tho' it is but four days ago, I writ to you, yet a ship being now about to go from Alexandria, I send this, as I may not probably have another opportunity a great while. I received this day letters from Europe, but not from Jerre [Jeremiah Milles] tho' two ships have come from Leghorn, where he was to write to me, & I might also have received a letter from you, addressed to Venice – you may let Mr Milles know that I received Mr Thomas's letter of the 27th of August directed to Cologne;[64] but I suppose Mr Hoare [his London banker] to whom it was sent, conveyed it to Mr Cutts Maydwell Turkey Merchant in London, at his house – & so you are to direct to me: – & I hope to hear from you once in six weeks, that is that you will write so often, but let Mr Thomas know that I did not receive his letter he mentions to have writ not long before, having received only one from

64 Revd Edward Thomas at the time held several positions in the Diocese of Waterford & Lismore, including that of Rector of St. Olaves. Later, after obtaining an LLD, he was appointed Vicar General of this diocese. See the various entries in Rennison, *Succession List*. Mr Thomas corresponded with Pococke during the earlier part of his second grand tour. See Finnegan, *Letters 2*, p. 136.

him since I left England in answer to mine from Leipsick.[65] I desire you from time to time to give Mr Milles an account of my motions, & the purposes of my letters, so far as is proper; & also of my intentions, & who I am to travel with &c:. I left a box at Venice to be forwarded to you, which has in it nothing but what belongs to me. Tell Mr Milles that my time is so entirely taken up, that I have not time to write to him, much less to write to my Lord, which takes up more time. I do desire of you to be chearfull & use exercise constantly; – I find Coffee raises the spirits much, & I believe is good for my constitution, which is moist, so I take it twice a day, – here they drink it – viz – the Turks & Jews 10 or 12 times a day, without sugar & only one dish at a time; & the Jews drink strong waters very much, & it does them no harm, they say, they use it at meals & at other times, & unless thirsty take it alone. God Almighties Providence is everywhere, & I have full trust that it will protect me; & I hope to by seing the world & gaining knowledge of mankind, to be the more able to do him service in the profession I am in. I have two tours to take, one of a day to the Pyramids five miles off, & one of two days, to those ten miles off to the south; – I saw all of them from St Georges Convent. Assure yourself there are not the dangers in travelling here, that there were formerly, – the Arabs are quiet & at Peace; & the worst they ever do, when they are not in peace, is to strip & rob, without hurting the person of anyone. To fill up my letter I will give you an Account of the dress on my head; a white skull-cap, & then a larger of red cloth over that, & a red Turban round of a coarse muslin, that is the distinction here of Franks; – other Xtians & Jews wear blew & white. The shirt is made like your old fashioned shifts, the neck is bare, for it does not come on it, then a wastcoat with cut sleeves, buttons not, but tyes at the side, then a short cassock of Aleppo [...] red & white narrow stripes, a piece makes two, which cost 20 shillings, then a long cassock of the same stuff, but different stripes, one of which takes a piece, these do not turn up at hands, but hang to the knuckles, with an opening, this is called a Doliman [overdress]; over this a garment called a Benish [robe] of same make, but open before the Doliman being hid with a sash. The breeches tye up on the thigh always red are worn by Gentlemen & come down near to the ankle like trowzers, to these are sown a short pair of yellow leather stockings, & these I wear with yellow slippers; – but Jews blue leather, and Xtians of the country commonly red slippers with hindquarters, under these of red they wear linnen Trousers, & on the feet linnen socks, the upper garment is furred in winter, & some wear under it another short one, furr'd, & then the upper one is not always lin'd with furr & with this short one, with sleeves down to the elbows many go within doors. When visits are paid to great persons, they wear a

65 Pococke and Milles were in Leipzig in November 1736.

Ferigee or gown like the Drs red gowns at Oxford, & not unlike a nightgown & here a cap turned up with furr, called a Carpack is a head dress among the Franks, when they visit great persons. None but Turks can wear white Turbans & none but Mahomets relations, green; one of those waited as a servant on us at Rosetto, he cannot be hurt with his green Turban on, but take off the green Turban, & a Turk may [hurt] a relation of Mahomet, as well as any other. The Grand Seignor governs here by his Bashaw [Pasha], who is changed once in two, three or four years. He & the Divan of 24 Beys, do all that the Janizaries will permit, of which there are seven bodies, & their officers meet in their particular Divans, & oppose what they please, & depose the Bashan in a regular sort of manner, & take care of the government in the interim; – this they do by ordering him to come down out of the castle, but his person is always secure; & this is often done: the 24 Beys govern the 24 Provinces in Egypt, & are put in yearly, but one of them is generally a leading man & does all, & when he has got that power, he tyrannizes even when he is out of office, like Pompey or Caesar, when they were not Consuls, & these help even the military power in order, who otherwise govern. It is cold raw weather now, & the sun has not appeared these two days. Mr Maydwell our Correspondent in London is the Consuls [Robert Barton's] cousin german & related to his Lady, who is a Lady extremely civil & obliging, her sister was married to Mr Pelham Ambassador in Sweden, but I think she is dead. Captn. Lisle left his Turkey ship in hopes of being Captn. of the South Sea Ship, when she goes out. I am recommended to Mr Lisle at Constantinople, to Mr Philips at Smyrna, & to the Consuls Brother Mr Barton at Cyprus, & to our Mr Pullinger at Aleppo.[66] Pray my most kind love to my sister. I think I ought to fill a sheet from this place, & you insist on my journal, & to tell you the most trifling particulars; one is that this country has no wine, tho' it had in the time of the Romans here. This city is very large, & above a mile from the river, & about it many magnificent monuments of great men, very

66 These can be identified as: Mr John Lisle (see Letter 2, above); George Barton, Consul at Cyprus until 1730; and Arthur Pullinger, formerly a merchant in Aleppo. See Wilson, *List of British Consular Officials*, pp. 42, 34 and 3 respectively. Pococke pays special tribute to Arthur Pullinger in the preface to *Description 2*, p. v, "for the many extraordinary civilities [he] received at Aleppo" from him, as well as for his "very extraordinary knowledge of the Greek medals… antient geography and history of those parts, and of Asia Minor, and Greece". (See Thomas Drew-Bear, Christian Naour and Ronald S. Stroud, "Arthur Pullinger: An Early Traveler in Syria and Asia Minor", *Transactions of the American Philosophical Society*, 75/3 (1985), pp. 1-80.) Pullinger's "Geographical and Antiquarian Notes Relating to Greece and Asia Minor" were bequeathed to the BL on 5 June 1767 by the late Bishop Pococke, and one of the "two hands" in the script (including the title page: "Antient Geography Collected By Arthur Pullinger Late Merchant In Aleppo About the Year 1740") is clearly that of Pococke, who may have copied part of it in the Levant.

fine mosques, & the environs very pleasant; when you read travels, put Queries to me of what you want to know. Pray my kind respects to the family at Whiteoak. I desire you again to be cheerful & talk of it & look on it as your happiness, that you have a son in a capacity to see the world, & that will not, I hope, come home entirely unimproved, by what he sees: – you see travelling was not with us an idle business, & my hands are more full now, that we are parted, insomuch that I can do litle more than write full notes of everything I see as yet, tho' I hope I may have more time on my hands; but I have a great deal to do in order to be at Jerusalem at Easter; before this comes to hand you will doubtless see Mr Jer. Milles; – let me know whether he has the man he took at Venice – or who? – & how he goes on? & pray remember me to my fellow traveller. I should be glad to receive a chearfull letter from my Lord Bishop [Milles], giving me an account of the confirmation of his good state of health. & Mr Thomas to my great joy informs me, on the 27th of August that he was wonderfully recovered. Let me know where Mr Masons eldest son [John] & Harry [Henry] Alcock are: desire Mr Milles to write to me once in two months, or if he can fill a sheet in six weeks, & let him know your time & write so as it may come to Mr Maydwell about the same time; – you must now get very large paper, & write particularly every day; but as the weather here & with you can have no correspondence, so you may leave out that, for more material things; & let me have deaths & marriages of great persons, & news of State, but not of the Grumbletonians [Malcontents].[67]

I am Dear Madam your most obedient & Dutifull Son Richard Pococke

My best respects again to all friends. – If Mr Milles would write me an account of the most remarkable occurrences as they happen he might fill a sheet once in six weeks; – & would by all means have you agree on a time to write so that both letters may come to Mr Maydwell at pretty near one time.

I seal my letter with a seal of onyx, but on 2d thoughts, twill be taken off, by being open'd; – it is a parrot with two cherries in its mouth; – given me by the French Vice Consul at Rosetto[68]

67 As the correspondence is one way, the reader has to surmise her son's responses what Mrs Pococke put in her letters.

68 See Letter 4.

GRAND CAIRO
5/16 December 1737
Pococke to his mother
Letter 6[69]

Honoured Madam,

As you insist on a particular account how I spend my time, & what I see, so I continue a particular journal to comply with your desire. – My last letter to you was of the 15th-26th of November.

16th-27th. I went & waited on the Procurator of Mount Sinai, who precides over the monks here belonging to that place, & does the business of the order; & sends them provision: – the Abbot of Mount Sinai is a Bishop, & has the title of Archbishop; there are 500 under this Monastery, they have estates in Cyprus & elsewhere, & collect charity in most parts, & have one in England for that purpose: – he was very civil, talked Greek, there being an interpreter between us; – sugar plums were brought, then a fine dram, sugar plums again & Coffee, & at going away the servant poured rose-water on my hands. They live in a large house round a court, & go to the Greek Patriarchal church. A greater treat before Coffee is sweetmeats, & the greatest compliment is, the addition of perfuming the beard. Father Francesco spent the afternoon with me, a most worthy man, late superior of the Franciscans of the propaganda fide in upper Egypt. Mr Congreve & Mr Vernon spent the even here.

17th-28th. Rid out to the hill over the Castle, from which I had a fine view; A Priest of the mosque, had us into the mosque where a great saint is buried, & his relations; – this they do with much civility in private places, but in hopes of some small reward; – he treated us with Coffee & biskit: We descended another [hill] & came into the town at the gate where Sultan Selim entered, called the gate of Victory [Bāb al-Naṣr]; – you should read Knollys's Turkish history[70] now. We met a marriage procession, 2 camels with men on them, beating kettle-drums, 2 youths finely dressed, one in Ermin, musick on horse back, a great boy followed on high stitches [stilts], & then women, some of them playing on musick. In the evening went with the Consul to see the propaganda fide Fathers, my fellow travellers Father Francesco &c: entertained with sweetmeats & wine. Mr Vernon & I went to see the ceremony, by giving a look into a mosque of a Sheke [shaykh] or Turkish priest, who has fasted 3 days & ye 3d night all the Shekes came to perswade him to eat. Messrs Congreve & Vernon spent the

69 BL, Add MS 22998.

70 Richard Knolles, *Generall Historie of the Turkes*, London 1603.

even with us. Went also to see the procession of the Sheke to his house, but he did not come.

18th-29th.Went out with Mr Upton to the shops, & to the place where they sell the Black Slaves. – In the even played at Quadrille with Mrs Barton &c:

19th-30th.Went a shooting with Messrs Congreve & Vernon, about some lakes just without the town to the north, these become dry about April; saw a Caioux or Colonel going out on horse back, with many slaves, throwing darts before him, as is the way, as a sort of exercise, & going to shoot at marks. We dined under a covered place in the garden of Apricots, because many standard apricots there, as well as oranges & Lemons; – some French also were there: – Mr Upton came out to us, we walked a litle about the country & came home at night. Messrs Congreve & Vernon spent the evening with us.

20th-1st. I went out with the Consuls Draggerman to wait on the Cophti [Coptic] Patriarch of Alexandria, the head of the Alexandrian or Cophti church, & the true successor of St Mark, as acknowledged even by the Romans, who give the Cophtis only the name Schismatic, – they hold nothing different from them, but the Eutychian notion of one confused nature in Christ of the divinity & humanity together.[71] He was cloathed in a brown silk cassock lined with white linnen, – over a pale blue gown lined with furr; – on his head a blue & white Turban as the Christians wear, & over that a purple piece of stuff like a hood somewhat, but going down under the upper garment like a scarf: – sitting cross legg'd, he had over his cap, a blew & white linnen cloth, & I once observed that his feet were bare. He treated us with Coffee. I went to see their church. Messrs Congreve & Vernon dined with us. Went with the Consul & all the English, to an entertainment given on account of a Jewish wedding: – seated on the Divan, Coffee brought musick on Tambour, Dulcimer & a sort of two stringed fiddle, & singing joined in sometimes, like the chorus of the country people; – on the carpet a stool brought, on that a large copper dish 5 feet diameter, on which were 30 dishes of sweetmeats, two dishes of a sort, sat at it cross-legg'd & wine was given round plentifully for half an hour. – taken away, a dance of three, consisting much in throwing out the legs & arms; & the Ladies who looked from above, gave their compliments & approbation of the dance, by a guttural trembling wise, something like doves. We had Coffee served twice. Aloe wood burnt in an incense pot, & last of all a dish of shalebe, I believe like our saloupe sold by Turks in London, made of honey & water & cinamon boyl'd. Messrs Convreve & Vernon sup'd with us.

71 Elsewhere Mrs Pococke has transcribed the word "Coptic" correctly, though in his book, Pococke uses "Copti".

21st-2d. Visited Father Francesco, spent the evening with Mr Upton at Mr Congreves.

22d-3d. Went out to some shops with Mr Upton, dined, drank tea with Mrs Barton. Father Francesco was with me. French Consul here till 8. Mr Vernon sup'd here.

23d-4th. Went & saw that great work of Joseph's Well; – then the coinage; after that Josephs Hall, a very grand thing, with 32 long pillars in it; went after into some vast ruinous buildings, where I saw the covering [Ar. *kiswa*], weaving & embroidery for Mahomets tomb which is a sort of thick black sattin, with figures of buildings on it; – it is richly worked in gold, with Arabick sentences out of the Alchoran; – I saw 6 frames & about 30 men embroidering it; – such a covering is carried hence every year to Mecca, with the Grand caravan. Mr Congreve & Vernon spent the even with us.

24th-5th. I was visited by Father Justo of the propaganda & a Cophti, went with the Droggerman & waited on the Cophti Patriarch, he was just going to dinner, so we sat down with him; – in a large dish, set on a stool was brought two plates of honey, one of dry'd Bakes, one of cold saltfish in oyl, a dish of warm cake in the middle; – when they had dined, a man came with a plate of young clover & threw down an handful at the feet of each of them, as they sat cross-leg'd, of which they eat in puris naturalibus [......][72] – so we dined in clover; – the Patriarch gave me a piece of a whiter cake laid before him, & put some dates before me, which was all I eat except robbing the Patriarch of some of his clover: – the Patriach said Grace, sitting, before & after meat, the two Priests standing, & making responses: – round the dish was laid as many cakes as it would hold; – no drink brought but Coffee, 2 dishes a piece; immediately I dined again. Went to the Propaganda, & then went with Mr Congreve to Bulac [Bulaq], to see him embark for Rosetto, to go to Alexandria; – went a litle way down the river as far as Heli & returned home;[73] on our asses with Messrs Certon [Barton?] & Vernon; – the latter spent the even here.

25th-6th.Went out to the shops to see & buy some things I wanted. Some of the Fathers came to see me. My whiskers are grown so long that they begin to be troublesome in eating, & now I have begun to let my beard grow on my chin which is proper, in order to go to Jerusalem, tho' not always done, so I begin to make a venerable figure.

26th-7th. Father Justo of the propaganda came to breakfast with me, went to wait on the Patriarch of Alexandria of the Cophti church to get two Arabick

72 Several words have subsequently been scored out by Pococke.

73 This village is spelled "Hele" in *Description 1* and is thought to be "some remains of the name of the antient Heliopolis, that was about five miles distant", p. 17.

letters from him to some convents of theirs I shall go by: – he was sitting crosslegd on a seat in the court, with his people about him: – his dress today was as at first, except that he had on a black gown, like a nightgown, which is the thorough dress. I carried him a present of 4 boxes of Prunellas, which my servt [servant] put by him, & then his steward came & took them away; he gave us Coffee. I went to the propaganda convent, & after to the Consuls Droggerman. Che dona, ben vende, si non e villano che prende – a proverb here.[74]

27th-8th. Visited Mr Vernon in the evening with Mr Upton, came home to supper; Signr Coggiat [Coggia, i.e. Yusuf Hoca], dined with us, late Ambassador from Tunis to England, France, Muscovy [Moscow].

28th-9th. Went to Old Cairo, saw in the way the slaves exercising; – two stand at 30 yeards distance & throw darts at one another, avoiding each others darts, this they do on horseback, one throws his dart & rides away, & then the other throws after him, & so on. I saw them riding full speed, & fire a gun at a jarr [jar] set up on a litle eminence; – the slaves here are much considered, & often succeed in their masters bed, that is, the head one, who is their Cashier, called the Hasnadar, & such are most of the Governours of the country, called Beys. I went to the top of an high building by the Nile, which has a Reservoir of water, conveyed from the Nile, & hence it goes on a fine Aqueduct of near 400 Arches to the fort of the castle hill, to the top of which it is conveyed by wheels. Went then by Josephs Granaries which are open enclosures with very high walls. Went to the Cophti church; where a chappel under ground; – they show a hole like an oven, where they say the Virgin Mary, our Saviour & Joseph lay hid, when they came into Egypt: – viewed a very fine old ruine here; & came home; – visited Mr Vernon for a quarter of an hour in the even, he being indisposed with a cold.

29th-10th. I rid out a litle about the town in the afternoon. The Government in & about this city is very well regulated, for altho' they are raising soldiers, yet there is not the least disturbance, & no disorders committed.

30th-11th. Father Justo came to see me. I call'd on Mr Vernon; – played at cards at Mrs Bartons Apartment.

December 1st-12th. I went to the Propaganda convent, & out into the Town, played at cards at Mrs Bartons Apartment.

2d-13th. Father Justo breakfasted with me, I went with him to the Franciscan convent of Terra Santa, saw their Refectory & chapel, but the convent is only several private houses joined together. Visited Mr Vernon. You may write to Jer. M. [Jeremiah Milles] to send an handkerchief a piece for for my two cousens, & also if you think proper, one of the best to Mrs Cooper of those Indian ones

74 "What can a woman well sell if it is not a rustic who buys it?"

at Waterford: – desire him also to send that cambrick, I had made up in a neckcloth 3 or 4 yards long, if you have the least occasion for it, because I shall not want it, & I hope you look on everything as your own at Newtown, table-cloths &c: & use them as you please, especially those coarse ones if you want them. I see the differences of our court in the news; – but write nothing that may offend, for letters may be seen. I spent the evening with the family.

3rd -14th. Drank tea with Consul & Lady at breakfast. Visited by Father Justo; dined. Visited Father Jacomo & Justo came to see me, & after the Superior & Procurator of Terra Santa Convent. I am in great hopes this letter will come to you before the Embargo takes place: & then you can't expect to hear from me again in all probability till towards Autumn. I writ to you from Alexandria on Septr 30th-the 6th of Oct. & the 24th of Octr & from Grand Cairo Novr 11th & on the 15th, some of which letters I hope have come to hand, to inform you that I was 3 weeks on voyage to Alexandria, 3 weeks in Alexandria, twelve days in Rosetto, a week coming up the Nile hither, & have been here now above three; everything succeeding beyond my expectations, & receiving the utmost civilities from the Consul &c: which makes my journey very agreeable. As we have a constant serenity of fine weather, & now just such weather, as may be in England, but cold mornings, sometimes tho' rarely foggy; – so I acquaint you always when it rains. Mr Vernon spent the Even with us.

4th-15th. [No entry]

5th-16th. Breakfasted with the Consul, & am to seal this letter before dinner. I desire you once more to be of good chear, for we shall meet I don't in the least doubt at Newtown; – use exercise, & have a full trust in God['s] Providence, that all things shall be for the best. Pray my very kindest love to my sister.

I am Dear Madam your most obedient & dutiful Son, Richard Pococke

I believe you will hardly hear from me, till I return from Jerusalem & so probably not till towards Autumn.

ACKMIM IN UPPER EGYPT
21st December 1737[75]
Pococke to his mother
Letter 7[76]

Honoured Madam,

On the 5th-16th December I dated my last letter from Cairo. I went to see Mr Vernon, & after dinner Father Francesco & Justo came to see me, & they Mr Upton & Captn. Forishal, &c: went with me to see me embark to go up the Nile to Ackmim [Akhmim], but the boat not being ready, returned.

6th-17th. Father Justo & Captn. Foresthill [sic] went with me again to Bulac, the great boat being gone up about 12 miles for fear of being pressed to carry soldiers to Rosetto; – we embarked on a smaller about noon, the great bark I was to go in belongs to the Prince of Akmim, & I went with his chief menager Soliman Elganany a Cophti Catholick man being strongly recommended to him by the Consul & Propaganda Fathers.[77]

7th-18th. In the morning we came up to our great Bark & set sail in it lying by part of the night.

8th-19th. Went ashore to see a convent of St Antonio the Eremite, found the priests drawing stones, who are married;[78] so that now it is not properly a convent, – showed me a place where St Anthony was; – here they say crocodiles can go no lower. To give you an account how we live aboard, – with the fine old Gentleman I am with [Soliman Elganany]; – drink Coffee as soon as up, they don't eat with it, as they keep Advent & eat nothing that comes from flesh: – so I live on mutton, beef & pidgeons pye, & brought fowls alive with me; – they eat commonly lentil soup, like pease soup, drink litle, but water, Coffee immediately after dinner & supper; – for supper sometimes they eat a sort of Treacle made of the sugar cane, in which they dip pieces of bread & eat also dates; with everything boyled I make nice soup, & commonly drink Punch, tho' I have a jar of wine, & the Consuls house sent a chest of Florence with me, as well as fowl, pye, Beef & mutton. It is not the fashion to go into naked bed in a boat, where we lye is just covered over & I lie on my mattress covered all over with a double blanket & over that a carpet to receive any dew that may blow in. We pass our time very agreeably, the Arabs sometimes chant out a prayer to all the saints from Mahomet to Adam & from Adam to Mahomet, & sometimes tell stories with an excellent accent, insomuch that it sounds like the best Italian

75 It is unusual for Pococke to forget to use the dual dating system.

76 BL, Add MS 22998.

77 This journey is described in detail in *Description 1*, pp. 69-81.

78 These were to be used for repairing the convent, see *Description 1*, p. 70.

preaching, & hence come the Arabian tales; another dish dipping bread into water & salt & a sort of seed pounded called simsim with oyl & onions shred on it. We lay by at night & have very fine weather.

9th-20th. We dined and my fellow travellers on the aforesaid dish, & they put it in the inside of oranges after we had squeezed them for punch; – they eat also a green called Belbe, gathered in the fields, putting their bread into pounded fennel seed & salt mixed. I went ashore & walked – the men drawing up the Bark, & the wind springing up we all went in, & got near as far as Fetne & tied to land. Next day-

10th-21st . We made litle way & fixed to an Isle opposite to Fetne, where I went after to see them gather the sugar cane, – where I saw the boys & girls half naked, with a brown piece of stuff only thrown on their bodies, skinning it with their feet & eating the pith as we eat liqourish, & one of them had the politeness to present me with a cane. These are Filaws [A. fellahin] or the true Egyptian Peasants. On this Isle I saw water-melon plants & others of that kind, with cane of Turkey-wheat set in the ground sloping over 'em, so as to defend them from the weather & so planted in rows about 6 feet apart.

11th-22d. We drew on & in the afternoon had wind & made some way, & passed a village called Aboulig [Abutig] a Bishoprick. This day began the Turkish fast, which holds all the month of Ramedam [Ramadan], when they neither eat nor drink nor smoak all day, but sit up & enjoy theselves most part of the night; – so we are to leave off dressing our provision at noon, & stay till towards even not to give them offence.

12th-23d. We had a litle wind & made some way. The country on the East side is sandy & inhabited by Arabs, the West mostly by Egyptians; – the Arabs keeping in the mountains on that side. You are now to imagine me like Old Tull excepting that I have a black beard, which is now half an inch long on my chin, & I have a vererable pair of whiskers [moustache]; & I have a lock of hair on my crown, so I am a perfect Arabian, & am trying a litle at Arabick. Our watermen live very much on thick Lentil soup, pour'd on bread which is very good, & not inferior to pease soup. I dress every day 2 fowls & soup of e'm with rice boyled in it.

13th-24th. I went out & walked, as the men drew the boat on shoar, which was here rocky & abundance of litle hills, which are in a manner rocks of petrified shells, cockles, oysters &c: which I picked out: – that would encline me to be of the opinion according to the tradition delivered by Herodotus, that all up to the cataracts [rapids], was a gulph of the sea like the Red Sea or Arabian Gulf. I then came to a fine flat country, where the Arabians were improving their land, for such of them as live in villages & improve their land are honest enough; – I gathered among the corn a fine sallad, of what I thought was

Endive, but proved to be Lettuce; I after saw them planting onions & [...] which is not good here, has no strength & is only used by the very poor Arabs who endeavor to have everything within themselves. We drew up above Samalou on the west, & here the rocky cliffs begin on the East side of the river, & at top of them is a Cophti convent. There is litle wind on the Nile in the winter, & we draw 6 or 7 miles a day when there is no wind. When we are below in the bark we sit on our beds & carpets after the Eastern manner, but spend most of the day above, sitting on the side of the ship. In the even we had wind & passed by Minio a town, & having wind all the night, made a pretty good run, & passed by Ensineh [El Sheikh Ibada], the old Antinoupolis built by Adrian [Hadrian] in honour of Antinous who was drowned here.

14th-25th. [No entry]

15th-26th. We had a fine wind all day, & stopping a while near Meloui, another town, I gathered a sallad of pepper – grew among the corn. We have now on the East near the river all along hills of Freestone, in which are the famous grotts of Thebais in these hills, which supplied stone to lower Egypt. When we stop at a village we buy eggs, fish, bread, & a litle meal, which is cheap; so we live very agreeably, & I have my health if possible rather better than on the land, so that every way it is a most agreeable voyage. We had a gentle wind all night, & passed by Manfalou [Manfalut] in the even, which is a Bishoprick, & soon after passed those hills of the Grotts.

16th-27th. We had a fine wind all day, & passed by Aboutik [Abutig] a Bishoprick this country to East & west is under a sheke Mahomed or Arabian head of such a clan; – South of Aboutik he was encamped, it looked like a fair, for the shekes go round their dominions constantly, & stay a month or less in a place, but generally have houses in every place, their tents some large some small like our European tents, & when they go thus about they collect their tributes, mostly in cattle & corn, & then they have with them 2-3 or 400 horsemen: this is the first country under a Great sheke we met with; – the other Arabs are under no Government & are the worst; we came to the East to the territories of the Emir or Prince of Ackmim & had a good wind all night.

17th-28th. We came in sight of Ackmim & about 3 in the afternoon landed there; abundance of people coming a mile out of town & walking on the bank of the river to see the great man I came with, who is the head of the Cophtics in these parts, especially such as are catholick, & many more came when we were landed among 'em two Italian Franciscans of the propaganda fide Hospitium here, & another catholick cophti priest of this country, bred at Rome, a very pretty [foppish] young man. I went with them & delivered my letter from Father Jacomo of Apostolical Prefects of the mission of upper Egypt, & my fellow travellers from Leghorn to Father Georgio President of the Hospitium aforesaid here; – who received me with much cordiality; appointed me the best room in the convent, which is a good one for these parts; but this Arabian city has but a mean appearance, the houses built of rambling unburnt brick, the streets dirty for it has rained here the President tells me but 3 times in eight years, & that not hardly enough to lay the dust: but this town is built on the ruins of an old one, nothing of which is now to be seen, & it is situated on a rising ground. I sup'd with the Fathers in their Refectory, – a dish of white pulse, broth, eggs, custard & another dish of some vegetables, in a sort of Ragou, each having his plate brought in to him of every sort, one after another; my servant eat at the same table, at the lower end, for that is the way here & all eat together that are strangers & others, the meanest & greatest of them, except the very serving people of the house.[79]

18th-29th. I went with the President & 2 Fathers, to visit Malim Solyman – & carried him a litle present of Florence wine &c: drank Coffee; – dined with the Fathers on Pilaw, a Ragou of meat with onions & pulse, fryed Liver, another sort of dish of roast chops; – no wine in this country, so we drank at table Brandy & water together, the former drink from Dates: – they came to my room as they did likewise last night, & took a pipe & part of a bottle of wine. The poor catholic Cophtics came many of them to salute me, & sat down on the ground & smoaked their pipes; – they spend the nights before Sundays & Holy days altogether in the church, singing Psalms till midnight, then repose an hour & then their long mass begins, which lasts three hours: – they are people of much simplicity & the catholicks are very good, but all strict in matters of Religion, fasting &c: having four seasons of fasting. This city two miles round perhaps is capital of the Emir of Ackmim's dominions, who resides here, he is an Arabian sheke, with the title of Emir, & lives about four months in the year in tents abroad, making a progress round his dominions. This Emir is a very good Prince, a favourer of the Xtians, especially the catholics & can carry 5000 men into the field. In this convent the Cophti catholic priest performs after the Cophti rites, & the Italians after the rites of the church of Rome; – of the two other Fathers, one is a Milanese Father Bona Ventura, the other of Lucca, Father Cyrillo. Difference between an Hospitium & convent is, that there must be a certain number 9 to make a convent. There are three of the propaganda in upper Egypt, this, Girge [Girga] & Farshut, in each of which are two or three Fathers. The convent here is built round a court, like those in Europe, part of burnt, & part of unburnt brick, & it commands the other west side of the river, tho' in the middle of the town, but without it has very few windows. I receive great civilities from the President & Fathers, & from everybody, which makes

79 Mrs Pococke has corrected the word "meanest" to "serving".

my journey very agreeable; – I shall be some time seing the Antiquities here abouts, & you may hear from me again once more before I return to Cairo, tho' it may probably come with this, but I fear the ship will be taken up by the soldiers, & that it may be long before your receive this, if it does not reach before a ship goes that was to carry my last. My wishes attend you all for many happy new years. The country here is much better than lower, & well cultivated, but no trees except about villages & those mostly date-trees. The afternoons here are warm. The Arabs have a very smooth way of talking, when a stranger comes they tell him it is an honour to their city; – & when anyone returns home of any consequence they say, it is a white day as milk, & you bring a Blessing along with you to the city. I went this evening to wait on the Emir or Prince of Akmim. Malim Soliman came first to see me & advise me of the time, & then went before to court. The President & one Father went with me; – I delivered my letter to him, from Osman Bey, who is in reality the Governor of all Egypt: – He received me with great civility, said he had a regard for the Franks, told me without the letter, he would have done anything for me, in his power, much more with it; told me he would send his own people with me whenever I pleased, & bid me be afraid of nothing, he would do me what service he could, both now & on my return, & offered me a letter to another person, & received very kindly the litle present I made of Venice glasss & 2 or 3 boxes of sweetmeats. There is a court with high walls, on two sides of the house, in it were about a dozen of curious sheep, black & white, & some all white, & the white of 'em all, was as white as snow, brought from ungela towards Ethiopia, & kept as a sort of Petts. Ascended into a room an L with arches supported by Pillars, & so all open from breast high; of these the Prince sat cross legg'd & a cousin german opposite to him, we sat down on the step to the Divan, by, & he being engaged in signing papers, we staid not long; – but he appears to be a Prince of great goodness; when papers are brought one puts ink in a pen on his litle [...], with this he blacks the seal on his finger & makes the impression on the paper. He was dressed in his Turkish & not the Arabian habit: – is about forty, has four wives & after custom gives them their nights in turn, & they live in separate apartments & never come together, & have separate households. The Arabians are very faithful when they give their word, & if they say they will carry you on your Head, which is a common expression, & they signify by putting the hand up to the head, they'll die in defence of you.

19th-30th. I went with the President & Father Bonaventura to Malim Soliman who treated me with Coffee & two dishes of hot sweetened water, boyled on cinamon. I went after dinner to see two old ruined churches, built of unburnt brick, – the walls almost entire; – the Fathers talk as if they had stood ruinous ever since the Mahometan Religion came here, but I rather think the cophtics were turned out of them [...][80] I then went to see some ruins in two places, & went into the place enclosed like a litle garden, where the Princes family is buried. The country to the north, a fine plain, is most beautiful, all in verdure. Malim Soliman came to see us. At supper a new dish, a sort of minced meat, dressed with onions, a breast of mutton stuffed with rice & forced meat.

20th-31st. I went out with Father Bonaventura & walked almost round the town, & viewed again the ruins of a Temple. I begin to have a venerable black beard, but you must not tell anybody, that I have a few white hairs in it; every body tells me I look exactly like one of the country. This day is windy, & for about a month to come, they say 'twill be cold in the mornings, & I find it so.

21st-Jany 1st. I walked out in the evening, Malim Soliman came to see me who is as a father to me, & indeed, tho' I litle deserve it, I find everyone exceeding civil, & everything far answering my expectations, & there are not those dangers here, there were formerly: – & I cannot but say I think myself very fortunate. Tomorrow I go out to see two monasteries, three or four miles off. Pray my most kind love to my sister, & due respects to Whiteoak, Greenham & all friends.

I am Dear Madam your most obedient & dutiful Son, Richard Pococke

FURCHUT IN UPPER EGYPT
3/14 January 1738
Pococke to his mother
Letter 8[81]

Honoured Madam,

My last letter was to you from Akmim in upper Egypt, dated Decr 21st – Janry 1st.

22d-2d. I set out with Father Bonaventura, from the Hospitium of the propaganda fide, to go 3 miles on the west of the Nile to see two convents: – at a village by the ferry a Christian invited us to take Coffee, would not be refused; – He went to his house & spread a mat in the street near his door, & then brought out himself on a shallow basket, dried dates, bread & meade to dip it in, – & then his relation brought us Coffee; – he said we should stay till Sunday Christmas day in the convent & then come & dine with him; – we went on: – & saw a great number of wild-ducks on the pools made by the Nile, & so tame

80 The text is corrupted at this point and in the line below by a large ink stain.
81 BL, Add MS 22998.

that one might go very near them; – a fine country; – came to the great convent under the hills, but the Priest not being there, I went on to the lesser convent; – both built in the same manner; – the outer walls in repair, but within in ruins: this consisted of a large church, the whole length of the building, with two rows of columns & at the end thus [diagram] in a cross: on the south side all along were the lodgings of the monks, this is now inhabited by one Priest, his wife family & relations, for there are not perhaps ten convents in Egypt that have priests unmarried: – having viewed everything well they laid us a carpet, in the Anti chapel of the church, where we dined on the cold meat we brought, & after had Coffee; – walked two miles to the mountains, I ascended almost to the top, to a hole which I believe is frequented by Eagles; – we tracked a wolf & a great serpent, which leave their tracks in the sand; – where the Nile does not come it is sandy, but East of the convent is a fine country, returned, & had a carpet laid for us sub dio [under the open sky], by the door of the chapel of the Virgin Mary, at one corner of the convent; here we sup'd on the rice we brought boyled in the soup of fowls & pidgeons, we had ordered to be killed; the cophtics supping on maigre [Lenten] rice by [nearby], & others sitting along with us, smoaking; – after two hours, we retired into the chapel of the Virgin Mary, where a carpet was laid for us to sleep on, & where I reposed very comfortably on the ground. We had at supper, for a candlestick, a litle cake with a hole made in it, for the candle, which was of wax. The next morn –

23d -3d. I breakfasted on cakes hot, made with butter, which I dip'd in honey, & after that a dish of Coffee; – & then went to the other convent, which I viewed well-built on the same model as the other; here are four married Priests, – they gave us a collation of Eggs buttered. We returned to Akmim, had a wind very troublesome by missing the sand, which appeared like a cloud, & it enters into our rooms & covers all with dust, & tho' it be shut up ever so close. Malim Soliman came to see me.

24th-4th. Malim Soliman came to see me, & I walked out with him to see some places. I went out to walk in the afternoon, & about 7 went to the catholic cophti church to see the ceremonies which lasted till 10; – went to bed for two hours, rose to see the cophti mass, which lasted till day; – the receivers go to the very table & holding their head over it, with a napkin in their hands, standing, the Priest puts the elements into their mouths, the Deacons only receive the wine, put into the mouth with a small spoon, & the receiver holds a napkin under his chin: – the bread is carried to the women, who put their heads out of the grate thro' a litle door from their place, which is separate from the rest, – & before the mass the priest is incensing all the people three times, puts his hand on the head of every person, & goes to the womens door for that end: – another ceremony is three candles on a cross in the priests hand, with which he makes some motions, & they are Emblems of the Trinity; – the priest has a yellowish woollen cloth over his head, but the old rite is with the bare head, in which manner one said mass. The priest has all the service without book in the cophti language, a corruption of the [...] & ye Greek,[82] which he does not always understand. Malim Soliman having invited me to dine, sent his servant to me, to come at 8 a clock; – I went – took Coffee – & went to sleep in his house; – we dined in an open place like a summer house; – on the floor a large piece of leather laid, – then on a stool a foot high, – an oblong square table, with 25 dishes on it; 8 on a row on each side, in the manner as our plates are put on, – but of several things there were four or five dishes, – mostly soups & sort of Ragous, roast lamb & pidgeons stufft with rice, boyld meat cut in chops; – we set round on the ground, – no knives or forks, no plates, help one another with the hand, & carve a fowl by tearing the joints apart; – I had a plait put in my lap & served myself with a knife: – a dram served first, & we had wine & water, which is a great thing here; – our company about 7 or 8, & my servant at the same table: – having drank Coffee, we went to the garden, half a mile off, where we walked, & had a carpet & cushions laid to repose, & Coffee. Sup'd at Malim Solimans, about 5 or 6 dishes brought on, one large dish, as usual on a stool, a leather carpet being laid on the ground, on which they threw their bones. After Coffee, I went with Malim Soliman to the Prince or Emir, who was sitting in a vast saloon at a window, we sat down on the outside; he said, he wondered that he had seen me but once since I came, & told me he would order men to move away some earth that I might see an inscription. His Cousin german told me of a piece of antiquity, where I am going & the Prince gave me among other letters, one to see that piece of Antiquity. At going away, I went into the saloon to salute him; & he desired me to make his house my own & command whatever I pleased in it; – & in a word behaved with all the politeness of an European Nobleman. Christians are much favoured here, – & a Turk who blasphem'd the Christian Faith he would not suffer to do his business he came about, till he had made full satisfaction.

25th-5th. [No entry]

26th-6th. I went to see the President Father Georgio embark for Cairo, many accompanying him. Malim Solyman carried me to dine with him, in the same manner as we had sup'd. I walked out most part of the Afternoon, called on Malim Soliman in the even. We have now but one Father, one being gone to Furchut. Christmas Eve, several brought 4 or 5 pound of wax candles, which are laid on the altar, & after carried into the vestry.

---

82 Pococke has added half a line above his mother's text but most of it is impossible to read.

27th-7th. A litle sprinkling of rain before day, which is very extraordinary. The cophti ceremonies are most like those of the Greek church: – a boy or two of ten or twelve years old received the sacrament. Went abroad with Malim Soliman who took us to dine with him; – we had a dish of the sort of treacle afore mentioned & butter boyled together, into which they dip a piece of bread & eats well enough; – drank Coffee & came to the convent; – & was employed in preparing for my journey further on.

28th-8th. I went to the river accompanied by the President & Malim Soliman & his eldest son in law. Malim Soliman presented me with a great basket of bread, a small one of cakes & a sheep: – & about 11, after having said a Pater Noster together, & wished we might meet to say another, – we sailed up the river with a fair wind. I went ashore to see the ruins of a convent, & after to see a convent opposite to the town of Menoufich, built of unburnt brick, but many stones finely washed, show it is built on the ruins of a magnificent one. Crossed to Menoufich, a Bishoprick but no church, all go to the convent; – went to view some ruins of the Old Town & a lake of salt petre. Visited one of the Governors here, who had been at the boat to see me, & left two letters for me, to his friends higher up; the Prince of Akmim having writ to him that purpose. He treated me with Coffee, & gave me many notices. I went up to see the captain of the bark, came [come] in from Cairo, who was sitting round a pan of coals with 7 or 8 Mahometans, treated me with Coffee, & a sweet hot ginger water, not spirits: – two or three sung a Mahometan song, beating time with their hands, & one playing on a tambour. This captain has four houses, at four cities on the river, in the space of 240 miles, this place & Cairo being two of them, & in each of them a wife. I sup'd, & lay in my Bark.

29th-9th. Left Menoufich about 9 or 10 a clock, – staid at a place on the East to take in wood, & arrived at Girge about 3 or 4 in the afternoon, & went to the Hospitium of propaganda fide, where Father Joanni of Trent is Superior, a very spiritual man of great simplicity & a stranger to the world. Father Arsenio a grave sedate young man of Vicenza, & I found here Father Gabriele alias Alphonso a Milanese, much practised in this country in which he has been 21 years & is a good Physician, & a man who lays himself out to do services, as I found. The missionaries are here in a worse condition than in any other place going under the name of Physicians & are not allowed a church, but have one, into which they do not suffer any Turk to enter. In this city resides the Bey of Girge or Governor of upper Egypt; a small part of the city comes to the river. The Fathers have a large house, & I sup'd in their litle Refectory, very well, but one dish, rice boyl'd with a few dryed beans, – the Fathers spent the evening in my room.

30th-10th. Father Gabriele went with me to wait on the Caimacam who is governor here, in the absence of the Bey, who is encamped near Manfalou to collect the tributes; – he is ill & Father Gabrieles patient, & keeps him by sort of force from going to Cairo. I presented him with 2 boxes of Prunellas, which I hear since, he thinks have done him much good; – & I think his expression was, "that I was sent by God Almighty to him to contribute towards his recovery." I showed him the letter I had to the Bey from Osman Bey of Cairo, who is the real Governor of this country; & he gave me a letter to a person where I am going, & sent his servant with me to another great man I had a letter to from the Prince of Akmim, who was not at home; – Father Gabriele had me to the Aga of the Janizaries, who I believe had been his patient, & received us with much civility, & gave me a letter: – after dinner I went to Mustapha Aga with the letter from the Prince of Akmim, who after sent me a letter I had occasion of. Father Gabriele is a man of great knowledge of the people of this country & of great discretion, & prepares all the medicines himself. I sup'd with them & they sat with me.

31st-11th. I went to the boat; returned – dined at the convent & about 2 sailed on with a Greek Aleppine, who goes with me to Furchout, we made a collation in the dusk of the evening on figs, lent-figs, dates, nuts & cakes, & then sup'd on fish, & drank Coffee.

Janry 1st -12th. I wish you all an happy new year & many. We sailed before sun rising & had a fair wind, & first saw the manner of floating down on Earthen jars; – they make a frame of date boughs tied together 30 feet wide, 60 long & four deep, – on this they put a layer of large jars in this figure [diagram] with the mouth uppermost; on this they make a floor of the date boughs, & then put a layer of smaller jarrs, so as to ballance the boats, leaving spaces for the people to walk between, & so they row down, as the floats of wood we saw in the Feltrine; they carry these even to Rosetto & Damietta & bring them from above colls, because they hold oyle, which the jars below do not, being the best earthenware, used also for butter, which they boyl in this country & it is soft & will keep a year, & is called mantoa. I intend to go a litle further than Furschout, & find it very pleasant travelling on the water; – I am well covered & sleep conveniently; & have letters of recommendation to all the Sheke Arabs & great people, & shall go no further than I can go with the utmost security. If this letter should come to you soon it probably will be long before you receive another, because all the ships will be taken up to carry the Janizaries to Constantinople. I am in these parts in the best of the year, there being no heat, tho' the sun in the middle of the day is warm, as May with us; – & this day going ashore towards noon, walking with my bare feet, I found the sand very warm: The men pick'd up wood & cut some green wheat for our sheep; & the men began to draw, for want of wind, – we had a collation about ten of bread & cheese, nuts, figs & Lent figs, after the way here, – tho' they generally have fryed eggs; & took a litle dram. I was 3 weeks in my Voyage from Leghorn &c:[83] – which account I give of myself least any of my other letters should have miscarried: I have writ to you at least every three weeks sometimes oftener. You will continue writing to me directed to Mr Cutts Maydwell. There are vast quantities of wild geese here, some are called Bau Geese, & are carried as rarities into England. We saw two crocodiles about [word missing] feet long, sunning themselves on the bank; – I had my gun to have shot at them, in case we had passed them, for I was desirous to pass them in order to see them, & just before we came up with them, they went into the water; & they say if the ball [bullet] had taken them under the foreleg, where they are soft it might not have killed them: – here the crocodiles begin to be plenty, but there is no danger of them; – they return again commonly to the same place & we saw them there soon after, & never come out of the water till towards noon in winter to sun themselves, there are but few below these parts. Saw another crocodile, they lye with their heads on the ground flat – for there are but few below these parts & the head next to the water, & never further hardly, but that they can go in at one spring. I beg of you exercise & be chearful; God Almightys providence disoposes everything & tho' we know not his purposes which are carried on & directed by particulars, – yet he has always wise ends, & when we are employed in laudable actions, which may tend to the perfection of ourselves, the good of mankind & the Glory of God; – we may look on ourselves as his instruments performing what is agreeable to his will & satisfye ourselves, that in such actions we do not displease him; – but that on the contrary he protects us & leads to wiser ends, than perhaps we had in view. As for my own part, I hope I am not doing ill, & cannot but acknowledge the greatness of God Almighty towards me, in making all things to work so much in my favour; to incline the people I meet with to be my friends & assist me, & indeed I look on myself here, just almost as if I was sailing in the Severn, & have more pleasure: therefore I beg of you to leave everything to Providence, & be not sollictous, but be chearful & divert yourself by company & exercise. In case this letter should come to you before the first of April, & anything of very great importance to me, should happen, which I cannot well conceive, – if you should have occasion to write on such an affair before the 1st of April, write two others to the same purpose, & inclose them to Mr Maydwell, & desire him to send one to Grand Cairo & another to Aleppo;

83 It appears that Mrs Pococke has edited this section, since it repeats what her son has already said in earlier letters.

otherwise he will know where to find my letters, which ordinarily you are not to enclose to him, but direct them to me at Cutts Maydwells Esqr. in London.

There is a tree in upper Egypt called Dar Dom[84] like the Date, except that the date has but one stem & makes no boughs as this does, the leaf of this also is like an hand screen [fan] in shape, the fruit is 2 inches long & 1 inch & ½ thick, on it the she[ll] is an hard substance which is eat & when this is taken off, it looks to me like a rough cocoa, in the shell is a kernel, so hard that when turned & polished, is smooth like a fine yellow marble, & they make religious beads of it. I have not yet eat the fruit, but eats like the dates, but hard & must be pounded with an hammer or pestle in order to be made a jamm to eat it. I begin to have a very venerable beard, & I believe it will be of the picked form; & I shall not cut it off till I get to Aleppo; – when I am shaved the barber combs my beard out. It is the fashion here to go with slippers, with naked legs & feet, or with leather socks on the feet, but the Trouzers come to the Ankles; & in Cairo short boots are sowed to the trousers; – so I have used myself to go without stockings but I wear the leather socks called Tertis, & a sort of slipper with kind of quarters called jeminiree [?] & now I shall take the Arabian shoes & slippers litle different from the others called merbhoube [?]. They are all of red, but in Cairo I wear yellow. In the boat I lye in my clothes, but do think to allow that manner [next two lines scored out]. There are some antiquities hereabouts which I entend to see, & so return to Cairo. We tyed [anchored] all night near a village, there being no wind, & in the –

2d-13th. The boat I have is 33 feet long, & 11 feet broad & yet it is of the smaller sort that sail on the Nile; – I need not tell you what degree we are in, you will see it is, as I believe between 26 & 27, only about 25 degrees south of you. We drew a litle way & mounted on asses to go to Furchout an hour off, without bridles, & only a rug girded over the Ass, a pityfull sight to see; [next line scored out]: – here we are in the country of the Sheke of Furchout, which extends to Assouan [Aswan]; so well governed that one may travel by day or night with a purse of Gold in ones hand. About a mile from the boat, the President Father Jeannio Tunn met me & mounted me on a horse, & so I went to Furchout thro' a flat country all green. 7 catholick families, Fathers here only a year, have built an Hospitium of The propaganda fide; – the Shekes secretary is a catholic cophti, & they are favoured by the Sheke, but pass as Physicians, & have not as yet a public chima. I am taken for a Physician, & have had good success, tho'

84 The Theban Palm (now better known as the Asian Palmyra palm). See Lane, *An Arabic-English Lexicon*, book I, p. 1861, on the authority of Forskål, who called it *borassus flabelliformis*.

I make no profession of it. Malim Gabrian came to see me; I found here only one Father more Bonaventura, who was at Akmim.

3d-14th. I went with the President & Malim Gabrian to wait on the Sheke El Arab, as he is called, the soverign of a great country here; – he was sitting in a corner with a pan of coals before him, a fat man, about thirty, his Father lately dead; – I presented my letters, & he said he would give me letters & a man to go with me, so that I might be in all security in seeing the places near: – but as I said this country is so well governed under this sheke that there is no danger; but it is necessary to have a man with one, when we go to view Antiquities. I walked in the shekes garden with the Fathers; – Malim Gabrian came to see us. Pray my very kind love to my sister & due respects to all friends. I desire you again to have a full trust in Gods Providence.

I am Dear Madam your most obedient & dutiful Son, Richard Pococke

GRAND CAIRO
28 February/11 March 1738
Pococke to his mother
Letter 9[85]

Honoured Madam,

My last letter to you was of the 2d -14th of January from Furchout in Upper Egypt.

4th-15th. I staid within all day.

5-16.Visited & drank Coffee with Malim Gabrian, he after sent a small sheep as a present to the convent to entertain me; walked almost all round the town with the fathers.

6th-17th. [B]eing Epiphany, I saw the ceremony of Crossing the forehead with Holy water, first blessed after the Cophti rites & then after the Roman, as they bless the water to wash the feet on Thursday before Easter.

7th-18th. Went out all day.

8th-19th. Malim Gabrian came & laid his carpet in the convent & treated us all with Coffee & cakes; – I prepared to go away, & about 10 made a light dinner with the Fathers, Malim Gabrian sending for his own collation of roasted eggs, sowr cheese like curds & bread, dining in another place by himself, & sending of also what he had to us: – He, the Fathers & Aleppine Malim Hadallah set out with me, we went 3 miles East to Badjoura, where a Christian catholick

85 BL, Add MS 22998.

entertained us with Coffee; we went 2 miles further to the river to my boat – I entertained them with Lent figs & cakes & a glass of wine. Malim Gabrian presented me with a large black sheep, sugar canes & bread for the men. I set sail & went 8 or 9 miles, lying by all night.

9th-20th. We sailed on till 2 or 3, tyed to a sandy Isle, where I had fine diversion, going out & shooting at crocodiles; – I saw 5 or 6 none more than 8 foot long; they walk gently into the water, & disappear in the water by degrees, but begin to march soon after you appear, & seem to see behind them or to hear & smell well; – tho' I surprised one or two & came pretty near, they never hardly go far from the water, but lye near it, unless a litle way by night so that they are in immediately; we sailed most part of the night & tyed at Dendrea, an old city, now an Arabian Town. I took a walk to it the morning of the

10th-21st. Ramedam the great Mahometan fast is now over, & this is the first day of three of their great feast Biram. We sailed about 10 & had wind all day & stopped opposite to Chous [Qus].

11th-22d. We went a litle higher by Meguade a Bishoprick, but a poor place & many Christians. Went opposite to it to the Port of Chous, & rid 2 miles to Chous, with two letters to the Arabian Sheke, went to the house of his Christian secretary, where a matt was laid in the yard with the horses on which we sat, bread, the oyl & seed mentioned, & a dry seed & raw onions brought to eat, with bread & water only, after we had Coffee presented him with a pair of 6d shoes & 2 pieces of Tobacco. Went to the Sheke who was in bed on the Divan, going to repose in a white shirt & blew one over it, a molotto, a fine large Eye; – he promised me a letter to Luxerein [Luxor]; – so I went to his scrivan [scribe or secretary], having presented the Sheke with some rice, Tobacco, soap & a p[ai]r of shoes: – he commands a part of the country on the East side. Saw the Cophti church, & a ruin of a small temple, drank Coffee & went to the boat, leaving my man for the letter. The Sheke sent for us to the scrivan designing I should eat with his scrivan, he was there, being to go tomorrow, 6 hours to it by land; – so I purposed to stay when I came there, instead of seeing it on my return.

12th-23d. My man came & I sailed about 9, & had litle wind. Che non ruol manda, he voue vada,[86] an Italian Proverb, to do ones business, etc. We had but litle wind, but arrived near to Carnac [Karnak] below Luxerein, here the Sheke was come, & the man he gave me going to him, he sent two men to guard the

86 The Italian expression "Chi vuole vada, chi non vuole mandi" conveys that if you want to get something done properly, you must do it yourself – here presumably a joke about the wind.

boat, who came with their spears, & made a fire on the bank, & treated them with Coffee & Tobacco.

13th-24th. I went with a letter from the Sheke of Chous & from the Sheke of Furchout, to Sheke Aboussin, who is as a justice of peace here, under the Sheke of Cous; – he made me sit down on a carpet in an open house, like our cart-houses; – I gave him my present of 3 pieces of Tobacco, 7 of soap, & a pr of red shoes; – he treated me with Coffee, & presented me with a sheep. I went 2 miles to Carnack [Karnak], where I saw this Sheke again, & the Scrivan Malim Boukter of the great Sheke staying with me an hour or two. I viewed the great ruins of a temple;[87] asking for some bread to eat, a man sent with me brought some fryed eggs & hot bread & pickled cheese curd – the latter not eatable by Franks. I continued viewing the great ruins this being the spot of Ancient Thebes, of which there are no other signs but the ruins of some Temples. The great Sheke of Cous being incamped near I waited on him in his magnificent Tent; he did not rise nor ask to sit, but only if I would not sup with his Scrivan, which I declined, being fatigued, but the scrivan came & eat some Lent figs & took a dram with me in the boat.

14th-25th. I spent the day in viewing the same great Temple & two others. Sheke Abouessains son came to me & had me to see an old building where there were women, & where otherwise I could not enter.

15th-26th. Went with him four miles, he bringing horses to view the ruins of a temple to the East,[88] we dined there on the cold meat I brought, & he came & took Coffee with me on the bank by my boat, & I presented him with a pair of red shoes & 2 pieces of Tobacco; & the steward of the G. Sheke with a pr of shoes & some soap; the day before I dined among the ruins, on soup & mutton, brought me by the Boatmen.

16th-27th. I reviewed everything again. Took my leave of the great Sheke, who was very civil, waiting in his scrivans tent, till he was raised from sleep, about sunset: the shekes came & drank Coffee; & presented me with a very fine Lamb, these creatures we keep alive till we have need of them: – the scrivan sent me a dish of Boyled mutton & boyl'd pidgeons, stuffed with pearl barley, &

87 It is not clear which temple he refers to here, particularly when he is more concerned about describing his picnic than elaborating on the architectural features of the antiquities; but judging from the account given in *Description 1*, pp. 91ff, together with the accompanying illustration (Plate XXVIII, entitled "The Temple of Jupiter at Thebes") he appears to mean the Precinct of Amun-Re.

88 This temple is further described in *Description 1* (pp. 96ff) and illustrated in Plate XXVII, sections B & E. Furthermore, two columns from this building (identified as "Temple East of Carnack") are illustrated in his Plates LXVI and LXVII.

excellent bread from the shekes kitchen, & so I set sail for Luxerein 2 or 3 miles high or, where we arriv'd in the night.

17th-28th. Sheke Abouessens son, who lives here, came to me & I went to the sheke of this place with a small present [of fryed Eggs & new milk & bread] & spent the day in viewing the ruins of a temple here, Thebes extending so far in all probability. [He] sent us to the ruins & came & took Coffee with me;[89] & at parting the sheke sent me a large kid & a basket of bread; – we sailed up higher than Erment.

18th-29th. No wind; – we drew 4 or 5 miles. My man shot at a crocodile, as he was going into the water, he open'd his mouth & grinn'd horribly, & leaped into the water, I believe he was wounded. We had a fine wind all night.

19th-30th. Stopped half an hour at Eme to buy some things in the morn; – & proceeded. Pray let the Bishop know I have nothing to write, – that it would take up much [time] to give an account of what I see, & be very difficult to convey an Idea by bare words, so I hope he will excuse me, as I must necessarily neglect the opportunity I shall never have again, of seeing considering & examining well into things which deserve attention. This to him: – & to you that it is my intention to write to him once a quarter – tho I see I shall not have time to write till towards May & shall only give him a general account of my travels.*[90] Going out under a hill I saw senna growing wild, – we went on with a good wind, – & lay by where several other barks [boats] stayed that night.

20th-31st. Sailed in the morn with a good wind, being within a day or two of Assouan [Aswan], the end of our voyage; – we cast Anchor at night near an Isle.

21st Febr. 1st [March]. I sailed with a good wind, saw at the end of an Isle, about 20 crocodiles, some of them I believe 20 feet long, the largest I ever saw; – they go into the water before one comes up to them, however we shot at one that staid but to no purpose. Passed by a small ruin of Ombos [Kom Ombo],[91] & after by the tent of the [...] of this country or governor under the sheke of Furchout: – I went ashore & waited on him & drank Coffee, & proceeded on my voyage, & in the evening arrived at Assouan, waited on the Governor with some letters, drank Coffee, he sent me a lamb for a present, & I returned to my boat.

89 The order of this passage is problematic, as Mrs Pococke has added lines between the existing text, presumbably having omitted some lines in error.

90 NB: See the particulars observation on the East Vol. 1. [...] 90 Chap. III Thebes. [This footnote is written in Pococke's hand.]

91 Famous for two temples dedicated to Apollo and Isis.

22d-2d. A Christian came to the boat to me, there being but two here;[92] I waited on the Governor, who assured me of his protection; & gave me a court & two rooms to be in, one open in front, & I had my things brought to this place; – being to sleep here the 4 or 5 days I stay here. I had the mousul who had assisted me with his advice &c: come to see me, & brought me a present of a dozen of pidgeons & some dates; – & some other Turks came to see me. I went & carried a letter to the Abedeen Odee Bashaw from Ali Surbage [ʿAli Surbaj][93] in Girge.

23d-3d. I went with the Governors brother in law & 3 Janizaries to see the Isle above the Cataract. I returned home; – we are here under the Tropick Circle, & this day I was most south I shall be, about 23-25 Latitude, Assouan being 23-30 exactly under the Tropick: The Cataract is 3 or 4 miles above it, & the Isle a mile or two below that. I saw them bringing Camel loads of Senna to Assouan which sells there for about eight shillings an hundred weight. I saw also coloquintida [bitter apple] growing, as I believe, like a calabash [gourd], but round, the fruit being like it. Saw barley in ear & a sort of Apples just ripe: at midday & afternoon near as hot as our summer.

24th-4th. I walked out with two Janizaries, to see the litle that is to be seen of Old Siena, which stood where the old city of Assouan was. I walked further to the East & came home, seeing cotton trees, which are a shrub like black currant. Staid within the afternoon. Had a supper ready dresst sent to me of baked mutton & hot bread; the Governor drank Coffee with me.

25th-5th.I went on the West side to see the ruins of a convent, that I believe was in Old Ethiopia, Egypt extending no farther than the Isle where Elephantine was on the west side, which Isle is opposite to Siena & we went to it, where I saw the ruins of a temple, & of a gate. All this country & the very bed of the Nile is Granite. I came to the other side & walked to see some quarries, which are all above ground. The Aga who is a great vassal, that is the Governor sent me goat, Pilaw & barley soup ready dressed for supper, hot bread also.

26th-6th. Had my things put aboard, went to the boat, but an high contrary wind kept us from going down on our return. Malim Daoud or David a Christian, the civil governors clerk, brought me letters & some money to carry to Akmim & presented me with a sheep, & I presented him with soap rice & a pair of 6 shoes.

27th-7th. We parted but could get no farther than a mile by reason of the wind.

---

92 According to *Description 1*, p. 116, these are the secretary to the Caimacam and a servant of the Aga's.

93 Cf. *Description*, 1: p. 166: "Serbajee, or captain".

28th-8th. We got a mile further in the morn & towards the even went on again till ten at night, the wind sinking.[94]

29th-9th. We got a mile further in the morn; the wind rising against us.

30th-10th. We made some way for about 4 hours in the morn & laid by all day.

31st-11th. We proceeded a litle all day, & I went ashore to see some things cut out of a rock, & stopp'd in the night at Etfou [Edfou].

Febr. 1st-12th. I went a mile to Etfou to see the remains of a Temple;[95] – I went with a letter from the Sheke of Furchout, to the Sheke of this village, whom I presented with some Tobacco, soap & a pair of shoes; – he was very civil, treated me with bread & Coffee, & conducted me himself to see the ruins; we dined with him, on a large wooden bushel bowl of bread made like pancake, broke in small pieces & mixed with that Treacle they make of sugar cane, which we eat with our hands. He went with me to the boat & presented me with a sheep. We went down the river most part of the night, & the

2d-13th. Went on with the stream & oars there being no wind; – about two we arrived at Etne. I went to Sherif Ahmed Aboulkerim, a green head & relation of Mahomet,[96] to give him three letters I had; – he went with me to see the remains of a temple; & then went to the boat. A Christian Cophti priest came to see me & two or three other Christians, there being two priests here & about 100 Xtians. I went with him to see their church, which is not so bad as I have seen; he after sent me a present of cakes made with butter. I went & delivered a letter from Ali Surbaje of the Janizaries in Girge,[97] to Sherif Ali Salim another relation of Mahomets, a very good sort of man & very civil. Sherif is as much as noble, & is the title here of Mahomets relations. Soon after I was in the boat he sent me a present of wood to burn, very scarce here, fresh bread & a fine kid.

3d-14th. The officers of the Governor, promised me in the even to go out with me this day. I went to review the remains of the temple. Sherif Ali Salim sent me another present of cakes & bread, & with him & the officers of the

94 Clearly, Mrs Pococke omitted this entry in error, as it has been added afterwards between the 27th and 29th.

95 The temple he visited and measured is described in his *Description 1*, p. 113 and illustrated in Plate XLVI. What Pococke neglected to tell his mother in the letter was the incident that occurred while he was measuring the ruin, when the Sheik's nephew ran off with his notebook and enraged his uncle. Pococke eventually got the notebook back in return for "about the value of a crown".

96 A "green head" of "sherif" was someone who wore a green turban to show that descended from the Prophet Muhammad. Also see Letter 5 for an outline of the significance of the various coloured turbans.

97 Earlier in this letter the name is spelled "Ali Surbage" and the correction here has been made in Pococke's hand.

governor, I went on horse back to see the ruins of a temple 2 miles north, & then went 2 miles S.S.W. to the monastery, called Helena, from the Empress Helena its Foundress, but it is known by a new name, that of the Convent of the Martyrs; it is a great place for burial; the two priests cannot marry, but their whole generation men & women live in it; – it is of brick, but I saw nothing of the grandeur, in which it was probably at first built, nothing seeing [visible]. The priests gave us bread, dates & Coffee – & I returned; visited the two sherifs ; Sheriff Ali Salim, writ a letter to me in the boat; – & we went away, but made no way, till the wind sunk at night, – then we went all night.

4th-15th. About noon we came to Erment, where I went to the sheke, was entertained with Coffee; – he went with me to see the ruin of a temple & a church. I returned to the Bark & we went on, & stopp'd almost as far down as Luxerein.

5th-16th. We went on a litle way about a mile lower to El-gurnou [El Qurna], opposite to Carnac, where I had staid four days to see the ruins as I went; at El-gurnou I delivered a letter to the sheke, from the Prince Sheke; & he went with me between the mountains 2 miles, where I saw, nine different ranges of Galleries cut under the hills out of the rocks, high & wide, which were the burial places of the ancient Kings of Thebes, & where I saw 2, or 3 of their tombs; – all most beautifully carved & painted, with Heroglyphicks, of which Strabo makes mention.

6th-17th. Went with the Sheke to see other subterraneous works; – ruins of a Temple & two Colossal statues, sitting, of one stone, 14 feet from knee to instep, of which I could say many things too long to mention.

7th-18th. Went again to these statues, & after saw some others tumbled down, three I believe of the same size, one less; – & very litle signs of the Temple, & other buildings, to which I suppose they lead. Saw ruins of another temple, & other buildings, & in the evening went down with the stream, & after midnight arrived at that part of the river nearest to Kepht, whence the Cophtics are called, the ancient Cophtos, – being here in great numbers, in the primitive times, & now I believe not one.

8th-19th. Went with letters to two Emirs, who showed me the remains of a small Temple, & some few other imperfect things: – they came with me to the boat & took Coffee.

9th-20th. I walked about noon to Emir Hassan to whom I had a letter from the Prince of Akmim, came to the boat, & went with me to see some very imperfect things; two other Emirs sent me a present, a kid, & a basket of bread: – I presented Emir Hassan with 4 pieces of Tobacco & 5 of soap; – & he after sent me a kid & two baskets of bread; & tonight we went on & came to the Port of Kena [Quena] [...] Emir here is the title of titled persons that command a village: in Turkish one of the Mahomets Relations, who in Arabick are called sheriffs.

10th-21st. I walked a mile to Kena, the Old Cenoboscillum;[98] no ruins, famous for materials of a light unglaz'd, earthernware for common use. Went in an hour to Dendera [Dandara], where I carried letters to Emir Omran & Emir Joseph, from the sheke of Furshout & Prince of Akmim; they entertained us with cakes & Coffee, & got the Governor to send his brother with us two miles to Amara to see some ruins; return'd, the Emir presented me with a lamb: & the Governors Brother came to the boat, & I presented him with some rice & soap, & departed going all night.

11th-22d. In the morn arrived at the Port of Badjoura & rid four miles to Furchout to the convent of propaganda fide, where I dined; – visited the great Sheke & his secretary Malim Gabrian & returned to the Bark, & went on.

12th-23d. Arrived about 9 in the morn at Girge, went to the convent, dined, went to the Beys camp near the town, who is the Governor of upper Egypt, having a letter to him from Osman Bey of Cairo; he was gone into the town, to the Harem, to the Ladies. I went with a letter to Ali surbage, – not at home, & returned to the camp & was well received by the Bey, in his fine tent, he ordered me letters for below, which were immediately drawn, & treated me with Coffee.

13th-24th. I rid out to see some things abroad, & then went to the boat accompanied by two of the Fathers; – we went away & saw a miserable convent, & some ordinary Grotto's to the East, & went on. I had slept in my cloaths till this night above five weeks. The great man of Furchout was eating beans boyled in the shells, when I came to him, & he gave me some. We arrived at Mensheh in the even, where the men would stay to see their wives.

14th-25th. Came early in the morn I arrived at Akmim; – went to the convent, then to Malim Soliman; dined; – he came to me. I took leave of the Prince, who gave me two letters I wanted in the way. Staid within.

15th-26th. [No entry]

16th-27th. Went to the two convents & some grotto's. Malim Soliman, the President & some others went with me to the boat; the former presented me with a sheep, the latter with a sack of bread; & a relation of Solimans, with a basket of fine bread, & so I departed for Cairo.

98 Several variations of this placename are used, as in for example *Description 1, p.* 87, where he cites Herodotus and Ptolemy, who identify it as "antient Caene or Neapolis". Elsewere it is called "Cainepolis".

17th-28th. At Raigny I went to see the Grott of two serpents, that are worshipped by the Turks, & they say work miracles;[99] – went on & saw some pillars of a Temple at Gaua[100], where we were detained by the wind.

Febr. 18th-March 1st. We passed Aboutig [Abu Tig] a Bishoprick, & were obliged by the wind to stop somewhat lower; – went on in the night to Sciout.

19th-2nd. Being detained by the wind, I went to see the town, which is large, we went not on till even, & arrived at Mansalouth, that night.

20th-3d. Went to see the town, & in the even arrived at Malloui, a town that belongs to Mecca.[101]

21st-4th. Went to the Governor in order to see some ruins

22d-5th. The Governor went out in Pomp, & I followed toward the ruins which I saw & return'd about noon to my boat.

23d-6th. Disappointed in seeing some ruins, staid within all day.

24th-7th. Expected in vain, & parted about 9.

25th-8th. [No entry]

26th-9th. Went on – & the

27th-10th. Arrived safe at Cairo, God be praised! after a voyage of three months on the Nile very happily finished, & now you are not any more to think I am in danger. I received your letters of Oct. 13th & Novr. 24th which gave me a mixture of pleasure & pain. I was 3 months without talking English. Arabick is the language universal here. The soldiers & white slaves speak Turkish, which by others is spoke as French in England. I am now setting out to be at Jerusalem at Easter... Damietta... – you see I have great fortune in everything, so be ...
R. Pococke.[102]

99 Pococke gives an account of this serpent-worship in *Description 1*, p. 125, in order to "shew the folly, credulity and superstition of these people", including the Christians. However, he does not mention it in his correspondence with the bishop. The grot is identified as Gabal al-Haridi, in Saqulta, just north of Sohag, by Christopher J. Kirby, "Gebel el-Harifi, in *Encyclopaedia of the Archaeology of Ancient Egypt*, ed. Kathryn A. Bard, London, 1999, 328-331.

100 See *Description* 1: 126: Gaua-Kibier, i.e. Gaw al-Kabir?

101 Pococke probably meant that this town was part of an endowment (*waqf*), the proceeds of which were used for the benefit of Mecca.

102 Parts of this sentence are illegible, as has been squeezed in at the bottom of the page in tiny script.

DAMATIA IN EGYPT
8/19 March 1738
Pocock to his mother
Letter 10[103]

Honoured Madam,

My last letter to you was from Grand Cairo dated Febr. 28th March 11th, upon my return, from a voyage of three months up to the cataracts, very happily ended: After Feb. 19th March 2d. I had room only to give you a very short account of what passed. My letter before was from Furchout in upper Egypt Jan. 3d-14th. & before that, one from Akmim in upper Egypt Decr. 21st Jan. 1st. before I writ three from Cairo & 3 from Alexandria Feb. 20th March 3d. I went to Manfalouth a small town & Bishoprick, church a league from the town. I carried a letter from Malim Soliman to a Christian, but as there was nothing to be seen I went on, & that even arrived at Meloui [Mallawi].

21st-4th. Went with a letter from Malim Soliman to a Xtian here, who was in Cairo; – went to some other Christians; the Sardar or Governor being abroad, & returned to the boat, – & in the even went to the Governor again with a General letter; – presented him with a basket of rice, & my knife of [...] consisiting of a pen-knife, knife, mohock & a sort of needle; with which he was pleased, said he would go with me tomorrow to see the ruins, that he might have a share with me in the treasures I found, & received me very well; & for the first time I saw a man bastinadoed [whipped] on the feet; – here in Egypt they lye on the belly, & the legs being in a pr [pair] of straps, 2 men hold 'em back, & two beat the bottoms of the feet with leather; returned to the boat.

22d-5th. Two asses came for me to the boat, near 2 miles from the town, went to town, & the Governours train were going out. I joined them; – first went the Banner of Mecca, this place being a peculiar principality belonging to Mecca, next followed about 30 soldiery; immediately before the Sardar, one went with a very long pike; – after him his great people, then followed two camels, with a man on each, beating kettle drums, & so we went 3 miles to Archmounain [El Ashmunein]; & to make up the glory of the cavalcade, I & my man on sorry asses, made you may be sure a noble figure among so many fine horses. – The secretary was sent with me to see the Portico of a Temple returned to the Governor, who was sitting on a carpet; – said we had given his

103 BL, Add MS 22998. Pococke sends a letter to his uncle in the same post, with identical address and date (see Account 6).

secretary part of the treasures to say nothing; – for that notions prevails among the Arabs, that Franks by Magick Art can draw away the treasures, when they have been on the spot; – Coffee was brought & I return'd to the boat.

23d-6th. The governors secretary who had disappointed me in seeing some other ruins, came to the boat & dined with me, having sent to him & promised to go with me tomorrow.

24th-7th. Expected him till the sun was two hours high, & sent to the secretary to let him know the Boat-men would stay no longer – & so departed. In our voyage a man swam to the boat with a green melon, 4 inches long; which they eat green as cucumbers, & eat 'em like the melon.

25th-8th. Came to Minio [Minya] after it was dark last night. Went all day, overcast, & litle wind.

26th-9th. Went ashore at Benizuaef [Beni Suef], a town where they make a striped coarse carpet, went all the way.

27th-10th. In the even arrived at Cairo, & went to the Consuls after 3 months absence, without speaking to an Englishman. It was like coming home; – & I received great marks of kindness & friendship; – Mr Congreve was there & sup'd with us; – & with this day I ended my last letter. I found here your two letters of Oct. 13th & Novr. 24th – your former letter of letters of your Hampton [London] expedition, never came to hand – I suppose directed to Venice. Your letters as I told you gave me a mixture of pleasure & pain; – you must not read travels, in which dangers are made or magnified. All sorts of meat-broth is bad here in sickness, & so I think is milk – as indeed it is with us; – so a water boyled & steeped on pounded Almonds is much used. But I thank God I had never my health better. I am very sorry my voyage has given you so much trouble; – but am glad to find your uneasiness wears off: – we are always under the protection of God Almighty, & if we do not ill, tho' we can't see the end of things, yet we may be directed by his Providence; – the universal Providence, does in reality consist, in a direction of every minute particular, & it is only considering them altogether, & the direction of them in relation to one another, that makes it a General Subject in his Providence, therefore let us trust in him, by whom the hairs of our head are all numbered, & without whom not a sparrow falls to the ground.

28th-11th. Went to the propaganda Fathers to deliver some letters; – Father Justo & then Georgio came to see me, & after the Father Prefetto, Fathers Francesco, Gabriele & Philoptio [...] was with the Fathers of terra santa to get Letters to Joppa & Jerusalem. Sat the even with the Consul &c:

March 1st -12th. Went to see the making of Sal Ammoniac [ammonium chloride], of the soot of cow-dung, & after to see the chickens hatch'd in rooms, which are kept warm by a constant succession of hot ashes or embers.[104] In the evening visited the Fathers of propaganda & spent it with [them happily].

2d-13th. I went to see the place where they observe the rise of the Nile, & after to see a college of Dervises [Dervishes]; busy in packing; – visited by Father Dr Maria of Lucca, my fellow traveller & after by Father Justo. Mr Congreve spent the even with us.

3d-14th. Busy in packing; – Father Justo call'd on me. Mr Congreve dined with us. He, Mr Upton, Father Justo & Malim Medas the broker of the house, went with me to the boat to Brulac [Bulaq] & staid there two hour & took a pipe, the boat not setting out till early in the morn for Damiatta.

4th-15th. Set out at day break & went on pretty well, obliged to stop in the night by the wind. As I had the greatest fortune in my voyage to the cataracts & back again, so I had good fortune at the same time in Cairo: – the English had put up by way of Lottery or raffling a Bureau, japan'd with gold on a blew ground, & with a Book case over with glass doors: – an hundred tickets at ½ a guinea each this fell to my lot in my absence; it is a beautiful piece of furniture fit for a Lady, made in England & so I have an opportunity of showing my gratitude, by making a present of it to Mrs Barton the Consuls Lady, as I intend to write from Damiatta: & indeed I have received most extraordinary favours from the house: – when I go abroad in this manner they send me wine & provisions abroad, & I give them not a litle trouble. The Consul has a younger Brother at Cyprus I am recommended to [George Barton]. In this short voyage I do not dress any meat, having pidgeon pye & roast beef, they sent aboard & also live fowls: – I have taken a part of the boat, & go very pleasantly, all the people, the Captn. &c: being very civil. And I am recommended at Damiata [Damietta] to the English Vice-Consul & everywhere. I take no money with me, but a Bill on the Reverendissimo Guardian of the Convent of Jerusalem: who will give me litle bills on all places, where I want them; – for I shall go to Acre, thence to Nazareth, Mount Tabor & the sea of Tiberias [Galilee]; – then I shall go to Sidon, to Damascus, Bareut [Beirut], Tripoli & Aleppo, & thence to Cyprus all very safe. A very curious voyage. I shall have no opportunity of sending letters till I get to Aleppo, so you cannot expect to hear from me, till towards Michaelmas. There are some Irish Fathers at Jerusalem. I have now a very becoming black beard near 4 months growth, & shall let it grow till it is proper to put on a periwig & hat at the English Factories, as is the way with the Turkish dress. It is a litle inclined to [be] picked, & I shall part with it with regret; but whenever that time comes, I shall send this dear part of myself, to my dearest friends as a present, on which I hope they will set a proportionable value. About ten miles

104 These incubators for chickens were earlier observed and described by Thomas Shaw.

below Cairo, the Nile divides into the Rosetto & Damiata branches, & on each side of this branch, the villages are very thick, a fine flat country, but wood only about the villages, mostly date trees. When I return from Jerusalem to Jaffa, I intend to go always by land, which is very safe, & by sea; no Frank ventures on any but Frank bottoms [boats], for fear of the Maltese who plunder Christians & set them ashore & carry the Turks into captivity. Drink a dish of Coffee immediatly after dinner, which will not affect your nerves & will chear your spirits, which you may try. If I had thought you would have taken my voyage so very much to heart, I would not have gone, but now that I am so far from home, I have a great desire to see everything this way, & make it a compleat voyage. As to the losing anything, but the spending of my yearly income, I well considered that before, & am very well assured, I shall be no loser in any other respect, & I believe no one will make any declarations to the contrary, which I shall look on as false, & take very heinously, unless by facts they are proved to be true; concerning which I have no hopes or fears, but this inter nos [between ourselves] & it is better not to touch on those subjects at no time whatsoever & to no person whatsoever;[105] – be the conduct good or ill, perpetual reproaches of ill conduct must be disagreeable; for I am sure my interest [prospects]will be no way affected, but rather the contrary; – & this voyage I undertook on very mature deliberation. I could give you many proofs that I shall be no looser [loser] by this journey, otherwise than by spending my yearly income & in other respects it may prove to my advantage: – & let those guard the treasure who are to have it; – this I am sure I shall enjoy a greater share of favour, than at present, tis true, our voyaging both together may have distressed, but I am not the person concerned; – it all regards the great person [Bishop Milles] & his heir [Jeremiah Milles], who should have taken care of their affairs: – he [Jeremiah] had leave to go on from Buda, received there, & if it had not been for me, would have been at Constantinople last summer; if I had not advised against it, he would have gone to Naples & perhaps Sicily, if I had gone as I had determined at Trieste: & if my advice had been followed [he]would have gone Post home from Venice. How am I concerned in all this, only in acting a part wiser than they, in advising & pressing him to take care of his affairs. At Vienna I advised him to return immediately home, very sincerely on another affair; – but I entreat you not to mention any of these things, which may have bad consequences: enjoy yourself & your friends [next two lines are scored out]. Felices si sua

105 The following text relates to church matters back in Ireland, where it appears that Jeremiah Milles was in trouble for taking so long to return from his second voyage, and that Pococke was being blamed. Pococke, who exonerates himself from such blame is careful not to mention any names, which are supplied by the editor in square brackets.

bona norint [Too happy, if they but knew their happiness]. I see the family at Duloe distressed certainly,[106] but why by my absence? was it by another? if he had been present would not all have been well? & was it not his duty & interest above all others to be present. If I had travelled alone all would have been well, so that I am not the person; & if I had been at home, it would have been depriving myself of the great advantage & pleasure of this tour, for the interest of others: It is more proper that those should be deprived whose interest it is. I could also inform you, as to some other facts not proper, that would please you in more respects than one, with regard to my interest [prospects], which I am fully perswaded this voyage will in every respect will promote. Stoick Philosophy is a sort of quietism, laying the passions, & waiting for the instances of the divine spirit, by which we may be lead to what is good without passion, & as to a thing purely good or agreeable to the Will of God; if you can divest yourself of passion & wait the events of things & not prejudice your health you will see we shall meet at Newtown, & you may see other things that will be agreeable to you. I would have you do everything to oblige the Soaring Bird [Jeremiah Milles], & say nothing that may disgust; for altogether he is a youth that is much to be valued & must be very dear to all that know him, & has few equals. I received no letter from him by fault of my German Bankers in Italy, who I am sure should not take in my letters, for fear of losing the postage.

5th-16th. We went on very well: In a separate partition of the boat is a Captn. with black slaves, ten or twelve of them boys & girls for Constantinople, he is very civil to me, & be not in any fear that he will kidnap your son, who say what you will is come to years of discretion, & is not in so great danger, for he finds everything mightily work[s] for his advantage & trusts it will be so. You say nothing what Dore's preferment is worth.[107] I have a bad memory as well as you; trouble not yourself to get things without book; – by exercise I have got a strange memory for remembering buildings, tho' I have to rely [on] occasion to use it, having measured even everything in Upper Egypt: Assure yourself I will go nowhere where the plague is, & will contrive accordingly. I shall in some time desire you to look out for a sightly boy for me, of good temper, especially if you should want one yourself, & then you would not refuse the trouble to take one & feed him up a litle for me, I would not have him younger than about

106 The family home in Cornwall of Jeremiah Milles, whose father (also Revd Jeremiah Milles) had been rector since 1704.

107 Revd Dore is not listed as a member of the Waterford & Lismore diocese, nor does he appear in *The Clergy of the Church of England Database* 1540-1835 and therefore he cannot be identified.

14, & if more it will be no disadvantage.[108] One of the black Ladies has 6 holes in her nose, for nose-rings, 2 on each side & 2 at top. Stop'd in the night.

6th-17th. Went on but obliged to stop by a contrary wind. As to what I said about my return home, as I have taken such a long Voyage, I am sure you would not have me return, without seeing what I took the Voyage for, & it would be a thing I should always blame myself for, as long as I lived.

7th-18th. Went on a litle last night & pretty well today, till at noon the wind rise too high passed by Manamura [Mansoura], where Louis IX of France was taken prisoner & Damiata was given for his ransom.[109] I am very glad to see in your last letter that you acquiesce & quiet yourself in a submission of Gods Providence, the only thing I thought proper to say on the subject of all your letters past, & so we shall conclude writing any more on this subject, which may only cause disquiet to one another. Pray present my very kind love to my sister & my best repects to White-oak, Greenham, &c:

8th-19th. I arrived at Damiata very early in the morn, & went with my letters to Seignr Fakor [Ḥanā Fakhr] a Christian Mercht & Factor here & Vice-Consul of England & all Europe here; we not understanding one another, he sent me to Captain Fieldings lodgings,[110] a very good sort of man & a country man, that goes between Constantinople, this place & Latichea [Latakia]; he told me he would not let me go among the crowd of Pilgrims, but would set me down at Joppa in his way to Latichea, & perhaps go with me to Jerusalem, you see I am very fortunate & tomorrow we are to make this voyage; – I write & have writ two letters to the Bishop giving a general account of my journeys from my landing at Egypt to this place,[111] & if you please you may desire Jerre in general, to send you my letters writ to the Bishop, or desire it of the Bishop himself, & that you will return them if desired, not that they are worth seeing, but because perhaps you may desire to see them & they will give you some general Idea of what I have seen, as these give you some account of what I have done. I dined with the Captain, you see how very fortunate I am. No one ever performed the journey of Upper Egypt with so much good fortune, & I can hear only of one company of Franks that ever were so far as the cataracts, & hardly

108 This is the first of several request for his mother to find him a servant to come and live with him in Ireland. However, on hearing of the news of his uncle's death, Pococke cancels the request.

109 During the Seventh Crusade, in 1250.

110 Pococke would meet Captain Fielding again in Latakia (see Letter 21).

111 The letters he had already written/sent to the bishop are as follows Leghorn, 6/17 September 1737 and two from Alexandra dated 30 September/11 October 1737, with a cover note for the latter two, dated 6/17 October 1737. The two letters he is now writing are those dated Grand Cairo, 1/12 March 1738 and Damietta, 8/19 March 1738. All are from the collection BL, Add MS 15779 and are reproduced in Chapter 5, Accounts from Egypt.

anyone has seen so far, & I am sure no one has seen so well, what they have seen; & what makes it more curious, everything is almost entirely unknown of that voyage. I walked about the town with the Captain, & sup'd with a top Greek merchant; – they can eat fish that has no blood in them, as shuttle, oyster all shell-fish, & the roes of fish in Lent, – but can't eat other fish. I went down with the captain to the mouth of the river, in order to embark for Joppa; so you see I am very fortunate. Be of good chear, we shall happily meet again,

I am Dear Madam your most Obedient & Dutifull Son Richard Pococke

PLATE 8 Frontispiece to *Description 1*
PHOTO DAVID KANE

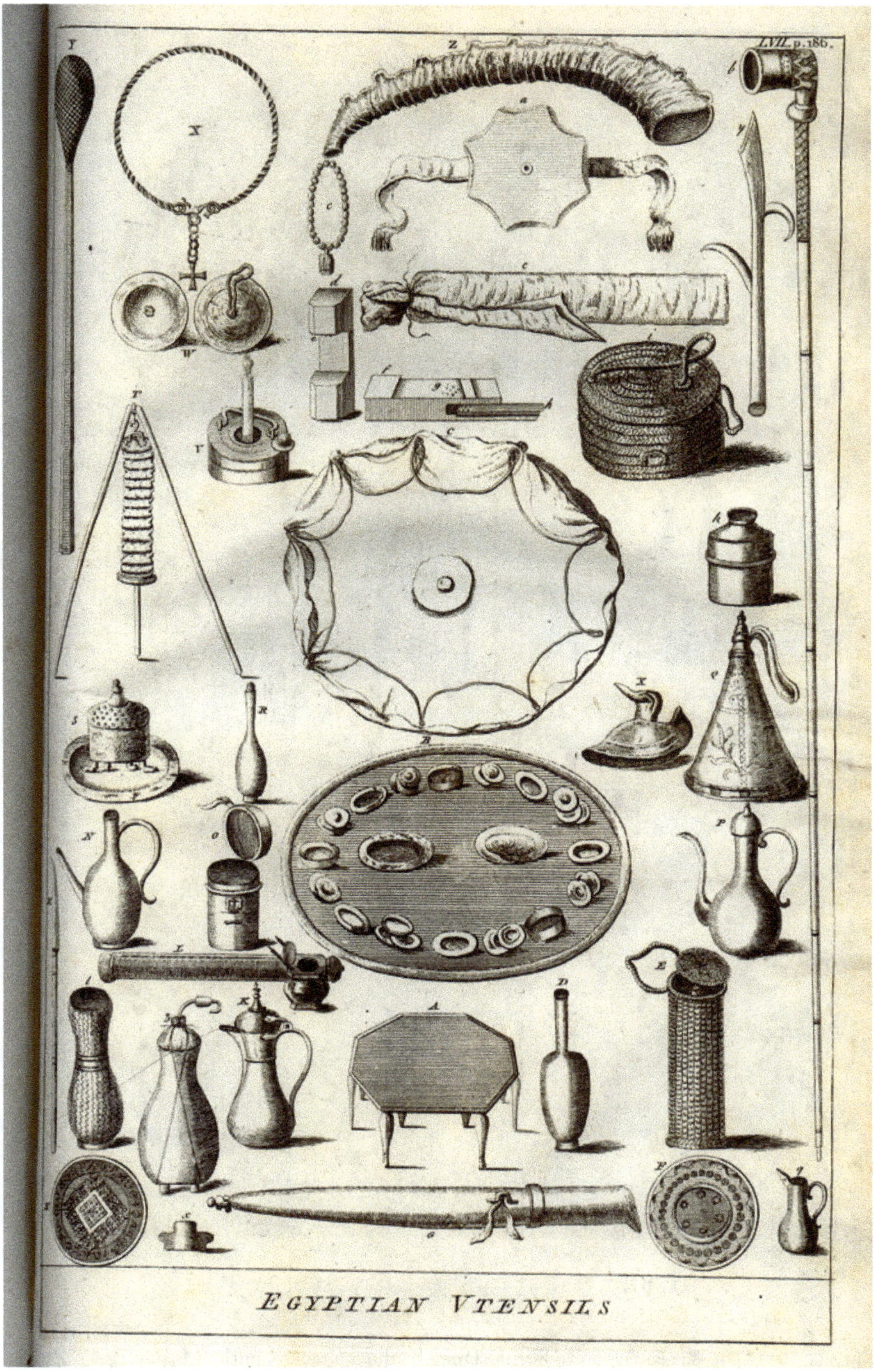

PLATE 9 Egyptian Garments, Litter etc., *Description 1*, Plate LVIII
PHOTO DAVID KANE

PLATE 10 Dresses Particular to Egypt, signed by the engraver, Charles Mosley, *Description 1*, Plate LIX
PHOTO DAVID KANE

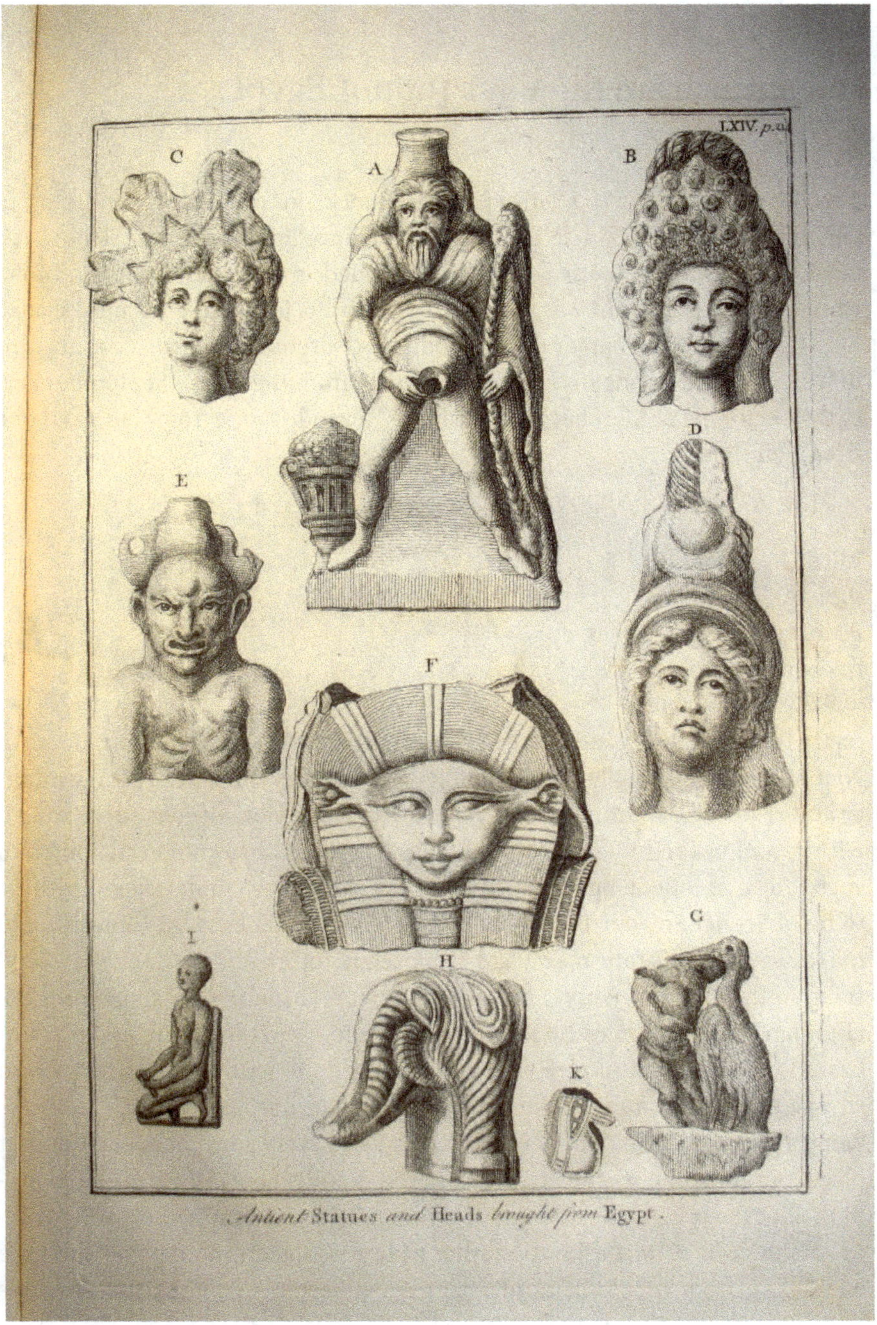

PLATE 11 Ancient Statues & Heads brought from Egypt, *Description 1*, Plate LXIV
PHOTO DAVID KANE

CHAPTER 5

Accounts from the First Tour of Egypt

This chapter begins with a letter sent by Pococke from Venice (Account 0), in which he informs his uncle of his plans for travelling to the east. This is followed by a note (Account 1) which is included in the bound volume of accounts from the east but was written from Livorno, just before Pococke set sail for Alexandria. The chapter also reproduces the remaining five accounts sent to Bishop Milles during his first tour of Egypt, from the end of September 1737 to the beginning of March 1738: three from Alexandria, one from Cairo and one from Damietta.

VENICE
19/30 August 1737
Pococke to Bishop Milles
Account 0[1]

As we thought it by all means adviseable that Mr Milles should return home to your Lordship, especially as we found it would be very difficult to get my mother and sister over into Ireland on any exigency, and not without some hazard with regard to their lifes [lives], or at least their health, as not accustomed to such long and fatigueing journeys, tho' they would have undertaken anything to be of service to your Lordship. And it being proper that Mr Milles should make hast[e] in his return, so I had proposed to return leisurely and see everything in the way. But when I came here I met a Firman or Passport from the Ottoman Porte sent to us by Sr Everard Falkener [Fawkener] the English Ambassador at Constantinople,[2] which encouraged me to undertake the Eastern voyage being well prepared for it in pursuance of our first design. And there being a ship here bound to Alexandria I disposed everything to embark for that place, but for some reasons chose rather to go to Leghorn [Livorno] and embark aboard an English ship for Alexandria, being determined on several accounts to go first to Alexandria rather than to Constantinople, and the time will fall out very well to be at Jerusalem next Easter, so that by the time this

1 BL, Add MS 15774. This letters from Pococke found its way into the volume of correspondence from Jeremiah Milles to Bishop Milles.
2 See Introduction, note 25.

 DOI:10.1163/9789004440050_006

comes to your Lordships hands I hope to be advanced far in my voyage. Mr Milles has with him a good servant, and I shall set out with [him] tomorrow, & accompany him as far as Trent [Trento], and thence make a little circuit to see some places I was not at before, and I shall give your Lordship an account of what places we did not see, and also from time to time of my Eastern voyage, which I hope will not be altogether unacceptable to your Lordship. And I shall beg leave to put into my letters some account of the antient Geography and history, and some other particulars of the places I see for my own satisfaction, which otherwise I should not think necessary or proper to be mentioned to your Lordship, as it may be the only account I shall take, or at least preserve of what what I see.

Whenever your Lordship is pleased to honour me with a letter be pleased to direct to me to be left with Cutts Maydwell Esqr Turkey Merchant in London, and it will be conveyed to me.

I earnestly beg your Lordships blessing and prayers on my voyage.

I am with the greatest respect and submission

My Lord

Your Lordships

LEGHORN [LIVORNO]
6/17 September 1737
Pococke to Bishop Milles
Account 1[3]

My Lord,

Mr Milles [Jeremiah] has set his face towards England from Trent, which place I left 28th of August old style, and travelled by Mantoua, Mirandola, Modena, Bologna, and Florence to this place, where I came on the 4th of September old stile, and found here the Mary Capt Parkins bound to Alexandria, and in company with two Franciscans from the Propaganda Fide, I hope to be in a fortnight in Alexandria; the ship being ready to sail so that probably I shall be in Egypt a fortnight before this comes to your Lordships hands.

If we sail immediately I shall not have time to give your Lordship an account of what I saw in the way to this place. But I shall from time to time forward to your Lordship a relation of what I observe in the East, tho it is probable twill be Midsummer before your Lordship will hear from me.

3 BL, Add MS 15779.

I earnestly beg your Lordships blessing and prayers.
And am with the greatest respect and submission,
My Lord your Lordships

ALEXANDRIA IN EGYPT
30 September/11 October 1737
Pococke to Bishop Milles
Account 2[4]

The sea south of Sicily was called by the Antients Mare Libycum or Africum, which sea comprehended that part of the Mediterranean which is on the coast of Africa, from the entrance into this sea at the Columnae Herculis, or the streights of Gibraltar, to the East bounds of Cyzeniaca where the Egyptian sea begun. This is commonly called the sea of Barbarry along the Africk coast and on the side of Sicily the sea goes by the name of the Channel of Malta.

When we approached Sicily I found we were sailing along the same coast by which Aneas made his voyage, and as I had a veiw of the citys and places on the shoar, I could not but observe the justice, and poetical beauties, of the descriptions, of the great masters of Latin Epicks.

As soon as we had doubled the South West point of Sicily we saw the city of Mazra, the old Mazara, from whence one third part of Sicily is called Vala di Mazara. Somewhat East of it, was the famous city of Selinus, which was destroyed before Strabo's time, the Poet makes mention of it as abounding in palm trees.

Teq[ue] datis linquo ventis palmosa Selenius.[5]

We after had a very plain veiw of the city of Ziacca on the side of a hill.

Sailing on I saw the city of Girgenti on the side of a hill being built up to the top of it, about four miles from the sea. This was the city of Agrigentum where the Tyran Phalaris resided. This city remained when most of the other citys on the south of Sicily were destroyed in the Carthaginian wars. It was first a Colony of Ionians, and after a Colony was brought thither from the cities of Sicily by F: Manlius the Orator. Under the greek name Acragas Virgil describes its eminent situation as well as mentions its having been formerly famous for a fine breed of horses

4 BL, Add MS 15779. A note in the bishop's hand states, "This letter to be placed after the other, of the same date."

5 Virgil's *Aeneid* 3, line 705.

> Arduus inde Acragas ostentat maxima longe
> Moenia magnanimum quondam generator equorum[6]

At the same time I had a plain veiw of Mount Aetna, which now among the vulgar goes by the name of Mount Gibelto, it is seen almost all along the South and East coast of Sicily. I discerned a very little smoak ascending from the top of it. This mountain so famous among the antients, is very beautifully described by Virgil, as seen by Aeneas from the coasts of the Cyclops, about Catana, where Ulysses had put in not long before, and where both of them met with such extraordinary adventures in relation to Polyphemus, according to the Poets.

I saw after Cape Lescate, at the mount of the river Salso, the old Himera, near which was a castle called Phalerium where it is said the brazen bull was kept. There is also a river called Rocetta, which runs into the sea North of Sicily the source of which is near the fountain of the Salso, and the Rocetta was formerly also called the Mimera. Which gave occasion to the antients to frame something very strange that these two rivers were one river called Himera, and that part of the river run North and the other part South, and that one part of it was sweet water, and the other salt of which Vitruvius gives the true cause that one part of this river or rather on of these rivers passed thro' places where they dug salt, for in the middle of the Island, about the source of the river Salso, are mines of rock salt which probably is the reason of the modern name of this river.

Further East I saw a city called Terra Nova, near a river of the same name, this is supposed to be Geta, with the river of the same name, mentioned also by the Poet.

> Apparet Camarine procul Campiq[ue] Geloi
> Immanisq[ue] Gela, fluvii Cognomina dicta.[7]

There is but one place more mentioned by Virgil, as also on the South of Sicily, which was in ruins in Strabo's time. The place where it stood is now called Camarana; the old name of the city being Camarina, a colony of the Syracusans.

We had a sight of Malta at a great distance, and at length, came up with Cape Palsaro, the old Promontorium Pachnum astisa peninsula, and the land very low to the West of it, so it appears at a distance like an island with a castle on it built to hinder ships from coming in there to lay in wait for other ships.

6 Virgil's *Aeneid* 3, line 703-4.

7 Virgil's *Aeneid* 3, lines 701-2

The ground of the Cape is very foul and ships cannot come to anchor there without danger of cutting their cables, so that it answers very well to the poets description of it.

> Hinc altas cautes projectaq[ue] saxa Pachigni [Pachyni]
> Radimus.[8]

Over the point we saw the high lands about Syracuse.

East of Sicily is that sea which by the antients was called first Mare Ausonium and afterwards Mare Siculum. It extended from the Fretum Siculum to the Promontory of Iapygia in Italy, to the mouth of the Adriatick sea, to the Sinus Ambracius in Greece, and as far as Crete, having the Mare Africum to the South. I do not find any particular name for this sea at present, but the mariners call all these seas as far as the Adriatick by the general name of the Mediterranean, as they call the Seas Eastward the Levant.

We lost sight of Sicily on the 21st Septr O:S the 2d October N:S in the evening and making a great run on the 24th-5th in the morning we could discern the high mountains of Candia north of us, the Island of Crete of the antients so renowned for the reign of Jupiter and the wise government of Minos, as well as for being the scene of many of the fables of the antients.

From Crete Eastward near to Cyprus the sea was called Mare Aegyptium, extending Westward on the coast of Africa to Cyreniaca, where the Mare Africum began.

On the 29th Septr. O:S 8th October N:S we came in sight of the coast of Africa about Point Solyman in the Kingdom of Barca just on the confines of Egypt which was that part of Marmarica, about Catabathmus parvus or the little valley, where was the famous Temple of Jupiter Amnon, Alexander went to with so much difficulty to consult the Oracle, and near it was a famous fountain of the Sun which they say was cold at noon, began to grow warm at night, and was very hot at midnight.

The next day we came in sight of the Tower of Arabia, and the day after of Alexandria, and as we approached it, we had a very agreeable prospect of the famous column, of the walls of the old city, of the country covered with Date trees which grow to a great height, and appear like tall pines pruned up, and also of palm trees rising up above the buildings of the city. And so on Thursday evening the 29th of Sepr 10th of October we happily arrived in the port of Alexandria after a very pleasant and agreeable voyage of twenty three days.

I am with the greatest respect and submission,

---

8 Virgil's *Aeneid* 3, lines 699-700.

My Lord

A ship going out this morning I take the opportunity to send these letters to Leghorn, and if I have time shall send by the same an account of my journey to Leghorn, but if not I shall send these letters by the next opportunity. The Consul of Cairo [Mr Barton] being here I beleive I shall go with him to Cairo, and thence after some time proceed on my voyage.

Any letter directed to me to be left with Cutts Maydwell Esq[r] Turkey Merch[t] in London, will be conveyed to me by first opportunity.

ALEXANDRIA IN EGYPT
30 September/11 October 1737
Pococke to Bishop Milles
Account 3[9]

On Tuesday evening September 6th O:S 17 N:S I embarked at Leghorn aboard an English ship the *Mary* Galley Captain Thomas Parkins, bound for Alexandria in Egypt. The next day we sailed out of the road of Leghorn. This sea is now called the Tuscan sea, and was formerly known by the name of the Mare Tyrrhenm, and Mare Inferum lying between Corsica, Sardinia, Sicily and that part of Italy which is on this side of the Republick of Genoua, the old Liguria. We sailed about two leagues from the Island of Gorgona, which is like a high rocky mountain, the cliffs of which are almost perpendicular all round, expect in one place to the East, where is the small port of Gorgona, which is the only entrance to the Island being a shelter for small fishing boats. Over it the Grand Duke has a fortress with about twenty soldiers in it, who by their situation are capable of hindering the landing of a considerable body of men. In this port live some fishermen, who cheifly are employed in catching Anchovies. We after sailed to the East of the Island of Capraia, the Capraria of Pliny. The Island is about two leagues long and one broad, being mountainous and rocky. We had a plain veiw of the only town in it of the same name of the Island which is situated on the high ground over the sea to the East, at the South of it is a large castle on a rock, and the town extends to the north to a small bay, in which there is a fishing village; the cheif support of this Island being a trade in fish which they carry to Leghorn. In the town is a Franciscan convent which belongs to the Province of Corsica, this Island being subject to the Genoese.

9 BL, Add MS 15779. There is a note at the top of this letter, in the bishop's hand, stating "This letter is to be placd before the other of the same date."

We after passed by the Isla of Elba, the Ilva of the antients. It is about five leagues long and three broad. Pliny says it was, a hundred miles round of which it may not fall much short, there being a great number of bays and creeks in it. The North part of this Island with the Port of Ferraro, and a Castle called Cosmopoli belongs to the Grand Duke of Tuscany, the South part (except Porto Longone) which belongs to the Spaniards and all along is within Canon shot of the fortress being subject to the Duke of Piombino. It is in the territories of the latter that the iron ore is found & tis said that having dug the ore out of the pits after leaving them for about thirty years they find Iron ore in them again which perhaps gave rise to what Virgil says of it.

> Insula inexhaustis Ilva Chalybum generosa metallis.[10]

And this also may be the reason of what Pliny affirms, that there had been more iron dug out of it in three thousand years, than the whole Island would contain. His words are

> Unde per tria annorum millia, plus esset ferri egestum quam tota contineret insula.

They have a vulgar notion that the iron cannot be melted here, which possibly may be owing to what Strabo saies of the Isle of Aethalia, by many thought to be Elba, who says they could not melt the iron on the spot, but carried the ore immediately to the continent melting or running, but it is more probable that they had not the conveniency of wood for their foundries in a small Island.

Three leagues South West of Elba West of Elba we saw the flat Island of Planosa called by the Romans Planasia. The land of it is so low that it cannot be seen at above four or five leagues distance. I am informed that ruins of houses and castles are seen on it from the sea, when they sail near it, that fishermen and others go there in the day time, but that it is not inhabited for fear of the Corsairs. This island belongs to the Duke of Piombino.

Four leagues South of Elba we saw the Island of Monte Christo, which appears like one high mountain, it is now uninhabited, and I have many reasons to think that it is the Αἰθαλία/Aethalia of Strabo, which has so much puzzled the Geographers, many of whom have taken it to be Elba, but as Strabo himself was at Populonium on the promontory of Piombino, and saw all these Islands himself from thence, we cannot suppose he could be mistaken, and he in another place mentions both Ilva and Aethalia, he also makes Aethalia equally

10 Virgil's *Aeneid 10*, line 174.

distant from Corsica and Populonium, that is three hundred stadia or thirty seven miles and a half, and Monte Christo answers exactly measuring on the sea charts about twelve leagues from each or thirty six miles. There is also no other Island except Capraia beyond Elba which can be seen from Populonium on this side Corsica and Sardinia: and Monte Christo being so near Elba, is most likely to partake of the nature of the soil of Elba producing Iron ore in the same manner, which grew again in the pits. The knowledge of which may be lost by reason that the Island is become uninhabited. Strabo mentions the port Αργοs/Argos in this Island which he observes (according to fabulous history) was said to be so called from Jasons touching these with the ship Argos when he was in search of the habitation of Circe, Medea being desirous to see that Goddess.

Four leagues South West of Giglio we could but just see the flat Island of Dianisa thought to be Dianium of Pliny called by the Greeks Ἀρτεμίσια and Ἀρτέμιτα.

We had a considerable time a sight of the Island of Corsica, and a plain veiw of the town of Batia on the East side of it, on the flat shoar.

We were several days East of Sardinia, having often a sight of that Island. Since we left Leghorn we were often becalmed, and met with contrary winds, so that we did not see Sicily till the thirteenth day from the time we set sail; tho' the voyage in other respects was by no means unpleasant, having very fine weather.

We were not a great way from the most Western of the Liparae Island, called Ustica, which I take to be the Isle Ευωνυμοs of Strabo, to agree with whose description of it the old Geographers in their maps have made an Island South East of the rest & call it Euonymus because Strabo says it is the farthest to the left sailing from the Isle of Lipara to Sicily, and that from thence it had its name. But for reasons I shall mention I apprehend that Strabo meant it was on the left sailing from Sicily to the Isle of Lipara, for he says this Island is farthest out in the sea of any of them, as he expressed it Πελαγια which could not properly be said of an island to the South East of the others, because that would be nearer to the land both of Italy and Sicily, and if it were not for this objection it might be an island called Volcanello South East of Volcano at a very little distance from it and a very small Island that has a smoaking Volcano, the other Volcano breaking out in flames. I must observe that all the maps especially those of the old Geographers are very false with regard to these Islands, and find the Sea Charts are most to be depended on for the number and situation of the Islands, tho' I find de Lisle's [Guillaume Delisle's] map only right in making two Volcano islands, a great and little one, who notwithstanding if I am rightly informed, is mistaken in placing the little one to the North, which ought to be in the South East. I could not but please my self with the imagination, that I was near the place, where the Romans at the Lipara Island gained their first sea victory in a most signal engagement with the Carthaginians under the conduct of the Consul Duilius who was not only honoured for it in a solemn manner but had a sort of triumph decreed him during his whole life; and the famous Columna rostrata erected to his honour, which is now to be seen in Rome with its long inscription, and is one of the greatest and most curious peices of antiquity remaining being about two thousand years old.

I saw Cape Gallo, which is very near the port of Palermo, and falling in with the West of Sicily we sailed between the Islands called by the antients Aegates; tho in all the maps we see particular names given them by the old Geographers; yet I cant find that the old names of these three Island are certainly known, that to the North East opposite to Trapano (from which town it is ten miles distant) is now called Levanzo. South of it is the Isle Favagnana ten miles round, it is a fine fertile spot of ground, being mostly a flat, with a high hill towards the North side, on which are three castles garrison'd by the King of Naples, in one of which the Governor of the Island resides.

This place was a great refuge for the Corsairs, from whence they came out and infested the seas till Charles 5th carried his arms into Africa. The third Island thirty miles West of Trapano is called Maritimo, it appears like a high mountain. To the North East of it is a rocky promontory which is a peninsula, and much lower than the rest of the Island. On this there is a castle built, in which there is kept a garrison. The insula Aegates are most famous for the second signal sea victory obtained here over the Carthaginians by the Romans, under the command of the Consul Lutatius Catulus, concerning which the historian says that after the battle, the whole sea between Sicily and Sardinia was covered with the wreck and ruins of the enemy fleet; and this total defeat put an end to the first Punick war.

The mountain of Trapano is one of the first things that strikes the eye on the West of Sicily, on the top of it is a castle: and at the foot of this hill to the West is a flat point of land that stretches into the sea, on which stands the city of Trapano, on the spot where the antient Drepanum stood the Illaetabilis ora, where according to Virgil, Aeneas left his father Anchises, and where after his return from Carthage he celebrated divine honours to this memory. This place is remarkable for actions, in the Punick wars as well as the small Isle of Columbaria opposite to it. The mountain of Trapana to the East, is mount Eryx so famous for the worship of Venus by Aeneas and his followers when he was about to leave behind him the women and the weak people to settle on the Isle as he expresses it

Tum vicina astris Erycino in vertice sedes
Fundatur venere Idaliae[11]

Strabo says that the town on the top of the hill was inhabited by women dedicated to the Goddess Venus by foreign nations as well as by the Sicilians, that in his time it was inhabited by men, and a temple served by preists, who lived in great poverty, the place becoming unfrequented: he adds that the Romans built a temple to this Goddess at Rome without the Porta Collina, called the temple of Venus Erycina, so that it is probable the devotion was removed thither.

South of Trapano I had a plain veiw of the city of Marzala built where Lilibaeum stood, the port where they usually embarked for Carthage. The promontory and town also are often mentioned in history, especially of the Punick wars. It is said the port was spoiled by the Romans, to hinder the convenient passage of the Carthaginians to and from this port in case they should recover it again, and it was entirely filled up by Don John of Austria in 1567. Augustus brought a Colony to this town. The sea coast being shoaly hereabouts answers in that respect very well to the description of Virgil in this verse, Et vada dura lego saxis Lilibeia caecis.[12]

I saw between Sicily and Africa the isle of Pantelera which is the Cosyra of the Romans and the Κόσυρα (Cossura) of Strabo which he saies is equally distant from Lilybaeum and the city of Aspis or Clupea of the Carthaginians, this is confirmed by the name the inhabitants of Africa now give it in the Arabick language which is Cosra. It belongs to Sicily and is made use of as a place of banishment. South East of this is the isle of Limosa, and a few leagues South of that a larger island called Lampidousa; which did belong to a Christian Hermit, and a Marabut or Turkish hermit, and served as a place, both for Christians and Turks to take in provisions, with agreement that neither of them should suffer from those of a contrary religion. The Marabut dying not long ago, the Corsairs seized on what was in the island, and carried the Christian away captive, of which great complaint was made by the French Consul who demanded the captive. Strabo in three places mentions the Isle Αιγιμωρος [Αιγιμορος] together with Cossura, and in one place speaking of several small Islands in general as near Cossura and Sicily, he only mentions Aegimurus in particular, and therefore probably it was the largest of them. The three Islands that are near Pantelera or Cossura are, Semetto, Limosa, and Lampidosa, and the last being much the largest it is probable, that it is Aeqimurus so called by the Romans

11 Virgil's *Aeneid* 5, lines 659-60.

12 Virgil's *Aeneid* 3, line 706.

also. On this Island in the first Punick war the Carthaginian fleet was ship-wrecked in the consulship of Fabius Buteo.

We thought we saw Cape Bona which is the North East promontory of this great bay of Carthage.

I am with the greatest respect and submission,

My Lord

ALEXANDRIA IN EGYPT
6/17 October 1737
Pococke Bishop Milles
Account 4[13]

I sent two letters to your Lordship giving an account of my voyage this place, dated 30th Septr 11th October, the day after we came into this port, and having an opportunity by a ship going to Marseilles I lay hold of it to do my self the honour to write to your Lordship tho' I have not time to give a regular account of any thing I have seen here

I am come here in a very good time of year, the spring & moderate weather now beginning, which continues to April when the harvest will be over, and all annuals will be burnt up with the heat, it is notwithstanding hot here in the sun during all these months which we call winter in Europe.

I have been very fortunate in meeting Consul Barton here who usually resides at Cairo, from whom I have met with all manner of civilities, and shall have the pleasure to go with him to Cairo, his brother is Rector of St Andrews London, and consequently the Duke of Montague is Patron of the family: so that every thing contributes to make this country very agreable.

I find great difficulties in examining the scituation of the old town, and the several parts of it according to the antient authors, the sea having left the shoar on some parts and gained on others; and I have seen modern authors that can give any light in this affair.

Begging your Lordships blessing. I am with the greatest respect & submission

My Lord

13 BL, Add MS 15779.

GRAND CAIRO
1/12 March 1738
Pococke to Bishop Milles
Account 5[14]

I always purposed, to do my self the honour to write to your Lordship a particular account of what I see, whenever I had any time not employed in viewing things, or committing to paper what I have seen. These indeed have so intirely taken up my whole time, as your Lordship will be convinced, that I have been obliged to lay aside the thoughts, of writing very particularly to your Lordship, having time only to send a rough, tho' I hope pretty exact account of every thing. I cannot however excuse my self from giving you from time to time, a general account of my voyage, not without mentioning some of the most remarkable things in it.

At the entrance of the port of Alexandria is an Isle, commonly taken for the Isle of the Pharos; but it was the island Antirhodos, so frequently mentioned in History, being now, at the East end, joined to the continent.

I saw the famous pillar, and when the sun shined on it, I could just discover some letters of a greec inscription on the pedestal; but tis so imperfect by the scaling of the stone, that it was very difficult, if not impossible, to make any think of it. I beleive the first word is Σὲβασιος[15] and am inclined to think that it was erected to the honour of some of the later Emperors, as Titus, or Adrian; because Strabo, who so accurately describes Alexandria, and was there, makes no mention of it. West of Alexandria are evident marks of the antient Necropolis; where one sees a great number of habitations for the dead cut out of the rock, and some of them in very regular apartments, one within another.

To the East I traced out, with difficulty the city of Necopolis, built by Augustus, in memory of his victory over Anthony, near this place. These are great remains here of a very large square building, or enclosure, with semicircular turrets all round built very strong. It is commonly called an Amphitheater. I am apt to beleive it was a Praesidium; there being such another building exactly at old Cairo, at the foot of that hill, which I conclude to have been the antient Babylon of Egypt, where was a Roman Garrison of one Legion. Farther on this shoar near Bickeern I saw the ruines of two Temples near one another. On is half a mile from the sea, and seems to have been a round building, and several of the red Granite Pillars are lying upon the ground. The other is almost covered by the sea; but there are several statues of Sphinxes, and of the Deity of

14 BL, Add MS 15779.

15 This word has been inserted in a blank space, possibly in the bishop's, hand and then corrected, rendering it difficult to accurately transcribe.

the Temple the later [latter] being of red Granite. It seems to have been [ ] and one of them, the temple there dedicted to Venus Arsinoe.

I saw at Alexandria the poor Patriarchal Church of St Mark. They pretend to have his head there, and even his body; tho' the later is more probably at Venice. They shew also his chair, and have a tradition that he as martyred, out of the gate that leads to the column; and the Greecs and the Latins say, that a very large Mosque, near that gate was formerly the Cathedral Church.

I had the pleasure to go a days journey to Rosetto, and up the Nile to Cairo, in company with Consul Barton, whose brother is minister of St Andrews London.

After about three weeks stay at Cairo, I set out for upper Egypt, a very curious voyage, as very little is known of the antiquities of those parts. About a degree and a half above Cairo, at Archmounain, near Maloui I saw a magnificent Portico of a temple, of twelve large pillars, with square capitals, after the Egyptian style; for they never received the five orders; tho' they have a variety of capitals. Alexandria indeed under a greec government, and some places, improved by the Romans, shew the utmost perfection, in fine Architecture, but in upper Egypt one sees many temples, which were built before the invention of the five orders. At this place are great signs of a large city, which was probably Hermopolis, capital of the last Nomos in the Heptanomis on the West. What I saw after was in the famous Kingdom of Thebees called the Province of the Thebais, when the whole land had obtained the name of Egypt.

About sixty miles higher, I saw such another Portico at a place, called Gava, but more beautifull, and every part adorned with Hieroglyphicks of men and other figures. In the ruins of the Temple are very large stones; of men and other figures. In the ruins of the Temple are very large stones; some of which I measured, and found to be thirty feet long, five broad, and about three feet deep. This place I concluded to be the old Antaeopolis so called from Antaeus, whom Hercules conqured by lifting him from the ground, his mother earth, and who as it is said, was called by Osiris into Egypt.

Higher up is Akmin [Akhmim], the old Panopolis, where I saw very imperfect ruins of two Temples. As to the Grotts of the Thebais which are in the mountains below from Manfalouth to Minio about fifty miles in length they were the places where the antient Egyptians deposited their dead, and to which they brought them from the citys and villages, which were on the West side of the Nile, these were the most Grottos, some of which lead by a sloped passage to other apartments, and in many of them there is besides that there is a perpendicular passage down; and when great persons deposited the bodys of their familys there, they adorned the stone with Hieroglyphicks, which very often is the cause that they are taken from Temples.

The places where the Egyptians deposited their dead were so sacred that Xtians in times of persecution, would not dare to approach them; they were so jealous lest any thing should be done that might tend any way to destroy those bodys, they had with so much art endeavoured to preserve. It is true indeed that there are two or three convents cut out of the rock, and there might probably be Hermitages near them, but all these must have been made in much frequented places, after the establishment of Christianity. Before that time, we find that the great retreats of the Christians were to the deserts which were, afterwards called from some of the most eminent of them; as that of St Macarius to the West, and the convents of St Anthony, and St Paul the Hermit in the deserts on the red sea. In these Grottos I saw several pieces of Mummys, & of the linen which they were wrapt in.

Opposite to Akmin are two of the most magnificent convents in Egypt, one of them without doubt was built by the Empress Helena, as I saw the Eagle cut in many parts of it.

Passing by Girge the present capital of Egypt I saw at a ruined place called Amara near Dendyra a Temple the most beautifully adorned with Hieroglyphical figures of men and animals of any I had seen, and which fall little short of the fine releifs of the Greeks and Romans. This by the capitals, which is the head of Isis, I concluded to be dedicated to that Goddess. And such a Temple Strabo mentions after Tentyra, which from the name I concluded to be about Dendyra. There are two or three other small Temples near this, and three very magnificent gates. Opposite to this on the East side is Kena, which I suppose is the ancient Chenoboscium.

At Kepht somewhat higher are signs of a very large city, but no ruins from which one may pass a judgment what the buildings were, except a small Temple and two bridges over the old canal. This was the antient Coptos, to which the merchandise of the East was brought by land from the Red sea, which is as they told me four days journey off. A trade was carried on in later times, but the Arabians have entirely put a stop to it by their robberys. The people find much of the refuse of Emeralds, probably thrown away by the Artists, and which were dug between this city and the Red sea. It is a stone not transparent, as the common Emerald, and yet is much valued. It is from this place that the Christians of the countrey were in derision called by the Turks Cophtis, as probably this city adhered firm to the faith, rather than from a concourse of Christians to the Grotto in the mountains which are not very near, and upon enquiry I could not find that there are any Grotts in these mountains.

Not much higher is Chous the old Vicus Apollinis where I saw the ruin of a small Temple dedicated to Apollo, and a Greek inscription on it mentioning its being dedicated to the Deity by King Ptolemy, Queen Cleopatra and their children. Higher up at Carnack, which was the East side of the famous city of Thebes I saw most magnificent ruins of the Temple of Jupiter of Thebes, the ruins extend for half a mile in length, and not much less in breadth, and are very perfect. Among them are six or seven grand gate ways, allmost entire. One of which is built of red Granite, and all the buildings inside and out, and also the columns are adorned with Hieroglyphicks, many of which were defaced; which was doubtless done in the time of Cambyses, the Persians being enemys to such representations. A consideration that raised in me sentiments of veneration for such noble peices of antiquity, which were defaced between two, and three thousand years ago, remains hardly to be equalled in point of antiquity, and only to be exceeded by those on the other side, where doubtless was the greater part of the city of Thebes; and when the city decayed being no longer the residence of the Kings it was inhabited only in villages, a town being built on the East side probably on account of the Temple of Jupiter, and this town was called for that reason Deospolis. An advanced ground on the West (North of a fine plain) extending to the hills both to the North, and also to the West from the river, is all cut out into Grottos. Up the sides and under the very hills themselves are most extraordinary works cut in the rock. These works below cut as on each side of so many streets, made me enclined to think that Thebes might be a city under ground; the dryness of the climate, and no ruins appearing over these works confirming me in that opinion. The works under the hills may be justly reckoned amongst the wonders of the world. They are several rooms like gallerys one within another cut out of the solid rock, and are from thirty to fifty feet long, about ten wide, and seventeen high. Every part of these is cut in most beautifull Hieroglyphicks of men, birds, and beasts, all finely painted, and the ceiling represents the firmament, and is cut in stars and painted likewise. At the end of each of these strings of gallerys is the large sarcophagus of some one King all cut out of one piece of red Granite. The way to some of them is between the hills, and there appears to be a grand entrance from the city under the hills, which probably had a communication with those between the hills, and in some that I was in were deep square holes, down by which I saw doors below, that lead to other apartments. Strabo saw these works and says they were about forty wonderfully cutt, and well worth seeing as indeed they are. I was in eleven of them, and saw the entrances to several others stopped up. I looked diligently for the Obelisks mentioned by Strabo on which there were inscriptions giving an account of the great riches and conquests of the Knigs of Thebes, but I could see no sign of them. Near the hills are great ruins of a Temple, and several Colossal statues are standing in it; and at the entrance are the ruins of one which seems to have been of one peice of red Granite. I took several measures of the parts of it, one of which may give your

Lordship an Idea of the enormous size of it, from the top of the head to the bottom of the neck measured eleven feet; but to lessen the wonder of this statue I must tell your Lordship that I have reason to think it was sitting.

About half a mile South West of this Temple are two sitting Colossal statues. I have several measures of the parts of them, but they are not now by me, but they seem to be little inferior to the other. One of them is entire, and of one stone, the other being broke (as Strabo says it was reported) by an Earthquake, is now built up from the middle with large stones, which are of an extraordinary spangling kind. The broken one was the statue of Memnon, and it was said that every day about the first hour a certain noise was heard to come from it, and the people have the tradition of it, and say it sounds now, but no body ever heard it, that I could meet with, but I copies from the statue several Epigrams, and a great number of testimonys of great persons in Greek and Latin that they had heard the sound some of them more than once. Strabo indeed was there, and heard the sound but did not think it came from the statue. Further on are the ruins of three, or four statues that seemed to be of the same size, and near the mountain some remains of the Temple I suppose they lead to, and the foot of the hill is cut away to make a regular Area for it.

I am my Lord

DAMIATA IN EGYPT
8/19 March 1738
Pococke to Bishop Milles
Account 6[16]

My Lord,

In my last letter I did my self the honour to write to your Lordship, I gave you some general account of my journey to the famous city of Thebes, and of the ruines, and wonderfull underground works of that city. Above it, on the East, at a place called Luxeran, are the remains of a very grand Temple, not less than a quarter of a mile long. At the entrance area two of the most perfect and beatifull Obelisks I ever saw; the red Granite being of the best sort, not scaled, and having the polish on it. There is an inner, secret room, on which much art is bestowed in Hieroglyphicks; probably some mysteries were performed in it; and seeing many Hawks cut in it, I concluded that a sacred Hawk was kept

16 BL, Add MS 15779.

here, under which shape Orus, or Apollo was worshipped, and that this was the city of the Hawk, mentioned by antient authors.

At Erment, the old Hermonthus, I concluded for the same reason that I saw the room in a Temple, in which the sacred Bull was kept, mentioned likewise by the antients to be kept there. I saw here likewise the ruines of a Christian Church, but so imperfect, that it was with much difficulty I made out the plan of it, which is one of the most beautifull I ever saw; being adorned within, with four rooms of fine slender pillars of red Granite. The plan pleased me so well and I beleive would hit your Lordships taste so exactly, that I wish I could see your Lordship bring it as it were out of its ruines, by putting it in execution.

At Esne, the antient Latapolis, is a temple entire. I looked among the Hieroglyphicks for the Latus fish, they worshipped, and should have thought by my not finding it, if I had been deceived in the place, if the Itineraries had not fixed it; for they have made the Latus so much like the Crocodile, that the people shewed it to me for the Crocodile, though this large fish is like other fish, and goes now by the same name in upper Egypt, though it has another in lower Egypt.

About four miles North of Esne, is a Temple which from the Hieroglyphicks I concluded was dedicated to Lucina, and that in this spot was the city of that Goddess. Amile South of Esne is the convent of Martyrs so famous in the primitive times, among the Egyptians, and from whose martyrdom they reckoned theirs. There are no marks of her devotion there now, nor of the antient regard paid to it except a great number of tombs in a very large cemitery, that here may be called magnificent, and which are built something like the Holy Sepulcher. One sees on tombstones the Eagles cut, and this motto which probably the Empress gave to the monastery ΕΙΣ ΘΕΟΣ Ο ΒΟΗΘΟΣ.

This is one of the seven monasterys where the Priests cannot marry, but they live here with their whole generation [of] men and women. Higher up the state of Christianity is very miserable; this being the last church to the South.

At Etfou the antient Apollinopolis are the ruines of a very grand Temple. Higher up is a strait of the Nile between two rocks. They have a tradition, that a chain was stretched across there, and shew the very nich in the rock where they say it was fixed. On the other side are some works cut in the rock and adorned with Hieroglyphicks, which seem to have been places of Pagan worship.

Above this I saw a small ruin of the city of Ombos and of a Temple on a small hill, which is now called Com Ombo, or the hill of Ombo. I saw on the Islands near it no less than twenty great Crocodiles together, which are probably the places, where the citizens of Ombo took so much care of them, and waged war in defence of their Gods. If the same superstition had continued, I should not have dared to have shot at them as I did, and I should have been in more danger from the religious people of Ombo, than I was in with regard to the Arabs.

At Assouan I was under the tropic. Here was the old Sene, where they had a well made to prove that they were under the Tropick which at noon of the Solstice day was all over illuminated, and cast no shadow. I will not say that I found out a fabric that was probably built over that well, or was made to try some experiment of that nature, or in some part of Astronomy, but hereafter shall beg your Lordships opinion on that subject. There is very little to be seen of old Siene among the ruins of the old town of Assouan, which was built of unburnt brick on rocks of red Granite. For here all the country to the East, the bed of the Nile, the Islands in it, and those rocks which make the Cataracts are all of Granite, mostly of the red sort, and I saw the Quarries both in the country, and in the Islands; and the places whence great columns were hewn out, and some marked out, and left in the Quarry; and in all parts the marks of the chisels; some of the Quarries being cut out in two or three steps are an instance of the ignorance of the people of this country, who say they are ruined Pyramids, and Paul Lucas, who delivers most things by hearsay, places Pyramids here in his map, which is as false as can be, in relation for upper Egypt.

Opposite to Assouan is the Isle on which stood the city of Elephantine, where I saw the remains of a Temple, and of a beautifull gate of red Granite. But it would be no recommendation to the art bestowed on the releifs of this Gate, in this place, to say they exceed the value of the materials, where there are no other stones, I looked in vain for the well cutt in the rock, by which they observed how the Nile rose. South of this city are several small Islands of high rocks of red Granite, of the best sort, which I saw had been worked for Quarries, and it is probable that out of one of these was hewn that wonderfull house of one stone, which Herodotus says was three years in carrying from Elephantine to Sais in Delta. This house according to his account was twenty one cubits long fourteen broad & eight high. I drank of the water of the Nile on the West side, that I might be sure according to the answer of the Oracle, and by account of the antient Geographers that I was in the old Ethopia. I went a mile along a valley to the West, opposite to the South part of Elephantine, which probably was the bounds between Egypt and Ethiopia; and if it was so I view'd the ruins of a monastery in Ethiopia on the South side of that valley.

Your Lordship may be sure I saw the Cataracts, which are but three miles higher. They are not to be described, and my curiosity was not satisfied till I went three or four miles above them farther South to the small Isle in which stood the city of Philae, inhabited both by Ethiopians and Egyptians. The whole Isle seems to have been walled round rising high above the water and where the walls make angles they appear something like a modern fortification. The

Temple here is exceedingly grand and allmost entire. In it was worshipped the Ethiopian Hawk, and I have reason to think that I saw the very room where he was kept, and a small enclosure in it which was the proper apartment of this Deified animal. Besides several rooms at the end of the Temple there are three Areas, one of which has a Portico on each side of ten pillars, another of near forty, and ruins of two Granite Obelisks; this Area goes to the end of the Island, where it is at least thirty feet above the water, and faces a long reach of the Nile, which is terminated by the hills of Granite, and alltogether affords a very beautifull, and Romantick view. Strabo who went to this Isle seems to have been a very superficial observer of natural curiositys. He speaks of large black stones, laid one upon another, in the way, which are all red Granite, and have indeed a brown cast, many of them being exposed to the weather: but it is very surprising that he should make no mention of these famous quarries out of which the Obelisks were cut, which were esteemed such great masterpieces of art in their kind, that they were afterwards carried to Rome with the utmost expence. He speaks also of going thro' a plain country to it, though a rougher face of nature can hardly appear, as on each side there is nothing to be seen but hills and rocks. It is true that it is a fine plain road, the crumbling of the Granite mixed with the sand, makes it like a gravel walk, and the authors words may perhaps be favourably interpreted as spoken only of the road, it we suppose πεδιον is an error for πεδον. In January I saw the corn here in ear, and some fruits ripe, and found the sun as hot in the afternoon as it is with us in May, for I was here in the degree of Latitute 23.24 and being South of the Tropick was in the Torrid Zone.

I returned to Cairo having performed this voyage in the three months time with the greatest good fortune in every respect.

Your Lordship may wonder that I say nothing of inscriptions, but I am very well assured that there are not ten in all Egypt, none older than the Ptolemy a few of the Romans, and all in Greek. There are indeed some which are certainly Hieroglyphical inscriptions, which before the government of the Greeks were allways used about their Temples; but I beleive it would puzzle the best Decypherer in the secretarys office to interpret them and would not be so profitable, as finding out a plot.

I shall not repeat to your Lordship, what has been observed by every traveller. All the great works, especially about Cairo, are attributed by the vulgar to the Patriarch Joseph, whereas Memphis was rather lower on the other side of the river, and it is supposed that the Pharoahs at that time resided at Tanis. But they say that the rise of this tradition is owing to a Sultan Joseph being promoter of most of these great works, who I beleive was Sultan Selim, who had also the name of Job or Joseph, and was the founder of that family of Sultans, which is commonly distinguished by the name of Jobile from him.

Coming to this place in order to embark for Joppa to be at Jerusalem this Easter, I saw on the East about twenty miles below the division of the Nile at a village called Benalhassar, great marks of an old city, laid out in a very different manner from any other Egyptian citys I had seen, which I beleive was the city of Onia; which Onias a Priest of the Jews in Egypt obtained of a King of Egypt, and built a Temple in it, and this city Ptolemy makes the capital of the Nomos of Heliopolis, which it probably was after the decay of Heliopolis.

I could not find out the city of Bubastis which was lower down and is judged to be the Pi-beses in Ezekiel mentioned by God Almighty in denouncing his judgments against Egypt.

I passed near Manoura near which the Christian forces of the Holy war were twice defeated. In the first action the Earl of Artois was drown'd, and the brave Earl of Salisbury died fighting on his knees, and the whole army that was in this action was cut to pieces. The French engaging afterwards with Good King Lewis [Louis] at their head had not better fortune: the King being taken prisoner, and Damiata among other things was given for his ransom. Mansoura seems to agree with the situation of Pacusa from which place the canal was cut from the Nile to Arsinoe, on the Red sea, a voyage of four days sailing, and if so the water mentioned in the actions between the Christians and Turks is probably the remains of that canal which was begun by Necus King of Egypt, carried on by Darius, and perfected by one of the Ptolemys, being called the Canal of the Kings.

Though some are of a contrary opinion, yet I cannot but think that Damiata is the old Pelusium, which is thought to be the Sin[17] mentioned in Holy writ, Magdolum was twelve miles from it, according to the Itinerary mentioned also in the prophets, and famous for the victory gained there over the Syrians by Necus King of Egypt.

I have a piece of news to inform your Lordship of, which is somewhat curious. A few days before I came to Cairo, a certain Count, and his Lady and a missionary from Rome set out for upper Egypt. The Count I am here informed is certainly King Theodore; he had (they say) engaged the Pope to send him with the Mission into Ethiopia; they went five days above the Cataracts where they narrowly escaped, and were obliged to return praeciptately, and so made hast[e] to Cairo. I parted from one place the moment they came to it, not knowing of it. They proposed to have gone to Ethiopia by a branch of the Nile, or rather by a river falling into it that rises to the South East called Asara, and if this did not succeed to go into Ethiopia by way of the Red sea, but it is said now that they are returning into Europe. When I left Leghorn it was said he was

17 Ezekiel xxx:15.

incognito aboard a Dutch ship, having been at Rome, and no one knew where he was, tho' he had embarked at Leghorn.

I am exceeding glad to hear of your Lordships better health, and when your Lordship has got rid of this gravel [gallstones] I hope (by abstaining from whatever might cause it) your Lordship will enjoy a second spring of health, and that we shall live to have a happy meeting.[18]

I beg leave to put in my kind remembrances to my fellow traveller [Jeremiah Milles] and am

My Lord

---

18 Bishop Milles died on 13 May 1740, a year before Pococke's return to the West.

CHAPTER 6

Letters from the Holy Land, Lebanon and Syria

This chapter reproduces the 11 letters from Pococke's tour of the Holy Land, Lebanon and Syria, as follows: one each from Jerusalem, Bethelem, Jaffa and Acre, one each from Sidon, Tripoli and Damascus, three from Aleppo and one from Latakia. The first letter, from Jerusalem gives no account of the voyage from Damietta to the port of Jaffa, except to confirm that it was pleasant and that "no one was sick".

In Letter 17, from Damascus, Pococke describes his visit to the famous Roman temple complex at Baalbek. As already noted, it is not known where Pococke obtained the original drawings for the 14 plates printed in *Description 2*, since no indication of this is given either in the text or in the correspondence itself. As also mentioned above, these beautiful engravings (two of which are reproduced here, as Plates 14 and 15) have been consistently ignored or dismissed since their publication.[1] The letters from Aleppo, in particular, confirm the view of other contemporary travellers to the Levant that the members of the English community there were eager to befriend any compatriots visiting the area.[2] For further details of Pococke's experiences in these parts, the reader is directed to *Description 2*, Books I and II entitled: "Of Palestine, or the Holy Land" and "Of Syria and Mesopotamia", respectively.

1 Friedrich Ragette for example briefly refers to the 11 engravings and scathingly declares: "Unbelievably inaccurate and unreliable, they do not merit much attention. Anyway, twelve years after their appearance they were overshadowed by Wood's great book, *The Ruins of Balbec"*. See F. Ragette, *Baalbek*, Chatto & Windus, London, 1980, p. 87. Wood's book of this title was published in London, 1757.

2 For a comprehensive account of the social lives and practices of the English and other foreign merchants in the Eastern Mediterranean, see Wood, A *History of the Levant Company*, pp. 229ff.

 | DOI:10.1163/9789004440050_007

JERUSALEM
26 March/6 April, Easter Sunday, 1738
Pococke to his mother
Letter 11[3]

Honoured Madam,

My last to you was of March 9th-20th from Damiata in Egypt, having writ to you a week before from Cairo, two months before that from Furchout in upper Egypt, which went with that from Cairo & about a fortnight before that, I writ to you from Akmim, both which places are in upper Egypt; having happily in three months made a very pleasant voyage to the Cataracts, & somewhat above them. The evening of March 8th-19th I went down to the Mouth of the river in the boat of Signr Hana Fakor [Hana Fakhr],[4] whom I before visited, Captain Feilding going along with me, we slept in the boat, the mouth of the river being 3 leagues below Damiata; as it was day we went to the opening to the sea, where we landed on the sandy point & waited all day to have tidings of the ship we were to go in, which was obliged by wind to go out to sea, there being no Port here. Dined with a Cophti Bp [Bishop] on the strand on bread & those seeds & Coffee, after, happening to pass by as they were sitting down on the strand, – having no prospect of going this day, we went up a litle higher in the river to be secure from the wind, & laid our carpet in a shady place & took some refreshment. Captn. Fielding returned to Damiata, & I remained with my other friend in the boat being fine weather.

10th-21st. Captn Fielding not being quite ready, I embarked with the Pilgrims, a large French Ship, Captain Tournier of Toulon, we sailed in the afternoon, & between calms & sometimes gentle contrary winds, we did not land at Joppa in Asia & in the Holy Land till –

14th-25th. but a most pleasant voyage & no one sick. Joppa is on a litle hill on the sea; & being landed I went immediately to the house which was the house of one Simon a Tanner by the seaside, which is the Hospititum of the Latin Fathers of Terra Santa; – that moment four fathers were gone in a small caravan to Ramah [Ramla], ten miles in the way to Jerusalem. I had a horse got ready to meet me out of the town & walking out & mounting I soon over took 'em. It is here about 18 miles from the sea, a very pleasant country & rich; what

3 BL, Add MS 22998.

4 Members of the Fakhr family, originally from Lebanon, settled in Aleppo, as well as Damietta, in Egypt. In all these places they became involved in international trade in one capacity or another.

is not sow'd appeared like meadow ground, anemonies & many uncommon flowers among the grass, & a great number of Tulips of the red & beautiful colours; – & anyone that considers the beauty of a bed of Tulips, may be enclined to think, that this is the Lilly, of which our Savour said, "that Solomon in all his glory was arrayed like one of these". Ramah is pleasantly situated, called Rameli by the Arabs, which is an easy corruption from Ramethi, as that is from Arimathea, for they say with great reason, this was that city of which Joseph was, who took so much pious care, to bury our Saviour; – & it is said Nicodemus had a house here, on the spot where the Old chapel of the Latin Hospition stands: here & in some other places, especially at Joppa are manufactories of excellent soap, which instead of Tallow is made of the oyl of olives in which this country much abounds, the olive trees being large. North of Ramah three miles I saw Lydda [Lod], where St Peter raised Eneas; – it is said our Patron St George lived there who killed the Dragon near Bareut, not far from Sidon. There is a church of the Greeks in Lydda, to him built or rather repaired, as it is said, by a King of England, I suppose King Richard. At Ramah I went to the Hospitium of the Terra Santa, – received with much civility & dined in their Refectory one day, with much neatness & after in my chamber, for I spent there 3 or 4 days without going out.

18th-29th. In the even set out for Jerusalem, about twenty miles, 8 of it fields to the hills but an uneven ground, sending forth much herbage of thistles &c: like garden weeds & rank. Nothing offers to viewesert but the castle of the good thief, the churches of the maccabres & of Jeremiah on the mountains – & near at the end of them, the valley of Elah, where David killed Goliah [Goliath] which is nowd called the valley of Terebinthus, by the Roman catholicks (as I suppose) from those trees growing there, for in the Vulgar Bible it is so called, & I supposed Elah is the name of that tree in Hebrew. I came to the gate of Jerusalem two hours before day.

19th-30th. Signs of a country well improved & very pleasant now for two miles before one comes to it. Sent to the gate for the Frank Droggerman, who came & paid what was proper; – went to the convent of franks, where I was very civilly received; went directly to the most holy sepulchre,[5] to see the fine function of Palm Sunday performed by the Guardian of this convent, who is called Guardian of Mount Sion & Custos of the Holy Land, is as a Prelate in the German convents, & has the title of Reverendissimo, & is always so called, but for shortness I shall write here Rev.d^{mo} a fine old Gentleman, & has a venerable look in the garments of holiness. It is not to be imagined by anyone that does not feel it, how the mind must be affected at entering into this most Holy place,

5 Pococke's drawings of this building are reproduced in the present volume as Plate 12.

within the walls of which in a manner, the whole scene of our Redemption was translated & effected.[6] The Christian with veneration beholds the stone on which our Saviour was embalmed & more surely the sepulchre, in which he was laid; but I will not speak of these things, till I have visited them all, in a more solemn manner, for these two are the only I have had time to attend to in the least. The present subject is Palm Sunday, our Saviours entering with Hosannahs in triumph into the city & Temple. The Prelate Guardian clothed Pontifically with mitre &c: gets with much solemnity into the Holy Sepulchre & there blesses the Palm branches laid on it, & after distributes them to all that approach him on the knee, first kissing his hand, when this is done a procession is made several times round the sepulchre, by the Prelate & Priests with the Palm branches in their hands, singing Anthems relating to the triumphal subject of the day. I dined in the Refectory, the Prelate at a table by himself at upper end, – I sat the second at the left hand table; not a word spoke to be heard, a more venerable Assembly eating I never saw, about sixty with their long venerable beards, & most of them Grey, portions of Scripture are read, & a French Jesuit said to me, we are among Angels, & indeed we are to expect something more here than ordinary among the keepers of the Holy Sepulchre; – after dinner was invited to the Prelates, sweetmeats, Coffee, a dram, the top [head] of the convent there. I delivered two letters, he received me with great Affability, & promised me all that lay in his power. He is of Bologna. Sup'd in like manner, a Lay man is assigned to take care of me, there being two for that office, of frank Pilgrims only a Jesuit from Aleppo & your Son. After supper went to my room, at dinner, Brocali [broccoli], rice milk, pickled fish, Tunfish, figs, raisins: at supper sallad, Brocali, pickled fish, fryed fish, the same desert; always a plate of each brought to everyone; – rained very much today & hailed.

20th-31st. Showery all day – drank Coffee, as every morn in my room – writ; the Procurator general of the Terra Santa came to see me, my attendant gave me notice, that the ceremony of washing my feet, was to be perform'd by the Reverendissimo[7] Guardiano this afternoon, a servant came to my room & washed my feet first with warm water; at five I went into the church with only my white skul-cap on, I put off my shoes &c: & sat dwon in a great chair, at the lower end of the church opposite to the high Altar the Rev.d^{mo} took a towel & girded himself & began to wash my feet & to wipe them with the towel wherewith he was girded, in warm water with dried rose leaves in it. He had about his neck a red cordon, & kneeling down, I put my feet into the bason & with both

6 In a letter to his uncle, however, he admits that he was disappointed (see Account 7).

7 Having said, in the previous page, that he would abbreviate this word henceforth, Pococke then gave the full term here.

his hands he washed first the left & wiped it clean & kissed my foot & then the right foot, wiping & kissing it & placing it on his left knee & covering the toes with a napkin, & a father covering my garments above the Instep with a napkin; – all the convent came & on their knees, kissed first the hand of the Rev.dmo that covered my foot, & then kissed my instep, all the while Psalms & Anthems were chanted. This done the Rev.dmo put a lighted candle into my hand & rose & put on my shoes & went with them in procession to the high altar, before which I kneeled, whilst they performed several devotions, with organ & singing, having a good Quire of boys; – I went likewise in procession to the two other altars & at last to the high altar again, where being incensed the ceremony ended, & returning to the lower end of the church & my candle being put out, I had it brought to me as a relique to be kept. This is a ceremony all pilgrims go thro', & as far as according to our tenets their devotions are agreeable to Scripture, I believe they may be innocently joined in, 'tho it may be looked upon by those who have no devotion as a complance of civility, the prayers indeed not being understood: for my part I meditated on the passage of Scripture which gives an Account of this instance of humility in our Saviour, & hope on this account I shall not have less part in the merest of Christ, & that I shall never be above doing the meanest [lowliest] acts of charity. At supper had only sallad, capers, olives, Hazle nuts without shells, & to me, first, this time only, & then to the Rev.dmo was brought a red Herring. After supper the French Jesuit who is my neighbour call'd on me, & I am like to have his company in all my voyages, to Aleppo, whence he came & to which place he returns.

March 21st-April 1st. Breakfasted on chocolate, the Superior having sent to me to know if I chose chocolate in the morn; they treat me here with great civility; – I am known to be a clergyman; – at Damiata the Greeks asked Captn. Fielding if he had seen the English Papas (Father) & aboard the ship the Cophtics gave me the title of their own Priests Abouneh [A. *abuna*, 'our father'] i.e. Father. We had on board a Cophti Bishop who was particularly civil to me; he came in the room of the vicarius natus of the Patriarch of Alexandria, who is Bishop of the north part of Egypt, as I take it a sort of sufragan to the Patriarch & this vicarious has the title of Bishop of Jerusalem, & comes always or his Deputy, to perform the functions of the season. Father Bonifacio I travelled with to Rama came to see me; – dined & sup'd, as usual; talked with some fathers after supper & the Jesuit sat with me a while. I heard the Miserere sung in the morning in the church.

22d-2d. I went along the via dolorosa, & saw every place, where every single thing was done relating to our Saviours passion, mentioned in the Gospel, from Pilates house to Mount Calvary, when Christ was led to be crucified; – went by the gate that led to Solomons temple & could see the area, which is very large, went out of St Stephens gate & saw the place where he was stoned & they showed an impression on the rock, which they say is of his body: – going down that hill of Moriah, came into the valley of Jehosaphat, which was very narrow, passed over the brook kedron, which has never water in it now, the foot of the hill on each side descends to it, & the bed is deep but narrow; – went down to the Sepulchre of the Virgin Mary by about 50 steps, a broad, fine descent, it is a tolerable church under ground, but no light on the side of the stairs; down are the Sepulchres of Anna Joachim & Joseph the Virgin Mary's husband & it is said Simeon lay here. Close to this place is the large cave, where they say our Saviour sweat as it were great drops of blood: – going south a litle way came to the garden of Gethsemane at the foot of the Mount of Olives; – there are 7 olive trees, said to be there in our Saviours time; – passed by the place where the Blessed virgin dropt her girdle & St Thomas took it up, where the disciples slept & where Christ was betrayed: – was shown the print as of steps in a stone, said, to be made by our Saviour when he was thrust along, after they had taken him: – I then saw what is call'd Absaloms pillar & the tomb of Zacharias, two fine things cut out of the rock, like two small temples, & to which originally there seems to have been no entrance;[8] near the former is the Tomb of Jehosaphat: – went on & passed by the village of Siloe [Siloah/Silwan], & before we had passed the fountain of the Virgin Mary. We then came to the Valley of Gehennon [Gehinnom] & the Mount of Offence [or Offense], which joyns to the Mount of Olives. Saw the place where Isaiah was sawn asunder; – & so coming to the end of this valley, we saw the fine deep well of Nehemiah, where they hid the fire of the temple, when they went into captivity, & not finding it, but water, they poured water on the sacrifice, which set it on fire. We then turned to the right into the Valley of Rephaim, having Mount Sion on the right, & the Jews burial places & Aceldama [Field of Blood] on the left; – saw that deep large Apartment into which they threw the bodies, & I saw the bones from the holes at top; before we came to that, saw the grott where the Apostles hid themselves, when Christ was taken; – passing along the valley & by the Aqueduct that conveys the water hither from Solomons Fons signatus near Bethlehem,[9] I entered by the gate of Bethlehem on Mount Sion & the tower which is said to be on the place of David: – & so came home & paid a visit to the Rev.dmo who is a very polite man & exceedingly civil: here is Father James

8 Both are depicted and described in *Description* 2, Part I, Plates VI & VII and pp. 22-23. Pococke also describes these monuments to his uncle and notes that he measured every part of the Sepulchre of Zachariah and took a rough sketch of it (see Account 8). However, he makes no mention of sketching of Absalom's Pillar.

9 The Pools and Sealed Fountain of Solomon are further described in *Description* 2, Part I, p. 44 and depicted in Plate VIII, reproduced in the present volume as Plate 13.

Quirk of the County of Galway, whom I met not till my walk this day, having been to the Holy Sepulchre; he is an ingenious man, & much more polite than the Irish priests commonly are;[10] – he took Coffee with me after dinner, & we went into the Holy Sepulchre not to come out in 4 daies, till Easter Sunday; for the doors are lock'd till Good Friday & then locked again till Sunday morning: The fathers towards twenty of them reside here in a straight convent built to the church & now more are come & the Rev.d^{mo}. I saw a ceremony performed & visited the Holy places, too long to give you a particular Account of them & you have it in books. Father Quirk sat a while with me after supper.

23d-3d. Spent the day in writing & in considering the Holy places. After dinner saw the Rev.d^{mo} wash the feet of twelve priests & Friars before the door of the Holy Sepulchre.

24th-4th. The Good Friday of the Latins. The Fathers rose early, & went into a large Apartment in the dark & said the Miserere & disciplined themselves with their cords, to which no secular is admitted. I saw some ceremonies in the Chapel, where Christ was nailed to the Cross. The fathers dined not in the Refectory, but in an apartment kneeling in two rows, & eating nothing but bread sallad & onions, & drinking nothing but water: – I would have made one of them, & found after I might, but was set to dine on fish in the Refectory. After dinner the doors of the church were opened for 2 hours & all that paid a few pence admitted; – & many Greeks, Cophtis, Armenians & Jacobites or Syrians & Latins both men & women & children, staid within, the doors not being to be opened till Easter Sunday. After supper began a procession of the Latins; – first one preached in Italian, in the Chapel of the Virgin Mary of the apparition in the dark, the cross being brought in a while to enliven the subject; then the Fathers & Pilgrims went with the Rev.d^{mo} two & two all with wax candles in their hands to the chapel where they divided, & cast lots for our Saviours garments, where one preached very movingly in French; then to the chapel where is the column on which our Saviour sat, when he was crowned with thorns & ill treated by the soldiers; – & one preached in Spanish. Thence went to Mount Calvary, to the chapel of the nailing to the cross, where one preached in French, & then to the chapel where Christ was crucified, where setting the crucifix in the very hole, where the cross of Christ stood, a Father held it there, & another Father made a very moving sermon in Italian & shed tears himself, & made not only some of the women to weep, but even the men; – when he had finished a priest on each side of the cross, first took off the crown of thorns & showed it to the people, who beat their breast & groaned, then taking the nails out of the hands, turned down the Arms, being made with joints, & they took also the

10 Pococke was to meet Father Quirk (or Quirke) again in Cyprus. See Letter 22.

nail out of the feet, showing each nail to the people, who expressed the same marks of sorrow; & the supposed body of Christ being put in a sheet, was carried by four priests to the stone of Unction [or Anointing], where the Rev.d^{mo} embalmed it on his knees, & a priest preached in Arabick; – then the body being carried into the Holy Sepulchre & laid there, a Priest concluded by preaching in Spanish between ten & eleven a clock.

25th-5th. In the morn I saw the Function of blessing the Fire & incense at the door of the church, & at the outside of the chapel of the Holy Sepulchre by pouring in oyl chrism [anointing oil] &c: & then the Rev.d^{mo} who wears the mitre & crozier, celebrated mass Pontifically, & at Gloria in excelsis, they threw down a covering of the front of the Sepulchre, & shewed a fine piece of French Tapestry, representing the resurrection, where the day before the St Sudaire had hung. After dinner the pilgrims went with wax candles in their hands in procession to all the Holy places in this church chanting Hymns, & the Priest at every one offering up a very proper prayer, & a ceremony performed of incensing the plates & the people. And tho' the church of England has a member here that payes his devotion to these Holy places, she has also a weak champion, who as occasion offers, gives a reason of her Faith & practice; my countrymen being very fond of dispute, & your son, as he hopes well grounded & firmly fixed in his principles. Saw the Procession of the Greeks & Armenians, for their Palm Sunday tomorrow. The Bishop goes with a cross in his hand, & two persons go backward before him, incensing him all the way; – but the Armenians were most grand. You see I am so happy as to come to Jerusalem to keep the feast of the Passover, the solemnity of which you shall have in my next. Pray my very kind love to my sister; I had not opportunity to send it till my return from Jordan & Jericho to this place on the

29th-9th. You must not expect to hear from me very soon,

I am Dear Madam your most obedient & dutiful Son, Richard Pococke

I go in four days to Bethlehem & return to Joppa & thence to Acre, Said [Sidon], Bareut [Beirut], Damascus, Aleppo & Cyprus, – & you see I am out of the way of Corresponding, – you must not expect to hear so frequently from me.

BETHLEHEM
8/19 April, Easter Sunday, 1738
Pococke to his mother
Letter 12[11]

Honoured Madam,

My last letter to you was dated from Jerusalem March 26th April 6th being Easter Sunday of the new style among the Latins. I was then in the church of the Holy Sepulchre; – last night when the doors of that church were shut, abundance of people staid in all night, & this morn the gates were opened & remained so till noon, & the church is so crowded, there is no going in to the Sepulchre. The Greeks & Armenians make a great noise, in imitation of bells, with candlesticks & a bar or plate of brass against a board. This day as soon as it was light, the Latins celebrated a Pontifical Mass, & all the convent, near 60 in their fine vestments went in procession three times round the Sepulchre, & to the Stone of Unction, the Rev.dmo being vested Pontifically, & three silver crosses being carried before the procession: then the Greeks began their procession of Palm Sunday round the Sepulchre, carrying banners before, & great boughs of olive like trees – & with much noise of the Hosannah, all having Palm branches in their hands, going round three times; – the Bishop went into the Holy Sepulchre & coming out, gave his blessing to the branches, & then all the people caught the boughs & broke off pieces. After this began the same procession of the Armenians; – followed by the Cophtis & their Bishop & the Syrians or Jacobites & their Bishop; – the two later [latter] a small number; – the Syrian Bishop no Mitre, but a sort of blew satin hood embroidered with gold & lined with red. The Cophti Bishop dressed in white, & a sort of coule [cowl] ending in a point, & somewhat resembling a Mitre. But the most grand Procession was that of the Armenians. First went the cross, & then about 36 deacons & subdeacons in silk garments & with stiff silk caps embroidered, & a small silver cross gilt in the middle at top, thus [small sketch]; after them followed about fourteen priests in their silk vestments, with a sort of covering on the head, like a broad sugar loaf or Grenadiers cap, but the point of it not upright, but declining behind; then came six very richly clad, with same cap as first, but very richly embroidered, & had small crosses in their hand, these men some of the head people of the church; behind them came six Bishops, vested in same manner only with stoles hanging down before; their caps were of silver plate, gilt: – then came their Bishop of Jerusalem, who has the title of Patriarch,

11 BL, Add MS 22998.

with very rich garments, and a very fine large mitre, a cross in his right hand & a palm branch & candle in left: – some development: – performed on the side of the sepuchre, & there these blessed the olive branches & Palms, & after having entered the holy sepulchre & the other Bishops the procession ended; & I returned to the convent of St Salvator in the city, where I first was, after having spent four days in the church of the Holy Sepulchre, the spot where our mighty redemption was [...]; dined & took Coffee, my countryman being with me. The Armenia[n] Bishop & the other six & the Bishop patriarch, have collars round about three quarters of a coule on their shoulders at a litle distance from their necks of silver gilt plate about four inches deep, & a man went in the procession & threw rose leaves among the people & others rose-water, the lat[t]er the Latins did; the first thing today, Eggs, custard in fine meat soup, boyled fowl, ragou white of veal, kid & Cyprus cheese 3 inches thick 4 diameter, kept in oyl, but very good, most like Cheshire, & wine of Hebron; & what we have commonly being of this place; – the convent bruising the grapes & make it themselves, all white, a litle on the sweet, but strong & good; at even hard boyled eggs, sallad, boyled fowl & mutton & cheese.

27th-7th.[12] At three set out for Jordan, with the caravan, walked down to the valley of Jehosaphat, where we mounted on horseback, the Rev.d^{mo} & about 30 of the convent, the Armenians followed us, we being stopped by a number of soldiers; the camels were gone on before, on which men, women & children rid in frames, one on one side & one on the other, & the camels going two & two was a comical sight. The Greeks then came followed by their Governor & his train; our train went on before, all we went on the south side of the of Mount of Olives, & came to Bethany, of which hereafter, in a mile more came to the fountain of the Apostles, so called, because they think Christ & his Apostles used to drink of it; going to Jericho about 7 or 8 miles further, passed through the field of Adomim (blood) because many were formerly rob'd there going down from Jerusalem to Jericho, & here it is supposed the man fell among thieves, who was assisted by the poor Samaritan, if that was not a meer [mere] parable: – after travelling over rocky hills, we came to the sight of the plain of Jericho & of the dead sea, to the south of it, when we descended into the plain, we turned away to the mountain to the north & came to the highest part of them two miles west of Jericho, which is called the Quarentana [Mount Quarantania], where they say our Saviour was tempted by the Devil. I went up three quarters of the way, many Grotts, & at the higher part ¾ up a chapel cut out of the rock, where it is said the Devil said to Christ, command that these stones be

12 The usual daily format of the journal is glossed over here, with three days being incorporated into this single entry.

made bread; at top of all where they dined – none go now for fear of the Arabs, is a chapel where it is supposed the Devil showed Christ all the Kingdoms of the world & the glory of them. The chapel we were in where they say Mass, was built or rather Arch'd by St Helena; returned & went a mile East to the fountain of Elisha, supposed to be that, which he from being bigger made small, & it is a very soft water & not very cold, & is at the end of a fine wood; we then went a mile further through corn & trees here & there to Jericho, now only a few ruined houses & uninhabited & one square tower called the house of Zacheus; – & they pretended to show the tree on which he got up to see Jesus; about a mile East of this, we went to the camp of the caravan, where the Governor was encamped & where the convent had a very large green tent pitched; – we spent the whole day here, & having eat at Quarantana & at Elishas fountain, we did not eat again till towards evening, sallad, fowls & cold meat, walked out a little & went early to sleep, a guard with a fire being set at certain distances, all round the camp, who call to one another at fix'd times. About two in the morn, we rose to go to the river Jordan about four miles East; the Greeks & camels set out first, then the Governor & his guard with kettle drums & haut boys & colours flying, & then the Franks with a particular guard, & the Armenians went in a body by themselves, being all lighted by a sort of Refircus wood put into a litle refin'd Iron frame, which is an exceeding good light being held in the air: – near the river we came to the ruined convent of John, thought by some to be the place where the voice came – This is my Beloved Son &c: & then descended to a lower ground & came to the river, deep, rapid & high banks, but some litle flats below the banks, which flats are often over flowed. The Fathers said six masses on the river side; & it was extraordinary to see the women strip to their smocks on the flats & to go to the brink, for the men to throw water on them, in memory of Christs Baptism. I washed nine times in the river, that is dipp'd myself over so often, – seven in memory of the Baptism of our Saviour, which they say was here abouts. The Greeks went to another part two or three miles south. The river may be about 100 feet over. We were there before the sun was up, & returned by 8 to our Camp; – the Governor was on a litle hillock a mile before we came to the camp, & we passed one by one by him & then the others, having all met here. I saw in this plain at a distance towards the South the large convent of St Jerom, where he lived many years, & I think translated the Bible, or at Bethlehem. This plain is in a manner bare, tho' it might be well cultivated, but it has no leaf of grass: – We began towards sunset to prepare for our departure & all the tents were taken down & about an hour after it was dark we set out, as we came out by another way, the first half thro' the mountains, we rested at the fountain of the Apostles, & about 3 arrived in Jerusalem; – slept till dinner time & then slept again & writ.

30th-10th. Went again along the dolorous way & saw Herods palace, & the place where Christ was arrayed in white &c – & then to Pilates palace, where the Governor resides, & saw the judgment hall & the place where Christ was crowned. From this Palace I had an exceeding good view of the octagon Mosque [the Dome of the Rock] which stands where the temple of Solomon did, in a very large Area, not to be enter'd by Christians. Saw the church of the nativity of the Blessed Virgin, – & saw the remains of a very fine Hospital built by Helena; – dined, – went to the top of the temple of the Holy Supulchre to see the Greek ceremony of washing the feet in the Area of the church; – the Bishop ascended the steps to the chapel of the Virgin & saying some prayers, disrobed, & twelve priests being placed down the steps, he went down & washed their feet, one pouring water on them, he wiped them, kissed the knee [of] the person at the same time kissing his ear, the last was Peter, who made the speech, but submitted to the washing; – the Bishop took a lettuce & dipping it in the water boyl'd on herbs sprinkled the people & then it was thrown on them. Went & saw the prison of St Peter, where was a church, the iron gate the Angel led him to; the church of St John & James of the Greeks. The church of the Syrians where St peter went when out of prison; the house of Anna the high Priest now an Armenian church, & an olive tree to which Christ was tyed; & a stone with a hole like a mouth, which they say, spake to confirm, if these held their peace. Went then to the fine church of the convent of St James, where he was beheaded belonging to the Armenians. Saw the ceremony of washing the feet there, which they do to all the pilgrims; – the Bishop kneeling, the pilgrim sitting, – & the Bishop wet each foot with a towel & puts a sweet pomatum on it, & then the person kisses the Bishops hand. The Procurator ordered us to his room, where we were treated with a dram, sweetmeats, – & had the honour of Rose-water to cool our beard: passed by the house of St Thomas, these places being near Mount Sion, & passed by the tower of David.

31st-11th. Writ. Father Quirk came & had me to a German room to see some Jerusalem works. Red herring & Brocali a dish here. Went for an hour to the Holy Sepulchre; paid a visit to the Rev.d^{mo} after supper the Jesuit sat with me.

April 1st-12th. Dined; – The under Procurator came to me about going to the Dead Sea. Went to the church of the Holy Sepulchre; where I saw what I should never see among Christians: – on Friday towards midnight the Greeks carry a coffin covered with a Pall three times round the Sepulchre in Procession, & as soon as midnight is passed, the mob begin to carry men wrapt in blankets round the Holy Sepulchre & then laying them down stretched before it, people going after & hollowing this & some other tricks continued all Saturday, till about 2 in the afternoon, boys imitating the men & some men standing on others shoulders were carried round, the country people run round the Sepulchre hollowing, the boys jumping on one anothers backs putting off each others caps, that it was more like a Bacchanal than something of true devotion, & they could not do more to Burlesque Religion. But what followed you have heard of, which is the ceremony of the Holy fire: The Greeks go three times in procession round the Sepulchre, & then the Bishop goes into the Sepulchre, & after him an Armenian Bishop, where being shut in, they expect as the Greeks say, the fire from Heaven to light the lamp, all the lamps being put out, here it is said they strike fire & lighting a bundle of wax candles, the Bishop comes out & holds them out of the door, the people striving to light their candles, & when one or two are lighted, it is very extraordinary to see all the rest processing about them to light theirs, with such mobbing & noise that the like was never seen, in no long time all their candles are lighted, & the church is in a smoke with them. They leave them to burn some time as they please, they say this fire will not burn the flesh or the beard &c:. After this was over the Armenian Cophti & Syrian Procession began; all these believe in the Holy fire, but the Latins laugh at it. Visited the Procurator General of the convent of Terra Santa & the Rev.d^{mo}.

2nd-13th. Was with the Rev.d^{mo} twice & set out at 3 for St Saba, going along ye brook Kedron towards the Dead sea: – St Saba is a greek convent, I veiwed it & sup'd with the Superior & lay there; it is 3 hours from Jerusalem.

3d-14th. Set out at 5 yesterday accompanied with 3 Arabs, & went 3 hours to the Dead sea, saw Antilopes in the way: – the water of that sea so salt, that it contracts the mouth like a strong alumn water, but it is limpid, tho' I saw an oyly matter in it. It is said very near to the place where the Jordan enters it is a salt as even here & it is supposed by some that the stream passes without mixing with it. I went into the sea, it is not cold, nor can it hardly be said to be warm – it bears the body up every way, when one lyes on the water without motion, & as long as you please, & so high that one that swims strikes his legs out of the water, & the water is so heavy that it is difficult to recover the feet.[13] The apples of Sodom are no more than this, all fruit that grows near as pomegranates, &c are dust & smoke within; but I saw trees where it chances to gain on ground (where they had grown) & it kills the trees. No fish in it, the air about it very malignant in the three hot months; but I could see no other steam or smoke more than on Jordan. Lots wife said to be this way, now is said to be beyond Hebron, but I can't find that ever anyone has seen it, since Josephus. The Bitumen is a pitch thrown on the shoar by great winds, which it is supposed by the motion of the sea break the rocks of Bitumen that may be caused

13 In *Description 1*, Part 1, p. 36, Pococke informs the reader that he spent a quarter of an hour in the water and notes "they have a notion, that if anyone attempted to swim over, it would burn up the body, and they say the same of boats..."

by subterranean fires, & is used as pitch. We returned to St Saba & dined there, & then set out for Bethlehem, 3 hours from it going out of the way to see the grott on the mountains of Engaddi, where David cut off Sauls shirt. Arrived at the Latin convent of Bethlehem, a litle after night. Sup'd in the Procurators room, a Spaniard.

4th-15th. The church of Bethlehem is a fine church, & adjoining to it are the Latin, Greek & Armenian convents & what is called the School of St Jerom; under the east part is the place where Christ was born & the manger in which he was laid. I went out & visited the grott where the Blessed Virgin hid; before she fled into Egypt; saw the Greek & Armenian convent, who repair St Jeroms school; I turned into a church, went out after dinner, & saw cast the foundations of an house cut in a rock, said to be the house of Joseph; – went farther East in the village of the shepherds & half a mile further to the field where the Angels appeared to them: here was a church above ground; now ruined, but the church under it & under ground still remains: – hence we went a mile north, to the place where the famous St Paula founded a nunnery; there is now a sort of a Mosque kept locked, which we entered; returned & went to the Procession they make every day in the church of the Nativity under ground, to all the Holy places chanting certain Hymns at every place.

5th-16th. I went in the afternoon to see some of the works they make of crosses &c & then went to the Well of David, to which the three men went through the enemy to fetch water for him, & because they ventured their lives, he would not drink of it, but poured it on the ground, it is half a mile north of Bethlehem.

6th-17th. Went in the afternoon to see places in the way to Jerusalem, as Rachels sepulchre, the house of Jacob, the field where it is said, the Virgin asking what they were sowing? they said stones, when they were pease or ask'd them to give her of them, & so the pease all turned to stones, & this field is full of litle stones like pease: – then went to the convent of Elias belonging to the Greeks, a fine situation & a large plain church; I do not know the greek history of the place; but before the convent is a place, where all say Elias lay & left the impression of his body; – a litle further are ruins of a church, in the place whence the Angel took Habacuk with the pottage to Babylon to Daniel, & beyond that the well, where it is said the star appeared again to the wise men. Saw a sepulchral grott near; – & went on to the olive tree planted in room of the Terebinth (the ash of this country) under which the Virgin Mary rested when she was going between Jerusalem & Bethlehem with the babe Jesus.

7th-18th. Went with the Jesuit as usual & some fathers 4 miles to the Pools of Solomon, the sealed fountains & the enclosed gardens;[14] – in the way on one side of the hill over a narrow valley is a village called Solomons; at the end of the valley is an hill, & on one side of it is the fountain sealed; – 3 springs arched over, under ground, at the foot of the hill are 3 very large ponds, one higher than the other, & run from one into the other, something like the mill-ponds but larger, & as I said not on a level, so that they are in a manner on the side of the hill; – This still was doubless the place where Solomon had his house of pleasure, & the enclosed gardens were probably about these pools, & also to the north of the foot of the hill, where is a fine litle plain. From the fountain & pools the water is conveyed to Bethlehem & Jerusalem round the hills on arches in large round earthen pipes set in stone, & is without all dispute the work of Solomon, goes round 30 miles, tho' but 10 from Jerusalem. We returned.

8th-19th. Went out under the conduct of Arabs 6 miles to Therma, the country of Habacuk & of Amos rebuilt by Reholoam, – a ruined church & a marble Font. Went thence to see a large grott, 2 or 3 miles nearer the Dead sea, dined near on a Lamb boyled in water & sowre milk, a la mode, Arabe. We then went to the Mountain Franceti or of Behulia, not that mentioned in Scripture – remains of a curious round fortification at top of a high hill, the upper part of which is sloped like a nut, held by the Knights 40 years after the last taking of Jerusalem & being a common weal of men only, caused its end; returned. Tomorrow to the desert of St John. Pray my very kind love to my sister.

I am Dear Madam your most obedient & dutiful Son, Richard Pococke

JOPPA
2/13 May 1738
Pococke to his mother
Letter 13[15]

Honoured Madam,

My last to you was from Bethlehem, dated Apr. 8th-19th.

9th-20th. I set out for St Johns, half way passed by the fountain of St Philip, where he baptized the Eunuch; – ruins of a church, they say this is the valley of Eschol & the place near where the spies took the bunch of Grapes carried between two. I should have observed near Bethlehem we passed into a vale where

14 Pococke's drawings of these are reproduced in the present volume as Plate 13.

15 BL, Add MS 22998.

they say, Senacheribs army was destroyed by the Angel in one night. We went on to the Convent of St Johns belonging to the Latin fathers, about a dozen monks here, & among them the Irish father: It is a new church & convent very neat; in the church they have an altar where St John was born; – after dinner went 2 miles to the desert of St John, an old merry Spanish father with us; it is said, Elizabeth fled hither when the infants were murdered, with her son, – that she lived with him in a grot we saw & where they say he lived till he was thirty years of age; & then went towards Jordan; – that his mother dyed when he was three years old, & was buried in a ruinous church we saw: – near at the grott is a very fine spring coming out of the rock, & over it a ruined church; – in the way saw a stone, on which they say St John stood & preached; – after supper we went half a mile to the country house of Zacharias, where it is said the Blessed Virgin met Elizabeth & staid 3 months, they show the stair on which they met, & the chapel of the house, & another church & other buildings it having been a convent, near St Johns is a fountain, formerly called St Johns but now the Blessed Virgins, because it is supposed she drank of it the 3 months she staid there.

10th-21st. Went to the top of a high hill, where they say the Macchabees defended themselves, against their enemies, but there descended to the valley of Elak, where David killed Goliah, passed over the dry bed of the brook, whence it is supposed he took the five smooth stones, & on the Hills on each side the Israelites & Balak were encamped, & they pretend to show the place where Balaams Ass spoke but the authority of both I much doubt. We then went towards Jerusalem & 2 miles from it stopp'd & refreshed ourselves at the Greek convent of St Cross a fine church there, & they show the place, where they say the tree grew of which Christs Cross was made. Went by the tower of Simeon & to Jerusalem. Waited on the Rev.d^{mo}.

11th-22d· I went out to see the grot where it is said Jeremiah writ his Lamentations, a print of his body where he lay, then went to the place, where he was let down with cords into the dungeon, a sort of dry cistern with no water in it; – we then went to the Sepuchre's] of the kings; – I believe to the Sepulchre of Helena Queen of the Adiabene. they are about seven chambers cut out of the rock of marble, with places in them to deposite the bodies & sarcophagus's some of which last are still there & one door of stone,[16] but I could see that it might easily be put in, & not out of the very place, which seems impressive

16 The tombs of the Kings were excavated by the French archaeologist Louis Felicien de Saulcy in the 1860s, and the sarcophagus believed to be that of Queen Helen was first exhibited in the Louvre and is now in the Israel Museum, Jerusalem. This and the other tombs in this complex are described in *Description 2*, Part 1, pp. 20-21 and depicted in Plate v.

unless there were another entrance. We went by the upper pool called Gihon to receive rain water from the hills, a channel from it to the water pool in the city, & came home.

12th-23d. We went out to Mount Sion, to the part now without the walls, to the place where Christ instituted the last supper, the Holy Ghost descended on the Apostles & Thomas might put his hand in our Saviours side, where the Latin fathers had their convent about 150 years ago, but now is a Mosque, no Christian can enter; before it is the well, where it is said the Apostles parted to preach the Gospel; – hard by is the house of Caiphas, & they show the place where Christ was kept, a litle chapel, where Peter denyed him, & a piece of the pillar, on which the cock crow'd. The high altar is the stone which was at the door of the sepulchre, this is an Armenian convent: – kept on round the walls, passed by the places where the Jews would have overturned the body of the Virgin Mary carried to be buried, for they show the walls of her house, near the Caenaculm [Franciscan Convent]; – then I saw the place where St Peter wept; by Porta Sterquinilia, into which they led our Saviour when he was taken, went into a garden, which goes to the walls of the Mosque, of the Presentation of the Virgin Mary, supposed to be the very spot where the temple of Solomon stood; – being near the mosque now called the temple of Solomon: in the same Area, there are vast large stones towards the bottom of this & of the old nunnery building, which probably were formerly used in the temple of Solomon: – continued round the walls, saw the larger stones at the corners of the walls of the Temple court, some above 20 feet long; – passed by the golden gate, through which Christ entered in triumph, now a fine gate, but walled up because they have a Prophecy that Jerusalem will be taken by that Gate.[17] Saw a pillar & window in the wall, whence they say, a rope will be stretched to Mount Olivet, & that Mahomet will come through the window at the day of judgement, & walk over it & all those who can follow him, will go with him to Paradise; made the whole circuit of the walls. In the afternoon went with some fathers along the valley of Jehosaphat.

13th-24th. I went with others up Mount Olivet, – saw where Christ wept over the city foretold the judgement, taught the Lords prayer a second time, where was a church; below is the church of 12 Pilasters, where the Apostles composed the Creed, & lower the large but rough Sepulchral Grotts of the prophets as they are called: At the top of the Hill, entered the round chapel of the Ascension, in it a foot step, as said, of our Saviour, the other being carried into the Mosque of the Presentation, it is a Mosque & the habitiation of a Dervis [the *zawiya* of Rabiʿa al-ʿAdawiyya]; – but the Christians the night before the

17 It was sealed up in 1541 by Sultan Suleyman the Magnificent.

Accension day celebrate, encamping in the Area. Over this chapel a large church was first, the lower part of the walls of which are round the court, near is the chapel; over the grot of Pelagia, the famous Harlot of Antioch, who retired & died there, in [next word scored out] a mans dress; – we went to the place where Christ mounted the Ass, & the round hill of Bethphage, whence they fetched it; leaving Bethany on the left, we went to the signs of the house of Martha, & the stone on which Christ sat, when she came to him, – it is a remarkable point of the rock; thence to the ruins of the house of Mary Magdalene; – went to Bethany & saw the sepulchre out of which Lazarus was raised; – 2 grots, one within another, the outer one a chapel, where they celebrate on certain days. Saw the ruins of the castle of Lazarus & of the house of Simon the Leper, & the place where the fig-tree was cursed; – & going to Siloe over the Valley of Jehosaphat, the place where Judas hanged himself: – went over the village of Siloe inhabited by country people, abundance of sepulchral grots & others in it, being on the side of the hill; returned as usual to dinner. Cold weather this season here.

14th-25th. I set out under the conduct of Turks towards ye N.W. viewed the sepulchral grotts of the judges,[18] 2 miles from the city, cut into the rock several of them 2 rooms, one within another, & one as beautiful as those of the kings – went on about 4 miles & came to the place, where Christ met the two disciples going to Emmaus [Imwas], on the hill near are the remains of a church, built in memory of it, then went to the high hill of Ramah [Ramot], where Samuel dwelt & was buried, & where they have now a place called his Tomb, in a mosque we could not enter [Nabi Samwil]: – saw a fine fountain of excellent clear water, filling a Gold heron cut of the rock & is called the fountain of Samuel; – hence we saw under the hill a very fine plain, here they say Joshua was when the sun stood still, & being told that a village I saw on an hill in this plain was called Geb [Jib], confirmed the truth of it, which without doubt is Gibeon, on which the sun was commanded to stand still. Went on about 3 miles further to Emmaus, on these mountains saw the large church said to be the house of Cephas, where Christ was known in break[ing] of bread; returned by the tower of Simeon, & came in the even to Jerusalem.

15th-26th. I went up Mount Olivet to the place called Viri Galilai, reviewed the chapel of the Ascension, went all along the top of Mount Olivet & then to the Mount of offence, where Solomon sacrificed to false Gods, & descending went to the Hill of evil counsel, near the road to Bethlehem, where it is said the Elders resolv'd on the death of Jesus.

18 Known today as the Tombs of the Sanhedrin.

16th-27th. The Jesuit went away yesterday; – I went out to the pool of Siloe, the fountain of the Blessed Virgin, & viewed very particularly those great & curious Antiquities, the Sepu[l]chres of Zechariah, Josaphat & Abrahams pillar or, sepulchre, & took another View [drawing] of the temple of Solomon, from the top of the Governours house,[19] & went to the Jews quarter at the foot of Mount Sion, & saw their miserable synagogue, part of which is underground, the glory of the Jews in Jerusalem, – many of them come to die & be buried here.[20]

17th-28th – 18th & 29th. Stay'd within writing & reviewing my notes. Jonas Korte of Altema [Altona] near Hamburgh a Lutheran arrived, his brother a Bookseller in Hamburgh, I believe he has formerly been employed as a supercargo, a good old man talks English was a week in Cairo & came on purpose from his country to make the Pilgrimage, & now he goes with me to visit;[21] – I had lost two days & cant find but one, the last ought to be the 19th-30th.

April 20th-May 1st. I went out with Myn here [Mein Herr] Korte, in order to see the old walls of Jerusalem, – went to the north & viewed the sepulchres of the Kings, – Jeremiahs Grot & prison, – Herod & the Governors palace, which was Pilates, & the Hospital of St Helen.

21st-2d. Went in the afternoon to the church of the Holy Sepulchre, which was open'd being the Eve of the Invention of the Cross, saw the solemn entrance of the Rev.d^{mo} which is always performed with great solemnity; & then they went in Procession to the chapel where St Helena found the cross, which is hung & adorned, & they say several Anthms with organ & good voices, & after went in procession to the other places; – doors shut about 5, & many of the fathers staid in.

22d-3d. I went for the last time to the Valley of Jehosaphat, unless we are to appear thereabouts another day, – it was to review the sepulchre, Mass was celebrated Pontifically by the Rev.d^{mo} in the chapel of the Invention of the cross, & then a Procession began of all the convent in surplices, – the Rev.d^{mo} carrying a small gilt Cross, – went three times round the sepulchre; I visited the Holy places in the temple for the last time, & myn heer Korte & a Ragusan sailor pilgrim, remained to stay there three or four days. I went to the Cophti convent. dined, took leave of the Rev.d^{mo} & about one, set out for Ramah, with a head Arab & historant [i.e. Korte] all of us on horseback. We passed through Emmaus & about an hour after night arriv'd at the Latin Hospitium of Ramah.

---

19 Represented by a "Plan and View" in Plate III, *Description 2*, Part I, p. 14.

20 He describes the plight of the Jews in more detail to his uncle (see Account 7).

21 Pococke meets up with Korte in Aleppo (see Letter 20). Jonas Korte would later publish his *Reisen ach dem gelobten Lande, Aegypten, Syrien, und Mesopotamien*, Altona, 1740; with Rhree Supplements, Halle, 1746.

23d-4th. Went 3 miles north east to Lydda, where they say our patron St George suffered as a confessor & some say as a martyr. I went to the ruined church ye East part held by the Greeks, the west by the Turks, as a Mosque, who have a great regard for St George; – I could see it was a very fine church of hewn stonework within as well as without, & handsome pillars all of great Architecture before it was quite corrupted, that is, of Justinians time, who is said to have built it, & a king of England, I suppose Richard repaired it. It happened to be St George's day & the Greeks were celebrating the sacrament: – here you know St Peter healed Eneas [Aeneas] of the dead palsie. We joyned a small caravan of about 20 of Magribines people of Barbary & others, & arrived at Joppa about nine in the morn, a small poor Hospitium, but said to be the house of Simon the tanner; only a President – curate here of the Catholicks & a Lay Brother Procurator. I went out with the servant to see some places. I left my man here to avoid the extraordinary charges of carrying to Jerusalem, which is about ten pound tax a head, & I was absent from this six weeks wanting two days; – having seen all the Holy places with great pleasure & examined them well, & I believe no one pilgrim for above two years past has seen so much.

24th-5th. Went a mile East to a place, where there is some very litle ruin of a building & as they say was a church, where the house of Tabitha or Dorcas was, when St Peter raised from the dead: – The Greeks yesterday celebrated the sacrament there, being the feast of that saint. I went into a garden or two, the soil here very rich; – I saw in the fields towards Ramah, Rue, chardon [chard/beet] & in some other parts Aders-tongue; – neither here or in Egypt are any trees that grow wild in England; they have abundance of standard Apricots. Jerusalem is a cool place, but it grew a litle warm before I left it; diverted myself with seeing the Dolphins play very near the shore.

25th-6th. I sit talking with the President always after dinner & supper & go to the the Coffee house after dinner. I wait for my things from Jerusalem – in the even went to see the old burial place of Franks, where Lady Gerard a Roman English pilgrim was buried in her return from Jerusalem 6 years ago.[22]

26th-7th. Spent these two days in writing of some of the most remarkable or unobserved things to the Bishop of Waterford in four Letters – least I should forget it hereafter.[23] When anything arrives for me from these parts, I desire you not to let any of my sketches in a pencil out of your hands, on no account,

22 She is mentioned in Walpole's *Correspondence* (vol. 1, p. 412 of the 1837 edition), where it is mentioned that she died in Jaffa.

23 These are Accounts 7-10. However, it appears from Account 19, a letter from Aleppo, dated 31 July 1738 (a hurried note that does not apply the usual dating method), that he did not send these immediately, as he tells his uncle that his account so far only covers his journey to the Cataracts and to Jerusalem.

for fear any of the figures or lines should be defac'd; – & so likewise what is writ with a pencil, & if Mr Milles should be with you & desire to see anything, pray turn them turn them [sic] over for him – but it is better to keep them unseen.

27th-8th.[24] I send now two letters – one dated from Bethlehem the other from Joppa with a Postscript I design from Acre, from whence they go to Cyprus & thence to Marseilles as I suppose. I go from hence to Nazareth, Sidon, Bareut, Balbeck [Baalbek], Damascas [sic] & Aleppo, whence you will hear from me again, it may be in three months after this.

28th-9th. After dinner drank Coffee in the Presidents room.

29th-10th. Went out with the Lay-Brother – who went to see carpets.

30th-11th May 1st-12th. Staid in writing.

2d-13th. My Portmanteau came from Jerusalem, – I was employed about going to St John D'acre [Acre];[25] youll see my other letter is two half sheets which I posted for a reason. A Father of Augsburg came from Jerusalem & we are to embark together for St John D'acre in the evening. Pray my very kind love to my sister.

I am Dear Madam your most obedient & dutiful Son, Richard Pococke

P.S. St John D'Acre May 6th-17th.

May 3rd-14th & 4th-14th – was on the sea with a contrary wind & May 6th-17th arrived about 2 in the afternoon at this place, – & went to the Consul Mr Wakemans,[26] with a letter from the Consul of Cairo; – who received me with great civility, a litle man, extremely obliging, something like Mr Whitfield & his manner.

St John D'Acre is 80 miles from Joppa, & they often sail in twelve hours, 'tho we were 3 days & 3 nights but fine weather; – passed in sight of Caesarea of the centurion, & by mount Carmel of Elias; – I find it not so hot in these countries as I expected & everywhere very healthy. I shall go to Carmel & Nazareth & return hither, & to [...] &c:[27] The President of the Terra Santa Hospitium & another Father came to see me: – Just as I arrive I have an opportunity of sending these two letters, that is, this is my last from Bethlehem.

I am yours as pr conta.

R. Pococke

---

24 Either Pococke or his mother makes an error in the original, stating the date as "27th-9th".

25 Pococke gives a lengthy explanation of the placename to his uncle, in Account 12.

26 This is George Wakeman, who was English Consul at Cyprus from 1741-53 (though perhaps earlier) and may have been at Acre immediately before that. See Wilson, *List of British Consular Officials*, p. 34 and Ali Efdal Özkul, "The Consuls and their Activities in Cyprus under the Ottoman Administration (1571-1878)", *Turkish Studies: International Periodical for the Languages, Literature and History of Turkish or Turkic* 8/2 (Winter, 2013), p. 256.

27 The text has been rendered illegible by Pococke's corrections to his mother's transcript.

ST JOHN D'ACRE
16/27 May 1738
Pococke to his mother
Letter 14[28]

Honoured Madam,

My last letter to you from Joppa with a postscript from St John D'acre was of May 2nd-13th & May 6th-17th. I informed you of the kind reception I met with from Consul Wakeman, on the letter of Consul Barton of Cairo to him. I settled my affairs, walked in the garden with the Consul & Mr Bo[d]dington his partner, when Mr Usgate came with his two nephews,[29] sons of Consul French who died about 6 years ago, to take leave being going with their mother to be educated in Geneva, she is Mr Usgates sister & he has another sister unmarried here, lately come, he being married to one that was a Jew, that became first Roman [Catholic], 'tho he is a Protestant, he is a Factor here; an Italian came to the Consul we sup'd, – he is a most obliging man, & has already presented me with several curiosities. I forgot to tell you, that Venice Treacle taken at night is a fine composer, good almost for every thing, taken as much as a nutmeg, & washed down with a litle wine, & when to operate immediately in acute pains, dissolved in wine. Balsam of Mecca also 3 drops in water a soverign remedy, & good to be applied outwardly to wounds: – but none of these to be taken often; the mithridate is also good specially for the Ladies. I sent a litle Balsam in a small vial from Cairo, use it if occasion, it must be warmed to drop & I believe dropt into warm water to mix with it.

7th-18th. I Breakfasted with the Consul on chocolate. The French curate I saw in Jerusalem came to see me, & after Mr Usgate & his Brother:[30] the Consul did me the honour to go with me to the President of their convent which is a very good one, being one side of a large lane where all the Franks live, except the Consul; – we then returned the visit to Mr Usgate & his Brother.

8th-19th. I prepared for my journey to Nazareth, the President of propaganda came to see me; dined. The Jesuit I met at Jerusalem, the President & another father came to see me, & about one I set out for Nazareth to the East, went over this fine plain, then going before fine fields of corn & herbage & trees came to Abylene, when the sheke would have set before me to eat, &

28 BL, Add MS 22998.

29 Thomas Usgate, who is recorded as having been sworn in as English Consul at Acre in 1723. See Wilson, *List of British Consular Officials*, p. 40.

30 Possibly Richard Usgate of Scanderoon. See Wilson, *List of British Consular Officials*, pp. 41 and 66. Pococke also meets this brother, Richard, in Larnaca. See Letter 22.

asked to stay all night, but I only accepted of Coffee: – four hours from Acre met a French merchant; – in two miles came to the fine plain of Zaboulon [Zebulun]; & in an hour ascended up the hills – another hour to Sappory [Sephor/ Tzippori], a ruined handsome church three, only spot where was the House of Joachim & Anna, being night went to the Shekes; had milk & fryed Eggs & Coffee & lay there.

9th-20th. Went an hour & half to Nazareth, on a rising ground at the foot of a hill, in a narrow valley, the hills rocky, the earth & stones white. – Went to the new built convent of propaganda, delivered my letters to the Procurator & after to the Guardian, who conducted me about the convent with much civility, dined, rid two miles to the precipice of the hill, from which they would have thrown our Saviour; – & thence to Jaffa, & saw the place where was the house of Zebedee the father of James & John. The Sheke here enterained us with Eggs fryed, soure milk & Coffee: returned by the ruins of a church where it is said, the Virgin Mary went after Christ, when they were going to throw him down from the precipice, & in Nazareth saw the synagogue where he interpreted that text of Scriptures of himself, which so enflamed the Jews. I then went a litle way up the hill & saw a great stone on which they say, Christ eat with his disciples; sup'd in the Refectory, the President coming to me.

10th-21st. I set out two hours before day, to go two hours East to Mount Tabor, & an hour going up, a fine hill covered with trees, the top about 2 miles round, fortified by Josephus, in forty days taken by the Romans, great ruins. Three altars in a grott in remembrance of the zeal of St Peter, who would have erected three tabernacles to Christ, Moses & Elias. A Benedictine convent was near, & a Greek convent where an altar now stands. This hill is in the plain of Esdraelon; from it you see the mountains of Samaria, Mount Gilboa, Mount Hermon, & at the foot of the later [sic], Nain & Endor; – descended to a [...] church [next two words scored out], where Christ bid the disciples not speak of what they had seen till he was glorified: – then to the foot of an hill to a village called Deboura [Deburah], & near a ruined church, where he left the 9 disciples. The Sheke would have given us Coffee. Went by the plain of Esdraelon through the village of Zal, where I saw many Jewish sarcophagus's, & graves cut out of the rock;[31] returned near Jaffa I saw before, having dined on Mount Tabor & taken refreshment at a well in the way back. Walked up the Hill west of Nazareth. Sup'd – viewed the church of Nazareth well, & the place where the house of Loretto stood, the grot within it.

31 Pococke suggests that this could be the ancient Xalod mentioned by Josephus, *Description 1*, Part I, pp. 65-66.

11th-22d. I set out at sun rising for the fountains near of the Virgin Mary, & near them a Greek church under ground, under which the water passes from the spring to the fountain: – at Mesheth 3 miles north, went into a Mosque, & saw the sepulchre of Jonah, this being his country, 'tho somewhat uncertain if his burial place: – when I came out the Sheke had brought into the Mosque yard fryed eggs, soure milk & Coffee, so we eat & went on a mile to Cana of Galilee,[32] saw the Fountain whence the water was taken turned into wine; & the church on the spot of the house where this miracle was wrought, & near it a Greek church. Near the fountain is the church of St Bartholomew, where I think they say he had a house; – 3 miles further passed the field where the disciples rubb'd the ears of corn on the Sabbath day, & about 4 miles further north, came to the Mount whence our Saviour delivered his sermon; – ruins of a small church on it; – went down to the West foot of the hill & dined near a fine spring at Haleri, where Jethro the Father-inlaw of Moses was buried, in a place over which now is built a Mosque. Thence returned up to the mountains in a line with the said Mount to the place where Christ fed 5000 with 7 loaves, – where the stone is shown on which it is said, he blessed the bread. Descended for 3 miles East to Tiberias to the sea of Genesareth [Galilee], a walled town but few inhabitants; – sent a letter to the great Sheke, who ordered his secretary to entertain me, – a carpet laid on the top of the house, sup'd on Pilaw & soure milk; & honey & oyl together. Lay in a cube of 7 feet made with reeds, on top of the house; a door place without a door, plaistered for 2 feet up from the bottom, this for coolness, & all, as soon as sun is set, go up to the top of the houses to be cool, tho I find it hot.

12th-23d. I went to the place where the sea of Tiberias runs into Jordan, saw abundance of wild boars; – returned by a hot salt Iron sulphureous Bath, as hot as that of Albano near Padua;[33] – viewed some runs of Old Tiberias & some grotts; din'd as I sup'd but below: – found the Procurator of Nazareth here. Went in a boat on the sea of Tiberias to fish, – threw the net myself, & tho' caught nothing, yet I hope I may serve God as a fisher of men; – the fishermen caught a good dish, which besides the usual fare, we had for supper. I saw the entire church of St Peter where it is said Christ gave him the power of the sheep, & told him he was the rock &c: The Latins come here on this day to celebrate: then went to see some Jewish Sepulchre's. Visited the great Sheke, desiring a man to go with me next day.

---

32 The precise location of Cana is disputed, with four possible places in Israel and one in Lebanon. However, Pococke's description of the various churches corresponds with the site of modern Kefr Kana, situated 5 km north-east of Nazareth on the road to Tiberias.

33 See Finnegan, *Letters 1*, p. 225, where Jeremiah Milles describes the springs at Albana as "one of the greatest curiositys we have seen in Italy".

13th-24th. I set out & went to Bethsaida 2 miles west of the sea, ruins of a Good church, then went north & passed by a large round fountain, in the west plain of Gennesareth, having first seen the ruins of magdolum at the sea, & I believe of a church there.[34] At the fountain refreshed ourselves. Ascended part of the high hill to Saphet [Safed]; – went to the Arabs tents, & were entertaine'd with eggs, soure milk which some eat with cucumbers, & we had Coffee: – went on up this hill, which is almost as difficult as the Alps; arrived at Saphet at the top of it, went to the Cadis house, to whom I had a letter, after Coffee, went to see the Sepulchres of some Talmudists of the 2d temple, which the Turks call the sons of Jacob, entered on to into the court, the entrance into the very Mosque not being permitted. Went with a letter from the Consul to Signr Cochorn an old head Rabbin [Rabbi] of the Jews, who have 7 synagogues here, 20 Rabbins & 200 families, expecting to reign here 40 years before they get to Jerusalem; – it has been a great place for the Talmudists: this [Cochorn] is a very knowing old man & Droggerman to the French Consul at Sidon[35], entertained with cake, sweetmeats & a dram: – sent his servant with me to see the very strong tho' ruined old castle, part of which was the [word missing] of the Jews: – this place was defended by Josephus, & is I believe what he calls Gamala. I sup'd & lay at the Cadis, an Aga being come there from Sidon.

14th-25th. I drank Coffee, went to see the great Rabbin; – their feast of Pentecost, – I staid till he came out of the Synagogue, had much discourse with him, drank Coffee & eat sweetmeats & cake, – & towards ten, set out, passed at the foot of the Fortress by the 7 wells which Isaac made, & were stop'd by the Philistines &c: – went north down the hill into that plain in which the sea or Lake, out of which Jordan rises called formerly Phiala, – went along the plain westward & met with the Arab soldiers, & the Sheke of Sanowata on their Patroul, reposing under shady mulberry trees. We sat down near [] & I gave the letter to the head of them they had given me to their great Sheke, the son did me the honour to come & dine with me, on what I had – boyled Eggs, new cheese like curds, bread & Coffee; – & the head sent 2 horsemen to wait on me: – we went to the Lake about six miles long & two wide. At the northern end of it are mountains where begins the country of the Druses & the Maronites, the water turbid, not wholsome; – at the end of it, where it flows into Jordan, are ruins on a small round hill, called the town of Jacob, where it is thought he lived after his return from Padan aram, but I know not whether it will agree

34 In his printed account, Pococke refers only to its "considerable remains of a very indifferent castle". See *Description 2*, Part I, p. 71.

35 The rabbi was undoubtedly a so-called honorary dragoman of the French consulate, not an interpreter on active duty. On honoray dragomans, see Maurits H. van den Boogert, *The Capitulations and the Ottoman Legal System*, Leiden 2005.

with Succoth: – a litle further on its Jacobs bridge, over which it is probable he passed & its thought Esau met him here; – on the east side of it, I went to a Kan the next day, lying this night in the tents of the Arabs near the Bridge.

15th-26th. I went to the bridge & about a mile lower to the ruins of Caesarea Philipi on a litle hill, two gates remain, & a sign of a fortress at one end; a fine situation over Jordan, which from the Lake runs about 6 miles between the hills with a noise over stones, with some litle falls, it is but narrow, & the trees on the banks almost meet, so that you hardly see the water: – & the shrub oleander that bears a red flower & is in our Greenhouses (especially much at Kensingon, being something like the blossoming with us) grows on the banks of it & the lakes, & appears most beautifully. I went along by the river; – when it comes to the plain, it runs gently for two miles into Gennesareth; – I went to the mouth of it, & then continued on about three miles on the west side of the Lake, to Caphernaum [Capernaum]. Great ruins particularly a small church of white marble, almost entire, supposed to be where Christ lived in this city; – a semicircular port near, where probably the fishermen commonly came. Near this we dined & went up the hill East & slaked our thirst with water at the Arabs tents, which in these parts do no harm, & only makes one sweat; – I kept on about 2 miles further, & came to the cistern, into which it is said Joseph was cast by his Bretheren; – it is in a court walled round, & the cistern made like a well; – within it is a Turks praying place, where we sup'd & adjoining a large Kan, this is in the high road from all these Northern & Eastern parts into Egypt. About 9 in the even we set out towards Acre, ten hours from it, when the moon near down, we slept two hours under a tree.

16th-27th. Set out before day & about 8 came to a village, & went to the Sheke a friend of the Consuls, entertained with Coffee, fryed eggs, rice boyled with soure milk, & milk boyled with butter & sugar, & Coffee again. About 4 I arrived at Acre, – met with a letter from Mr Pullinger my friend at Aleppo,[36] with letters enclosed of recommendation to Damascus & other places I intend to go to. Mr Bodington went with Mrs French to Sidon;[37] – I go with the Consul tomorrow to meet him, in his return about 4 hours off, – & having an opportunity now of sending this letter, I conclude it; – designing to go in 4 or 5 days to Sidon or so proceed on, in my journey through Syria; at Sidon I shall be with the French – there being no English there.

Pray present my kindest love to my sister & due respects to all.

I am Dear Madam your most obedient & dutiful Son, Richard Pococke

[Two lines of a postscript are scored out]

36 See Letter 5, above, Chapter 4, note 66.

37 This is presumably the widow of the late Consul French, mentioned above.

SIDON
30 May/10 June 1738
Pococke to his mother
Letter 15[38]
(Miscarried lost) recover'd from Ireland Oct. 7th 1746[39]

Honoured Madam,

My last letter to you was of May 16th-27th from St John D'Acre, the day I returned from my most agreeable tour to Nazareth, The sea of Tiberias &c:

17th-28th. Went with the Consul, Mr Usgates brother & 2 merchants to meet Mr Bodington, went about 8 miles up to a hill whence we saw Tyre & almost as far as Sidon: Met Mr Bodington, Mr Usgate & 2 French merchants, & dined under shady fig-trees by a spring; – rid back half way to Acre & spent an hour in a garden & returned.

18th-29th. Went with the Consul & Mr Bodington to dine at Mr Usgates by invitation, [met] his Wife, Brother & a young sister. Soup with cucumbers stewed in it; – Ham & fowls & chicken, Boyl'd Beef, kidney beans stewed with the Giblets of the fowls, a pidgeon-pye, roast green Goose, & fowls – & a desert – apple-pye, a green Almonds Custard & tongues. Took Coffee – visited with the Consul, the Procurator of Terra Santa & 2 french Gentlemen that had been to see me. Erdel chock in Arabick, is the Thistle of the earth thence comes Artichoke – sperage is Asparagus – Coss is the lettice of Egypt. Sat in the Garden &c: over a sneaker of Arrack Punch. Sup'd.

19th-30th· Prepared for my Journey to Mount Carmel; & set out about 2 in the afternoon; – went round the Bay & came to Caiphas, – where I delivered a letter to the Aga, who sent 2 men with me. I went up to the top of Carmel to the Carmelite convent; – well received by the President on the Consuls Letter: saw an old convent higher on the Hill, sup'd – reposed.

---

38 BL, Add MS 22998. This letter marks Pococke's entry into what is now Lebanon. For further particulars of this course of his journey (as described in *Description* 2, Book II, "Of Syria and Mesopotamia", pp. 88ff) see Bartlett, "Richard Pococke in Lebanon, 1738", who identifies the placenames mentioned in this section of Pococke's book.

39 The quality of this letter is particularly bad and much of it is illegible, especially towards the end, where Mrs Pococke has squeezed a lot of text into the end of the page. This is probably because she had to insert the letter after the completion of the manuscript, i.e. 10 days after it was recovered from Ireland, presumably having gone astray en route to Marseilles. Probable meanings and omissions, etc., are indicated by the editor with square brackets.

20th-31st. Set out before the sun – went a league between Carmel to the South, to the fountain & garden of Elias, where was a convent; – then went on a league & ½ to a village, there was a fine old house & entertained by the Sheke. Thence went to a [ ] – one of the strangest, fine built places I ever saw – & seems to be the building of the Greek Emperors.

21st-1st – Set out about 2 in the morning southward, passed by Tentura & about 8 arrived at Caesarea on the sea side, where Cornelius sent for St Peter: the walls almost entire built by Lewis IX; an old castle with many fine columns built into it & a church; reposed here for two hours, where there are only 4 or 5 poor families; – returned, reposed a while by a river & passing by Tentura, the Sheke would not let us pass by, but we were obliged to drink Coffee, & eat Pilaw, with small pieces of boyl'd & roast mutton in it; arrived at Castle Pelegrini.

22d-2d. Set out early & dined at the Carmelite Convent, saw an old church at Caiphas, & the wind being fair took a boat, & crossed the Bay to Acre, to save 4 hours riding – the Consul being rid out I visited Mr Usgate.

23d-3d. I staid within all day.

24th-4th. After dinner rid out with Mr Bodington to see the source of the river Belus; – where we took Punch & bread & returning had sport with the wild boars & Jackals.

25th-5th. Set out at 3 with Mr Usgate, the Consul & Mr Bodington to go 4 hours, to a delightfull castle, in English the Enchanted Castle; going to it an hour through a narrow vale by a rivulet; – staid here 3 or 4 hours, returned, reposed by a water & got home by 10 at night; having shot at a wild Boar.

26th-6th. Went out a walk to review the ruins about the Town.

27th-7th. Rid round the old walls – in the even visited Mr Usgate the President of Terra Santa & some French – the latter returned my visit.

28th-8th. Mr Usgates Brother, & Scotch Captn. Anderson of a ship here came to see me: after the French Vice Consul & 2 others came to see the Consul. I set out about 4 for Sidon. The Consul, Messrs Bodington & Usgate & Captn Anderson accompanying me a league; I lay about 5 hours from Acre at a strange encampment made of a sort of hurdles of boughs, with places to sleep on of the same 3 feet from the ground; I slept without the court by these lodgings, on my carpet most soundly.

29th-9th. I set out early; saw a ruin, passed the famous road made by Alexander, on the side of an hill like Penmanmour, & came to the fountains of Solomon, said to be made by him for him they are two & from them an Aqueduct to Tyre a league off. Went round about to Tyre, where are great ruins of the walls of a church, churches &c: staid here 4 or 5 hours, – & went on passing over the Casnun the old Eleusherius, where the Emperor Barbarossa was drowned, lay at a village on the mountains by the sea in a court.

30th-10th. Set out early, went to Serphant on a hill , the old Sarepta; drank Coffee, & near the sea entered a Mosque, where it is said the widows house was who entertained Elias. Came on to Sidon & copied a Milliary [Roman milestone] in the way. Went to the convent of Terra Santa, where I was received with the utmost civility, by the French President; – dined, – slept. The President went with me to deliver a letter to Monsr. Audimar[40] & after one to Messrs Sequier & Giraud[41] French merchants from Mr Usgate, & after to Monsr. Gautier with a bill. Monsr. Sequier & Giraud returned my visit & invited me to dine next day.

31st-11th. I rid out round the Town, saw what the Turks lately call the Sepulchre of Zabulon, a Mosque into which I went, & into another of some saint of theirs, both in a garden: – & the Mosque where it is said the woman said Blessed is the womb that bare thee: – Went up the high hill of Sidon & half way up into a litle Mosque; – there are many places here called from Elias. They have here as nut the Almond. Apricot 'tis round, has a reddish cast to the sun, & eats like a very fine plum, is reckoned as wholesome as the others are unwholsom. The gardens are rather orchards here of figs, Apricots, mulberries, oranges & Lemon trees are most delightful, & the ground under cultivated like a kitchen garden. I saw a castle in my return & the ruins of the old Port, & went to dinner with Messrs Sequier & Giraud in company with many French; Frank Coffee at Mr Audimars room: – came home & writ; – & went with the President to visit the French Consul & Lady; – am going out to the gardens with the French & am to sup with Mr Audemar, & tomorrow morn, set out by land to Bareut & Tripoli: & find it every way most agreeable travelling, & have my health exceeding much, & do not find it as yet more hot, than I have found it sometimes in England, having this opportunity of sending directly to Marseilles. I close my letter. I met with most extraordinary civilities from Consul Wakeman &c: & believe few people have made this voyage with the pleasure & good fortune that I have done, from Tripoli, Balbeck, Damascus & Aleppo; at Tripoli is Consul Carew a very obliging gentleman & exceeding gentle to strangers: [...][42] Love to my sister & due respects to all friends.

I am Dear Madam your most obedient & dutiful Son, Richard Pococke

40 In the 1720s one Jean-Baptiste Audimar was a member of the town council of Marseilles. It is not clear wether this Mr. Audimar – presumably a merchant – was the same man, or perhaps a relative.

41 Two Frenchmen called Giraud and Sautet made a trip to Baalbek and Palmyra from Tri[poli in Syria in 1705. Perhaps the Mr Giraud mentioned by Pococke was the same man, or possibly his son. See Paul Perdrizet, "Les dossiers de P. J. Mariette sur Ba'Albek et Palmyre", *Revue des Études Anciennes* 3/3 (1901), 225-264.

42 John Carew (c. 1670-1747), a native of Wales, was the English Consul at Tripoli for 50 years. His epitaph is noted by George Robinson in his *Travels in Palestine and Syria*, vol. 1, London, 1837, p. 70.

TRIPOLI IN SYRIA
11/22 June 1738
Pocoke to his mother
Letter 16[43]

Honoured Madam,

My last letter to you [was] from Sidon May 30th June 10th. That evening I went with Messrs Sequier & Giraud & other French, about a mile to a garden, where we sat down by a stream, entertained with Ham & Wine & had an Apricot tree, shook over our heads, of the fruit of which we eat; returned to Monsr. Audimars room, – sup'd about nine & took leave.

31st-11th. I set out early, viewed some sepulchral grotts near the Town & went about 4 hours to the Wall of Jonah, where they say he was thrown out by the whale; staid a litle there, – went on to a river where we dined under a tree, & about 6 arrived at Beyreut, the old Berytus, – first passing through Faccardines pleasant grove of pines.[44] The country about this Town very beautiful, – went to the capuchin convent; walked to see the Palace, stables & pleasant garden of Lemmons of Faccardine a famous Prince of the Druses, about 100 years ago;[45] passed by a Mosque, that was a fine Gothick church, & where is a crucifix of which 'tis said St Athanasius relates some miracles – sup'd in the garden.

June 1st-12th. Walked round the Town & set out; – saw some sepulchral grotts, & came to a Mosque very pleasantly situated, called St Georges because here abouts he killed the Dragon or rather serpent. Went on 2 or 3 hours & came to the River Kelp [al-Kalb], the old Lycus; – a fine way cut on the side of the mountain & an inscription of the Emperor [Alexander] who made it cut in the rock; we dined under the large Arch of the Bridge: – where we met the Christian maronite sheke of a Town near with his four sons, fine youths, we interchanged our Provisions, & passed an hour agreeably. I went on & ascended the mountains of Antilibanus, now called Castravan, & came to a handsome maronite convent, where I met with a priest bred in Rome that talked Italian, who showed me all & entertained me with a glass of wine. Went & saw a good Parish church, for all these mountains are christian, have cells & are all

43 BL, Add MS 22998.

44 The pine forest still exists in the form of a municipal park known as Horsh El Snoubar or Bois de Pins.

45 Faccardine or Fakhr el-Din (Emir of Lebanon in 1591-1635) was put to death at the orders of Sultan Murad IV. In accounts to his uncle and his printed work, Pococke refers to him as Feckerdine/Fecherdine. See *Description* 2, Part I, p. 90.

Catholicks. Then went to a new founded nunnery of greek catholicks of the Order of Francis de Sales, in which are ten young Ladies of Aleppo the Founders of it, who are performing in it, the second year of their novitiate; was entertained here with cake, raisins, conserve of Roses, dram & Coffee. Went to the Jesuits near under whose direction they are; – here I staid all night, now their seminary higher on the hill, to prepare youths to send them to Rome.

2d-13th. Set out early & travelled 3 hours to the convent of Refond [Rayfoun],[46] almost rocks that look like a ruined town; – they set before us fryed Eggs & a soup of pulse, & Coffee & excellent white wine, which these mountains produce; most of the convents have a nunnery adjoining, who cook & wash; the priests being employed in cultivating the land, the great produce of these hills being mulberry trees for silk. Went on 3 hours to a convent [in Eirskeen] & waited on the Patriarch of the maronites, who was sitting near the convent on a carpet under a tree, with two Bishops; – he received me with much civility, one of the Bishops talked Italian; we had wine, Coffee, & a dinner prepared within, of Pilaw, fryed Eggs & sour milk; – the Patriarch sat near but did not eat – being friday, a day when they can't eat butter or eggs or milk: – we then went out & had Coffee, was much pressed to stay, & the Patriarch was almost angry, when I would not.[47] This place at the top of the mountains very cool, others East being covered with snow. I travelled about 14 hours to Gusta where is the Patriarchs family; I found his Brother under a tree & sat a while with him & invited to stay all night; – he is one of the Principal Shekes of the country: – Went on an hour to the convent of Terra Santa finely situated on a high hill over the sea, a very polite young Italian the President.

3d-14th. Returned by same way, went a litle down to a catholick Armenian convent; – was entertained by a Bishop here (who has fled from persecution) with cake, Pistachio nuts, morellas & dry sweetmeats, wine & a dram. Went up the hill again to Gusta [Ghosta], drank Coffee with the Patriarchs Brother who pressed me to stay. Went on South West in very bad road, on side of hills, 3 hours to the Capuchin convent [at Ghazir], but no fathers there; – I dined in the garden & the Sheke came to me, who travelled 9 years in Europe & talks Italian; – he carried me to his house, entertained me with Coffee & then we dined again on roast fowl, Eggs fryed, Pilaw &c: Went to see a convent near, entertained with wine & Apricots, on which they live in a manner in the season, eating them with bread; made a litle supper & reposed, being pressed to stay all night.

46 Of the Order of St. Anthony.

47 In his book, Pococke notes, "and he seemed almost angry when he found I was determined to go." *Description* 2, Part 1, p. 95.

4th-15th. Set out & left these mountains & travelled along near the sea 3 hours passing over the river Adonis to Eisbele [Esbele/Gibele/Jbail]; dined at the latter under a tree, the Sheke to whom I had a letter coming out to me.[48] I went to see the town full of ruined houses, a good church ruined & castle, it was the old Byblis [Byblos]; went to Patrone [Batroun/al-Batrun] the old Botrus. Saw the ruins of two churches & went on between Mount Libanon to a small single high rock, on which is built a castle, reposed a while; a maronite priest coming by I warmed him with a couple of glasses of wine, & he carried us 2 hours to his village, where we sup'd, & lay on the top of an house.[49]

5th-16th. Went on 3 hours & staid near a fountain & in 2 hours more arrived at Tripoli; – went to Consul Carews with letters from Consul Wakeman & Barton;[50] – was received with much Politeness; dined. Settled my affairs, slept, sat with the Consul, sup'd.

6th-17th. Rid out with Mr Crouch his [Consul Carew's] nephew & partner,[51] a mile to the Port, & to an old Town at the point, that now runs out into the sea; returned; writ. In the even went a mile to see an Aqueduct that crosses a valley from one hill to another with Messrs Carew & Crouch. The Consul is about 70,[52] has been 40 years here, & was so in Aleppo, a fine old Gentleman & a Batchelor; his Brother was in trade with him, but now is not; – his nephew is Mr Crouch & is his parner & are the only English in Tripoli; the gardens here are very pleasant; & all round about full of Almond, Apricots, some of the finest & most wholsom fruit I ever eat, & the kernel eats like a Philbert [filbert/hazel nut].

7th-18th. Staid within all day.

8th-19th. Mr Carew walked out with me about the town; – we went to a pleasant garden, & thence to the [convent of] Dervises, built over the river on

48 Not wishing to cause his mother anxiety, Pococke refrains from relating the tale that he later includes in his book (*Description* 2, Part 1, p. 98) whereby the Sheik demanded his pistols and on refusal apprehended them. Pococke walked away from the scene, threatening to report the incident to the Pasha of Tripoli, and when it was clear that he would not accept the offer of 10 dollars, his pistols were eventually returned to him.

49 Further described in his book as, a "poor cottage". *Description* 2, Part 1, p. 99.

50 The English Consuls at Acre and Cairo, respectively.

51 A Thomas Crouch was a correspondent of Alexander Drummond, later English Consul at Aleppo, who in 1750 published his *Travels* and founded the first Freemason Lodge in Izmir. See Maurits H. van den Boogert, "Freemasonry in Eighteenth-century Izmir? A Critical Analysis of Alexander Drummond's Travels", in Maurits H. van den Boogert (ed.), *Ottoman Izmir. Studies in Honour of Sander de Groot*, Leiden, 2007, p. 115. Pococke was to meet Mr Crouch again in Latakia (see Letter 22).

52 Pococke was not far off, as Mr Carew was then 68 years of age.

a valley, half a mile from the Town, only one in it; a pleasant place; – went to see the Greek church, which the Bishop showed us & his own handsom house. Came home – waited on the Consul to the Pasha's, the Governor of this country, to compliment him on his return from meeting the caravan of Mecca with provisions, as he always does half way; – after we came, he came in with a Gown lin'd with unspotted Ermin, a knife in his girdle, on the end of the handle of which was, a very fine large Emerald; – he sat cross-leg'd at the corner of the Divan, a cushion set up on each side & one above, behind him; – the Consul sat on a stool at some distance before him, the Droggerman stood on the right, & we on the left; sweetmeats, Coffee & sherbet; – the perfumed water & incense only to the Consul. We all had our hats on going as usual in these parts with Hat & wig. From the Pasha we went to his prime minister, where we all sat on the Divan, & had all sweetmeats, Coffee, shirbet [sic] the perfumed water & Incense; but before great persons it is not respectfull to sit cross leg'd, but you kneel, sitting back ony our legs, which is very uneasy. In the afternoon Messrs Carew & Crouch went with me, mounted on the Consuls horse, to the south upper hills, to see the Greek convents of St James & Belmont finely situated over the sea; entertained with glass of wine at the former, & with fryed Eggs, cheese & wine, at the latter where the Consul often goes to spend some weeks. In Tripoli is a Terra Santa convent, Jesuits, Carmelites & Capuchins.

9th-20th. Went nowhere.

10th.-21st. Went with the Consul Messrs Crouch & Carew, about an hour & a half east of the Town, near the foot of Mount Libanus to spend the day in a fine shady place, on a rivulet amidst pleasant Gardens; – the country being very beautifull: – we had a collation of Bologna sausages, diverted ourselves by walking &c. dined on Pilaw, boyled mutton, mutton pye. – Mulberries & Apricots; after dinner left the Consul & went to see the head of the stream, half an hour off, at the very foot of Mount Libanon, coming out of the hill in many cold streams & immediately makes a very considerable rivulet; – returned, took Coffee & about 5, setting out came to Tripoly [sic] before sunset, after a pleasant ride.

11th-22d. Being Sunday I read prayers in the morn & even; & after went to see the Fathers of Terra Santa. Tomorrow morning I set out to see the Cedars of Libanon, thence go to Balbeck & from Balbeck to Damascus, & so to Aleppo, where I propose to stay till the heat of the summer is over, which I have not yet found very extraordinary; – there is almost always an air in this country. Arabick is the language in all these parts. I travel on horse back always, & am well recommended to every place I go to, one of the handles of the knives I sent you from Egypt, is not mounted; – you may send to Mr Milles & let him know

I bought it for him, made of an Egyptian pebble, & I will send you another.[53] Pray present my very kind love to my sister & due respects to Whiteoak &c:

You see I am now going inland a litle, so you must not expect to hear so often from me.

I am Dear Madam your most obedient & dutiful Son, Richard Pococke

DAMASCUS
1/12th July 1738
Pococke to his mother
Letter 17[54]

Honoured Madam,

My last letter to you was from Tripoli in Syria June 12th-22d [11th-22d]– 12th-23d. I set out early for Mount Libanon & in 3 hours arrived at it riding through groves of olive trees & near it through vineyards; ascended for half an hour & reposed a while, & then ascended for an hour; the mountain is covered with Pines, Cypress & Ever-green oaks &c: – descended into a narrow valley; & dined by a spring under an olive tree; – went through a Romantic narrow vale to the convent of St Anthony [Casieeh] on the side of a hill, & mostly cut out of the rock; – a monk that talked Italian showed me the convent & treated me with a dram & strong wine, which they have here very good. – I went 3 hours to such another convent called Canobani [Cannobine], the general residence of the Patriach of the maronites, whom I saw in Anti-libanus: His nephew a Bishop entertained me at supper on Pilaw, soure milk, fryed Eggs & olives.

13th-24th. I returned by the same bad way up a very steep mountain. I had travelled in a Turban, – but now put on an hat & wig; going through a village a maronite priest took the rein of the horse I was leading, had me under his tree, laid a mat, & treated me with sowre milk, fryed Eggs, olives & wine; & was not above taking about the value of a shilling. Came on saw many fine streams, running through the rocks with a noise, & often in beautiful cascades, came near a village called Priherch [Bisturch], & sat & dined under a tree near a stream: – went a mile further to the Carmelite convent under a hill, & most of

53 See Letter 27, in which Pococke gives alternative instructions to his mother about this handle. Langford, *A Catalogue of a Large and Curious Collection of Ancient Statues, Urns, Mummies, Fossils, Shells, and other Curiosities…*, p. 3, lists Lot 18 as "Four knife handles of Egyptian pebble, 2 ditto (1 broken) of serpentine marble, a cornelian pendant, a small knife handle of jasper, ditto of green jasper, and a toothpick case of plumb pudding stone."

54 BL, Add MS 22998.

it cut out of the rock, especially the chapel; – but the most beautiful place in all Mount Lebanon. 2 fathers & a lay Brother who are very civil & please me better than any of the Monks – sup'd on eggs in several shapes, sallad, walnuts, cheese, olives.

14th-25th. Ascended up the hills & in 2 hours came to the Cedars of Libanon; now only a small grove of Cedars & pines, about ½ a mile in circumference, but they commonly grow as if it was two or three trees together, & soon shoot out into very large branches which are flat at top with litle shoots springing up perpendicularly, as do the apples the point of which are to the heavens & in shape like a tun & close, not opening like pineapples, but grow something like a very close Artichoak. There are 15 large ones, & one faln [fallen] down, the wood is like deal, but has the smell of the W. India Cedar; we ascended up the highest mountains, passing by the snow & sometimes over it, & had a fine refreshing air; descended on the other side near to the Lake Lemoun [El Yammoune], dined by a stream, went two hours between low hills to the plain of Balbeck, about 9 miles wide, extends a great way north & many miles south & is what is called Coelesyria; we stop'd at a village; – & went on not arriving at Balbeck till an hour after night. Went to a kane where I reposed in a poor room on my Carpet: – you must know, the passing the Alps is nothing, to the passing Mount Libanon; – it is not I believe in a strait line 12 miles broad, & it must be a very long days journey to cross them, without stopping.

15th-26th. I went to the Pasha's seraglio to find his Christian secretary, to whom Consul Carew procured me a letter; – he had me to his house, we had a collation of rice-milk, Eggs &c: – I waited on the Pasha with a letter Mr Usgate gave me, he made me sit, & sweetmeats & Coffee was brought; – he told me when he was called from his Pashaship in Bosnia about 3 years ago one of Ragusa had his & his litle boys picture, who was by, drawn: – he is a non-cheearfull good sort of man & particularly civil to Franks, inviting them to him 'tho they have no letters. I dined, & went with an officer the Pasha ordered to show me everything. Went first & viewed the most beautiful Temple which is beyond any piece of Antiquity remaining: – measured & examined it well;[55] – then viewed a fine colonade & other buildings, & sup'd in Fresco in the secretaries court, adorned with trees & a fine bason of water in the middle, into which the

55 This is what he continually refers to in his book as "the famous temple" and further describes as, "a most equisite piece of workmanship, on which the utmost art has been bestowed". See *Description* 2, Part I, p. 108. This is known today as the Temple of Bacchus. He says surprisingly little about this monument in the letter but devotes more than a page of text to it in his book, as well as representing it in a full-page plan and four beautiful line drawings (see *Description* 2, Part I, Plates XII-XVI). He notes in the letter that he viewed, measured and examined this monument, represented here as Plate 14.

water runs by several pipes; had roast fowl, mutton & broth, stewed cucumbers stuff'd with forced meat – & after fruit.

16th-27th. I went & viewed a grand Entrance & two grand courts, with most beautiful buildings on each side, being the avenue designed to that Grand Temple, of which only a most magnificent colonade is seen. After dinner viewed a small old circular Temple turned into a Greek church, but is a most beautiful piece of Architecture:[56] – then went to see three great stones in the Town walls, each above 60 feet long; – went round part of the walls & got some bits of Inscriptions, & within the town saw a fine Tuscan pillar thought to have been used in some conveyance of water.[57]

17th-28th. I went out on horseback 2 hours to see a large Corinthian Pillar in the middle of the plain.[58] Saw some quarries in return. A Turkish Sepulchral Dome of red Granate Pillars, – & then a quarry near the Town, where I saw a stone larger than those in the walls, cut out every way, except that it was not quite seperated [sic] from the rock at the bottom. Went to a garden & dined on bread, Almond Apricots mulberries & cherries; rid round the walls which are not less than four miles, & found on the stones some few words, & at a fountain half a mile East of the Town, a greek inscription; returned; the secretary told me the Pasha had asked two or three times why I did not come to see him, & bid him bring me to him after supper; – showed me a young Tyger that was brought to him that day; talked about the affairs & Princes of Europe; – told me the Grand Signor [Mahmud I][59] had sent Ragotskie [József Rákóczi] into Hungary with 30,000 men, that Boneval [Count Bonneval] was Captn. General of 3000 Bombarderers. Coffee was brought, – he told me I should be welcome to stay 3 or 4 days longer, or as long as I pleased, & in a word treated me with all the politeness imaginable.

18th-29th. I went out with the Oda Bashee [supervisor of a building], to view for the last time those magnificent ruins that are not to be equalled in the world. Bashaw is the title of every Janizary, Oda Bashee is a Janizary like a servant, head of twenty, – & a Pasha is the great Governour of a country, whom we wrongly call a Bashaw. In the evening walked out with the secretary, & sat an hour under a tree by a river; – being a place where all the people go out in the evening to be in the fresco.

---

56 Again, his commentary on this monument is limited, given the amount of attention he gives to it in his book, where he presents two full-page views with plans (*Description* 2, Part I, Plates X and XI). These are entitled respectively, "a temple at Baalbeck" and "the semicircular temple at Baalbeck" (see Plates 14 and 15 of this volume, respectively).

57 This is depicted in Plate XVII of his book (*Description* 2, Part I.).

58 Depicted in the same plate (*Description* 2, Part I).

59 Mahmud I was the sultan of the Ottoman Empire from 1730 to 1754.

19th-30th. The caravan not going to Damascus till tomorrow, I began to write letters to the Bishop, having seen everything.[60] In the even went to the fountain half a mile out of Town, & lay there with the caravan.

June 20th – July 1st. At one in the morn the caravan set out for Damascus, we dined 6 hours from Baalbeck & lay by greatest part of the day; – & went on 2 hours further & lay by a rivulet.

21st-2d. Set out at one in the morn, went mostly on the side of hills & about 7 came within 2 hours of Damascus, ascended up the hill which commands the plain & city of Damascus, which is about 2 miles from it; – out of this hill runs the Barrady [Barada]. East to Damascus through a green sod called the field of Damascus, out of the earth of which they say, Adam was made, it is narrow, & on each side of it, the green tents of the Bashas camp are pitched, & every way you see the fine gardens of Damascus & a view of the city in that great plain, which appears more magnificent without than within; – being built from about 4 feet from the ground of unburnt brick; but without the minarets, the costly Mosque of St John, & some Cypress trees in the city overlapping the houses gave it a very fine view. I enter'd the city & went to the convent of Tera Santa where I was well received by the President & fathers who are Spaniards; – dined – slept & writ.[61]

22d-3d. The Pasha's secretary was with me to de some business for me; & I went out with the President to see the house of Ananias a sort of cellar where was a church now a sort of private grot; – then went to St Pauls Gate, where he entered after his vision, & thence to the place where he was struck blind. Some signs of a church near, & here the Christians are buried; – returned by the place where St George Porter of St Pauls Gate was martyred & buried, having favoured his escape, I suppose when he was let down in a basket; returned, dined, writ.

23d-4th. Went out all day about the town with Signr. Nicola, one who assists strangers; – he showed me at the gates the great Mosque dedicated to St John

60 The letters he had written to the bishop up until this point (though failed to mention) are as follows: Nazareth, 11/22 May 1738, Tiberias, 13/24 May 1738, Tyre, 28 May/8 June 1738, Sidon, 31 May/11 June 1738, Bayreut, 1/12 June 1738, Tripoly, 9/20 June 1738 Cannobine on Mount Libanon, 13/24 June 1738; and the letter he was writing at the time must have been the one dated Balbeck, 16/27 June 1738. All are from the collection BL, Add MS 15779 and are reproduced in Chapter 7, Accounts from the Holy Land, Lebanon and Syria.

61 There is a discrepancy in dates between his journal letter and his printed book. In the letter he notes that he "enter'd the city" of Damascus on 21 June, yet in his book he claims he arrived there the following day. The mistake may perhaps be attributed to the fact that Pococke was disturbed by an incident in which stones were thrown at him by a drunken soldier who had picked a fight with his servant – an incident which he was careful not to relate to his mother but recorded in his printed work. See *Description* 2, Part I, p. 114.

Baptist or St Zachary as the Turks say, but the Christians to St John Damascus. Saw the castle, dined at a Pastry cooks, where they have Pasties of forced meat & others with a sort of cream, or white butter between, on which they pour honey, both of fine puff-paste & very good. Went to a fine Coffee house, with an Island behind planted & accommodated for the people, a most pleasant place, in these Coffee-houses they have musick 3 times a day; – walk'd to some other parts, & sup'd at a cooks-shop on calves feet soup.

24th-5th. I rid out with two of the fathers, a Janizary & Nicola 3 miles to the Sepulchre of the 40 Martyrs half way up the hill to the West. It is a Mosque, which we were shown & the place where, they say, they were buried then went to the place of the famous seven sleepers, on the side of the same hill, & came back ½ a mile to Selheia [Sahnaya] a most pleasant village, where are the most pleasant Gardens, in one of which we dined, by a stream of running water; – went to see some others & were entertained with Coffee in one by a Mussul man [Muslim]. We then went to see the seven rivers that run out of the Burrady to Damascus or the country near; returned.

25th-6th. Signr Antonio came to see me yesterday, of the Greek church, 'tho educated in Rome, a young man that talks Italian well; he came also today & showed me the Greek, the Armenian & Syrian churches; – from the later I went to see their Bishop, who was very civil & treated us with Coffee; dined; went abroad with Signr Antonio & to his house, where he treated me with Apricots & a glass of wine; – Went 2 miles to Jobar, where Elias anointed Hazael King over Syria, & where they say falsely he was fed by a raven, 'tis a synagogue.

26th-7th. I set out with Signr Antonio & went about 4 miles to the south when Abraham overtood the five kings & delivered Lot, & they say the 5 kings are buried there; near is a mosque, in which the Turks showed me a cliff, here they say Abraham was born. We took our mornings collation of eggs, & went in between the mountains & travelled north west 2 hours to a nunnery, & lay in a building without, on purpose for strangers. The abbess showed me the church & several ruined chapels about the village. Greek nunneries are but litle different from Hospitals; – the 20 nuns work in the fields, as well as the Abess, who showed me her hands hard with labour, they are particularly employed in the silk worm business, have great plenty of excellent strong wine.

27th-8th. I set out early to go 3 or 4 hours to Malouca [Maaloula], where there is the convent of St Thecla, disciples of St Paul, who was martyred here, & is the female Proto martry, stop'd at two places; – & came to the convent,; the church is within a very large Grotto, in which we lay. I went up the hill & say many sepulchral grotts cut out of the rock, & another convent called St George now uninhabited, the other having but one monk in it.

28th-9th. Returned to Marah near Sydonaia [Saydnaya], breakfasted at the convent of St Elias, where there is only one Lay-brother to conduct to the Grot of St Elisha, where Elisah anointed the prophet to be his successor, which we went to see, & going to Sidonaia saw the ruined convent of St Christopher, that of St George where ther is but one monk: – saw the catholick Greek church in the Town, sup'd, & lay at the nunnery.

29th-10th. Set out & breakfasted at Menin, went 2 hours eastward to see the river Fege come from under the hill at once as a great river: Dined; rid about the place; – & went two hours up an high hill, where, they say Cain slew Abel; a very fine small church there; returned.

30th-11th. Returned to Damascus, stop'd at Dummar & went by the Barrady [Barada] & saw the division of the waters, & the rivers running on the side of the hills; – a very fine sight; returned to Damascus whence I propose to go in 2 or 3 daies towards Aleppo. This afternoon I went out in the town with Signr Antonio. I desire you to present my very kind love to my sister & due respects to all my friends.

I am Dear Madam your most obedient & dutiful Son, Richard Pococke

ALEPPO
31 July/12 August 1738
Pococke to his mother
Letter 18[62]

Honoured Madam,

My last letter to you was of July 1st-12th from Damascus, I spent all that day in the city with Signr Antonio, pretty much among the silver smiths, who are Christians & work together in one great hall, got a few silver medals among them; dined at a Cooks-shop, & went to a Coffee house, where there is musick at noon, bought some things I wanted, & came home, & writ.

2d-13th. Sigr Antonio came to see me, I writ most of the day: – went out with Sigr. Antonio to hear the musick at the Coffee house, & took a walk.

3d-14th. Went out with Signr. Antonio to see the city & do some business, for an hour or two, in the morning & afternoon.

4th-15th. Went two miles south out of town with Signr. Antonio, to see a Mosque – where they show what they call the footstep of Mahomet in a stone, who seeing Damasacus thence would go no farther, saying, there was but one

62 BL, Add MS 22998.

Paradise for man. Saw a Mosque where they say Moses was. Went on & passing by Deriah, reposed 2 or 3 hours, in a shade, by a rivulet & went two hours further to Deiret Caucab[63] & lay at a village near it; they look on me, in the character of a Frank Physitian [physician]; – Sigr. Antonio understanding Physick is wonderfully respected.

5th-16th. Went 3 hours So. South West of Betimie under the mountain, din'd there by the rivulet, went a mile further to see the sepulchre of Nimrod, only 2 or 3 great stones of the building remain; near saw remains of a fine building called his Castle; – returned to a kane in the road to Jerusalem, being but a days journey from the parts I was in when I went from Acre to Tiberias; lay at the kane.

6th-17th. Went again to Deiret Caucab, near which I saw the Grot where St Paul hid the day after he made his escape by the wall in a basket.[64] Came on towards Damascus, & dined in a garden at a village called, Mezy [Mezi] near it. – Went on to Dummar 2 hours west of Damascus, where I lay.

7th-18th. I went to a village called Estropi [?], where the Barrady comes our from between the mountains, to see the process of an Aqueduct, cut through the mountains which I saw near Fege [?]; returned to Damascus.

8th-19th. Signr. Antonio came to see me. I went out with him after dinner to the Coffee house.

9th-20th. Went in the morning with Signr. Antonio to my Turkish Merchant a brave man & serviceable:- Went to the Coffee house; after dinner Signr. Antonio was with me twice. This day the Pasha here removed, & the Pasha of Tripoli comes here. The children here are suprizingly fair & beautiful; – but their Bagnio's [bath-houses], their vices & beards make them ugly when grown up. I know not whether I know more of this place, than of any other I have seen, – but it appears to me, to be the most wicked place in the world, for some private vices not to be named, as well among Christians as Turks, the former being only ashamed of them, as the others are not. It is said my Lord Sandwich has freighted a ship a Messina for the Eastern voyage, but whether to further than Constantinople I cannot tell; – hear he is returned by Greece.[65]

10th-21st. Writ all day to the Bishop. Sigr. Antonio with me twice.

11th-22d. Signr. Antonio came to me – I writ all the morn; about 3 went with him to see my Turkish Merchant Binny, & then to see the Chancellor of the

63 Dayrat Kawkab, possibly the village Kawkab in the district of Hama.

64 In present-day el-Tell (or al-Sahra), half a mile south of Damascus's Eastern Gate (Bab Sharqi).

65 As noted earlier, Lord Sandwich (1718-92) embarked on his grand tour in 1737, setting sail for the Levant in April 1738. His movements are assiduously followed by Pococke, who refers to him again in Letters 22 and 43, below.

Greek Patriarch of Antioch, who is a Lay man. Green Pistachio nuts are esteem'd very delicate, but to me, eat no better than young green pease, – the shell is covered with a rind, so that they look something like an exceeding small Katherine pear, & grow in bunches, the kernel is green as a pea throughout, but long.

12th-23d. I went out early with Signr. Antonio to visit the Patriachs Chancellor who received me very civilly, & promised me all the assistance he could give me for my journey, & with him I drank Coffee, took a walk to the Leper Hospital of the Turks, & saw the tomb of Gehazi. Here they say was the house of Naaman the Syrian; at 4 went to the Coffee house with Signr. Antonio, & then to Signr. Binny, & in search of the Chancellor, whom I could not meet.

13th-24th. Went to Signr. Binny: dined at a cooks shop. – & then went to the Dervises convent, & saw their extraordinary dances, & whirlings round, performed to music. Visited the Patriarchs Chancellor, & went again to [Autinias] house,[66] a litle behind to the east of the street, called the Royal Street – Desbnet.

14th-25th. Took a walk with Signr. Antonio round part of the city. Went to the Patriarchs Chancellor who is my friend, thence to Signr. Binny; returned; dined; at 2 went to Signr. Binny then to the Chancellor, & came home.

15th-26th. Went early to the Aleppo kane, din'd early with Sr. Antonio at a cooks shop & set out with the Aleppo caravan about noon & went 2 hours to Touma [Duma] & laid on a bulk in the street.

16th-27th. Set not out till near sunrising, A camel being gone astray: about 100 persons in the caravan; I ride on horse back, went 7 hours passing over the hills into another plain & came to Kteiphe [Adra], & reposed in a very fine kane, built like a cloister; – raised places in the Piaches [?] where the people lye, & within the Portico round are the stables, all built of Hewn stone.

17th-28th. Set out at 2 in the morn & travelled 8 hours to Rekte [?] a kane with chimneys, & got some boyled mutton for dinner: – I like the caravan travelling very well: – when I come in sleep for two hours, dine, – write & read; & the caravan never stops, till at the end of the days journey, about 8 hours or 16 miles.

18th-29th. Went out before the sun & travelled only 3 hours to Coraw [Qarah] – an officer of the Governours a civil man came & drank Coffee with me, & soon after a Christian; writ.

66 The original word was "Judas's" but this has been scored out and the word above it is difficult to decipher.

19th-30th. Went out at day-break & travelled 6 hours to Hasseah [Hisyah] a sorry place; – In the road my merchant Binny in Damascus his son came up with the caravan & we joined company.

20th-31st. Set out 2 hours before break of day, travelled along in sight of the plain of Balbeck & passed only by one place yesterday & this day, being in a manner through a desert, no water last night, but that of a cistern; – saw the Lake of Arse, under the Ancients, & in about 8 hours. Arrived at Hems [Homs], the Old Emesa, waited on the Governor with a letter I had, went to see the sepulchre of Caius Caesar who dyed in Asia, & the castle on a high hill, shaped into a mount.

July 21st-Aug. 1st. Set out an hour before day, & passed by a village, called Restoun [al-Rastan], probably the Old Arethusa, a large convent, was in it quite ruined; passed over the Arse [Orontes] here; & went on 4 hours more to Hamah as I conjecture the Old Epiphania situated in a narrow valley almost water & trees, a large Town, supplied with water by wheels & Aqueducts 40 feet high, & others in great numbers in a most curious manner, but the work of the [] not so well executed; dined, – went into the Town & walk'd round it.

22d-2d. Rise early & walked about, went to get a few medals at the silver smiths all Christians as in Damascus; – went again into town to the Coffee house, & in the Even went out to lie with the caravan.

23d-3d. Set out 2 hours before day & travelled 8 hours to Shehoun [Khan Shaykhun] a village & kane there, writ till 5; – then set out & went five hours to Marah [Maarat al-Numaan], the Old Arra & lodged in a very [ ] Kan[e].

24th-4th. Staid all day at Marah, walked out with my friend, no Antiquities, tho' it is the Old Arra, received a letter from Mr Pullinger.

25th-5th. Set out about one in the morn, & travelled 7 hours to Surmeen [Sarmin, 9 km southeast of Idlib?], a days journey from Aleppo, where Mr Pullinger to whom I am recommended came out to meet me,[67] & conducted me to his tent, after dinner we went two hour S.W to Reah [Ariha?] & saw some sepulchral Grotts, & pitched our tent on the side of a hill, where is a fountain; the Governor of the Town below treated us with Coffee. We took a walk. This day received your four letters of Decr. 31st – Feb. 12th – March 2d & April 16, which gave me a great deal of pleasure.

26th-6th We went out about 9 hours to see the ruins of two magnificent villages called Burruss[68] & Kuph[69] – full of country seats of rich Primitive Christians, a litle after Constantine, I suppose; where they retired, & built near

67 See Letters 5 and 15, above.

68 The "Dead City" of al-Bara?

69 Kufr Nabl?

sepulchres, of which I saw very many magnificent ones, the cross being a sign that they were Christian. Went 2 hours to Firihay[70] such another village; – where I got 2 or 3 Greek inscriptions mostly Christian, returned to Reah.

27th-7th. Went after dinner 3 hours to Roncat [?], a place, more magnificent than the others & of the same nature, here we lay.

28th-8th. Went 6 or 7 hours North East to a place called Old Aleppo [Qinnasrin] by the Franks, which I take to have been Chalcis, we pitched our Tent on a Hill over it; in the Even rid to see the ruins.

29th-9th. Went early to Rumousa [Rambuta] within an hour of Aleppo, pitched the tent in a garden – Mr Aide, Mr Hammon [Hammond], Mr Shaw[71] & the English doctor came out to dine with us;[72] after dinner the Consuls Chancellor Mr Hayes[73] & his Messenger & Janizary came to meet me with complements [sic] from the Consul & to invite me to sup with him, the la[t]ter I declined as being fatigued; – & after came Mr Edwards the Chaplain, who was Chorister of Corpus [Christi, Oxford] & left it in 1717.[74] Mr Pullinger had sent for his fine State-horse with exceeding rich trappings, on which he mounted me & we set out for Aleppo; – I was met by Mr Fitzhugh, Mr Bigg [Bigge] [75] &

---

70 Furayka? Drummond's 'Furkia'; Drummond, *Travels*, p. 289.

71 Jirjis 'A'ida was an Ottoman Christian merchant who also served as consular interpreter to the British consulate in Aleppo. Hammond and Shaw were English merchants based in Aleppo, of whom only the latter two can be further identified. Peter Shaw, named as the signatory to the Last Will and Testament of Lister Bigge, below (dated 16 November 1742: see TNA, Prob 11/759) was at times in debt to the Radcliffe House trading in Aleppo (the English merchants often borrowed from each other) and died in 1752 (see Davis, *Aleppo and Devonshire Square*, pp. 220 and 94). William Hammond belonged to a long line of Hammonds trading in the Levant. He was also in debt to the Radcliffe House and in 1752 complained of the inconvenience of having to borrow large sums of money at interest. Davis, *Aleppo and Devonshire Square*, pp. 66 and 220-21.

72 Named in Letter 19 as "Dr Duke", undoubtedly the Levant Company physician in Aleppo.

73 There are several officials by the name of "Hayes" or "Hays" (see Wilson, *List of British Consular Officials*, p. 60) but their dates are not compatible with this official. This may be the Aleppo factor David Hays whose correspondence 30 years later is mentioned in Davis, *Aleppo and Devonshire Square*, pp. 222 and 226.

74 The details given by Pococke identify him as Edward Edwards who, according to Foster's *Alumni Oxonienses*, was from Bedford town, matriculated at Corpus Christi College in 1711, took his MA in 1717 and became Rector of Holwell, Bedfordhire in 1721.

75 William Fitzhugh, son of Captain William Fitzhugh of the East India Company, went out to Aleppo in 1735, aged 18, to work for David Bosanquet, a member of the Levant Company's factory there. He later became factor to a relative newcomer to Aleppo, Jacob Chitty (see Davis, *Aleppo and Devonshire Square*, p. 245). Mr Fitzhugh is recorded by Alexander Drummond as accompanying him and William Hemming, in 1747, on a "journey into the desarts [sic] of Arabia, and other countries inhabited by savages". See Christine Laidlaw, *The British in the Levant: Trade and Perceptions of the Ottoman Empire in the Eighteenth Century*, Tauris Academic Studies, 2010, pp. 99-100. He left Aleppo in

Mr Hooper [Cooper]; – & alighted at the Consuls Nevil Cox Esqre.[76] to whom I went to Mr Hydes house where Mr Pullinger eats & supp'd there & lay at Mr Pullingers house.[77]

30th-10th. Breakfasted with Mr Pullinger & Mr Shaw who lives in the same house, we are always to breakfast together: – they all sup, breakfast & sleep on the tops of the houses for coolness, in little beds with curtains, & the air is so good, no one catches cold. The Consuls Chancellor [Hays] & Messenger came from him, with his complements [sic] to know how I did: & after Mr Bossington [Boddington],[78] Mr Cowper [Cooper],[79] Mr Clerk [Clarke],[80] Mr Free,[81] Mr Shelton,[82] Mr Fitzhugh & Mr Edwards came to see me, & Mr Sherman,[83] & a Dutch Gentleman that talks English viz Mr Bites [Sites?][84] after I was gone to

1751, a wealthy man, at the age of 33. He was licensed to trade at Aleppo in 1742 and admitted to the Company in 1754. London directories list him as a Turkey merchant at Garroway's Coffee House, Exchange Alley (1759, 1763, 1765, 1767) and at the Sword Blade Coffee House, Birchin Lane (1760, 1763). For further details, see Wilson, "More Notes", p. 2. Lister Bigge, another English merchant, made his Will in 1742 and died four or five years later, since it was proved by the Perogative Court of Canterbury 4 January 1747. The will was attested by Consul Arthur Pollard and Cancellier (or Chancellor) Nathaniel Free, the latter of whom is mentioned in the same letter.

76 Henry Neville Coxe was the English Consul at Aleppo, c. 1727-39. See Wilson, *List of British Consular Officials*, p. 36.

77 This was probably Herbert Hyde (cf. Pococke, *A Description of the East*, Volume 2, part 1, p. 152: "I saw in Aleppo a bronze statue of Minerva ... This statue belongs to Mr Herbert Hyde, who has it now in London"). The son of Captain John Hyde, he was admitted to the Levant Company in 1726 and entered into a partnership with Peter Lupart at Aleppo in 1735. See See Wilson, "More Notes", p. 3.

78 Probably George Boddington (one of a line of George Boddingtons), who was consul at Smyrna from 1722-33 and died in 1759 in London (see Wilson, *List of Consular Officials*, p. 15).

79 Mr Cooper was a member of the English factory in Aleppo in 1746 and is recorded as being in partnership as Cooper & Free in 1747 (see Davis, *Aleppo and Devonshire Square*, pp. 156 and 110). A death notice in 1774 refers to a "William Cooper, Esq; a Turkey merchant, and one of the directors of the Bank of England, in Winchester street" (see See Wilson, "More Notes", p. 2).

80 Presumably of the English firm at Aleppo, Clarke & Bosanquet.

81 It is not clear if Mr Free is Cancellor Nathaniel Free, who signed the will of his two colleagues Lister Bigge and Rowland Sherman (see note above) or John Free, who died in Aleppo in 1755. Either is possible, as each was in Aleppo around that time (Davis, *Aleppo and Devonshire Square*, p. 219)

82 The unidentified 'Mr. Shelton' is mentioned only once. He was almost certainly not a merchant at Aleppo, so perhaps he was a ship's captain.

83 This is Rowland Sherman, who sailed to Aleppo as a factor in July 1688, was employed by March's in 1738 as their factor and was consul at Aleppo in 1743. He wrote his Will in 1748 and died at Aleppo the following year (see TNA, Prob 11/771).

84 Possibly the Dutch merchant Giacomo Seitz, who was in Aleppo at the time.

the Consuls to Mr Edwards's who had engag'd me to preach for him, which I did, an handsom arched room, neatly fitted up for the chapel. The Consul's country is near Cirencester, a very genteel man & a most obliging manner: – he desired me to make his house my own, with all the civil things that could be said: – we had a good English dinner of two courses & a desert. Came away at two, & returned after all my visits, met the Dutch Consul,[85] him & the French Consul,[86] I am to visit tomorrow. No lights in the air & no Astronomers in these parts, but we are to have here on the 4th an Eclipse of the Sun. The summer here remarkably cool. Never received your letter of your Hampton journey; – but all writ since came into these parts. On Board Ship I paid the Captain: – but the Turky [officials] are exceeding civil, & above taking a present unless it may be something accidental. The Phoenix seems to be a fable.[87] No chaplain or service at Alexandria whilst I continued in Consul Bartons family. No rain in Asia from April to October. Tell Mr Milles the registry of the License of Absence, must be copyed from the original & not from a copy, & attested by the Register, as a true Registry, if that cannot conveniently done, he may take out a new one & have it Registered. The sulphur of my letters is the purification of Leghorn, no plague in all these parts. Rosso Egypto I have seen none, few natural curiosities in these parts. The Lady you met at Bath is Mr Uptons mother, he has a sister married to the same name in the north of Ireland a Parliament men. I have the servant I took with me who is very faithfull. They [the Pyramids] are called Josephs Granaries, but [are] not his work. Pharoah lived probably then at Tanis in Delta, & sometimes perhaps at Memphis, west of the Nile & Pyramids East of them the sepulchre of a king. I shall write to Mr Milles to congratulate him on his preferment, which I know was designed for him & I have writ to my Lord to give me the next.[88] [Next two lines are scored out]. Arabick [is] spoken from Aleppo to Tangier – North of Aleppo the Curdeni [Kurds] who talk Curden [Kurdish] & Turkish. I talk Italian to my servant, who

85 This was Jan Jacob van Liebergen, in office 1733-1740, the business partner of Hendrik Abraham Heirmans, who later became consul in Aleppo. Schutte, *Repertorium*, p. 353. Unlike their Western counterparts, the Dutch consuls were allowed to conduct trade on their own account to supplement their salary.

86 Léon Delane was the French consul in Aleppo from 1735 until his death in 1742. Mézin, *Les Consuls de France*, 232-3.

87 Many of the following comments are Pococke's responses to questions that his mother must have put to him in her most recent letter.

88 This preferment must have been the Precentorship of Waterford, which was given to him in 1737 (see Rennison, *Succession List*, p. 40). Pococke lost no time in writing to the bishop and this letter follows immediately.

is my interpreter; if not too much trouble continue to abridge & send to Ireland.[89] Pray my very kind love to my sister & respects to Whiteoak.

I am Dear Madam your most obedient & dutiful Son, Richard Pococke

ALEPPO
12/23 August 1738
Pococke to his mother
Letter 19[90]

Honoured Madam,

My last letter to you was from Aleppo dated July 31st Aug 11th – that day I was visted by Mr Clark & the son of Signr. Binny, my merchant in Damascus. When at dinner our Consul sent to let me know, that he had appointed me, to wait on the French Consul [Léon Delane][91] at six, & he invited me to supper, which I excused being to meet Mr Sherman at our table; – this gentleman is the oldest in the Levant, understands Arabick perfectly, has translated much into that language, & given many Psalters & new Testaments, being a Greek advocated for the Greek church, against the catholicks, & spends much this way.[92] At 5 I went to the Consuls, & waited on him to the French Consuls; he goes out in state, first his Messenger with a long staff & silver head, then 2 Janizaries with their poles, & after them the Drogerman. We were treated with different sorts of wine punch & Coffee, when we return'd, I sat a while with the Consul, went

89 Though this is the first mention of such a request, it becomes a recurring feature of the correspondence, since Pococke claims that he has insufficient time to write to Bishop Milles himself. The request provides a further indication of Mrs Pococke's central role in the management of her son's personal and professional affairs.

90 BL, Add MS 22998.

91 See note 86 above.

92 The fact that Pococke mentioned the French consul's proficiency in the native language shows that it was unusual, and this view is substantiated by other contemporary observers such as Russell (cited in Woods, *A History of the Levant Company*, p. 236) who notes that, "few of the factors took the trouble to acquire any knowledge of the Turkish tongue; and they had little or no social intercourse with their hosts". It is odd that Pococke fails to mention his distant relative Edward Pococke (1604-91), who had served as chaplain at Aleppo from 1630-35 and whose famed oriental scholarship earned him a place as the first Professor of Arabic at Oxford (1636) and later of Hebrew (1648). However, perhaps he assumes that his mother, who married into the Pococke family, would already be aware of this.

to supper, Mr Sherman & Mr Purnel[l] a servian of Alexandria, the son of the late Consuls sup'd with us.[93]

Aug. 1st-12th. The Consuls Chancellor, Drogerman & Janizary came & went with me to wait on the Dutch Consul, he was not at home, but his partner Mr He[i]rman[s] was; treated us with a dram of visney, which is a sort of fine Morella [morello] cherry brandy of Smyrna;[94] at 12 went & dined with Mr Sherman; Messieurs Pullinger, Hyde, Dr Duke, Purnel[l] & Edwards there, a very plentiful English dinner, staid till 2 drinking Coffee, saw Mr & Mrs Eatons picture there. Mr Pullinger to whom I am recommended is a Gentleman of exceeding good nature & easy behaviour, is a great medallist, & has presented me with several valuable ones. I rid out at 5 with Mr Pullinger round the Town, which is about 4 miles round the walls, 2 suburbs being great, 300,000 people in it. Went with him to see Mr Fitzhugh by invitation, treated with Arrack Punch. Supp'd as usual, Mr Cooper being with us,[95] what they call Goosberry pye made of Grapes a usual dish here.

2d-13th. Mr Edwards came at 6, & we walked to see the Greek, Maronite & Syrian churches & the 2 Armenian; visited the Armenian & Greek Bishop, entertain'd by them with Coffee & sweetmeats, we went to see a fine house built by the Pasha here for his own private use, where he would be glad to retire with his money. The Consul came & sat with us. When I came home the Dutch Consuls Drogerman & officers came to pay the compliments of their master in return to my visit. Mr Pullinger is of Hampshire near Petersfield & Mr Hyde is our county [Hampshire] originally but born in London, a man of Learning & skill'd in Astronomy; about 11 I rid out with Mr Edwards to a summer house in a garden to dinner, where came also Mr Free, Mr Clark, Mr Bigg, Mr Cowper. Pilaw, mutton steakes, boyled fowls, & fruit; [next few words scored out] [played] at Quadrille; – the Consul came & we walked in the garden, & took a ride round about the Town – I supp'd at the Consuls.

3d-14th. Mr Hide [Hyde] breakfasted; busy all day in making a catalogue of medals; at 7 went to Mr Shermans concert, he played on the organ, & there

93 John Purnell was English Consul at Aleppo from 1716-26 and died in 1743. See *The Scots Magazine ...*, vol. 5, 1743, p. 428. Other Purnells are listed as consular officials for the period, including a William and an Edward, though it is not clear which of these (if any) is the consul's son (c.f. Wilson, *List of British Consular Officials*, p. 36). Pococke was to meet a relative, Mr Purnell, in Larnaca (see Letter 22).

94 A recipe for this cherry brandy is given in a contemporary publication by Richard Bradley: *The Country Housewife and Lady's Director in the Management of a House, and the Delights and Profits of a Farm*, London, 1732, p. 61. Perhaps Mrs Pococke or her daughter Elizabeth had a copy of this book.

95 Mistakenly named as "Hooper", in Letter 18 (see note 79 above) and sometimes named "Cowper".

were 3 fiddles, a base viol, several there; – Arrack punch went round. We went at 8 & sup'd at Mr Clarks with 6 or 7 more.

4th-15th. Walked round the castle with Mr Pullinger & Mr Hyde; observed the Ecclypse [sic], which began here at 45 minutes after one & ended 45 after 3 about 4 digits or one third of the sun being obscured, but it was hardly visible to the naked Eye; we made some observations concerning the macula solares, one of which it just covered, the nearest but one to the outside of the sun, which spots may be seen half an hour before sunset without hurting the eyes, with a common unblack'd spy-glass. Mr Purnel son of the late Consul Purnel came to see me; – & after Mr Edwards to invite me to go out to the gardens to-morrow, with the Consul & others; – they drank tea, Mr Hide came to observe the sun. We went & took a bowl of Punch, which is made small & drunk out of the bowl, at Mr Cowpers by invitation; had grapes & pistachio nuts & walnuts. Sup'd at Mr Shermans, who was playing on his organ, with the Italian Musician, who has a double flute of Milan lately invented by a Musician, it is two flutes of different keyes, cut in one piece of wood.

5th-16th. Went out at 11 with the Consul to the garden; – dined; – plaid at ombre; went to see a fine situated convent of Dervises, came home, sup'd at the Consuls.

6th-17th. The French Consul had sent word he would return my visit at the Consuls house.

7th-18th. I went & drank tea with Mr Edwards, & paid a visit to the Consul.

8th-19th. In the afternoon went to see a fine collection of medals, of the Kings of Syria &c:

9th-20th. The Doctor called on me, Mr Fitzhue [Fitzhugh] dined with us; – at 5 went to see the rest of the collection of medals, & after visited Mr Free. Mr Hammond & Mr Shaw sup'd with us.

10th-21st. Mr Edwards dined with us; – at 5 rid out with Messrs Pullinger & Hide; went to Mr Shermans concert, & sup'd at Mr Clarks as they do every Thursday night.

11th-22d . Messrs Clark, Cooper & Hyde breakfasted with us; we sup'd with Mr Sherman.

12th-23d. As usual we are to sup at the Consuls. The weather is not very hot & I shall be moving soon towards the sea-coast. This [letter] goes by Marseilles & may arrive before my last. Love to Mr Milles who is [...] in England. You may open it & write to him in it. My very kind love to my sister & respects to [...].[96]

96 The text is corrupted at this point, as Mrs Pococke has crammed the last few lines into a small space with minute script.

ALEPPO
18/29 September 1738
Pococke to his mother
Letter 20[97]

Honoured Madam,

My last letter to you was dated Aug. 12th-23d from Aleppo. – I sup'd that even with the Consul.

13th-24th. Went to Mr Edwards & to church & preached, the Consul invited me to dine, but I was engaged to Mr Hyde; – at 4 went to prayers; drank Coffee with the Consul & after tea with the Chaplain with other company; the Chancellor [Hayes] sup'd with us.

14th-25th. I went to see Mr Edwards & wait on the Consul, the latter asleep. Messieurs Hyde, Shaw & Hammond & the Dr [Duke] drank tea at our house; & about 5, I mounted & they with me to make an excursion north of Aleppo for a fortnight. I took leave of them & went an hour farther, & lay with the caravan, which was going to Antab [Gaziantep] & from thence to Ezroun [Erzerum] in Armenia, I being bound to the former place.

15th-26th. Set out 3 or 4 hours before day, & travelled 3 or 4 hours after day, & the caravan stopping I went into a house, where I was very well received, they brought me grapes & I treated them with Coffee: – here they all speak Turkish north of Aleppo; writ.

16th-27th. Went 8 or 9 hours as the day before & sat down by a rivulet, under a spreading walnut tree by a village & writ.

17th-28th. Arrived at Antab in 3 hours, about day break: – I sent a Letter I had to an Armenian Merchant, – went to sleep. – the Merchant came to see me & drank Coffee; – writ – & bought a few medals brought to me. This place was Antiocha ad Taurum in the litle kingdom of Comagena; on the other side of Taurus is Capadocia [Cappadocia], of the kings of which many medals are found here; – but mostly of one king. The merchant came to me with two people going that even to Romcalia [Rumkale] on the Euphrates, to go in company with them: – about 5 I walked to see the castle & the town & the Armenian church of which there are here about 2000, & no other Christians there being few others in these parts; – was had to the merchants house, where we had a collation of sweetmeats, cake, sugar plums & fine large Imperial pears & a dram, & at night a very splendid supper of boyled fowls, rice soup, several sorts of broth, – Ragous, mutton &c: – drank Coffee, went to the kane, & about 8

97 BL, Add MS 22998.

mounted, – & about 2 or 3 in the morning, came to Hayam, where those fine pears grow; went to sleep in two hours.

18th-29th. Rose, took Coffee, & went to some shady trees near a rivulet, where I dined & writ & about 4 in the even set out & passing over rocky hills covered with ever-green oak, & some parts with Pistachio nuts. About night came to an open encampment of a village, where I lay for in these parts they leave their houses & bed under tents, or houses made of boughs for coolness. A man played as a Curdine [Kurd], here on the Tambour & sung with it a Curdine song.

19th-30th. Went in about 2 hours to Romicalia on the Euphrates, being North east of Antab: – It is an old Greek castle on a hill, which I saw, & in it two churches, one of which turned into a Mosque. Here the famous Ionam Cogia Captain Basha[98] or Admiral of the Grand Signr. is continued, having had ill success about 3 years ago in the Euxine sea against the Muscovites, & suspected to be bribed by 'em; he cut off the head of the chief rebel that deposed the late Grand Signr. & continued to head a tumult after this sultan was set up, & is reckoned a very great man.[99] I crossed over the Euphrates here which is about as wide as the Thames at Richmond, but is now low, a whitish green water; having crossed the Euphrates I was now in Mesopotamia called in scripture Padan Aram, ascended up the hill covered with Pistachio nuts, a tree somewhat like the walnut, but does not grow so large, the leaves thick. Went about an hour & ½ to an Armenian Christian village, reposed under a shady tree, the people came out & sat round & the two priests: – After dinner I went to see the church, & in the churchyard treated the Priest with Coffee, & went 3 hours farther East a stony rocky country & no wood over all Mesopotamia, not even about the villages; – lay at a summer village, the houses or huts made of large stones covered with boughs; travel now in a hat & wig.

20th-31st. Travelled 4 or 5 hours & dined under the Tent of the Rushouina Curdines [Reşwan Kurds]; – went on about 4 hours more & lodged among the same sort of people in the open air. They imagined that there was an Ecclypse of the moon which I did not discern but they were told so [I do not know but this was a misapprehension & that it was only from devotion on seeing this full

98 Canım Hoca Mehmed Pasha, three times Ottoman Grand Admiral (Kapudan Paşa), whose third term in office had ended in 1736.

99 He further explains in his book, "This place has been made use of by the Turks as a place of banishment of great men in disgrace; and when I was there, it was the fate of Ionam Cogia to be confined in it, who had been captain basha or high admiral of the grand signor's forces, and was an old experienced officer of great abilities." *Description* 2, Part 1, p. 157.

moon],[100] & after the mahometan manner they tinkled their pans, & there I heard one making a long prayer for the moon as they sd [said].

Aug. 21st-Septr 1st. Came in about 3 hours day break to Oursa [Urfa], the old Edessa & they say, the Ur of the Chaldees, which the Jews say, it certainly is, out of which God brought Abraham. Went to a kane, slept, sent a letter I had for a Turkish Merchant & to a Christian officer of the Pasha's. Went out among the silver smiths, but could find no medals. The Christian had sent for me to bring my things, & make his house my home. I went to him to his office; he presented me with a few medals, carried me to his house, where we had a collation of mutton chops & forced meat, water melon, grapes & Coffee. At 5 I rid out about the town & to the castle. A fine canal below it, said to spring out when Abraham was here, & full of fish, which the turks permit not to be caught, a fine Mosque on it, the portico & yard of which is open to it, with a fine colonade. In the castle high wood pillars. Here Abganis lived, who sent a letter it is said to our Saviour & they have a miraculous cistern relating to that affair. I had a splendid supper at the Christians, on [the roof of] his house, where I lay; here Caracalla died.

22d-2d. I set out at day-break, rid round the town, & went to the church of St Andrew, belonging to the Armenians, an Altar in a court, & a church under in a grott, where they showed me the sepulchre of Ephrim [Ephrem] Core [?]; – which I take to be Ephraim senis [the elder] who was a Deacon of Edessa; proceeded on in my return having been here at the most Eastern point of my journey & dined near the caravan – went on betwixt, 2 & 3 hours & lay at an encampment of Turk-men.

23d-3d. Travelled south west about 6 hours to Beer [Birecik] of the Euphrates, a town on the side of the hills over the river, having set out early; – Went to the kane, dined, sent a letter I had to the Governour, to have leave to see the castle; which I went & saw, & a room of old bows & arrows, & other arms before the inventon of Gunpowder. Went over on the other side of the river in the Ferry boat & lay in a kane.

24th-4th. Went with 2 Turkmen guides south to a ruined Town called Jeribolous [Jarablus] from the Syrian God of that name, to whom there was then a Temple; in 5 hours came to this place, view'd it, & went on an hour & dined under a Rushoueins tent & going on came to the sejour [Sajur[101]], & lay near it at an encampmt. of the relations of Mahomet, descendants of Sultan Ahmed of Boursa [Bursa], of the former race of Kings, before the Ottoman Empire was founded.

100 A note in Pococke's hand appearing on the reverse of this page.

101 "Village Sadjour on the river of that name". Drummond, *Travels*, p. 289.

25th-5th. Went an hour to another encampment of them, where their head resides, a comely man dressed in a fur-garment of green silk, lined with a yellow silk, but a little faded [next half line scored out], invited me into his tent; – but I desired to see the ruins of Hierapolis or Bambych now called Bambouch & by the Franks Bambech:[102] he went with me saw part of the walls, great Aqueducts under ground &c: – I was given to understand a present was expected by the Emir or Prince, & having nothing – he was so mean as to accept of a piece of money. I went on about 2 hours & dined in the tents of some Turkmen, & going on discharged our 2 Turkmen & went on & slept on the top of an house at a village of Turkmen in the way: saw here 3 or 4 Begdelies[103] who are Turkmen on horse-back that live by plunder; – this place is in the high road from Beer to Aleppo.

26th-6th. Set out about midnight with the Aleppo caravan, & leaving Bashe[104] went to Tedis[105] about 4 hours, a village finely improved with mulberry trees like the country between Chantilly & Paris; – dined in a garden – went on 3 hours to a village called Shirbey[106] where I was well received, going south had a good supper prepared.

27th-7th. Went 3 hours by sun-rise to the valley of salt about 5 or 6 miles long & 2 or 3 broad. 5 hours from Aleppo. In winter it is over flowed with water & the soil being salt when the water dries away it leaves a crust of fine white salt on the earth about ½ an inch thick, which they take off with Instruments for that purpose & put in heaps & sell it. I went on towards Aleppo, stop'd within a mile of the town & about 2 arrived there. Saw on this side the Euphrates a large bird call'd in Italian the Wild Turkey, keep [kept] together two & two are very shye; – but when the cock crows is at the time of coupling 'tis said he sings till he is blind, or rather shuts his eyes when he sings, & they are the Bustard kind, & that they have 2 or 3 sorts of flesh of different taste [Of this I was informed by one of the English, but have been since told that they never perch on a tree having but three toes & all plac'd forward].[107] Mr Hyde & the Dr came to see me; – I visited the Consul & Mr Edwards: – rid out to see the encampment of

102 The ancient ruins of Hierapolis Bambyce or Hierapolis Euphratensis are located 20 km south of Manbij in Syria.

103 Alexander Russell, *The Natural History of Aleppo*, London 1794, vol. 1, p. 390: "Begdelies, a tribe of Turkmen."

104 Half an hour's travel before the village of Bab. See *Description* 2, Part I, p. 168.

105 Tedis was a town "seven leagues east of Aleppo". Clement Cruttwell, *The New Universal Gazetteer*, vol. 3, London 1798 [no pagination; under T].

106 One league from 'Ayn Dhahab. See *Description* 2, Part I, p. 168.

107 This is a note in Pococke's hand appearing on the reverse of the page.

the Pasha of Oursa, who is going against Solly Bey the Rebell about Smyrna; Mr Vernon sup'd with us.

28th-8th. Jonas Korte of Altenan near Hamburgh, but a Saxon born, whom I met at Jerusalem, came to see me.[108] I had rose at 5 a clock & walked out to a garden with Messrs. Pullinger & Hyde to see the Pasha of Oursa [Urfa] go away; – many led horses & soldiers, & He with twenty or thirty footmen round him on horseback, & about 12 Led-Horses, with silver shields, with ornaments of Gilt on the right side of each of them made a fine figure, musick &c: – Mr Edwards came to see me in the even. This day busy in ranging some medals I had bought rejoycings for some victories of the Turks taking Orsoua [sic] &c: saw the rockets fired; Messrs. Sherman, Cooper & Purnel supp'd with us.

29th-9th. Visited the Jesuit lay brother & Mynheer [Mein Herr] Korte, both whom I had met at Jerusalem.

30th-10th. Breakfasted at Mr Hydes, & he, & the Dr went with me to the fine house of Salviari a great Jewish Merchant, he showed us a very fine manuscript of the Bible about 500 years old, & a beautiful case to put the Roll of the Law in; – & after showed us their old synagogue in which they say is an inscription of the time of the 2d Temple; – & I saw some work about that must have been near the time of Constantine. We went to Mr Hydes, where we staid a while & came home.[109] Mr Hyde called on me & I went with him to change & dinner, drank tea Mr Shaw being here; rid out with Messrs Pullinger & Hyde to see some Mosques. Mr Shaw sat with us & took some refreshment, & He & Mr Hammond sup'd with us.

31st-11th. You see the plague is warm in Constantinople, but you knew the winter always puts a stop to that; – & you may be sure I will go to no place, where the plague is; – It is in no other parts of the Levant. Visited the Consul at 11 who invited me to dine; sat half an hour with Mr Edwards, much company at dinner; at 5 I walked with Messrs. Pullinger & Hyde to the English Burial place, & saw the Epitaphs of Mr Maundrel [Maundrell] & Mr Harley, Lord Oxfords Brother, a Turkey Mercht. here, who refused to go home when his Brother was in power, being a man of great Honesty & integrity, & not liking a court.[110] Went & walked in the Pasha's garden, rid home, & went to Mr Shermans concert.

---

108 See Letter 13.

109 It is not clear from the letters where Pococke stayed while in Aleppo, though he notes that on the first night he stayed at Mr Pullinger's house.

110 On Maundrell, see Introduction, note 32. Nathaniel Harley was the brother of Robert, 1st Earl of Oxford and arrived in Aleppo in 1685 aged 20, where he stayed until his death 35 years later. He is said to have dealt in exotic commodities such as Arabian horses, falcons and ancient statues. See Laidlaw, *The British in the Levant*, pp. 114 and 204.

Septr. 1st.-12th. I went & drank tea with Mr Edwards, Mr Solesby Chaplain to the Bishop of Winchester who died lately not in the best circumstances for his family, was chaplain here.[111] Went into the Bagnio with Mr Pullinger & sat an hour at Mr Hermans [Heirmans],[112] to hear Musick.

2nd-13th. I rid out at 4 with Mr Hyde to the gentlemen who had dined in a garden & returned with them.

3rd-14th. I preached at the chapel, invited to dine with the Consul, but was engag'd to Mr Hyde – went to prayers & took Coffee with the Consul, tea with Mr Edwards, visited Mr Sherman, Mr Fitzhugh called on us, & the Chancellor [Hayes] sup'd with us.

4th-15th. Messrs Clark & Hyde called on me; – after dinner the capuchin father from Diarbeck came & sat a while; at 5 rid out to the fountain of fishes, with Mesrs Pullinger, Hyde & Purnel; – 2 later came & sat with us, & Messrs Sherman & Purnel.

5th-16th. Went out before sun rising & copied Mr Harleys & Mr Maundrels Epitaphs; dined with Mr Herman being invited. Mr Edwards invited me to drink tea. Mr Cooper supped with us. Quince pye – & plum sauce with Goose without the stones.

6th-17th. Mr Hyde breakfasted with us; – Mr Purnel came & went with us to see a Mosque, that had been an old church. Mynheer Korte came to see me. Mr Shaw sup'd with us.

7th-18th. Mr Edwards came to see me, & I visited the Consul & dined at his house; rid out with Mesrs. Hyde & Cooper; – went to Mr Shermans concert, & sup'd at Messrs. Hammonds & Clarks & we had the first rain & terrible thunder & lightening, which seldom does harm.

8th-19th. Went to see some medals, in the afternoon rid round the city with Mr Hyde. Saw a young Tyger to be carried to the Pasha of Bagdat [Bagdad], sup'd at Mr Shermans.

9th-20th. Went again to see some medals, & rid out with Mr Pullinger, & dined with the greatest part of the Gentlemen at the gardens.

10th.-21st. Visited the Consul & Mr Edwards, Mr Hayes supped with us.

11th-22d. Nothing remarkable.

---

111 Revd Joseph Soley, who had been English Chaplain at Aleppo from c. 1716-23, may have been known to Mrs Pococke, as he had held various posts in Hampshire up until his death, which took place on 25 November 1737. See Laidlaw, *The British in the Levant*, p. 228.

112 Hendrik Abraham Heirmans was a merchant in Aleppo as early as 1733, when he was also appointed Treasurer of the Dutch consulate there. He was promoted to consul in 1740, an office he held until his death on 7 October 1747. Schutte, *Repertorium*, p. 353.

12th.-23d. Mr Hyde called on me, went with him to dinner, rid out with him & Mr Vernon at 5, the latter & Mr Cooper sup'd with us.

13th-24th. I rid out in the Even with Messrs. Hyde & Vernon: after went with them to see the feast of the circumcision of the son of one of the Pasha's Agents, in the Pasha's private house; – sweetmeats & Coffee brought. Saw the Rockets & other fire works. Messrs. Hammond, Shaw & Clark supped with us.

14th-25th. Went to Rose & bought a few medals. Mr Hyde called on me; at 4 He & Mr Cooper & I rode out, when we returned we heard there had been a very great Earthquake, not usually perceived by people abroad in motion; as many of the English were, it was at ¼ after 5, continued about ¼ of a minutes, & as near as I could learn, consisted of 9 motions, backward & forward, there being a stop between every three, or rather three motions in all, the houses & every thing with them seemed to move, & the water in basons, waved backward & forward, it is said some walls fell, & I saw where there were cracks in houses, much dust & dirt faln [fallen] down, it is the greatest they have had here above 50 years part; – since there was no danger, I should have been glad to have perceived it. If there is a war with Spain, you may be assured I shall embark only on ships that are friends with Spain, such as the French. The plague is abated at Constantinople. I supped at Mr Hammonds.

15th-26th. I rid out in the even with Mr Hyde – & supped at Mr Shermans.

16th-27th. I visited the Consul & Mr Edwards; rid out to the gardens with Mr Hyde & dined there with most of our Gentlemen, play'd at Quadrille. Supp'd at the Consuls.

17th-28th. I preached at the chapel; – In the even after prayers, visited the Consul & Mr Edwards, & went with the Consul to take leave of the French Consul. I visited Mr Free & Mr Korte.

18th-29th. I visited all the Gentlemen, English & Dutch, dined at the Consuls, visited Mr Hyde & then a father of Terra Santa; – they all came to return my visit; & the Consuls sent their chancellors & officers to wish me a good journey. Thus having spent these few weeks very agreeable [sic] among my countrymen, from whom I received all manner of civility & friendship; I am setting forward & in 3 weeks expect to embark at Latichea [Latakia], for Cyprus. My kindest love to my sister & due respects to all friends.

I am Dear Madam your most obedient & dutiful Son, Richard Pococke

LATICHEA THE OLD LAODICIA
9/20 October 1738
Pococke to his mother
Letter 21[113]

Honoured Madam,

My last letter to you was of Septr. 18-29th from Aleppo.

19th-30th. I rose at 5 to leave Aleppo, waited on the Consul who went out with me with all his officers, as did Messrs. Edwards, Vernon & Purnel, about half an hour out of Town & then took leave: – I went about 6 or 7 hours mostly through a stony desert, north west & by west to the ruins of the magnificent convent of St Simon Stylites[114] who was so called from living here many years on a pillar, the lower part of which I saw in the church: – went an hour & half & lay at a village in the way to Alexandrette [Alexandretta/Iskenderun] : they brought me grapes & a nose gay of Basel [basil].

Septr. 20th-Oct. 1st. Set out early & saw a relief & Greek Inscription in a house; ascended the high hill, called Sheke Baraquet[115], found the remains of very ancient buildings & three Greek inscriptions, I never heard of before; – descended south into the plain crossed the Alexandrette Wall, in which are no accommodations, & went an hour more south to Daina [al-Dana] in the Antioch road from Aleppo & about 8 hours from the latter, saw some antiquities there, & took a litle refreshment & went on westward about 3 hours to Shesick [?], where I went to an house, sup'd on boyled fowl & rich soup, a great dish in my voyages. About 8 at night set out in company with an Aga & his retinue & travelled about 5 hours, to the bridge over the Orontes, where I slept till day.

21st-2d. Went on through the plain & arrived at Antioch, where I staid in a garden & sent a letter the Consul gave me to a Greek merchant, – who sent & invited me to his house close by, I went, dined, went abroad; this city is on the Orontes, partly on the plain from which the walls are carried up in very high hills like those of Genoua; – the whole about 5 miles round; – the present town ill built along a road on the side of these hills, saw an extraordinary bridge built from the one to the other, called Iron gate; the walls are suprizingly built up along the precipices, which are almost perpendicular: – saw near it upon the side of the hill the ruined churches of St George & St John; the latter used by

113 BL, Add MS 22998.

114 Known as Qalʿat Simʿān ('the Fortress of Simeon') or Dayr Simʿān ('the Convent of Simeon') this site is located 30 km northwest of Aleppo.

115 Shaykh Barakat, south of Qalʿat Simʿān.

the Greeks the former by the Armenians; & half way up the hill in the walls at the north end, saw the east end of the church of St Peter standing & probably the Patriarchal cathedral, & in all likelyhood built where St Peter lived, who was the first Bishop of this place; – went home on the west side of the walls.

22d-3d. I rid round about 4 miles to have an easy ascent to the high part of the old city called the castle, – went the whole length of it; saw the ruined castle & descended the steep part, & went 4 miles south to the supposed Ancient Daphne, where that person is said to be turned into a bay tree – saw two springs in the way to this heighth [sic], where the chief spring, which is said to be the fountain that comes from the castalian spring near what I suppose was the temple of Apollo Daphnaeus, which was in the middle of a wood of cypresses 10 miles round, planted by Seleucus first King of Syria of the Greek race: no cypresses now, nor the Laurus or Bay; – but much of the latter within the walls of Antioch; – these hills covered with myrtle & box & Ilex &c: I stop'd at the fountain & took some refreshment & returned.

23rd-4th. Went to see the ruins of what they call the Emperors Palace & then went over the Orontes by the bridge, & mounted to go to Scanderoon or Alexandretta [Iskenderun] , travelled along the plain northward, & near the great Lake of Antioch, which is 3 hours from the city, went about 5 hours to Mount Amanus as it was called formerly; just at entrance; dined at a village called Caramout;[116] – ascended other mountains 3 hours & came to Bertztown a place much frequented formerly by the franks even from Aleppo, for a cool retreat, & now by those of Alexandretta, situated between the mountains: – here are the gates of Syria Alexander pass. Went on two hours more to Alexandretta, to the Vice Consuls Mr Brand;[117] sup'd, Captn. Pridy in company. I went & slept aboard his ship, the air here being exceeding bad.

24th-5th. Came ashore, breakfasted, writ, rid out in the even with the Captain & Consul: – sup'd & lay aboad.

25th-6th. Walked about this poor litle town with the Captain to the Terra Santa Convent, – to the Greek church, – & saw the English burial place & among them, the Tomb of Edmund Sawyer a Merchant of Aleppo, who died, 1686 & I thought might be Sr. Robert Sawyers Brother.[118]

---

116 'Caramoot' in Russell, *Natural History* (1794), vol. 1, p. 314. Caramoot was a three-hour ride from Alexandretta at less than 2.5 miles per hour, according to Alexander Drummond, *Travels through different Cities*, London 1754, p. 287.

117 Possibly John Brand Kirkhouse, who later moved to Aleppo.

118 The Sawyer family from which the merchant Edmund came were lords of the manor in Kettering, Northamptonshire, until 1723, when brought down by the South Sea Bubble crash. Edmund the merchant was the son of Edmund Sawyer (died 1680) and left £600 in his Will for the foundation of the hospital or almshouses in Kettering. See William Page

26th-7th. I rid 8 miles north to Baias [Payas] in Cilicia, the old Isus [Issus], a litle Bay north of it, which makes the north east corner of the Mediterranean, before I got half way, went out of the road to what are called Jonass's pillars, & they say Jonah was last out here by the whale, tho' another place is shown for it near Sidon; – when I came to them I soon concluded them to be a Triumphal Arch, erected in memory of Alexanders victory over Darius, which must have been in the narrow plain just north of it. The top of the Arch is broke down; – all built of fine grey & white marble, the corners adorned with Pilasters, & in the south front on each side a Pillar on its pedestal; descended into the plain, through which runs a river, I suppose the old Pindus or Pindarus, ever which Darius crossed to meet Alexander; & where he was so fatefully defeated & all his family taken;[119] had a shower of rain I went to the kane at Baias. Here are abundance of Gardens of Oranges & Lemon trees, Aleppo being supplied from hence: – in my retrun overtook an Aga with 60 soldiers, going as he said against rogues, & desired me to stay till they went on, sat down with him, offered Coffee & a Venetian slave to buy, went on came to Scanderoon [Alexandretta]; – sup'd at the Consuls – went aboard ship.

27th-8th. I set out by the sea side to travel south, met the soldiers driving away the cattle of poor people; – in about 5 hours reposed by a rivulet on a pleasant green, encompassed with fine, shady Alders: – went on about 2 hours to a village of honest Turk-men, finely improved with Mulberry gardens; sat down in a place encompassed with fig-trees & vines twining round them by a stream. Staid all night.

28th-9th. Set out at 7, keeping on the plain of Arsous [Arsuz], & came to the low hills that are at the foot of the high mountains of Rossus [Nur/ Gâvur Mountains], as they were formerly called, adjoining to the old Seleucia South of it, which is near the Orontes, & but 5 hours from Antioch; called at a village where they were afraid of us, but seeing we were honest people, sent a man to put us in the way; – went on to another where I stop'd, & then to another near the entrance from the top of the hills to the mountains, where we found it convenient to stay, a pleasant place, round a litle hollow full of Orange & Lemon trees & pomegranates; saw over to the coast of Cilicia, much where about Tarsus is, which from this shoar is not above 16 or 17 leagues. I writ. We had rain in the night, & this day after I came to the village, thunder & lightening & rain: – the Head of the village came & sat with me, & in came 8 dishes of Wheat

(ed.), *A History of the County of Northampton*, vol. 3, 1930 [online], footnote 47. Pococke mistakenly thought that the merchant was the brother of Sir Robert Sawyer (1633-92) of Highclere, whose daughter married the 8th Earl of Pembroke.

119 Though earlier believed to have been the Deli Çay River, it has more recently been identified as the Payas River.

boyl'd with a very litle meat like our buttered wheat, & 2 dishes of sort of Pumkin [sic] or Gourds boyled in slices, which they eat with the wheat: – I sat down with them eating with the hands as the way is, taking out the victuals with a piece of bread. Writ. Supp'd on my fowl.

29th-10th. Set out & passed over part of the mountains, the Montes Rossii of Pliny, & took some refreshment at a spring & approached the high mountain close to that which was called Coryphoeus,[120] & also Mons Pieria, passed part of it, & having rid near ten hours the worst mountains I ever passed, stopped at a house & lay at the top of it, being well received, & had wheat dressed with mushrooms we brought, & the Pumkin sliced in the same manner, Grapes &c: –

30th-11th. Set out early, & travelled about 4 hours still up the mountains, & passed by a ruined church & then by a Christian Armenian village & ruins of a large church & convent near. Came to another Armenian village where I staid & received great civilities, dined on bread dipp'd in Treacle & olives, & my bread & my [next word scored out]. Writ; – & whilst I was writing, they saw there was an Earthquake, I look'd up, it was only one shock, & one does not perceive unless one looks against the wall. Went 2 hours to Kepse, the last place among the mountains, & near the sea [next few words scored out], being the old Seleucian Pieriae;[121] – went to the chief of the village, in which all are Armenians, the people came to look at a Frank, & brought me a few medals, which I bought, sup'd & writ. Saw on this mountain, Horn beam, an oak with an acorn with a feathered cup, Ilex, Bay, Boz, Myrtle, Platanus, Laurel, Yew, & many trees you know not. The women in these Christian villages, have to their head dress a thing like the end of mobs, & to that pieces of silver pieces tyed as big as a silver 3d, & at the end a piece of money as big as an half crown.[122] Make oyl for lamps of Bay berries which are a nut with only a soft black skin, which they boyl, the oyl coming from the skin, & dryed are good for the cholic. They eat here myrtle-berries, the broad leaf produces black & white, the later the better.

---

120 After Zeus Coryphaeus, god of that mountain.

121 Pococke describes this city in great detail in *Description 2*, Part II, pp. 182-86 and Plate XXV represents a plan and view of Alexander's Arch, and a plan of Seleucia. 'Kepse' is now Çevlik, near Samandağ in the Hatay Province of Turkey.

122 This rather curious description repeated in his book, as follows: "I observed a particular fashion among the women of Kepse; they wear a sort of cap of silver money, fastened round in rows by holes made in them: Among these there are many antient medals of the Syrian kings, and of the city itself, which are often found here; so that the head of a lady of Kepse is often a very valuable piece of antiquity." *Description 2*, Part II, p. 186.

1st-12th. Went to their poor church, & saw their Armenian service, copyed an inscription or two; – went out & viewed the antiquities of this place, great numbers of sepulchral grotts, an old port, & what is most curious, a way cut through the rock & under a mountain, from the strong town between the hills, to the sea near; reposed ourselves at 2 or 3 springs & met with excellent white figs. A fine plain to the south & about 4 miles off the river Orontes; came home towards night; a boy came in carried away water from the spring where we sat, in two large leaves coulhas, 21 inches long, 15 broad, one he tyed up gathering it round, the other carried in his hand, it is like coltsfoot, – they eat the rest, but it causes ye stone [gallstones].

2d-13th. Walked about Mount Coryphoeus, near a mile from the Town [Kepse], saw the ruins of a convent on it, over which is a sort of strong situated castle, to retire to on the very top of an high hill. Took some refreshment at an Herdsmans cot, who brought us some bread & [next word scored out]; – turned round to the west & went to the side of the hill over the sea about 3 miles to another ruined convent & some hermits cells & oratories in ruins, returned & saw another hermitage 40 feet high in the perpendicular rock to which there is a ladder; & another opposite called St George, as this is coarillus [?]; returned about 3 & writ; they have Porcupines here & I saw one [someone] picking up their Quills.

3d-14th. Went out & encompassed the old city & saw a sepulchral grott, with reliefs & returned about 2; dined & writ.

4th-15th. I designed to go to Antioch but it rained hard, as it did early in the morn the day before yesterday, but clearing up, set out about 11, & went to the present port of the Orontes, – & saw the signs of the old noble basin or Port; went 4 or 5 miles towards Antioch & laid at a village.

5th-16th. Went up the hill called the thousand churches, saw the ruins of a fine old church of St Simon Stylites, lately begun to be repaired by a Patriarch, & by the Turks of Antioch both ye old & new pulled down.[123] In about 3 hours arrived again at Antioch, at the house of Cogia Elias the Greek Merchant & priviledged Drugerman of the English nation;[124] rained in the morn an hour, & before I came here very heavily, but my umbrella that had sheltered me from the sun, defended me also against the rain.

6th-17th. Raining, went not out – dined on excellent spitch cock'd [roasted] Eel of the Orontes went out in the afternoon to see an old Aqueduct, & the

123 The Convent of St. Simon Stylites, who lived on a platform on top of a pillar in Antioch for 37 years, is represented in his Plate XXIV.

124 This was Ilyas Fakhr. On his career, see Bruce Masters, *Christians and Jews in the Ottoman Arab World: The Roots of Sectarianism* (Cambridge, 2001), p. 76.

house of St John Chrysostom & his Father & mother, who they say, being chosen Patriarch of Constantinople, the people here did not care to part with him. [next word scored out] it seems to have been a chapel to him, probably in the place where he lived. Sup'd.

7th-18th. Set out towards noon for Latichea; – viewed a fine Aqueduct in the way, something in the nature of that of Pont de garde near Montpelier, went to the water called Brahema about 5 miles south of Antioch, where was the middle of Daphne & without doubt the famous Temple of Apollo, & afterwards a church built by Galeas in which he deposited the body of Bacylas Bishop of Antioch, & of some other Martyrs, the ruins of which I saw; – went 4 hours further & stayed all night.

8th-19th Travelled ten hours good way over the mountains & through Ourden a Greek village, part of Mount Cassius, & lay under the trees.

9th-20th. I went 9 hours to Latichea, the old Laodacea, pleasantly situated in a very fine country. I went to Consul Parnels [Purnell's] house, by whom I was received with all civility;[125] dined at 3. Captns. Fielding & Long partners here in ships & no other merchants of England;[126] – & meeting a ship just going to Cyprus, I send this. I am three days from Tripoli, where I go to see some places in the way, & then in a week, hope to embark for Cyprus. I thank God I have my health very well, & am successfull in everything. My very kind love to my sister & due respects to all friends.

I am Dear Madam your most obedient & dutiful Son, Richard Pococke

---

125 Either John or Edward Purnell, the latter of whom is recorded as having been English Consul at Latakia from 1750-59, though he could have had an earlier appointment too.

126 Pococke had already met Captain Fielding in Damietta and travelled with him to Jerusalem. See Letters 10 and 11.

PLATE 12 The Holy Sepulchre at Jerusalem, and the Church of Bethlehem, *Description 2*, Part I, Plate IV
PHOTO DAVID KANE

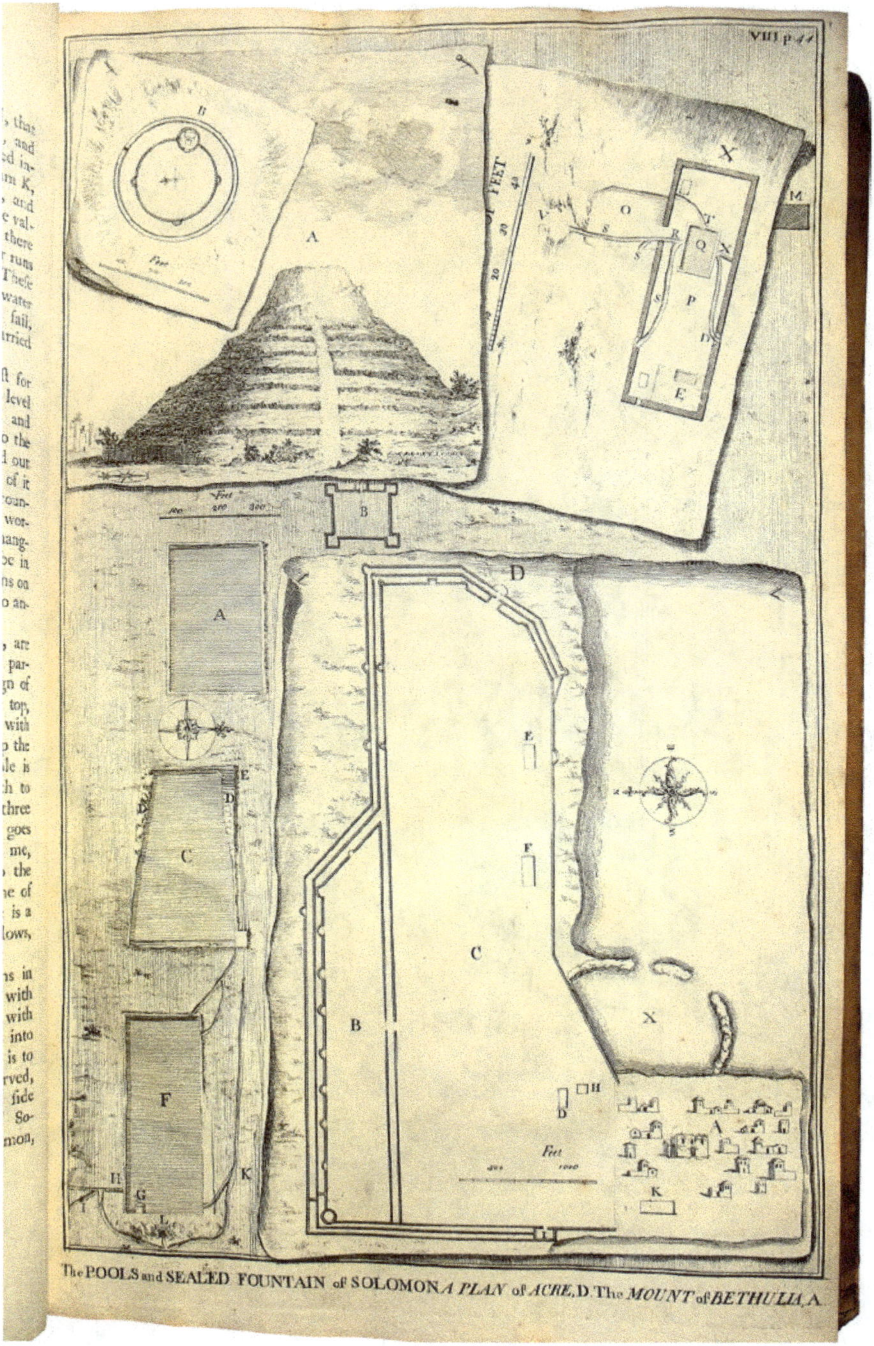

PLATE 13 The Pools and Sealed Fountains of Solomon, and the Mount of Bethulia, *Description 2,* Part I, Plate VIII
PHOTO DAVID KANE

PLATE 14 The Temple of Bacchus at Baalbek, *Description* 2, Part I, signed by the engraver, Samuel Wale, Plate XV
PHOTO DAVID KANE

PLATE 15 The Semicircular Temple at Baalbek and the Great Stone Wall, *Description* 2, Part I, Plate XI
PHOTO DAVID KANE

CHAPTER 7

Accounts from the Holy Land, Lebanon and Syria

The 13 letters in this chapter (Accounts 7-19, inclusive) comprise detailed accounts of the biblical towns of Jerusalem, Bethlehem, Jaffa, Acre, Nazareth, Tiberias and Safed, together with accounts of Tyre, Sarepta (Sarafand), Sidon, Cannobine, Tripoli, Mount Lebanon and Baalbek. The final letter, written from Aleppo, is a short personal note to Bishop Milles in Ireland requesting ecclesiastical favours.

The period of this correspondence spans four months, from the beginning of April until the end of July 1738. However, there is such a gap between Letters 18 and 19 (sent from Baalbek and Aleppo respectively), it is likely that a letter from Damascus is missing. This would explain why the correspondence in this collection lacks an account of the famous temples of Baalbek, which feature so prominently in Pococke's book.

JERUSALEM
2/13 April 1738
Pococke to Bishop Milles
Account 7[1]

My Lord

When I landed at Joppa from Damiata I went to the house of Simon the Tanner by the sea side; which according to tradition, was where the Hospitium of the Latin Fathers now stands.

Joppa is situated on the side of a hill, on the sea. It is no harbour for ships only for small boats, which are shelter'd from the weather by a ridge of rocks that run along, at some distance before the Quay. The great vessels anchor a league off, and in bad weather are often obliged to go to sea.

They have a great trade here in Soap, which is made at Ramah, Lydda, and Acre, as well as at this place, and they export a great quantity of cotton in small barks to Acre, to be shipp'd off for other parts, and these barks often return loaded with soap in order to be shipped off here. It is that fine soap, which the perfumers sell in London, by the name of Sopa of Joppa, and is made of the oil

1 BL, Add MS 15779

 | DOI:10.1163/9789004440050_008

of olives, in which these countrys abound, and which makes it so much excel the manefactory [sic] in our parts. All Egypt is supplied with it from hence.

The most curious thing in Joppa is the manner in which they are supplied with water, which is by digging wells along the quay close by the sea side, where they have fresh water that never fails. The story of Perseus, and Andromeda is related of this place, and the grave St Jerom says that the ring remained in the rock in his time, to which she was tied. This town belongs to the Kisler Aga [Kızlar Ağası] or head of the Grand Signiors black Eunuchs, who sends a Governor to this place, who has a soldiery under him, but they can be of very little use in the country against the Arabs, because if they happen to kill any one they are obliged to pay eleven hundred Piasters; which is about £200 and sixteen changes of raiment for the blood; which the soldiers of the Grand Signor are not obliged to.

A mile Eastward of Joppa on a rising ground are some signs of the foundation of a church; which they say was built where the house of Tabitha stood, who was raised form the dead by St Peter, and on hier festival day the Greeks celebrate the sacrament there.

Ramah, called by the Arabian Ramely, is three leagures distant from Joppa, in the way to Jerusalem; and it is generally agreet that this is Arimathea, the city of that Joseph who took so much pious care of our Saviours body, and laid it in his own new tomb. There the Hospitium of the Latin Fathers is, they have an old well built chappell where the say Nicodemus had a house. There were two large churches here with well built square towers; one dedicated to St John, the other to the forty Martyrs. I have e t think that this place flourished and as much enlarged in time of the Holy war, and there are great ruines of well built arched houses.

A league North East of Ramah is Lydda, where St Peter cured Eneas. Here St George the Patron of England suffer'd as a Confessor, and some say as a martyr; tho' I think authentick history is somewhat dark in that respect. The city at that time, was called Diospolis, and Justinian built a church here to St George, which I could see by the ruins was a very fine one, and built before the good architecture, was corrupted by the Gothick taste. A King of England is mentioned to have rebuilt or more probably repaired it; which doubtless was Richard the first in the time of the Holy wars. The Greeks have the East part for a church, and the Turks the West for a Mosque having a great veneration for this Saint. The East end of the nave and of one Isle, which are in a semi circular form remain with the arches over them, and one pillar in the body of the church. It is all of fine hewn stone of excellent masonry both within and without. The pillars are large, square, and of the Corinthian order, with a round pillar of the same order against each side of it, and on each side of the entrance of the semi circle is a round pillar with a Corinthian pillaster behind it. In a line with the Captial is a very particular Cornish [Cornice] round the semicircle, being a sort of continuation of the work of the Captial of the pillar; the upper members projecting and under each of them is a sort of volute some thing like that of the corner foliages in the Corinthian capital. Under that is a large leaf; and the lower part of that is covered with two small leaves, the Capitals as usual having only one small leaf to the great one by reason of the lessening of the Capital, which in these pillars is only of two such foliages. I have here endeavoured to describe the manner of this Cornish. [diagram of cornice]

I happened to be there on a festival day, when the Greeks were celebrating the Sacrament.

Lydda is a small town but in a very fine rich country, as is all this plain to the mountains which are at about ten miles distance. Ramah and Lydda are indeed the gardens of Jerusalem; from whence that city is supplied. After having passed this plain the remainder of the way to Jerusalem for ten or twelve miles is over rocky mountains.

Jerusalem and Bethlehem and the places very near have been so often described that I shall not trouble your Lordship with a repetition of those things, but only make a few observations on some particulars in Jerusalem, and give your Lordship an account of the places I saw at a distance from that city.

I examined as well as I could into the situation of the old city before it was destroyed. The Hill Acra on which Jebus was situated is at the North East part of the present city, as it ws of the old city before the suburb of Bezetha was added to it. It is the highest part of Jerusalem, and now inhabited by mean people, and on the North, which is the highest part of it, the city wall is built a furlong. Tho' the North part of the town, on each side of this heigth [sic] was not strong by nature, yet that defect was supplied by art; a very deep fossee [earthworks] being cut down in the rock, which they are continually filling up with the dung of the city, so that now this fossee is only about a hundred yards long. The North, which expect the foresaid heigth was the weakest part of the city according to history, was defended by three walls, there being a deep valley, as a natural fossee on the South East, and West, and the wall was built round on the edge of the hill. Mount Sion was afterwards added to the city by David, which as I take it had its particular walls round the top of the hill; thourgh without doubt the built a wall along the valley from one hill to another The same may be said of Mount Moriah; the temple having its particular fortifications, the pool of Bethesda being one part of the strength of it to the North, which now appears as a very wide and deep fossee, at present extending near half the length of the Area of the Temple. The lower pool also at the foot of Mount Sion might contribute to the defence of the city, which is like a long well fossee, receiving its waters from the upper pool or pool of Gihon, which without the city is a receptacle for the rain water that comes from the mountains on the North East.

Mount Acra and mount Sion appear as hills within the city; but Mount Moriah where the Temple of Solomon stood is now a flat, and commonly called a hill with regard to the valley on the East side of it. Bezetha was a large suburb North of the city, and at length taken into it. I traced the foundations of the walls on the West side, but being all rock to the North, and it being natural to suppose that the stones were carried away, I could see no signs of a wall but suppose it went along an advanced ground; over a shallow valley to the North and so on the East over that valley, which is in a line with the valley of Jehosophat, and as the North part of it extended half a mile beyond the present walls Westward an further Eastward, so it was near as large as all the rest of the city, but was inhabited by Tradesmen and Artists, who doubtless had open places for the conveniency of their arts, and probably there were in it many gardens. Bezetha in the Eastern languages signifying [sic] a garden, from whence it might have its name.

To give a better Idea of the present state of Jerusalem and what I judge to have been the old circumference, I have made this rough plan the single line is where the old wall went the interrupted lines where the new walls are only supposed and is double lines where the new and old walls are supposed to have been in the same place. [Diagram of Jerusalem with main buildings and gates denoted A-G and 1-8 respectively]

The present walls were finely built of large hewn stones, about two hundred years ago, after it was taken by the Grand Signor; but they [there] are no fortifications in these times. The circumference of them is about three miles.

Though the sentiments raised by seeing the Holy places is not to be express'd, yet I cannot but say, that I was in some measure disappointed; for I expected to see the very place where our Blessed Saviour lay, or at least the very sepulchre, only adorned or cover'd over with marble; but it is a place like a sepulcral [sic] grott built above ground, and I suppose exactly like the true one, which they say is in the rock under it. St Helena seem[s] to have had a wrong taste, in covering every thing with ornaments that related to sacred history; for it would have been much more affecting, to have seen the very tomb, in its natural state, and all the other places, and only to have built churches over them.

The area of the Temple of Solomon, not commonly to be entered by Xtians [Christians], takes up near a quarter of the city. The middle part of this Area is raised about eight feet above the rest, and in the middle of it is a fine Octogon [sic] Mosque, called the Mosque of the Temple of Solomon. It is cover'd with a Dome, the Diameter of which seems to be about half the Diameter of the whole building. In each of the eight sides appear to be seven oblong windows semicircular at top, like the most modern church windows in our countrys. I conjectured that they might be about five feet wide, and that their heigth was twice and a half as great as their breadth. But upon a further view I found there were in reality but five windows, on each side, the windows next to the corners being false ones. The peers between them seem to be about three feet wide, and they go down like a Pillaster to the ground. The Dome and the flat part of the roof of the Mosque, are cover'd with lead and the walls as low as the bottom of the windows are cased with a beautifull sort of tile of green and other colours, that looks at a distance like mosaick, and which over the windows are placed like three frieses [sic], and round the sides of the Dome they are set in square compartments with lozenges within them. Below the windows the walls are cased with white, marble, and I am informed that it is exactly the same within as without. There are four doors into it, and before the Eastern door is a Dome supported by an Octogan Colonade. The advanced Area on which this Mosque is built is ascended to by two flights of steps every way, but to the East, where there is only one, and at the landing place, from each of these are Colonades from three to five arches with their entablatures all of the Corinthian order. Their not corresponding to the doors of the Mosque, nor being in any way regular, with regard to it, made me conclude that they were probably erected as grand entrances to the Area of the Temple of the Lord which the Xtians built after the vain attempt of the Jews to build their Temples.

At the South end of the Area is the Mosque of the Presentation of the Blessed Virgin which was a church under that title, and is supposed to have been the very spot of the Temple of Solomon. I was informed that it is built on the model of the church of Bethlehem, with two rows of marble Corinthian pillars, on each side in the style of the Temple of Balbeck. Near the place towards which they pray is placed in the pavement one of thos[e] impressions of a foot, taken out of the chapel of the Ascension on Mount Olivet, which is said to be the impression of our Saviours foot when he ascended up to heaven; the other I saw in that chapel. By a garden I went to the back part of this Mosque. In the walls of it and of an adjoining building that was a nunnery are several rows of stones upon one another of a very large size, and under it are arched works, which were formerly enter'd by strangers but now are shut up, and in the walls East of the Area of the Temple are very large coin [coign/quoin] stones between twenty and thirty foot long.

My curiosity led me to see the misery of the Jews in this city, were they once reigned in so much splendor. Instead of the glorious Temple of Solomon in which they worshipped with so much solemnity, as the people of God, they have now only a poor Synagogue half under ground. The women, as the greatest contempt they can put upon them, are obliged to go unveiled, and as it is a general custom among the Easterns not to wear their hair, but to shave their heads; so the men among them as a sort of ignominy, are obliged to wear their hair in such a manner, as that a lock may appear on each side before. This custom of the Easterns in shaving their head may help to explain a passage in the 11th chapter of St Pauls first Epistle to the Corinthians and the 14th verso, where he says "Does not even nature itself teach you that if a man have long hair it is a shame to him," which tho' it does not seem to be true in our part of the world; yet it is very much so in these parts, where that Apostle was born and educated.

I am my Lord

BETHLEHEM
6/17 April 1738
Pococke to Bishop Milles
Account 8[2]

My Lord,

My last letter to your Lordship was dated April 2d-13th form Jerusalem; in which I have your Lordship some account both of my journey thither and of the city itself.

The two most remarkable things for art in the valley of Jehosophat East of the city, are the tombs of Zachariah and Absalom, which latter is commonly called the pillar of Absalom. The sepulchre of Zachariah is a square tomb cut out of the rock like a little Temple with a Pyramid at top, the base of which is of the same dimensions as the tomb. It has a very simple etablature [sic] without any distinction of friese [sic] and architrave. The pillars, which project about three quarters or a circle, from the building, are of the Ionick order two on each side with a quarter pillar, and pilaster in a line with it, that mkes the Angle; they are finished only in the front next to the valley, there is no visible entrance into this sepulchre, though I suppose the way to it was under ground, as probably it was in Absaloms, the perpendicular rock is about six feet distant from it on three sides as it is likewise about Absaloms; tho' there the rock is not so high. There are holes broke into Absaloms tomb into which I enter'd; the lower part of it is in the same taste as that of Zachariahs, only it has a friese [sic] of Metopes, and Patinae, and the capitals of the Ionick pillars are all finish'd. It is a square sepulchre cut out of the rock for some little heigth above the room

2 BL, Add MS 15779

and entablature, over which one si tiers of stones laid one upon another without any other stones or building within them. About six feet above the entablature is a Cornish, and three feet above that another above which the building is round, and about three feet higher a half round from whence it slopes so as to make a Cone for the two upper teers [sic] of stone, all which is crowned with one stone about ten feet high, which lessens so little that it looks like a round pillar at the top of all, is an ornament resembling a flower of a flame.

I was in the inside as high as the top stone, where it is all rough above the room, and I measured every part of it, and have such a rough draught, as that I hope to be able to lay before your Lordship an exact design of it, tho' I could not forbear giving such an imperfect description of this curious piece of antiquity. It is certain that there is a heap of small stones raised so high about it, that it greatly disfigures this monument. Within the rock, out of which this is cut is the sepulchre of Jehosophat consisting of six rooms cut out of the rock, and the entrance is adorned with a very beautifull Pediment. Round these tombs the Jews bury, and there are a great number of grave stones over them, they being very desirous when they are old to come and dye at Jerusalem, in order that they may be buried here.

I went all over the village of Siloe, which is a little father South on the East side of the hill over this valley. There are many sepulchers there cut out of the rock, one particularly is like a square house, and a little North of the village is one work cut out of the rock wich the Arab told me was now called Gethsemane, that village probably extending from this place Northward towards the place which they shew for the garden of Gethsemane half a mile from it, and about the spot where this village might stand or between it and the garden they shew the place where Judas hanged himself, and where there grows a single tree to distinguish the place.

North of the city near a mile are what they call the sepulchres of the Kings. One first enters a large square court which is cut out of the rock, uncoverd at top and about ten feet deep. One enters the sepulchres, by a sort of Portico, over which is a very fine entablature, and on each side of it I could see there was a Pillaster, which I take to be of the Dorick order. It is a rich friese of Metopes and Patens, and some variety in the middle. One thing in it is very particular; below the two plain members of the Architrave, is a third which projects and is enriched with reliefs of flowers and foliages. The entrance from the Portico is by a low small door into the common room which leads into four others, cut all out of the rock with great art, and with small apartments round them for depositing the bodys, and some have within them small square ones, as for young people. Two of these rooms are in the same manner, except that from one, there is a descent into a small room, in which are three benches cut in the rock with the top made archwise for depositing three bodies; a third has at the further end onl three steps of the width of the room on which they put the Sarcophagus's. This is within one of the four which is in a different manner from the rest, there being a door in the middle of each side, as into so many narrow rooms on each side of which is a bench or nich for depositing a boy, and a square place at the end and on each side of these is a little apartment or cell for single bodys. A fourth has in it only three long niches for depositing bodys. In some of these are Sarcophagus's of stone with semicircular tops, both adorned with releifs [sic] of flowers, all with the covers taken off, and some broke.

What is mentioned as the most curious thing in these sepulchers is the stone doors, but it is impossible that they could be cut out of the same place, unless the workmen could be on both sides and then it would be very difficult to cut the hinge above, which is above to receive it, and this also made it difficult to account how this part on which it turns above could be let in. But examining the hinge of the same nature on which it turns below, I could easily account for this piece of art, for the hinge below is not an inch deep, so that by cutting a channel in the floor, the depth of the upper hinge to let in the door, and to bring the hinge under its soccet [sic] they had nothing else to do but to raise the door and put in a stone very nicely to fill up this channell. The doors are left rough within and cut without like pannel work.

Some antients having mentioned the caves of the Kings somewhere, may be the reason why authors who have wrote of Jerusalem call these the Sepulchers of the Kings which were on Mount Zion, and at the same time they are at a loss to find out the sepulchre of Helena Queen of the Adiabeni, which was this way also. But I think I can make it appear, that this is her sepulchre, and she being buried here and one of her sons a King, might be the reason why it might be vulgarly [commonly] called the Sepulcher of the Kings, in case they have that name from tradition. The Sepulchers, which Queen Helena built for her self and her family are described as having over them three Pyramids; and Villalpandus speaking of these sepulchers of the Kings mentions one Pyramid, as standing over them in his time; the other two having doubtless been destroyed before, as the third has been since that author wrote.[3]

All pilgrims who are at Jerusalem at Easter take the journey to the mountain, where our Saviour was tempted; to the Fountain of Elisha: to Jericho; and Jordan. The Caravan does not go to the dead sea. I took that journey after, I saw

3 Juan Bautista Villalpandus (1552-1608) was a Spanish Jesuit famous for his commentary on the book of Ezechiel, which he co-wrote with another Spanish Jesuit, Hieronymus Prado (published in three vols from 1596 to 1604).

the Greek Convent of St Saba over the brook Kedron, half way between Jerusalem, and the dead sea. The situation of it is very extraordinary. It is a triangle two sides of which are bounded by two perpendicular rocks, and the third is over the brook Kedron; from whence one looks down from a high precipice. In this retirement in cells cut out of the rock, lived John Damascenus, Euphemius, and Cyril the Monk and at one time they say above ten thousand monks.

The water of the dead sea is very clear, but looking close to it, I could plainly see an oily substance mixed with it. There is likewise a sulphur in it, as I judged from a smooth white incrustation on the stones, which are on the bank, but what prevails most in it is salt, to that degree, that it seems to burn the mouth, and has also the same effect as a strong allum water. I went into it, it is not cold; the water bears up the body every way withut motion, and when one strikes with ones legs in swiming [sic], the water bears up the body so high, that one strikes with ones legs out of the water. They have a notion that if any one attempted to swim across it would burn the body. I think it cannot be more than three miles over, though some say that it is ten, and the high mountains on the other side may in some measure deceive the judgment. It is as salt on its banks near the place where the Jordan falls into it, as in other parts, and it is the common opinion that the waters of Jordan pass through without mixing with it. And indeed I thought I saw some marks of the stream passing through it. The Bitumen which is like pitch, but has an ill smell is thrown on its banks after great winds, which tis thought, by the motion of the sea breaks those rocks of Bitumen, which are supposed to be made at the bottom by subterraneous fires. It is remarkable that at a Mosque called Moses, about three hours from it, is found a stone like our stone coal but hard, and burns, and has the same smell as this Bitumen. Where the water overflows into pools, it makes a very strong salt. They take it out in very large pieces like rock salt, and it is this which is used at Jerusalem, and all over this country. The trees and other vegetables near it grow very rank, but if the sea happens to gain upon them, it kills the trees: I took of the dead wood of some of the trees that were in the very sea, and under the waters, I saw no fruit trees, but they say all fruits as Pomegranates Olives &c that grown near are full of dust and smoak within. When I found the leg of a Crab on the banks, I though that it was a proof that there were fish in it, contrary to the common opinion; but they certainly have crabs in the rivers here, and no sort of fish lives in this sea. They say that large birds, which fly high can fly over it, but that birds that fly low would be killed by the vapour which comes from it, especially in the three hot months, when the air about it is so malignant, that it is dangerous to go to it, and the Arabs then put some thing before their faces to prevent their falling in the Malignant air. I could not find that the bottom of the sea is hot, nor could I see any smoak or vapour over it, more than over the river Jordan which I observed both near, and at a distance. To be able to judge the better of the nature of this sea, I have taken with me both of the water, and of the salt, and also of the stones incrusted, and of the Bitumen.

Enquiring about the pillar of salt into which Lots wife was turned, they now say that it is twenty or thirty miles further South, though it has been always mention'd as towards this end. The Greeks talk of a stone pillar on the other side, but I cannot find any body that has seen it, since Josephus, not even St Jerom[e] who was so many years in these parts.

Going from Bethlehem to St Saba I passed by the mountains of Engaddi and the cave or grot where David cut off Sauls skirt.

The church and sanctuary of Bethlehem were in the possession of the Greeks; when the French Embassadour [sic] had audience of the Grand Signor to acquaint him with the birth of the present Dauphin, the Grand Signor asked what he could do on that occasion which would be agreeable [sic] to the King; on which the Embassadour desired the church of Bethlehem for the Latins, which was granted, and the fathers made a partition between the cross and the body of the church, and would have permitted the Greeks to have celebrated on a side alter, but as they could not have the privilege of the high altar, they have entirely forsaken the church, keeping to their chapel in their convent adjoining to it. The Armenians have also a convent here and have repaired the fine room of the Schol [sic] of St Jerom and turn'd it into a church.

I am my Lord

JOPPA
23 April/4 May 1738
Pocock to Bishop Milles
Account 9[4]

My Lord

My last letter to your Lordship was from Bethlehem, dated April 6th/17th. From that place I went seven or eight miles South to Thecua, a city often mentioned in Holy Scripture, and was rebuilt by Rehoboam son of Solomon. It was the country of the Prophets Amos, and Habbacuc [Habakkuk]. It is situated on a hill and there are very great ruins of the old city, many walls, and arches of houses still standing; particularly a large building which seems to have been a

4 BL, Add MS 15779.

Castle, and a church, about which there are some broken marble pillars, and a great number of bases of pillars, and a very fine deep Octogon [sic] font of red and white marble with some releifs on the sides. About a mile further on the hill is the ruined church of the Pantaleon. Three miles East is a very large grott called the Labyrinth. The entrance is difficult, at the side of a high hill allmost perpendicular there being a terrace to the grott. A narrow passage leads to a very large high room, with several natural pillars supporting the rock, and between are cavities in the ceiling like so many domes. From this go several narrow passages a great way in, no one venturing to go very far, least they should loose [sic] their way. There is a tradition that all the people of the country to the number of thirty thousand retired into this Grotto to avoid some bad air, or it may be a hot wind, which is sometimes very fatal in these countrys. I am very much enclined to think that this is the Cave of Adullam, to which David fled from Saul, and whither all his family, and those that were in distress fled to him, to the number of four hundred men; though they shew another place for this Cave.

About five miles South East of Bethlehem is a mountain called now by the Franks Montagna D'Francesi, though its true name is the mountain of Bethulia, from a village of that name near it. It is a very high single hill, and the summit of it is so formed I suppose by art, that it appears like a very large mount on the top of the hill. This place was held by the Franks, the Knights of Jerusalem, forty years after the taking of Jerusalem the last time. It must have been a very strong fortress. The hill is all laid out in terraces, the lowest of which is about thirty feet above the foot of the hill, and the ground from it very steep. A strong wall surrounded the foot of the hill, and to the East (where half way up the hill there was a very gentle ascent) a deep fossee is cut down in the rock, which separates the hill from that part. At top the fortification was round, with a round tower built to the East, and three semicircular towers round the great fortification, pretty near equally distant from one another. At the North foot of the hill are great ruins of a church, and other buildings, on a hanging ground, on the top of which went the outer wall, and West of this a cistern and a fine square pond, with an Island in the middle, on which are ruins, as I suppose, of a pleasure house. There were two Aqueducts to this place; one from the Fons signatus or sealed fountain of Solomon, and another from the hills South of that Fountain. There are great signs of improvements about this hill, and the fortification is so beautifull a thing, that I hope to lay a plan of it before your Lordship.[5] It was so very strong that it might have held out, as they say, longer,

5 While there is no corresponding illustration in his book, Plate VIII of *Description 2* presents views of "The Pools and Sealed Fountain of Solomon", described below (see Plate 13 above).

but it was Populus Virorum which they assign as the cause of the loss of it. From the top of this hill I was shewn in a plain to the South East the place where was the famous garden of balsam trees, which Cleopatra inspight [sic] of Herod transplanted into Egypt, having the leave, and being secure of the favour of Anthony.

Four miles West South West of Bethlehem is the sealed fountain of Solomon, his famous pools, and the enclosed garden. The way to it is along the side of a hill over a narrow valley; at the end of which is a hill, and between the valley and hill is a very gentle ascent, or rather plain, of about half a mile long. In the part next to the valley are the three fine pools, one higher than the other, and the water flows from one into the other. Between them, and the hill is the remaining part of the plain, now a very fine feild [sic], and also to the North of it is a plain about three miles round, enclosed with hills, on which is a Greek convent called St George's in repair, but uninhabited. These plains were probably the enclosed garden, and the hill at the end opposite to the pools laid out in the terraces, was doubles place where Solomon had his Palace for retirement and pleasure; the top of it commanding a view of Bethlehem, and of all the country round, and of those small plains, then without doubt cultivated, with the utmost art, and in the most beautifull manner, and above all it had before it those three fine large pools or ponds of excellent water, which must make it a most delightfull place in the heat of summer. This is a situation worthy of the wisdom of Solomon, and not a little ruined village on the steep side of a hill over a narrow valley, with no prospect but that of a dismal hill on the other side, and even out of sight of this water, which is generally called the village of Solomon, and thought to have been his Palace, and the place he seemed so fond of.

The sealed fountains are three in number. They rise in two arched rooms underground, at the foot of th hill first mentioned, towards the North East corner of it. The streams flow plentifully out of the rock into an oblong cistern in the floor of the first room, out of which the water runs into one pipe by two holes one over another. This pipe divides into three. One goes to a large square castle on the North side of the first pool, another goes to the first pool, and a third goes to Bethlehem, and Jerusalem. These pools have been lately [recentlu] repaired. They are surrounded by a low wall, which is built very strong at the head, where the weight of the water comes, each row of stones battening in about two inches on the outside. They are said to be very deep in the middle. On the South side the rock has been cut away in order to make the pools, and though the figure of them is now irregular, yet I do not doubt that they were at first made regularly. The entrance to the sealed fountain seems not to be the old one; but I saw two holes in the top of the arches, which probably had doors to them, that were sealed and only opened to take out water for Solomns own use.

I took the best account I possibly could of all these things, and a section of the pipes by which the water goes to Jerusalem. It is by around [a round] earthen pipe of about ten inches in Diameter laid with a fine cement on stones which are cutt in a semicircle to receive it, it is covered with such another stone at top which rests on the lower stone; and the whole is cover'd with large rough stones cemented together, a wall being built along as a foundation either higher or lower, according to the level, and sometimes the rock is cut away; but it is no where built on arches except across the valley, which is West of the walls of Jerusalem, where are only nine low arches, the pipes being placed all round the hills, for thirty miles about, though the fountains are but ten miles from Jerusalem. These run into the pools, and these pipes, two other fountains, from the hills to the South; and a pipe as I observed, went to the mountain of Bethulia. Bethlehem is supplied and its cisterns filled; and then the hill being cut across, the pipes are laid under ground, and go between the church and town of Bethlehem, and in all other parts, for so much as I could observe, it goes by the sides of the hills, and when it approaches Jerusalem goes round the hill of evil council, crosses the valley, and winds round the side of Mount Sion, and so to the Temple; supplying allmost all the places that are not on higher ground, as is above half the city where they build cisterns under all their houses as we build cellars, and the houses being all arched over, and terraced at top, these cisterns receive the rain water, and supply the greater part of the town.

The Convent of St John is six miles North West of Bethlehem. About a mile from Bethlehem is a valley, that way, where they say the army of Sennacherib King of Assyria was encamped when the angel slew one hundred, and eighty five thousand men. About two miles further, and a quarter of a mile out o the road to the left is the fountain where it is said St Philip baptised the Eunuch. It falls very plentifully from a high ground, against which is built a very handsome wall, and arch adorned with Corinthian Pillasters, out of the middle of which the water flows. This is supposed to be the work of the Empress Helena as well as a church above, of which there are great ruins. The common way to it now used does not seem practicable for a chariot, but on the other side of the valley is a plain road crossing over at this fountain. This is said to be the valley of Eschol [Eshcol], where the spies took that bunch of grapes, which was so large, that they were obliged to carry it between two men and they shew the side of a hill exposed to the West which one ascends to go to St Johns, and whence they say the grapes were taken: it is inded a very rich spot of ground.

The Convent of St John is built on a low hill among the mountains. It is a new building very neat, and a plain handsome church with an altar finely adorned with releifs; in the spot where they say St John Baptist was born. The Latin fathers have not been quietly established here above forty or fifty years. This was, as they say, the house of Zacharias in the city; and about half a mile from it are the ruins of a church, and convent, a little chappel, and stairs down to it, which they say was the country house of Zacharias, and the stairs on which the Virgin Mary and Elizabeth met, and where the Blessed Virgin abode [lived] three months. In the way to the desert is a stone on which they say St John preached, and in the desert on the side of the mountain by a fountain over a deep valley is the Grott where Elizabeth and St John are said to have lived together, till he was three years old, flying thither to avoid the cruel decree of Herod. Over it remains a small church, and there are some ruins of a Convent as I suppose higher up; and not far off is a small church partly in ruins, where they say Elizabeth was buried, dying when St John was three years old, who is supposed to have staid in this wilderness till he was thirty years of age, after which he went into the wilderness by Jordan to preach and to Baptize. There is a tree here called Caroub [carob], which bears a fruit like a bean, the seeds of which are small and flat. They now eat the shell of it, which is very good, and is supposed to be the locusts on which St John fed, and not what I have seen shewn for it, which is cassia [Chinese cinnamon] and does not grew in these parts.

I am my Lord

ST JOHN D'ACRE
6/17 May 1738
Pococke to Bishop Milles
Account 10[6]

My Lord

Returning from St Johns convent to Jerusalem I went out of the way to the North to see the valley of Elah, where David slew Golia[t]h. This valley is a very proper place for a battle, as the hills on each side are for the disposition of two armys. I went to the bed of the brook, out of which David is supposed to have chosen five smooth stones (which like most others in the Holy Land) has no water in it except after rain.

6 BL, Add MS 15779.

In the road from Bethlehem to Jerusalem at the sepulcher of Rachel I examined to see if it were built on twelve stones as it is said representing the twelve Patriarchs: but being a burial place of the Turks the ground is risen up; though it might very well be built on twelve stones being a Dome built on four arches supported by four square peers, each of which might be built of three stones of no extraordinary size. These arches are lately built up by the Turks who do not like that the Jews should resort to places, where they pay their devotions, as they do here, having a place without it to pray in.

Half-way between Bethlehem, and Jerusalem is the Greek convent of St Elias. In it is a plain lofty church, and the situation is very delightfull, commanding a view of two very sacred places, Bethlehem and Jerusalem. Near it is a ruined church dedicated to the prophet Habbaccuc, who they say was taken from that place to Babylon, by the Angel.

From Jerusalem I took the journey from Emmaus which is about eight miles distant. Two miles from Jerusalem in the way there are a great number of sepulchers cut out of the rock like the sepulchers of the Kings, and some few of them as beautifull as those, but only of two, or three rooms at most. One I entered, which was unfinished, and saw the manner in which the[y] worked out the stone in one piece from the top to the bottom. Where these sepulchers are I have reason to think that they had their country houses, as I observed before; most of them have a sort of Area cut out in the rock, on which I suppose their houses were built. One great use of Grotts in these countrys was for depositing their dead. Another use of large Grotts was to keep their cattle in by night, in the winter, as they do at present, familys living there with them, as many do all the year round, and it is very probable in those times when they lived under tents, that the familys in great rains which they have here in winter, retired in general with their cattle, into these Grotts, which accounts for the great number of them. And there is still another use of Grotts which is for cisterns to which they make channels from the hills to convey the rain water. And as some of these cisterns are turned into Grotts for cattle, so some of these sepulchers of the Judges are turned into cisterns. They are supposed to be so called, because some of them might belong to certain ruling persons of the city, and not because they were the sepulchers of the antient Judges of Israel. All the way to them is by pleasant feilds [sic] of corn, planted with Olive trees Fig trees, Almonds, and Apricots, and is by far the most pleasant spot about Jerusalem, and most inviting for a country retirement. Here the Jews come out on their Sabbath day, to pass their time in ease, and diversion; as they go on the other side of the town amongst their sepulchers at other times, when they would employ their minds another way.

Half way to Emmaus on a hill a little out of the road are the walls of [a] well built church, in memory of our Saviours meeting the two Disciples there after his Resurrection. I took the way to Emmaus which is by the high hill of Ramia, where was the dwelling place and sepulcher of the Prophet Samuel, which at present is called Samuele by the Turks. At some little distance, before I came to it, I saw, by the way a very fine large open cistern, cut out of the rock, with an entrance to it, and at the top of the front of it, cut out of the rock like a wall is a small cistern [..] through on each side for watering cattle. Just at the top of the hill on the right hand of the road, is a sort of mount, with regard to the lower ground, tho' on a level with the top of the hill, and shaped in a semicircle. One who is fond of conjectures might imagine that possibly Samuel might sit there to judge Israel, it being a very convenient situation to have a great number of people round as auditors. The very middle summitt [sic] of the hill (which is a rock) is cut down regularly seven or eight feet deep, for a large space, which was without doubt for receiving the rain water. The church which was built over the sepulcher of Samuel is turned into a Mosque, not to be entered by Christians. It is on the top of the very high hill and is seen all the country over. A Turk informed me that there is no sepulchral Grott, but only a raised tomb covered over with a silk covering after the manner of the tombs of their saints. History gives us an account that Arcadius carried the bones of this great prophet into Thrace. Just below the summit of the hill on the North side is what they call the fountain of Samuel. You enter an open Grott, from which is a small door to another Grott cut out of the rock, in a regular manner, and sunk down some feet below the door. This inner Grott is always full of most clean excellent water, which springs from the fountain in it. North of this hill of Samuel, is a very fine plain or valley encompassed with hills. It is said that Joshua was here when the sun stood still. There are two hills in this plain; the Western hill has two summitts, it is finely improved and a delightfull situation. On the Northern summit is a large village, and when on asking the name of it, they told me it was Geb I was convinced that this was the place where the sun was ordered to stand still on Gibeon, and the moon in the valley of Ajalon, and that on this hill was the antient Gibeon and that this was the valley of Ajalon.

About three miles further on the same chain of mountains, but towards the West part is Emmaus, where are very great ruins of the old city. The church which was built on the spot where the house of Cephas stood is a long solid building. In it is a very large round shallow vase of white stone, which probably served for a font. The large Cemitery [sic] was encompassed with a fine wall of hewn stone, and within it are ruins of several arched buildings which might have been the habitation of the Priests.

All these countrys are under the Bashaw of Damascus. Jerusalem, and a district round it is under a Governour sent by him, and approved, and confirm'd (if he has interest) from year to year to Constantinople. The present is the best they ever had, and is a particular Protector of the Franks, who support his interest at Constantinople, at no small expence. There was a nail in the Temple, of which they had a prophecy that Jerusalem should be taken by the Franks, whenever it should drop from the ceiling; which happened not long ago, on which a tumult arose, and they were about to destroy all the Franks. But the Governor sent to them to keep the gates of the Convent shut and dispersed his soldiery about to reason with such as talked on this subject, and if they could not silence them to dispatch them forthwith, which effectually put an end to the tumult.

There are in Jerusalem about fifteen hundred Greeks four hundred Armenians and as many Roman Catholicks. Bethlehem is in a manner a Christian town, having in it, a thousand of the Latin church, and about two or three hundred of the Greek and Armenian. A village near called Botes halak is entirely Greek. In the small village of St Johns there are only a few Roman Catholicks, here and in Bethlehem, they live by making breads and crosses.

The Greeks are proud haughty and furious, and the aversion that there is between them and the Latins, is only exceeded by that which is between the Greeks and the Cophtis; though the latter have much the same ceremonys, and were once in a manner but different Patriarchates of the same church, but the Cophtis can never forget the intolerable insolence and Tyranny of the Greeks, when they had the upper hand in Egypt.

The Armenians are the people that seem to stand the fairest withall, the least with the Greeks, being a quiet silent people, and civil to every body but withall of the deepest subtlety and Policy. About a hundred years ago they united with the Latin church, and continued so for some years, and obtained of the Latins, by an order from the Pope, several chappels and privileges in the church of the Holy Sepulcher, and soon after they separated from the Latins, and kept all that they had got from them. But I must observe that the Armenians are the most pompous and decent in their Religious ceremonys, especially their processions which exceed any thing I ever saw of the kind. The Latins I beleive [sic] exceed them in all their other Pontifical functions.

As to the state of this country, with regard to those of the Mahometan Religion; it consists of Turks, Arabs, and peasants natives of the country, and improvers of the Land. These have not power, but are worse in their nature than the Arabs themselves, who are remarkably hospitable on all occasions and they even take it ill, if you refuse to eat with them. I have had great experience of them even in their tents, in which I spent a night, and a day, being carried to them out of fidelity and a particular case of me tho' I could have excused their super abundant kindness. Their fidelity is most extraordinary, when they take any one under their protection, or make an alliance or friendship. A Sheke of a village to whom I was recommended, would (I am well assured) have killed his own nephew, if he could have caught him, because he affronted me before his face, and money being exacted of me unknown to him, when it came to his knowledge he obliged them to return it, and sent it after me. When they rob they never murder if no resistance is made. In these parts they are not under great heads as in Egypt, but one is over a certain village, and the tents in a certain district about it. They are chosen out of the family of the deceased sheke, and approved by the Government, to whom they are obliged to pay the tax which is fixed on their territory. With these Heads you go in safety thro' their country, and the country of their allys [sic], paying only a small Caphar [duty] in any village you go to, which is pretty much fixed by custom; but has regard to the dignity of the person. They are often at war with one another, when allys go to the assistance of the Principals. In Egypt I saw an Arabian battle. They have a great aversion to the Turks, and the Turkish government, but at the same time they are the only people they fear, apprehending the consequences of injuring them, if it should be known of what part they are of. They do not care to come among fire arms, their bravery depending more on the weakness of others, and their love of pillage, than on any extraordinary innate courage.

I am my Lord

NAZARETH
11/22 May 1738
Pococke to Bishop Milles
Account 11[7]

The voyage from Joppa to St John D'Acre is always perform'd by sea; the land journey being impracticable, by reason that the parts near Joppa are infested by the Arabs. This computed to be about eighty short miles. Ten miles from Joppa, I saw on the cliff over the sea, a round fortress that was very strong towards the sea, the cliff being laid with stone, after the manner of a modern rampart. They say this was a fort, held by the Venetians, during the time of the holy war. Near it they embark wood for Egypt brought from the neighbouring mountains, where are probably those large woods mentioned by Strabo between Caesaria and Joppa. I could learn nothing of Antipatris to which place

7 BL, Add MS 15779.

the soldier brought St Paul by night from Jerusalem which the Itinerary[8] places onely [sic] ten miles from Lydda, and it was about forty from Joppa. Half way between Joppa Acre is Casaria by the sea side, or Caesarea Palestina, as it is distinguished from Caesarea Philipi, on the river Jordan near its rise. T'was rebuilt by Herod and called Caesarea in honour of Augustus Caesar, the old name being Turris Stratonis. Having assisted Vespasian against the Jews, he made it a Roman Colony free from taxes, and the governour of Judea from that time began to reside here. It was from this place that Cornelius the Centurion sent for St Peter to Joppa, when he was admonished by that famous vision to go to the Gentiles; and it is said that Cornelius was the first Bishop of this city, in which Philip the Apolstle lived with his four sisters Prophetesses, and here St Paul was kept in Herods palace, and pleaded before King Agrippa and Felix when he appealed to Rome, and on his departure Agabus prophesied of his future persecutions by binding himself with Pauls girdle, and declaring, that so should the man be bound, who owned it.

When we lay at anchor near Caesarea, the passengers spied a sail, and a Janizary aboard, thinking it was a Maltese cruiser resolved to leave the vessel; dressed himself in the worst clothes he could get, that he might not be plundered by the Arabs, and left his money with an European aboard. A panick fear seized the others who did the same, and went a shore to Caesarea returning to us the next morning, when they saw there was no danger. On these occasions the Turkish Captain makes a Christian master of the boat he always had with him for that purpose, and the Cruiser cannot touch the vessel nor the Captain with his twelve men he is allowed.

The danger there is of the Arabs prevented my going a shore at Caesarea, but seeing great ruins, I made a journey to it by land from St John D'Acre with the head of that country, but was much disappointed. The walls are of small hewn stone and almost entire being about a mile round but they are not of great antiquity no more than any thing within them, being built by Lewis 9th of France during the time of the Holy War. From the wall on the South side runs a point of land on which are the ruins of a very strong castle of the same date, being full of peices [sic] of very fine marble pillars, some of which are Granite Cipolino [sic], and a most beatifull grey Alabaster. North of this there seems to have been a small port, the principal harbour appearing to have been to the South, well shelter'd by the point from the North winds, and by some rocks from the South. On the shore to the East I saw some signs of a wall with three towers from the disposition of the ground and a point of land running out from

8 Pococke later mentions the "Jerusalem Itinerary" (see Accounts 12 and 16, below) and all references seem to relate to Maundrell's *A Journey from Aleppo to Jerusalem at Easter*.

it to the South of the harbour a little way into the sea on which there are some very small ruins; and South East of this on a rising ground on the shore the side of a large square castle with a tower on each corner and a fossee on three sides of it. Some peices of wall are seen above ground, but I much doubt if they are of any great antiquity. Within the walls of the city are a great number of ruins of arched houses that probably were built during the time of the holy war, but the whole place so overgrown with briars and thistles that t'was impossible to go to any part where there was not a beaten path. This place is a remarkable resort for wild boars, there being only four or five poor familys in it, who live in perpetual fear of the Arabs, against whom their poverty is their only security. There is no ruin of any note but a large church without doubt the Cathedral and probably built where the house of the Centurion was who they say was first Bishop of this place. It appears to have been strongly built, and to have been destroyed as the castle was by war. By what I could judge it was built in the style of the Syrian churches three naves ending to the East in a semicircle, in which were there altars: tho' I could not see within the city the east remains of Roman antiquity, yet to the North of it on the shore I saw the foundations of the walls of the old city extending for near half a mile to two Aqueducts by which water was conveyed from mountains some miles distant to the North East. One which is the most Eastern, and probably the oldest, is but low, built along on the ground, the channel for this water is five feet six inches wide and arched over, the other is about forty yards nearer the sea. The arches are about then feet wide, and fifteen high, built on piers that are five feet wide, the side next [to] the sea is of rusticated hewn stone tis the greatest part of it covered over with sand. I observed the East side was plaister'd with a very strong cement which is unusual in Roman buildings, that are not made to contain water: tis probable the sand continually blowing against this side, twas contrived to keep the Aqueduct from being damaged by the wet sand after rains.

Ten miles North on the sea is a small village called Tortura [Tantura] with a little port for small vessels who often put in there on stress of weather, when every passenger is obliged to pay a Caphar of about half a Guinea a head. This must be the place called in Hebrew Dor in the sacred writing and Dora by the prophane writers who place it nine miles from Caesarea. This present village is in the plain on the bay, to the North of which is a promontory on which is a ruined castle where probably was the ancient town. Between Tortura and Caesarea are two small rivers, that nearer Tortura is called the Coradge, the other about three miles North of Caesarea is the Zurka. I was not informed till I returned to Acre that in the latter are certainly crocodiles, which I could not have believed, if it had not been affirmed by very good authority, some of them having been brought to that city which I found attested by all the Europeans there;

and I no longer doubted of the truth of it, when I looked into Strabo and found him mention κροκδαλωπολις among some cities destroyed between Ptolemais and Caesarea, the names of which he says then only remained. I was informed the crocodile are but small, about five or six feet long, but that that they had been known to take some young cattle that had been standing in the river. Tis probable this is the river Keresos [Kerseos] of Ptolemy placed four miles South of Dora. Ten miles South of Dora. Then miles North of Tortura is a very extraordinary old fortification called by the people of the country Athleti which seems to be the greek name from its great strength, by the Franks it is known by the name of Castello Pellegrini probably from the Pilgrims landing and embarking there in the time of the Holy war as being a more secure place than Joppa. And I find another name I suppose of the middle age which is Petra incise probably derived from its scituation [sic] on a rock and having a fossee cut in the rock on the East side, so as to make it an island, before that fossee was filled up with sand as it is now. It is a most magnificent castle scituated on a small promontory with a little bay to the South of it, to the East and South East are the fine waters of the ancient city almost entire, being built only on the South and East sides, as it had the sea to the North, and they [the] bay and castle to the West, the South wall is built into the sea, the bay probably making further in than it does now, all being choaked up with sand in such a manner that the very waters of the city are almost covered within. The wall is about half a mile long from North to South, and a quarter of a mile from East to West, seven feet and a half thick, built of very large rusticated hewn stone in the same manner as the castle. Within it are no ruins that appear to be of any great antiquity, every thing that is ancient being buried in the sand. The castle it self is about three quarters of a mile round and is encompassed with two walls fifteen feet thick. The inner one on the East side, cannot be less than forty feet high, and within it there appear to have been some very grand apartments. The back part next to the sea seems to have served for many offices for the fortress, among them I could not but take notice of the oven which is eighteen feet diameter. In this castle are the ruins of a fine lofty church of a light Gothick architecture. It was a figure of ten sides with three altars, one opposite to the Grand entrance, and one in the sides adjoining to the East end, which three sides by what I could judge appear to have been built out like chappels with five sides. This castle was probably built under some of the last Greek Emperors, that enjoyed these countries, before they were invaded by the Saracens, and the whole is so magnificent and so finely built, that it may be reckoned as one of the things that best deserves to be seen this way.

North East of Caesarea begins a chain of hills that are called mount Carmel, running Northward to the bay or Acre, and then Eastward to the plains of

Esdraelon. It is the North part of this mountain that is famous for the residence of the Prophet Elias. The way to the Grott where that holy man lived is by a narrow valley before one comes to the North West corner of the mountain that leads near a mile in between the mountain, where they also shew his fountain cut out of the rock. Here are the ruins of a Convent built by that Brocardus second General of the Latin Carmelites, who has writ some account of the Holy Land. On the top of the hill over it is a spot of ground they call Elias's garden, because they find many stones there resembling pears olive stones, and as they imagine water melons, these latter being broken, are hollow and the inside are incrusted with a sort of fine crystalization in spools some of which appear like Diamonds. One part of this spot they call the garden wall, appearing like a foundation being of loose stones about eight feet wide, and near a quarter of a mile long. I could not learn what this really was, but conjecture that a long bason might be dug here to receive the water for the common use of the Convent which is at some distance from the spring, and that they filled it up with stone that it might not be observed and destroyed by the Arabs, as it probably has some channel for the water to go down to basons I saw cut in the rock at the convent full of water. All along this supposed wall are a great number of stones that have this crystalization on them. But the place now most regarded, is the North West corner of Mount Carmel, reckoned about ten small miles from Castel Pelligrino [Pellegrino], coming near to the sea at the South side of the opening of the bay of Acre. At the top of the hill are the ruins of an old convent probably inhabited by the Greek Caloyers of the order of St Elias who had possession of these parts before the Latins were established here. Near it is a Chapel in a Grott where they say Elias sometime lived, this is resorted to with great devotion especially on St Elias's day, by the Turks as well as by the Jews and Christians. Under the ruined convent is a small one, now inhabited by two or three Latin Carmelites. Great part of it particularly the church and old Refectory consist of Grotts cut out of the rock, but they have lately improved it with some small buildings. There has not been a Convent here a great many years, the monks living like hermits in separate Grotts about the mountain. After their convent was destroyed, and when they lived in this manner, it is said they used to meet and confer in a Grott towards the foot of the hill, which is one of the finest I ever saw; being like a grand saloon about forty feet long, twenty wide and fifteen feet high all cut out of the rock, and is now turned into a Mosque.

I am with the greatest respect My Lord
Your Lordships

TIBERIUS
13/24 May 1738
Pocock to Bishop Milles
Account 12[9]

My Lord

The North part of Mount Carmel comes near to the bay of Acre, there being only a plain between about half a mile broad. This bay is three leagues over and two deep. On the South side of it, and North East of the Convent of the Carmelites, is a small town called Caipha. A modern writer mentions that it was called Porsina and Porphyria (I suppose in the middle age) and that it was a Bishoprick, and had the name it goes by at present from Caiphas the high preist to whom it belonged, and it was probably inhabited by Jews: for between it and Mount Carmel are several sepulchral Grotts cut in the rock in the Jewish manner; and at the town is an old church very well built almost entire which might be the Cathedral. This might be Calamon, which in the Jerusalem Itinerary is placed twelve miles from Ptolemais, which places Sicuminos three miles nearer Jerusalem mentioned also by Ptolemy in the same latitude as Mount Carmel, which I conjecture might be on a rising ground at the point of land which makes the bay of Acre on the South side. At Caipha are the ruins of a large building that probably was a castle; and they have built two forts against the Cruisers; this place serving as a port to Acre, the ships lying off this place very safely at anchor, which they cannot do in the winter on the side of Acre.

The way to that city is round the bay, and near the South East corner of it one passes over the river Kishon which rises from several springs in the plain of Estraelon that are near Nain at Mount Hermon, and also towards the South side of the plain. This is the river into which Elias threw the bodies of four hundred and fifty of the prophets of Baal, after he had caused them to be slain; and over it to the South East of this road is that part of Mount Carmel, where was the famous sacrifice when the preists of Baal cut themselves and cried from morning till evening to their false Gods and where the fire descended from heaven on the sacrifice at the instance of Elias, and they pretend still to show the stones of the altar. This part was so infested at that time with Arabs that I could not go to it, but they say it is the pleasantest part of the mountain and beautified with many sorts of fruit trees.

On the North side of the bay, not above a mile from Acre, the river Belus falls into the sea. It is said the sand of this river gave the first occasion in the time of

9 BL, Add MS 15779.

the ancients for the invention of glass, probably by some accidental melting of it. It is a small stream, that rises out of a lake about six miles East, which probably is the Palus Cendovia mentioned by Josephus near which and the river Belus he speaks of the sepulchre of Memnon of which there is now no account.

At the North West opening of the bay stands the city of Acre in a very large and fertile plain, bounded on the North at about twelve miles distance from Acre by the mountains of Saron, and to the East by those fine fruitfull hills of Galilee about ten miles distant which seems to have been the bounds between the tribes of Zabulon and Nepthali from the tribe of Asser [Assar] never possessed by the Israelites, and consequently may be esteemed the bounds of their Kingdom, as the river Kishon and the bay of Acre was between the tribes of Asser and Isachar [Issachar]. The ancient name of Acre was Ace, which it has probably even since retained with a little alteration among the people of the country, the Arabs now calling it Aka. It was called Ptolmais by the ancients, and the Europeans seem to have corrupted the old name to Acre, and when it was inhabited by the Knights of St John of Jerusalem it obtained the name of St John D'Acre, tho' now it is only called Acre. It may be considered as three citites, the old city, the city of the Knights and religious, and the new city. The old city is that which is the present town on the Southermost point, which has the sea on the South and West, and to the East a small bay which made the old port now almost filled up, tho' there are great remains of the old water of the mole, into which small ships come and take in their loading in the summer. On the North side of this town there was without doubt a strong wall, to defend it on the side of the land, but now there are no signs of it. This present town is near a mile round, without walls nor will the Arabs suffer them to build walls, by which they might [exert] their power or influence over the city and themselves be shut out of it.

North of this city, and of the port and strand East of it, is the city of the religious, extending from West to East three quarters of a mile, and a furlong in breadth, at the West end of it was the palace of the Grand master a large building which was afterwards repaired and inhabited by the great Feckerdine. Adjoining to it are the ruins of a very grand saloon, and a room of the same architecture at the end of it which probably served for a Chapter house. South of the palace is a very well built noble chapel of the Grand master, the walls of which are almost entire. Towards the East end of this town was the house of the Knights, and a very strong built church adjoining to it dedicated to St John. What remains is a low massive building over which it is probable there might be built a grander church, in the ceiling of this is a releif of the head of St John in a charger [large serving dish]. Between this and the Palace of the Grand master was a very large and magnificent Nunnery of Clares, some of the lofty walls of the house are standing, and the church almost entire. When the city was taken by the Turks, the Abbess and Nuns after the example of Air [St Ebba] of Scotland, cut off their noses, to secure their chastity, and were unhumanly butchered by the soldiers. North of this town is Fossee and some signs of an old wall, and North of that but not extending so far to the East was what I call the new city probably built when the soldiers of the Holy war and other Christians retired here as their forelorn hope. Here one sees the remains of the most beautifull old fortification of the Knights made with two fossees lined with hewn stone, and two walls, to the inner wall were round towers at certain distances, and the outer fossee was defended by semicircular Ravelins [triangular fortifications] faced likewise with hewn stone which fortification was continued on to that part of the middle city which was not covered by this and to the East end of it, so that in this place seems to have been the first invention of that sort of military architecture.

I have great reason to think that the river Belus was brought along through the Fossees, because in the siege of the city it is said that such a body attacked the city from the bridge over Belus to the Bishops palace. And if so it made the city an island, and a place of so great strength that it would not have been so easily taken had it not been for the divisions and the iniquity of the besieged. There is nothing of antiquity in the old city, but very little remains of the magnificent lofty Cathedral church of St Andrews, it appears to have been a very fine Gothick building encompassed with a Cloyster, near it probably was the Bishops palace, and at a like distance to the North West are the remains of a very strong building called the Iron castle, from which there seems to have been a double wall by the sea side and other buildings as appears by several cuts in the rock as for foundations of walls. There is in this city a Greek Bishop, a very good old church and a convent belonging to it. The Maronites and Armenians have their churches, and the Latin Fathers of the Terra Santa their Hospitium.

From Acre all strangers go to Nazareth which is about twenty four miles East South East of it, the way is across the plain of Ostere which produces great abundance of excellent Cotton and corn, and what is very extraordinary the Grand Signiors country supplys his enemies the Maltese with corn, which is carried from this port to Malta every year, Ascending the low fruitfull hills from this plain, I passed thro a village called Abilene, and after thro' a narrow valley of that name, which with the hills on each side of it is one of the richest and most beautifull spots I ever beheld. This leads to the plain of Zabulon which is about four miles over, extends Southward to the plain of Esdraelon, and for seven or eight miles North, having a communication to the East with another plain of the same name, from the former we ascended the hills and came to a village called Sepporch, on the spot where stood the famous city of Sepphor, called by the Romans Dio Caesarea, which was made capital of Gallilee when it was given by Nero to Agrippa Junior Tiberius having that honour before. Here was held one of the Conventus's of all the land of Judea and this city being the head of a sedition among the Jews it was destroyed by order of Gallus. I have two medals of the city, with a Palm tree on the reverse and this inscription SEPFORHNOWN. One is of the city itself and the other with the head of Domitian. Over the present village on a higher part of the hill is a small ruin of a Castle very finely built of hewn stone. Little more of antiquity is to be seen except several sarcophagus's about the village and in it a ruined church that seems to have been built before architecture was quite destroyed.

It is said to stand on the spot of ground where formerly was the house of St Joachim and Anna the parents of the Blessed Virgin Mary, and consequently to be dedicated to them. And some antient church historians mention the chair there in which the Blessed Virgin sat when she was saluted by the angel.

About five miles East South East is the village of Nazareth is scituated on a rising ground at the foot of a ridge of hills, and over a very narrow vally [sic]. It is a white soil with a very soft white stone out of which the Grott of the holy family is cut, behind the place where they say the Santa Casa stood. The descent to them is by several steps, and over it is lately built a very handsom church. In the convent of the Franciscans of Terra Santa, there are some few marks of a very large antient church, said to be built over it by the Empress Helena, which is very probable from what I saw of this architecture, for one sees several capitals and bases of pillars lying about and some other antient peices of work not in a bad taste, and over a gate of the Convent is put up a peice of a very fine old alto relievo of Judith cutting off Holofernes's head. The Fathers have lately built a large handsom Convent, their accommodations as to a church in the Holy Grott, and in all other respects being before very straight. What other things they shew here are some signs a very little to the North of the convent of a church where was the house of Joseph, in which it is supposed he exercised the trade of a Carpenter. West of it is a strong small arched building which they say is the synagogue in which Christ explained that text of Isaiah of himself; which so irritated his countrymen. This in the middle ages was called the church of the forty martyrs. They show also a large natural stone in a sort of hollow or quarry, on which (from what author I know not) they say Christ eat [ate] with his disciples.

I am with the greatest respect My Lord
Your Lordships

TYRE
28 May/8 June 1738
Pocock to Bishop Milles
Account 13[10]

My Lord

About a furlong North of the town of Nazareth is a fountain, which has an arch turned over it, and runs into a fine marble sarcophagus. A little further is a Greek church under ground, where it is said the Angel Gabriel first saluted the Blessed Virgin, which is mentioned by an author of the seventh century, and one of the twelfth century speaks of the cave in it which I saw, out of which the water runs to the fountain above mentioned, which was formerly nearer to the church, where he says the Blessed Virgin used to draw water, and where the Angel Gabriel first spoke to her. There is some difficulty in relation to the accounts of these places, the authors of the first and middle ages mention this church of the fountain, and only mention another church, where they say the house of Joseph was, which probably is that built over the sacred Grott, whereas that which is now called Josephs house, where perhaps he might exercise his trade of a carpenter, might be only a Chapel within the precinct of the other, and so not be mentioned; the greatest devotion being paid to the holy Grott, which might be the place of the women, at some distance according to the Eastern custom. About two miles South of the town, and very near the plain of Esdraelon is a high rocky steep ridge of hills, where they say is the praecipice [sic], down which they say they were about to cast down our Saviour, and shew some holes in a stone which they tell you were made by our Saviours fingers. On the side of the hill somewhat lower, is a kind of an altar cut into a rock, and some remains of a Mosaick pavement. About half way to this place on a rising ground are ruins of a church called the Madonna of Fear, to which place they say the Blessed Virgin followed them when they went to cast their son down from the praecipice: West of the praecipice is a village called Jaffa. In the gardens South of it is an altar made up with loose stones where the Latin Fathers celebrate on St Johns day, there having been a church there dedicated to St John the Evangelist, on the spot, as it is said, where was the house of Zebedee Father of James and John.

About five miles East of Nazareth is Mount Tabor, a very high hill, and one of the most beautifull I ever saw, being a rich soil, producing excellent pasturage, and is finely adorned with variety of trees, the ascent is about half a mile

10 BL, Add MS 15779.

long and near a quarter of a mile broad. The North is the lower part, and there was a wall all round, which Josephus built in forty days, there was also another along the milled of it, on that ground where the South plain of the hill commands the North part which is below it and is called the Meiden [Meidan] or riding place; being probably used for exercises, when there was a city here, which Josephus mentions by the name of Alaburion. On the inside of the North walls are several deep fossees, out of which it is probable the stones were dug to build the walls, and which served also to preserve the rain water, there being a great number of cisterns under ground for that use; to the South where this ascent to the hill or approach to the walls was most easy are cut fossees without and the access render'd more difficult. There are many remarkable things in history concerning this mountain, but what has made it more famous than any thing else is the transfiguration of our Saviour, when Moses and Elias appeared as talking to him, in the presence of Peter James and John. On the East part of the hill are the remains of a strong castle, and within the precinct of it is the Grott, where are the three altars in memory of the three tabernacles St Peter proposed to build, and where the Latin Fathers always celebrate on the day of the transfiguration. In this place was formerly a convent of Benedictine Monks, and on another part of the hill was one of Greek Basilians [Order of St. Basil], where the Greeks have an altar and perform their divine service on the same day according to their style. Coming up the hill they shew a church in a Grot, where they say Christ charged his disciples not to tell what things they had seen, till he was glorified. Mount Tabor is not only most beautifull in itself but also in its prospect, especially as one sees from it so many places famous in sacred writ. To the South the mountain of Samaria, and the hill of Engeddi; to the East the hill of Hermon, and at the foot of it Nain and Endor, and North East of that the mountain of Gilboa so fatal to the family of Saul, and at the South West corner mount Carmel; to the North the mount on which our Saviour delivered his sermon to the people, and near it the place where he blessed and miraculously distributed the loaves to the multitudes, and the sea of Tiberias, and further North Saphet on a very high mountain and within a days journey of Damascus a higher [mountain] called Gebelshene which is always cover'd with snow, at the foot of which the river Jordan rises: not to mention the whole plain of Esdraelon extending as far as mount Carmel. At the West foot of the hill on a rising ground is a village called Debocra, and a ruined church where it is said Christ left his other disciples. Any one who looks into the fourth chapter of Judges may see if it is not very probable, that Barak and Deborah met in this very spot of mount Tabor with their forces, and thence went and pursued Sisera, which on this vary account might have its name from that great Prophetess who then judged and governed Israel. I returned from

Mount Tabor by the plain of Esdraelon, going South by a village called Zal about three miles from it, it is on a rising rocky ground, near it are a great number of Sepulchers cut in the rock, some like sarcophagus's above ground, others cut into the rock like graves, having some of them still the stone covering, over them, so that formerly it might be no inconsiderable place and perhaps was Ξανως the great plain, as the bouns of tower Galilee to the South. Turning West I passed near the famous precipice, and ascended the hills near Jaffa and returned to Nazareth.

From Nazareth I went Northward to Mesheth [Mashhad]. This the Turks say is the country of Jonah, and here also in a nich in a Mosque they shewed me what they told me was his sepulchre. What foundation they have for this latter I know not, but as to the former St Jerom in his preface to the prophecy of the prophet Jonah mentions Geth two miles from Saphorim in the way to Tiberias as the country of Jonah, and the place where they show'd his sepulchre, and this village is about three miles from Sepporeh – so that probably the name of it is changed since his time, and there still remains some similitude in the names.

About two miles further is Kepher Kenna, where the Latins say our Saviour wrought his first miracle of turning water into wine at the marriage. On the South side of the village is a fountain out of which the water they say was taken that was turned into wine, and near it are the ruins of a church dedicated to St Bartholomew and said are almost entire, whether it was a house or a church I could not well judge, but they say here was the house of the marriage. Near it is a large new Greek church. It is certain this situation so near to Nazareth makes it very probable to be the place, but the Greeks have the tradition of Rana on the West side of the plain of Zabulon about three or four miles North West of Sepporeh, and I know not how it thus came to be allowe'd [believed], that the water was carried from this fountain four or five miles from Rana. Which ever was the place it was a matter unsettled about the beginning of the last century, when a writer of the Holy land endeavoured to fix it here as the most probable place. About three miles further is the feild where they say the disciples plucked the Ears of corn.

Twelve miles North North East from Nazareth I came to the mount of Beatitude where our Saviour delivered his sermon. It is about ten miles North of mount Tabor. From the plain to the South it appears as a long low hill, with a mount at the East and West end. When you come to it the hill appears to be rocky and very low, but on the East mount is a level surface covered with fine herbages. And here it was that those blessings proceeded out of the mouth of the Redeemer of mankind. The mount is about ninety paces long and sixty wide, about the middle of the East side are the foundations of a small church twenty two feet square on a ground a little elevated which may probably be the very place where our Saviour was when he spoke to his disciples. West of it is a cistern under ground which might serve for the use of the preist who had the care of the church. About two miles East near the brow of these hills which run to the sea of Tiberias are several large black stones. Two of them stand together, and are larger than the rest, and on them it is said Christ blessed the loaves when he fed the five thousand, when he made to sit down on the grass. These hills tho' they appear very low for the South yet they are high with regard to the plain of Hain North of them to which I descended and went to the village of Hutin,[11] at the West foot of the mountain of Beatiduudes. This place is famous for some pleasant gardens of Lemon and Orange trees, and here the Turks have a Mosque they pay a great vereration to, as having they say a great Sheke buried there they calle Sede Ishab [Sidi Ishaq?], who (as a very learned Jew assured me) is Jethro the father in law of Moses. Two miles North East of Hutin, and North of the plain of Hutin is a narrow pass called the valley of Doves, which is a dscent between two rocky mountains into the plain of Gennesarreth, which plain is West of the middle part of the sea of Gennasareth; the plain of Hutin being also West of that sea. The mountains are full of sepulchral Grotts which probably belonged to the towns and vilages near, and on the North side of the hill, over the plain of Gennereth is a castle a considerable heigth up the hill, with a great number of Apartments cut into the rock, said by some to be the work or at least the improvement of Fecherdine. The reason of my mentioning this pass so particularly is because that South of it in the plain of Hutin, add about two miles West of the sea of Tiberias are the ruins of a town or large village which is now called Bailsida [Baitsida], and must have been the antient Bethsaida of Galilee so often mentioned in the Gospel. I cannot find that this has been observed by any authors and antient Geographers finding there was a Bethsaida East of the sea of Tiberias or of Jordan in Gaulontis have much doubted whether there was another West of the sea, and consequently if our Saviour did not speak of that on the East; but as that on the East had its name changed to Julias by Philip Tetrarch – before our Saviour frequented these parts, it may easily be concluded that Eastern place was never intended, but always this town which is in Galilee, and tho' two miles from the sea or lake, yet may be said to be by the sea of Tiberias: among several ruins here as of a large cistern and other buildings, there are particularly great ruins of a church and a very fine worked door case to it of white marble, and some columns. Three miles East South East is the town of Tiberias situated on the sea of that name,

11 Presumably Ḥiṭṭīn, a village 8 km west of Tiberias, which was known for the shrine that Sunni Muslims venerated as the tomb of Jethro.

at the North end of a plain about half a mile broad, and in some part less, that runs along by the sea of Tiberias and farther South by the river Jordan, the present town is encompassed with indifferent walls about three quarters of a mile round, lying open to the sea, there are remains of a very large castle in it, and the Sheke has lately built one on the hill North of it, it is now but a village, the houses being built scattering; at the north East corner of the town is an oblong square arched church dedicated to St Peter it is mentioned by antient authors, and at present they say it is built on the spot where Christ gave St Peter the power of absolution &c and the Latin Fathers come hither from Nazareth every year to celebrate on St Peters day. As to the old city said to be built by Herod and named in honour of Tiberias, it is not known whether there was any town here before, or if there was of what name, tho' some falsely think the town of Kanareth was here, it is said by some to be built by Tiberius himself, it extended about half a mile further South than the present town where one sees a great number of imperfect ruins, and particularly remains of the tower in the sea where the South wall ended, and I could see the suburbs extended further South, near the present town are ruins of a church and further South some signs of a large square building about which there lye several pillars, which might be the house of the Government, this being the head city of Galilee, till afterwards that dignity was removed to Sepporeh as above mentioned Justinian repaired the walls of the old city. When Jerusalem was destroyed the Jewish Rabbins came and lived in this place, and continued here, till the eleventh century, and on the North side digging for stone to build the castle when I was there they found a great number of built sepulchres under ground in which the[y] said the Jews were buried, but whether they are of so great antiquity or no I will not venture to affirm, tho' the Jews have left the place but about eight hunderd [sic] years.

I am with the greatest respect my Lord
Your Lordships

SIDON
31 May/11 June 1738
Pococke to Bishop Milles
Account 14[12]

My Lord

A quarter of a mile South of the walls of old Tiberias are hot baths. The waters seem to be impregnated with salt and sulphur, and by a red settlement on the stones I concluded they have something of the chalybeate tho' not to be tasted, a salt and bitterish flavour prevailing. They are very hot and are used for bathing being esteemed very good for the Gout. By authors they are commonly called by the Hebrew name for baths, Emmaus, but they are now called by the Arabian name Hamam.

The sea of Tiberias is a very fine lake, the mountains coming close to it on the East, the countrey [sic] on that side being all a desert, and the hills are covered with wood, from whence Tiberias is supplied. On the West side is this plain of Tiberias, the high ground of the plain of Hutin, the plain of Gennesareth, and the foot of these hills, by which one ascends to the high mountain of Saphet, to the North and South is a plain country. Josephus accounts it to be eighteen miles long and five broad, tho' I think it is not above fourteen or fifteen miles long. The water is esteemed very good and has something of a sweet taste. It has Crabs in it, as there are in all the rivers of Asia, and a great quantity of fish, which they catch with a casting net. I went on the sea in a boat, and fished in it. A learned Jew, I discoursed with at Saphet, lamented he could not have an opportunity when he was at Tiberias to go in a boat to see the well of Miriam in this lake, which he said according to their Talmundical writers having accompanied the children of Israel thro' the wilderness was fixed in this sea, and that at the place the water may be seen continually rising up. There is nothing known of any place, on the East side of the sea. I went to the place where the sea empties itself into the river Jordan, the lake is very narrow henc[e] not above two miles over, and runs into the river Jordan rather nearer to the West side, Jordan first runs South for about a furlong, and then turns West for about half a mile, in this space between the river and the sea is a rising ground which seems to have been improved into a fortification by art, and on the West side of it are some signs of buildings where there is a very long bridge or causeway built with arches over a marshy ground, and under this the sea flows into the Jordan, when the waters are high which makes the above

12 BL, Add MS 15779.

mentioned site of an antient town or fortress an island, and by making a cut they might always have a stream there which would make it a very strong place, being out of the reach of ordinary canon from the Western hills, except from a small hill which formerly might add to its strength by defending the pass, there being on it some little signs of an antient building. I find the old Geographers place Sennabris here, because it is mentioned by Josephus to be thirty stadia from Tiberias in the way of Scythopolis. This is a place very little frequented, and on the other side of Jordan I saw very large herds of wild Boar, and several of them on the same side lying among the reeds by the sea, that narrowly escaped us. On the East side of the sea towards this end is a narrow plain where some Geographers place Hippos mentioned as thirty stadia from Tiberias, and I was assured that thro' that plain runs a river from the narrow vale between the hills, and continuing its course along East of Jordan, falls into it four hours below this mouth, and that below that is a bridge over the river, probably where the antient Scythopolis stood. This river is called Skener Mousen (the Jordan of [Moses]) and I was informed it is as big as the Jordan when the waters are high, and that it rises three days off in the country of Tauran. This seems to be the river Jabok [Jabbok], that divided the kingdom of Sihon, the portion of Gad and Reuben, from Galaad and the kingdom Bashan, which was the half tribe of Manasseh beyond Jordan; and if so it is that Jabok over which Jacob went when he had wrestled with the angel, and near which Esau came to meet him. Going North of Tiberias the first place I came to at the South East corner of the plain of Gennesareth on the sea was Magdol, where there were pretty perfect remains of a sort of a castle of very indifferent building. This does not seem to be Magdalum mentioned in the Scriptures, because that is spoke of with Dalmanutha which was East of the sea. This plain which is a fine spot must be what Josephus calls the country of Gennesareth, and describes as thirty stadia broad from North to South, and twenty deep from the vale of Doves to the sea, which appears to be very just. About the middle of it or rather towards the North side is a very fine fountain enclosed with a round wall about a hundred feet diameter, from whence it is called the round fountain. It runs off in a stream in a stream [sic] thro' the plain into the lake, and is probably that fountain mentioned by Josephus by the name of Cesaina, as watering this plain, which is a very warm spot of ground. But I dont find that they have ripe fruits there all the year round as some have said, except it be a sort of little apple not disagreeable growing on a thorn tree, of which they say they have all the year round. This spring I have mentioned seems also to have been called the spring of Capernaum, and from thence one may suppose that Capernaum was at the lake where this rivulet fell into it. Capernaum is mentioned as on the borders of Zabulon, and Nephthali which tribe[s] were probably divided by the brooke

Sernon which runs thro' the vale of Doves, and so thro' the plain of Gennesareth into this lake, now indeed they commonly shew for Capernaum, another place at the foot of the hills North of the plain of Gennesareth, where I saw the ruins of a small church of white marble with some remains of pilasters about it. The ruins are very extensive along the sea, and I could plainly observe a round port for small boats to put in, so that this without doubt was the antient Tarichea which Josephus describes, as situated under the hills like Tiberias, in which it very much resembles it, but is farther than thirty stadia, not being less than six miles distant at least. The ruins extend all along very near to the North West corner of the sea for three or four miles. It was encompassed with a wall by Josephus on the part that dos [sic] not lye on the sea and I saw here signs of a wall along the West side of the ruins. I enquired for Chorazin, but could find nothing like the name except at a village up among the hills West of the supposed ruins of Tarichea which is called Gerasi. Opposite to Tarichea was Gamata a strong place famous in the history of Josephus. Coming where it runs into it after it has taken its course for near two miles, thro' a fine plain, on the East side of it at its entrance into that plain is a hill, on which I imagined I could discover some sign of ruins. It is called Jelouwy which seems to be a corruption of Julias which was the antient Bethsida in the Gaulonitis and must have been here, there being another Julias in Peraea on the East side of the lake, which before was called Betharamysta [Betharampta], both having their name changed to that of Augustus's daughter in compliment to the Emperour. From the lake Samochonitis to this place the river Jordan runs about ten miles except the two first running over the rocks with much noise, its passage is between the hills almost hid by the shady trees the Platanus &c that grow on each side of it which makes it a most delightfull veiw. After having travelled about four miles on the side of the Western hills, on a plain spot is a mount on which I saw some ruins, but could not judge whether they were of any great antiquity. Some writers of the Holy land place Lakum here about, I suppose because it is mentioned in Joshua as the bounds of the tribe of Nephthali, and it is also spoken of as on the river Jordan. They place also along this part of the river Jabneel and Thelia. And about a mile and a half South of the lake Samochontis is a bridge over Jordan called Jacobs bridge, as it is said because Jacob passed over here returned from Padanaram. On the other side of the bridge is a large lane being the high road to Damascus from Palestine. On this side of the bridge Baldouin the fourth King of Jerusalem built a fortress against the Saracens on a rising ground probably on this very spot. I went over this bridge into that part of Arabia Petraea called Petraea, and that particular division of it is known by the name of Gaulonites, which was part of the Kingdom of Bashan, and after, the half tribe of Manasseh beyond Jordan. A small mile below the bridge is an oblong square hill shaped by art so as to be very strong against an enemy. All round at top are the foundations of a strong wall and at the South end, and on the East side I saw some remains of two very handsome gates of hewen [hewn] stone and at the corners, round turrets. At the North end is a heap of ruins as of a castle, the whole being about half a mile round, with some signs of a suburb to the South on a lower ground which seems also to have been fortified, the place is now called Kaisar ateran [aterah], and it seems to have been an improvement of the Romans, but what place it could be I cannot conjecture, unless it were Thelia. A mile above the bridge is a mineral water, which seemed to be of Sulphur and Iron, it is walled in as if it had been formerly frequented. Half a mile from the lake and this place, is a little low hill with some ruins on it which they now call the town of Jacob, some by conjecture place Harosheth here, the city of this lake. I came to the lake Samochonitis, called in scripture the waters of Merom which is placed one hundred and twenty stadia from Julias, tho' I think it cannot be above ten or twelve miles at the most. It is situated on the East side of a large uneven country extending about five miles to the mountains of Nepthali being East of the Gaulonitis. Josephus says it is seven miles long but it is not above two miles broad, except at the North end, where it may be something of the nature of morass water, which partly owing to the brooks on the West being stoped [sic] to water the country, so that the water comes thro' the banks into the lake, and partly to the bed of it being muddy. For when the snows are melted and the waters falen [sic] it is only a marsh thro' which the river Jordan runs, and yet notwithstanding the water passing thro' the rocky bed towards the sea of Tiberias settle and purify, and become very wholsom. Its waters forsaking it towards the latter end of summer, is the reason why it was generally called Pulus Samochonitis. I observed two rising grounds on the West side of it and a third towards the North West corner on which might probably be some of those antient towns mentioned on it, particularly Saanaim placed by some Geographers here. It was in this plain at these waters that Joshua smote Jabin King of Hazor and all his allies. North of the lake was the tribe of Dan, and near the waters of Merom on Jordan was Traconitis which was a hundred and twenty stadia from Paneas, the name it still retains, but it is called in scripture Caesaria Philippi being near the rise of Jordan, but this is a very dangerous country to travel in.

About six miles South of Jacobs bridge in the high road to Jerusalem is a cistern which they call Jacobs well, and suppose this to be the country of Dothaim where Josephs brethren which any one will easily be persuaded is an error, and this has introduced others, particularly Dothaim in the book of Judith being mentioned as near to Bethel or Bethulia, they have thought Saphet was Bethulia besieged by Holofernes about six miles north of it on the hills. Jacob when he returned from Padan Astram went South from Sechem near Samaria to Bethel which was then called Luza. In the history of Joseph we find he was sent to this brethren to Sechem, and when he came there he was told they were gone to Dothaim, probably to the East of Sechem, for it is said Holofernes's army extended in bradth [sic] from East to West from Dothaim to Belmaini and in length from Bethulia to Cyamon, which is over against Esdraelon. And tho' this may be said to be another Dothaim in the above mentioned place, yet the great distance makes it to be unlikely to be the place where Joseph went to his brethren, it being four ordinary days journey from Sichem, and could not be performed in less than eight with the cattle they were charged to feed. It is possible this place might be called Dothaim, as a Jew told me that place was said to be called Dothaim, which might occasion this mistake.

I am with the greatest respect My Lord
Your Lordships

BAYREUT IN SYRIA
1/12 June 1738
Pococke to Bishop Milles
Account 15[13]

My Lord

From the places South West of the waters of Merom is an ascent up very high hills to Saphet, in the tribe of Nepthali. This place is not mentioned, by name, in our translation of the Bible, but in the vulgar Bible Tobias is said to be of the tribe and city of Nepthali in the upper parts of Galilee by the way that leadeth to the West having on the right the city of Saphet. The city of Nepthali is said to be a mile South of it.

Saphet is mentioned by several writers of the middle age. Its situation is very high, and commands the whole country round. On the summit of the hill are great ruins of a very strong old castle, particularly of two fine large round towers. The Jews think part of it to be as old as the time of their prosperity. The Christians had possession of it in the time of the holy war; and I saw on a building in the town a relic of the arms of the Knights of St John of Jerusalem. It was surrender'd to Saladin Sultan of Egypt, and afterwards came into the

13 BL, Add MS 15779.

possession of the Ottoman family with all these countrys. The town is a little lower down on three sides of this heigth of the hill, on which the castle stands, and is not small; it having been formerly the residence of the Pasha of this country which was thence called the Pashatick of Saphet, and by that name that whole territory now goes, being called the country of Saphet. But the Pasha resides at Sidon; and a Cadi from Constantinople resides here. Many Jews live in this place; it being a sort of university for the education of their Rabbins, who are in number about twenty or thirty, and they have no less than seven Synagogues. Several Doctors of their Law, of the time of the second temple are buried here, particularly three in a place which is now turned into a Mosque, and the Turks say, they are three of the sons of Jacob. The Jews have a notion that the Messiah will reign here forty years before he will take up his residence at Jerusalem. At the North foot of the hill of the castle are several wells which they say Isaac dug, and about which there were such contentions between the herdsmen of Isaac and Gerar in which they were dug being far off on the other side of Jerusalem.

From Acre I set out towards Tyre, going thro' the great plain near some ruins, and particularly an old bridge of one arch. This is said to be Azhzib mentioned in the scripture. North of the plain I passed over the mountains of Saron, on them is a small square tower, and I saw several others Northward, and it is said they are seen as far as Tripoli. It is probably they were to give notice by fire, and some even say, that they were built all the way to Constantinople, and that by them the Empress Helena gave notice to Constantinople, according to the appointed signal, that she had found the Cross. But I take them to be of much later date, and probably were built by the Christians at the time of the holy war. About four miles further North on the side of the hills is a ruin of a building which I judged might be about thirty feet square, and about it several broken pillars two feet diameter. Two of the pillars are standing, and I could see by the defaced capital of one, that they were of the Ionick order well executed; so that I concluded it was an antient temple. There is a most beautifull road that goes five or six miles along the mountains being green sod shaded with trees, and mostly an ascent. This I find is called by the Franks the Tyrian ladder, and must have been the Scala Tyriorum, as being the most usual ascent from Tyre up these mountains. About four mile[s] further is the famous road, cut by Alexander the great, along the side of a mountain over the sea, after the manner of those in North Wales. As its first ascent is a small tower called Acanderette, the road is mostly thirteen feet wide, tho' in some parts it is only six; where the hill would allow it there is a battlement cut out of the rock, and in other parts of the breadth of one hewen stone, one foot four inches thick. The whole road is a mile long.

About a league and a half South West of Tyre are the famous fountains of Solomon said to be made by that King at the time when he cultivated an alliance with Hisam King of Tyre to facilitate the building of the temple of Jerusalem, and supposed by some to be the well of living waters mentioned by him in the Canticles. They are three springs about half a mile East of the sea, which rise so plentifully, that they make a considerable stream that turns several miles. These springs are enclosed with very strong walls, about fifteen feet high, by which the water was raised to a sufficient heighth so as to be conveyed by an aqueduct to Tyre. That nearest the sea is of a multangular figure; from this the water runs about a hundred paces on the aqueduct Eastward to the other two, from one of which there was a channel also into the grand Aqueduct. These are oblong squares, and the water runs from one into the other, the buildings being contiguous. It is said by some that these fountains are of a very great depth, but they have been found to be only about four or five fathom deep. The aqueduct which is a very fine one takes its course in different directions, but mostly Northward, near to a small hill called Smashook, on which is a Mosque, and a town, and this by some has been thought to be old Tyre, which as on other accounts so more particularly being a league from the sea is very improbable. Hence the aqueduct is carried Eastward to Tyre. Near the North East corner of the great bay South of Tyre, is such another fountain enclosed, as the others, but the wall not so high, and I saw the foundations and remains of an aqueduct, that seemed to have been low and not built on arches which probably went to old Tyre, which seems to have been in this corner of the bay, and that the […] rather because near this spring is a little hill, which in all probability is the very mount Nebuchadnezar [sic], raised in order to take that city, which he destroyed as described by the Prophet Ezekiel; and it is no wonder that there are no signs of the antient city since Alexander carried all the ruins of it away in order to joyn new Tyre to the continent; and being a sandy shoar the face of every thing is altered, the great aqueduct in many parts being almost buried in the sand. The Tyrians returning to the Island near, made so great a stand against Alexander the great, that tho' it was half a mile from the land he joined it to the continent and made a peninsula of it. If it was so far from the land it must have been a very small Island and a work of very great expence to joyn it to the land. I saw all across this peninsula a sinking of the ground, and the higher ground West of it was probably the East part of the Island. There are some little remains of the walls all round and a port enclosed with walls on the North Side. At the East end also are ruins of two great square towers very strongly built which seemed to have serv'd for reservoirs of water from the Aqueduct to be distributed from thence all over the city, there being foundations of a thick wall from one to the other. The East end of this city appears to have been defended by a double wall, and double ditch. There are great ruins of a very large church built inside and outside of hewn stone, after the Syrian manner with three naves each ending in a semicircle in which was an altar. There are also very perfect remains of several buildings North of it probably belonging to the Archiepiscopal palace, some Granite pillars, which they say are the remains of a church dedicated to St John, and near it part of the ruined church of St Thomas is repaired and turned into a church there being here two or three families, only of Christians and few other inhabitants except some Janizaries in a poor castle, near the port. There are also ruins of two or three churches, but nothing that carrys any great marks of antiquity in it. The harbour North of this peninsula is so good that all ships whose business in the winter leads them to have to do with the merchants of Sidon are obliged by the contract of Insurance to go into harbour here, where they take in their loading. Near the aqueduct without the town, I saw a ruin which probably is the place where according to tradition, now lost, they said our Saviour preached when he came to the parts of Tyre and Sidon, and on this coast it was that he cured the Canaanitish womans daughter, which possibly might be at a village called Kana on the hills South East of the fountains. It was at Tyre that St Paul staid, when they disuaded [sic] him from going up to Jerusalem. Two miles South of Tyre in the road is a spring that had been enclosed with a wall, the waters are not very good, having in them a very nitreous [sic] salt. I observed from it the foundations of a wall to Tyre by which the water may be convey'd to the city for common uses.

Two leagues further is the river Casriny, the people of the middle age called this the Eleutherus which was beyond Tripoly, this must be the reason why the Historians give an account that the Emperor Frederick Barbarossa was drowned in the river Eleutherus falling off his horse, as he was pursuing his enemies, and sunk under the weight of his armour; but as they also call it the Casimy, that determines this remarkable peice of History to this river. There is a bridge over it of one very large arch and a small one; it is probable the arch was then broke down to prevent the pursuit of the Christian forces; and that the Emperor took the river, and then it is no wonder that he lost his life; for it is a deep rapid stream in so much that travellers dare not water their horses in it without dismounting. On the other side of this river the hills approach very near to the sea, and in them are cut some spatious sepulchral Grotts, and a city called Ornithon mentioned between Tyre and Sidon might be here, it being a place easily to be defended, having the river to the South, and the hills to the North, thro' which there is a narrow pass into the plain in which stands the famous city of Sidon. And here I cannot but make a conjecture that these hills were probably the bounds between the sites of Tyre and Sidon, as the Southern bounds of the former were probably these hills over which is the road made by Alexander on the mountains of Saron, as the Northern bounds of the latter, was probably that river running under the hills about four miles North of Sidon, and also of the tribe of Ashar [sic] of the Holy land; and tho' their territories might extend some way into the mountains, yet it naturally leads to the reflection how great any state may become by commerce, when neither of these plains are above twenty miles long or five broad, and yet these Republicks make a very extraordinary figure in antient history, and one of them first on the Continent, and then an Isle gave Nebuchadnezar, and Alexander the great two of the most potent monarchs that ever were in the world, more trouble than any other place in all their wars.

Soon after the entrance into the plain of Sidon is a village on the mountains called Sharphant [sic], supposed to be a corruption from the old Sarepta. There are great marks of improvements about the hills, and at the foot of them are cut a great number of fine sepulchral Grotts into the rock. The house of the widow who received Elias, and was so miraculously supplied during his stay with her, is said to be where there now stands a little mosque by the sea side into which I enter'd. In it is a little closet where they say the prophet lay. The old Sarepta was most probably here, where I saw some foundations of walls, and those sepulchres must have belonged to the people of this town. About a quarter of a mile North of the Mosque I saw some little ruins of a very antient building as I concluded from a pedestal rounded off and projecting about a foot beyond the pillar, after the manner of the very antient architecture I saw in upper Egypt. About four miles south of Sidon is a fine spring called Elborok, with a wall round to raise the water, as to those of Tyre; and I could see the foundations. The plain of Sidon is not above two miles wide, but East of it are fine fruitfull hills, whereas the plain of Tyre is four or five miles wide but the hills are high, and do not appear to be capable of any other improvement than that of wood in which they abound. But Tyre seems in every respect to have had the advantage of its mother city, and makes a figure in history accordingly.

I am with the greatest respect

My Lord

TRIPOLY IN SYRIA
9/20 June 1738
Pococke to Bishop Milles
Account 16[14]

My Lord

The city of Sidon is situated on a rising ground having the sea on the North and West. The present city is mostly on the North side of the hill; the old city seeming to have extended further East, as may be judged from the foundations of a thick wall that are seen over the sea extending Eastward. On the South side it was probably bounded by a rivulet, the large bed of which might serve for a natural fossee, as another probably might on the North side, if the city extended so far, as some seem to think, that is to the foot of the high hill East of Sidon, about three quarters of a mile from the present town, the space between being all laid out in gardens or orchards, which appear very beautifull at a distance. On the North side of the town are great ruins of a fine port, the walls of which are built of stones of a great size, being all twelve feet long, the thickness of the wall, and some eleven feet wide and five deep. The harbour is now choaked up; and it is said this, as well as some other moles on this coast, were destroyed by Fecherdine, that they might not harbour the Grand Signor's gallys, to lend forces against him. I could find only one of the milliary stones of the two Mr. Mandrel [Maundrell] mentions, which is exactly copied by him, and is in the road about a mile South of the town, on a pillar of Grey Granite.

In a garden South of the town is a small mosque called Nebi Sidon, where the Turks say the Patriarch Zabulon was buried, but they are mistaken, his bones being laid in Sichem. In another garden to the East is such another mosque called by the Mahometans Zalam, who pretend also to have some holy person buried there. The Europeans call it La Cananea, being as it is said the place where the Canaanitish woman cried out, Blessed is the womb that bare thee, and the paps that gave thee suck, which building has some appearance of an antient chapel. On the high ground, on the West side of the town, is a large old church turned into a mosque. On the highest ground which is a little hill on the North side of the city are great remains of an old castle said to be built by Lewis the ninth, of France, but on the very top is a work of an older date, a square castle of hewn rusticated stone, and remains of a wall which is part of a circle so that probably it was a round castle of the work of the Greek Emperors, and repaired or rebuilt by St Lewis.

14 BL, Add MS 15779.

Sidon is the place of residence of a Pasha, and there are in it a great number of very handsome new built houses. The trade here is carried on entirely by the French, the export being cheifly raw silk, cotton, and corn. Their Consul obliges them to live all in one Kane, in which the Jesuists, Capuchins, and the Fathers of Terra Santa have their respective convents.

Going out of Sidon I saw several sepulchral Grotts at the foot of the hills, some of them cut out with pilasters and painted in a very handsome manner. About three or four miles North of Sidon is a considerable rivulet called Elouly very near the hills which bound this plain, which river is that which I take to be the bounds of the Tribe of Asser and of the Holy land, shaping over these hills, which are called the mountains of the Druses from those people that inhabit them. I came to a bay about four miles long, in the middle of which is a village called Fee and a mosque near the shoar with a well by it called the well of Jonah; and here they say the prophet Jonah was thrown out by the whale. After some time I arrived at the river Damour which may be the river mentioned by Ptolemy as four degrees South of Berytus, which he calls Λέων tho' it do's not answer to that distance, but there is no other river nearer to that city. A few miles further we passed by a village called Camey. At a well near it I saw an antient sarcophagus, a fine peice of entablature, some large hewn stones, and two round vases of red and white marble. Going on at some distance, from this place on a rising ground are a great number of sarcophagus's cut out of the rock with large covers and beyond them I saw the remains of a wall twelve feet thick which was continued along on the East side of them. These might be Heldua mentioned in the Jerusalem journey [Maundrell] as twelve miles from Berito tho' this is not above six or seven which may be a mistake, as the distance in that Itinerary between Berito and Sidon certainly is, being made twenty eight miles, tho' it is not above twenty. As I approache'd near Bayreut I found the country exceeding pleasant being a rich soil and a finely improved, and at about two miles distance from the city I passed thro' that fine grove o tall pines which it is said the famous Fecherdine planted with his own hands. A finer scituation cannot be imagined. It is a green sod, bounded on the East by a hanging ground over a beautifull valley, thro' which the river of Bayreut runs, and at the North end of which it commands a veiw of the sea, and to the North West of the fine gardens about Bayreut. This city is the antient Berytus it is said Augustus was so delighted with the scituation of it, that when he made it a colony he called it after the name of his daughter with the epithet of happy, naming it Colonia Julia Felix. It is scituated on the sea on the North side of a broad promontory a little on a declivity, the gardens appearing very beautifully on the rising ground over it, to the sea are rocky cliffs and the old port or mole is a sort of a little by that was well defended against the sea by strong peers which were destroyed by Fecherdine as above mentioned, for the had possession of this city, and his successors the [princes?] of the Druses have most of them been made governors of it, till of late years. The town may be near two miles in compass surrounded with a very indifferent wall, that on the West side of hewn stone, with small square towers, may be remains of the antient wall, but it has been very much shatter'd. At a small distance West is a little bay where I saw some signs of ruins but nothing I could make any thing of. It is possible the Theater built by Agrippa might be here, and might be built so, as to have the advantage of the hill, as those of Pola and Frejus, and the sea may have gained on it; some mention an Amphitheater, which may be the same. In the middle of the city is a mosque that was a very large well built church, supported with Gothick pillars and dedicated to St John, there are seen about the town several Granite pillars, and particularly six or seven in one part of grey Granite, some standing and some lying on the ground. What this city is now most famous for, is the improvements of Fecherdine that prince having got some taste in architecture during his stay in Italy at the court of Florence. His Seraglio now only a nich of a building has the air of a Roman palace. Water was conveyed in channels in the walls all over it and in the middle of it, is a garden of very large Lemmon trees. The stables are truly magnificent built with several rows of square pillars, and down the sides are niches with a stone manger at the bottom of each for the horses provender. There are several other unfinished buildings that have a still greater air of magnificence in them, and look more like the remains of antient Roman buildings, than unfinished modern ones

About a league East of Bayreut is the place where they say St George killed the Dragon that was about to devour the King of Bayreuts daughter. Here is a mosque which was formerly a Greek church, near it is a well, out of the hole, which is now the mouth of it, they say the dragon used to come. In this mosque I saw an extraordinary ceremony performed on one of the Turks that was with one, who sitting down on the ground, the religious person who had the care of the mosque took a piece of a small marble pillar and rowled it backward and forward on the back of the Turk for a considerable time, which they judge to be of an extraordinary virtue.

Soon after I left this place I turned South and travelled about two leagues along the shoar and came to such another road as that of Alexanders near Tyre, cut along the side of a steep mountain, over the sea. This road was formerly called via Antoniniana. The ascent to it is difficult, and in the rock is cut that inscription of which Mr. Maundrel has given a copy. Before I came to it I saw some figures cut in the rock very much defaced by time being of single men in different compartments. One I observed had a particular cap like Phrygian.

I thought it most probably to be the Persian habit and may be as old as the time when the Persians had possession of these countries. Under this road runs the river Kelp, the old Lycus both words of the same signification, this river being so called form the statue of a dog formerly here that is said to have uttered oracles. On one side of the road there is a ruin something like the pedestal of a statue, and under it the sea. At the mouth of the river is a large stone, which the people shew for the statue of this dog, from whence the river they say was called, and in the rock at the end of the bridge over the river is a releif much defaced by time which seemed to me to have represented a a [sic] dog. Over the river is a bridge of four arches one of which is very large being built by Fecherdine. On the other side of the bridge is an aqueduct brought four miles along the side of the hill of so great a height that seventeen arches about twenty feet high are built against the hill near the bridge for the water to run on, being the work of one of the successors of Fecherdine, now known by the name of Castravan extending Northward near to the antient city Byblus, now called Esbele, where also the dominion of the Prince of Druses ends which extend to the South near as far as Sidon, being only on the mountains. This part called Castravan is inhabited solely by Maronite Christians, the other parts being inhabited by the Druses and Christians promiscuity. Some Franks call the mountains of Castravan Antilibanon, in which they are greatly mistaken, for this must be parts of mount Libanon itself tho' now the mountains East of Tripoli are only called by that name extending northward to Laodicea Cabiosa near to the antient Emera on the Orontes. Strabo saw there are two mountains equally distant called Libanus, and Antilibanus, between which is Calesyria; he says Libanus begins near Tripoli at Θεσπροσωπος which is the promontory about twenty miles South of Tripoli, from which it must be observed running Eastward for some miles it extends Northward near to Emesa, and Southwards I suppose near to Sidon, either where the mountains bound the plain of Sidon to the North, or to the South, which latter best agrees with the account we have of Hirams furnishing King Solomon with the Cedars of Lebanon for the building of the temple, because in that case, this part of Mount Libanon must have been the bounds between the territories of Tyre and Sidon, and consequently part of it belonged to the Tyrians. He says Antilibanus begins near Sidon, which would incline one to think that Libanon ended North of Sidon and that Antilibanon began South of it at the river Casmir, but supposing Strabo should not be so very exact, and that these mountains south of Sidon should have been the end of Libanon or even those North of it, Antilibanon might begin South of Tyre at the mountains of Saron, which seems to agree best with the course of the mountains to have between them that plain in which Balbeck stands which I take to have been properly Calesyria, tho' the plain of Damascus and most of the plain country of Syria Southwards seems also to have gone under that name in its largest acceptation, in which Strabo seems to take it speaking somewhat confusedly afterwards of Jordan as in Calesyria, tho' at first he speaks only of Calesyria proper. And this account of Mount Libanon ending North of Tyre and Antilibanon beginning South of it best agrees with this authors account that Calesyria is two hundred stadia wide at the sea or twenty eight miles, tho' the distance is not quite so much, and it must be observed that the plain of Calesyria does not extend in any equal breadth, for the plain of Tyre being bounded to the East by the mountains, Calesyria must only be in this narrow vale in which the Casmir runs and so extend Northwards probably a very narrow vale to the plain of Balbeck. Strabo says these mountains end near the Arabian territories, beyond Damascus, at the mountains of Traconitis, which must be meant of the mountain only of Antilibanon considered as running Eastward to the mountain at the rising of the river Jordan the country East of that river being reckoned a part of Arabia; tho' in speaking of the end of these mountains I should rather have understood it of their ending Northwards towards Emesa, and the desarts [deserts] of Palmyra to the East which might be fruitfull high grounds, when well water'd by aqueducts as it is thought they formerly were.

I am my Lord

CANNOBINE ON MOUNT LIBANON
13/24 June 1738
Pococke to Bishop Milles
Acount 17[15]

My Lord

Having given some account of the scituation of Libanon and Antilibanon, on my going to see the Castravan mountains, East of the road that leads to Tripoli: I shall now give you some relation of these mountains which were a part of Libanon. They are very high rocky mountains but the ascent is not very difficult, the highest parts are almost all the year cover'd with snow. It is surprising to see such barren hills so well inhabited and improved, cheifly with mulberry trees for the silk worms, and also with vineyards which produce excellent wine, far preferable to any other wines of Syria. A considerable way up in these mountains, I saw the rocks for about two miles, in such extraordinary figures,

15 BL, Add MS 15779.

rising above the ground, that at a little distance they appeared like a ruined city, like castles, towers, and houses, and some of them like men. It is such a scene as this which probably gave rise to the fable of a petrified city and men, beyond Tripoly in Africa. These mountains are altogether a Christian country, where they suffer no Turks to settle nor the Pashas themselves to come. The Prince of the Druses plays a certain sum for his whole country to the Grand Signor; and the people pay for their lands to the Prince. It is a place of refuge for Christians from the tyranny of the Turkish Governors; and especially for those unhappy wretches who to escape a punishment having denied their faith, repent of it and become Christians again. Every village has a well built church and there are almost as may monasteries as villages, and to all their churches they have a bell, which is an extraordinary thing in these parts. As I observed they are all Maronites, acknowledging the Pope, who dos [sic] not think fit to exercise but very rarely any immediate power here the Patriach of the Maronites being a sort of Legatus natus. He is chosen by the Bishops about ten in number, and the Christian Governors of the country; a certain district having over it a Sheke or head. His usual residence is here at Cannobine, but I waited on him at a lillte [little] convent near the top of these mountains, having come here on some disturbances in the country where he usually resides. He is of a family of one of the heads of this country, having been married, and has children, but being a widower he became a Monk, and was chose to this dignity. The Bishops have their Sees at some cities near, many of which are ruined, so they mostly reside in Convents here. The Monks are of the order of St Antonio Abbate, many of the convents have been built within this fifty years, and those are mostly of this order reformed by a Monk of Aleppo from which they are called Aleppines. Many of these convents have a nunnery adjoining, but they have usually only poor old people for the service of the convent. The monks both preists and others (as in all the Eastern churches) are employed in taking care of their land being persons of no manner of learning, and so perform their long offices of devotion by night, which are in the Syrian language that the do not understand, but being used to that character both they and the Syrians, or Jacobites with their mother language, the Arabick, in Syrian characters, which is called [blank] of which there are several books published in Assmans [Joseph Simon Assemani's] Bibliotheca Orientalis, that Prelate being a Maronite of Mount Libanon, and was sent by the Pope to buy up all the M.S.S he could get in these parts, and a second time lately on some affairs with regard to the Maronite church and Patriarch, from whence he proceeded into Egypt with some commissions to the Copti Patriarch of Alexandria, whom the church of Rome is labouring to bring over to it with his whole Province of Africa. In the reformed convents, the superior is chose every three years, whereas in the others they are during life, and if I mistake not they take the vow of poverty, which none other of the Eastern religious do, the inconvenience of which in that station of life, appears very much, in a people who have naturally such an exorbitant love of money. There are also some few nunneries that are not dependant on the monasteries, they are very rare in the Eastern countries, and are rather like hospitals for the aged and decrepit, and if any young women are in them they rather continue in a state of probation, rarely taking the vow, and so can change their life when they please, which perhaps might be an improvement on that kind of institution. I was at a nunnery on these mountains of Greek Catholicks, very lately founded by some young Ladies of Aleppo on the rule of St. Francis de Salles, under the direction of the Jesuists [sic], who have their convent near, these Ladies are retired to it and are performing their two years of probation. There is also an Armenian Catholick convent, founded by a Bishop I saw there who was obliged to retire to these parts on account of some distresses. The Latin Fathers, those of the Holy Sepulchre, the Jesuists, and Capuchins have handsome convents here, and the Jesuits have erected a Seminary both to prepare the youth for the education at Rome, and to fit those in a better manner for the Maronite church, who cannot be sent thither. The Maronites are esteemed the most honest simple and least ingreigueing [intriguing] of any other of the Christians of the East. If any account can be given of the original of the Druses it is, that they are the remains of the Christian armys in the Holy war, and they themselves now say they are derived from the English, they are esteemed men of courage and of greater probity than any others of these Eastern parts. As they and their Prince, are protectors of the Christians that live among them, so they seem to have the best opinion of and greatest regard for Christians, tho' they are in reality of no religion at all and only upon occasion profess themselves Mahometans, and go as seldom as possible to their mosques, to enjoy the privileges of the established religion; and I have been informed that in some of their books that chanced to be found, they both blaspheme our Saviour and speak evil of Mahomet, they have among them a sort of religious they call Akel that is [blank] these drink no wine, and will not eat any thing that belongs to the Prince, because they say it is rapine. They have most private places under their houses, for their ceremonies or worship, where no persons enter but these religious themselves, and it is said if any person should be too curious he would certainly lose his life. Here they keep their books, but the people joyn in no sort to religious worship, nor do these perform any openly except as I was told reading out of their books over the dead, before they are carried to burial, tho' in this I must doubt my authority. These religious people meet together in their private places, they seem to be something like the wise men or philosophers of old, rather then [than] the chief persons of any religion in a community that has none. I rather think if these in particular have any that they are worshippers of nature. I was indeed told that by some accident the statue of a calf had been seen in their retired places: but if I can depend on any thing it is that these persons have a small silver box, so closed as not to be opened, and that many even of them know not what it contains, that they pay a sort of worship to it, and that there really are in it the images of the nature of both sexes.

From these mountains I descended into the road by the sea side which leads to Tripoli, and passed over the river Ibrim which is the antient river Adonis. Of this the travellers observe that the river being red from the soil it runs through after great floods, which having happen'd about the time of the feast of Adonis, the antients said the river run [ran] with blood for the death of Adonis. Beyond this I came to Esbele called by the Franks wrongly Gibele. It is the antient Byblus, which Mr. Maundrel supposes to be the country of the Giblites mentioned in Joshua. Here Cingras the father of Adonis had his palace, and the city became famous for the temples and worship of Adonis. The walls of the town remain about a mile round and at the South East corner is a very strong built castle of rusticated work; towards the foundation are some stones that are twenty feet long. There are very few inhabitants in the town, but the ruinous houses are standing, which show that it has been well inhabited and that probably within an age or two. There are the remains of a beautifull church, which seems to have been the Cathedral. It is of the Corinthian order, and appears plainly to have been built before the entire corruption of architecture, probably about the fourth or fifth century.

Some miles further is Patrone, the antient Botrus. It is a Bishops see, and gives title to one of the Maronite Prelates as well as Esbele. There are remains of a large church which was probably the cathedral, and buildings about it which might serve for the religious that belonged to it. There are ruins also of a smaller well built church. There is nothing to be seen of the walls and not so much as a village on the spot. The rocky low cliffs to the sea have been much work'd with the tool, and I observed a sort of a canal, cut between them from the sea, which probably might be to draw up boats and small vessels for security from the weather, as it is an open port without any points of land to shelter it. Four or five miles further we enterd in between those mountains which stretch Westward to the sea and make that head which was called by the antients Qesuproswpos and where Mount Lebanon ended at the sea. Here those Arabs and Iturean robbers, who infested the country had one of their strong castles, which with many others belonging to them were destroyed by Pompey. Shaping this part of Mount Libanon, I went along the narrow plain that runs to Tripoli, and going near a small town called Enty on the sea, where they say are remains of a well built large church, I came to some ruins that seemed to have been an antient temple and several heaps of stones for some distance round it. These might be Trieres mentioned by Strabo between the mountains before mentioned, and Tripoly; which may be the Tridis placed in the Jerusalem itinerary twelve miles from Tripoly. On the right or East is a low ridge of mountains that extend almost as far as Tripoly inhabited by Greeks who have some convents on it, particularly the large monastery of Belmerit which is delightfully situated, and another of St James here the Europeans are usually buried that dye at Tripoly.

Tripoly is situated under the hills to the East of a low promontory that runs about a mile and a half into the sea, not being above half a mile broad. On this promontory were the three cities, colonies from Tyre, Sidon, and Stradus, a furlong distant from each other, and which seem at length to have joined in their suburbs and to have made one city thence called Tripolis. One of them probably the first built which might be that from Tyre was at the end of the point, and so could be easily fortified by building a wall across the promontory, on the East side of this city; of which wall, about fifteen feet thick there appear great ruins and it looks as if it had been thrown down by force. It was cased with hewn stone[s], which are now carried away. About the middle of it was the gate of the city, about which are several pieces of fine large pillars of grey Granite. The second city, it may be the colony from Sidon, might be at the angle made by this promontory to the North, where the river falls into the sea that runs from between the hills and passes thro' the present city. But there are no signs of either of those other two citys, most of the land being turned into gardens, and all being a sandy soil moving with the wind has coverd what has not been removed of those antient citys. I thought the third city, the colony perhaps from Stradus, might have been at the South angle, where there is good water and a small stream, but it being all a rocky ground and exposed to the South West wind which is the most dangerous, and so could not be a port. It is most probable the third city was in the middle between the other two on the North side, where there is a tolerable good harbour used at this time, being shelter'd by some Islands and rocks to the West. Along the North side of the promontory are six large towers about a quarter of a mile distant from each other, whether they might be built with any wall that might formerly have been raised for defence along the shoar or only to protect the place against Privateers, for which they serve at present I cannot say. On this side where the wall went across the land is a little town called the Marine, inhabited mostly by Greeks. Here they have the Custom house, and all goods are landed in small boats from the shops, which lye out at some distance. The Greeks some years ago built a fine church here, which soon after was pulled down by a Pasha.

I am my Lord

BAALBECK IN SYRIA
9/20 June 1738
Pococke to Bishop Milles
Account 18[16]

The city of Tripoli is about two miles round. It stands long; and the river after the great rains, sometimes overflows in such a manner as to do much damage to the city through which it passes. There are some buildings up the side of the field and over the South East corner of the city is a large castle on a hill, thought to be built during the time of the holy war, for there is a mosque in it that was a church dedicated to St John. In this city are five or six mosques with square towers, which they say were churches, and particularly one with a large square tower after the European manner, but the finest mosque has an octagon tower and was a church dedicated to St John. There are a considerable number of Greeks here, who have a handsome Cathedral church near which the Bishop resides. The Maronites also have their church. Many of their Bazars or streets of shops are made out of old convents and nunneries which one may easily see by the manner of building. The Fathers of Terra Santa, the Jesuists, and Carmelites of Mount Lebanon have their convents here, the latter residing in this city only during the excessive cold weather in winter on that mountain. This is the place of residence of the Pasha of Tripoli, from which city the whole Pashatick is denominated. The river of Tripoli runs through a most delightfull narrow valley to the city from the East: about half a mile above the city on the side of the hill over this vale and river, is a convent of Dervises [Dervishes]. It is one of the most beautifull situations I ever saw, is adorned with several water works, having a command of water from an Aqueduct that runs through it. This aqueduct derives its waters from the foot of Mount Libanon about eight miles distant. It runs along the side of the hills in a canal made for it on the North side of the river till it comes within a mile and a half of the city, when it runs on an aqueduct of four arches across the valley and over the river for one hunder'd and thirty paces, and seven feet eight inches broad. As it serves for a foot bridge, so it has the appearance of a bridge. The two middle arches which probably had been broke down are Gothick, but the other two are true arches and are probably very old. There is a cross on the aqueduct, which is said to be built or rather repaired by Godfrey of Bulloign, and upon that account it is probably called the Princes bridge. From this the water runs on the South side of the vale by the Dervices [sic] convent. The trade of Tripoly consists cheifly in exporting raw silk to Europe, and the cotton and silk manufactures of Damascus to the different parts of the Levant. They have also a manufactory of soap made with

16 BL, Add MS 15779.

the oyl of olives for which they were formerly more famous than Joppa which now rather has the fine reserve [?]. There is only one English house here, but several of the French nation.

From Tripoli I went Eastward to Mount Libanon, and ascending in about four hours I came to the Maronite Convent of St Anthony. The large church is a Grott cut out of the rock, and so are several other parts of the convent. There is also a large natural Grott that extends a great way under ground, and there are in it what they call petrifactions of water, which being hewn appears as a very fine white alabaster like that in the Grotts of Carliola. In [a] dark part of this Grott they discipline mad people, this convent being as they say famous for the miraculous cure of the disorder of the brain, and being brought here they commonly come to their senses in three or four days, or a week, and rarely continue there longer, and even sometimes are cured in their way to it. They bury the monks in a sort of vault above ground in their habits: in which they appear like skeletons, and one I saw whose skin seemed to be uncorrupted, who they say was a holy man. This place is famous for excellent wine, which they preserve as they do in all these parts in large earthen jarrs close stopt down with clay, but being sent to distant places in skins, it receives from them a strong flavour that is disagreable.

From this place I went towards Canobine, the convent where the Maronite Patriarch usually resides. The descent to it is down a very steep hill by a narrow winding way. It is a convent cut out of rock into Grotts. About three quarters of the way down the hill, below it runs the river that goes into Tripoli in a valley little wider than the bed of the river having on both sides two very high ridges of mountains covered with pines. The situation is the most extraordinary and refined I ever saw, there being only one way to it, which makes it a very secure retreat, probably the reason of the Patriarchs having taken up their residence here, tho' now the mountains where I saw the Patriarch are much securer. The church is a fine large Grott and has three bells hanging in a window of it. The Bishop of Mount Libanon was there, who generally resides with the Patriarch, and is a sort of vicar to him. Near the convent is a Grott which is a chapel to St Marina, without [outside] the chapel are descents to two vaults, in one the Patriarchs area buried, and in the other the monks. I returned by the same way and going Eastward passed near a village called Eden, which is reckoned one of the most pleasant villages in the world on account of the veiws it affords, its waters, and the improvements about it. I after came to the convent of the Latin Carmelite Fathers, which is a most delightfull retirement in the summer, the beauty of the opposite hills, the several Cascades and streams of water, and the perpetual freshness of the air in these high regions make the place ever agreable whilst the heats in the plains are almost intolerable, but in the winter the Fathers reside in Tripoly. From this convent is a gentle ascent for about an hour to a large plain between the highest part of the mount; towards the North East corner of it are the famous Cedars of Libanon. They are a thin grove about a mile in circumference, which consists of some large Cedars that are near to one another, a great number of young Cedars, and some pines. The great Cedars at some distance look like very large spreading Oaks. The bodies of the trees are very short making out into three or four very large branches, and some seem in a manner to begin from the very roof appearing like so many bodies united in the manner of Gothick pillars. One that has the roundest body, tho' not the largest, measured twenty four feet round, and one with a sort of triple body in somewhat a triangular figure, measured twelve feet each side. The limbs of the trees after they are grown to a considerable heighth spread horizontally, and make a flat surface at top. What is very particular, the Cedar apples grow perpendicular up with their ends to the skies, and so do the buds of their trees. The fruit is in the shape of a Tun [beer cask], smooth, something like a very close Artichoak, and never open like the pine apple. The young Cedars are not easily to be known from young pines. I observed the[y] bear the greatest quantity of fruit. The wood differs not from white deal in appearance, nor does it seem to be harder, it has a fine smell, but I think not so good as the Cedar of America, as it falls short of it in beauty in which I cannot be deceived because I took a peice of the wood from a large one blown down by the wind, which is left to rot. Besides that there are fifteen large ones standing. As the Christians of the several denominations near, come here to celebrate the Festival of the Transfiguration, so against several of the large trees are Altars built of large hewn stone, where they administer the Sacrament. These trees are about half a mile North of the road, to which we returned, and from this plain the mountains, ascended about three hours up to the very highest point of Mount Libanon passing over the snow which was frozen hard. These mountains are not inhabited above the Carmelite convent, nor all the way down on the East side, which is very steep and a barren soil; I observed the only trees which grow up towards the top are Cypress which being niped [nipped] by the cold grow not spirally but like small Oaks, and one may conclude that this tree bears cold better than any other. From the top of Mount Libanon one sees the beautifull parts of Mount Libanon below, and all away to the sea beyond Tripoly: to the East lake Lemoun at the foot of the hill about two or three miles round which nearly retains its antient name Lemone, and that great plain which was Calesyria proper on the East side of which I saw Balbeck. Tho' all the people about Mount Libano drink of the waters that run from the snow melted on this mountain, yet there is no such thing as those swelled necks which the people have who drink of the snow water that comes from the Alps.[17] I know not what reason can be given for this, unless that the snow in these climates may not be charged with so great a quantity of nitrous particles, and that human bodies in these climates which are so much hotter enjoy a greater perspiration to carry off the matter which occasions that excrescence. It is observed on Mount Libanon that in the spring when the snow begins to melt the waters of the rivers rise but the fountains continue as before. After a certain time the fountains – plentifully flow, and the waters of the rivers diminish, and then the fountains continuing to flow the waters of the rivers encrease [sic] again, which periods I have some where committed to writing. The reason seems to be, that when the snows first melt, the waters fall down on the surface of the snow without soaking into the ground to feed the springs, and so runs all into the rivers, and when the snow is melted towards the lower parts of the mountains, the water begins to be drunk up by the earth, and consequently to encrease the fountains, and that when the earth is full as it were with water, and consequently do's not imbibe so much of the liquid, it runs again more plentifully into the rivers, continuing still to feed the fountains. At the foot of the mountain of Libanon is a narrow valley in which on the South of the road is the small lake Lemone before mentioned, the road is to the East for two hours between two hills covered with wood, and leads into the plain of Balbeck which is about eight miles broad, extends a considerable way to the South, and farther to the North; so that plain which has to the North those high deserts which run away Eastward about three days journey to Palmyra. This plain between Libanon and Antilibanon is Calesyria proper, the Calesyria in its largest extent is all the vale country of Syria from Seleucis to Egypt and Arabia. Towards the North end of this plain rises the river Arse which is the Orontes of the Antients. It is a very barren unimproved red soil insomuch that it hardly answers the expence of tilling and watering, the crops produced are so very poor, and they cannot sow it two years together, this part having no water but what is brought from a little stream that rises plentifully half a mile South East of Balbeck, runs thro' the city and is lost in the feilds and gardens. Coming out from the hills into the plain on the left is a small village inhabited by Maronites, where is a ruined church repaired that seems to have been built after the model of the temple of Balbeck without the Colonade round and must be of great antiquity. Crossing the plain I arrived at Balbeck on the East side of it.

I am my Lord

---

17 Pococke noted this phenomenon in a letter to his mother from Milan, during his first grand tour. See Finnegan, *Letters 1*, p. 261 and note 179.

ALEPPO
31 July O:S: 1738
Pococke to Bishop Milles
Account 19[18]

My Lord

I was exceedingly glad to be informed by the Chantors [Jeremiah Milles'] letters, that tho' your Lordship had had some painfull returns of the Gravel [gall-stones] yet that your Lordship was well recovered, and I trust in God, and have a strong persuasion that your Lordships health and strength will continue many years, to which I don't doubt all your Lordships friends tell you, proper diet and exercise will very greatly contribute.

I believe your Lordship is persuaded that the reason why I do not write more frequently is the full employment I have, so that as yet I have been able only to give your Lordship an account of my journey to the Cataracts and to Jerusalem: since that I have been at Nazareth and the places near it, and have travelled all along the coast of Syria to Tripoli, and from thence to the Cedars of Libanon, Balbeck, and Damascus, from which city I came to this place in eleven daies. I purpose now to step to the Euphrates two daies journey from this place, and to see everything unseen in these parts with all convenient speed and then move Westwards.

The short time I shall stay at Aleppo I find will be very agreeable, as I receive all manner of civilities from the Gentlemen of our nation, who in every respect answer the character of their Preceptors, and make one of the most agreeable and easy societies in the world. I am in a more particular manner obliged to Mr Pullinger, to whom I am recommended, who met me a daies journey from this place, and also to the Consul Nevil Coxe Esqr, a Glocestershire [sic] Gentleman, from both which Gentlemen I have received all manner of favours.

I am exceedingly obliged to your Lordship for your good wishes, and the kind concern your Lordship is pleased to express for me.

I flatter myself I have no need to mention the following request to your Lordship, that in case a stall should become void in the Cathedral church of Waterford, that your Lordship would be pleased to confer it on me, but only take the liberty to mention it in case I should not be at home in time, tho' on advise [sic] fom your Lordship of a vacancy I hope to be able to be with your Lordship before the half year is up, but in case I should not, Mr Milles may be

18 BL, Add MS 15779.

removed into the vacant stall, by which another half year will be gained, which I humbly request of your Lordship.

Most earnestly wishing your Lordship a continued state of good health and begging leave to put in my kind remembrances to the Chantor, my fellow traveller [Milles]

I am my Lord

CHAPTER 8

Letters from the Second Tour of Egypt and from Turkey, Asia Minor and Greece

This chapter reproduces 27 letters from Egypt, Turkey and Greece, as follows: two from Larnaca, four from Cairo, one each from Mount Sinai and Suez, one from Alexandria, two each from Canea (Chania) and Smyrna, one from Ephesus, two from Cora in Samos, one each from Aydın and Ankara, four from Constantinople, one from Bursa, and a series of short letters (one from each place) sent from Salonica, Thebes, Athens, Corinth and Messina. This part of Pococke's voyage, which included his home journey, spanned a period of almost three years, from November 1738 to his arrival back in Dover at the end of August 1741.

His account of Crete earned him great credit;[1] and likewise with his exploration of the north-west Troad, for which he is recognised as being one of the first modern Homeric scholars to identify the locations for Ilium and Troy.[2] For further details of the places described in the correspondence from Pococke's second trip to Egypt, the reader is directed to *Description 1*; and for the other places *Description 2*, as follows: Cyprus, Part I ("Of the island of Cyprus"); Crete, Part I ("Of the island of Candia"); the Greek Islands, Part II ("Of the Greek islands of the Archipelago"); Asia Minor, Part II ("Of Asia Minor"); Constantinople and mainland Greece, Part II ("Of Thrace and Greece").[3]

1 As noted by Peter Warren in relation to his findings on this island, "Pococke well exemplifies eighteenth-century spirit, sympathetic, curious for polymathic observational detail, lightly moralistic in social and religious comment ..." See "Early Travellers from Britain and Ireland", in Davina Huxley (ed.), *Cretan Quests: British Explorers, Excavators and Historians*, British School at Athens, 2000, p. 6.

2 See Introduction, above.

3 Readers with access to a three-volume set of *Description* should turn to the third volume for the Greek islands onwards.

 | DOI:10.1163/9789004440050_009

LARNICA IN CYPRUS
10/21 November 1738
Pococke to his mother
Letter 22[4]

Honoured Madam,

My last letter to you was of the 9th-20th of Octr. from Latichea, the old Laodicea in Syria, the day I arrived there; that city is pleasantly situated, on a rising ground on the sea, being a port; it is now but a middling town, tho' formerly so famous a city.

10th-21st. I went out with the Consuls clark, to see Captn. Fielding, I met at Damiata & his partner Captn. Long, the former married to a Bristol woman in the West Indies; – they have four ships: – met Mr Usgates brother there of Acre,[5] I met them near our house coming to see me; went to the key [quay] & so home; at 2 all that house came to see me. The Consul went out to show me some Antiquities, a piece of a Portico round a Temple, a Triumphal Arch, & an Inscription or two, imperfect or difficult to be taken [copied]; returned, found Mr Usgate at our house, who sat a while. I writ. Sup'd; they have here the turnip cabbages, the seed of which I sent you.

11th-22d. I set out with Mr Usgate northward to see the ruins of Possidonium, about 20 miles north, near the Bay of Antioch; dined at a village on what we carried; lay near the sea in a salt ware-house.

12th-23d. Saw what remained of Possidonium; – returned about 10 miles toward Latichea.

13th-24th. Found out the ruins of Heraclea at the sea; dined at a village in a garden under a Lemon tree, & after dinner eat a Citron from a tree with people in the garden. It is almost all peel, which is what one eats; returned to Latichea

14th-25th. Walked out about the Town & Port, visited Captns. Long & Fielding, found Mr Crouch Consul Carews nephew of Tripoly returned from a journey, he had taken to Aleppo.[6] The Consul went with me to copy out a long inscription.

15th-26th. I went to compare the inscriptions. We all Fdined at Captn. Longs & rid out in the even 2 miles to copy out an imperfect milliary inscription. Supped at Captn. Longs.

4 BL, Add MS 22998.
5 Brother of Thomas Usgate, who appears in Letters 14 and 15.
6 Pococke had met Thomas Crouch in Tripoly. See Letter 16, above

16th-27th. About 9 I set out with Mr Crouch for Tripoli, – the Consul, Captn. Long & Mr Usgate went with us two hours, we dined together by a stream, & parted; arrived at Gibele [Jeble] the ancient Gabala; – went to the Aga's house; – Mr Crouch & I went to see Sultan Ibrahims Mosque where he is buried,[7] & the Grotts by the sea where he lived, & prayed, & after to see the ruins of a Theatre.

17th-28th. We set out at 3 from this poor Town, & in 4 hours came to the ruins of Balanea, a Bishoprick now called Baneas [Banyas]. Ascended up the hill to Meska castle where we were kindly entertained by the Aga, & shown the large castle, said to be built by the Franks:[8] – we left the place & went on & dined by a stream, & towards even arrived at Tostosa [Tartus]: – & went to the Aga's to whom I had a letter, & he was an acquaintance of Mr Crouches.

18th-29th. Mr Crouch went on to Tripoli, & I viewed the fine old walls of the Town & castle, & the church in the latter, – & a most magnificent entire one in the former of about the 5th century; got into a boat & went a league, to the Isle of Road [Ruad/Arwad], the ancient Aradus. Went aboard Captain Crispin an English ship, who invited me to dine: – I went to the Aga with a letter, who set before me a collation; – & sent one with me round the Isle, which is about a mile in circumference, all a rock, & was encompassed for the most part by a very strong double wall; dined with the Captn. on fish & returned to Tortosa.

19th-30th. I set out & viewed the Antiquities of a fossee, with three stairs on each side for seats. I take it to have been an Hippodrome or some place of sports; – then went to an open temple cut in the rock, with a throne in the middle; 4 sepulchral towers, & the grotts under them all mentioned by Mr Maundrel, & proceeded on & lay at a sort of an encampment of Arabs.

20th-31st. Set out, & two hours from Tripoly was met by Mr Crouch, who showed me the way a second time to the Consuls house [Purnell]; – where I dined & writ.

21st Octr-Novr. 1st. Captain Crispin dined here, who was with his ship come from Tortosa. the Fathers of Terra Santa came to congratulate Mr Crouch on his arrival. I writ all day.

22d-2d. Captain Crispin dined with us & Captn. Breton who came in yesterday, going to Damiata. I writ all day; – walked out in the even, to the marine or Port a mile off.

23d-3d. I rid out with Mr Crouch in the even to see a fountain of fishes.

24th-4th. I walked out by the sea shoar; dined; Mr Crouch & the Consuls droggerman rid with me to the port; & I went on Board the *Richard*, Captn.

7 Cf. Karl Baedecker, *Palestine and Syria: Handbook for Travellers*, London, 1876, p. 239.

8 A Crusader castle now known as Margat or Marqab.

Crispin for Limesol [Limassol] in Cyprus, the same ship that carried King Theodore from Tunis to Corsica;[9] about midnight we got out.

25th-5th. Made litle way.

26th-6th. Put into Bayreut [Beirut] where our supercargo had a mind to go, he only went ashoar & returned.

27th-7th. About 1 in the morn we sailed again, with a good wind.

28th-8th. In the even we arrived in the bay of Limasol [Limassol], & anchored about ten at night.

29th-9th. Went ashore to the house of a Greek, who acts for the English Consul; dined; then I set out on a mule for Larnica; where the franks reside; – went Eastward along to the Bay, & in 2 hours came to the ruins of old Amathus, a Bishoprick now, called old Limisol – & lay an hour & ½ further at a Christian village, as most are here, being Greeks.

30th-10th. Set out before the sun, & travelled mostly along by the sea, dined by a stream & after sunset arrived at Larnica, a town a mile from the sea, it is 2 or 3 miles round & every house has a garden & yard, & looks something like a village or perhaps like Moken, I went to the Consuls house, Mr George Barton, brother to Consul Barton of Cairo,[10] whose Partner here is Mr Grovesnor; another English house here, Mr Luport [Lupart] with whom lives Mr Purnel, brother of those in Syria,[11] who formerly was in business being recommended here by Mr Maydwell & particularly by the Consul at Cairo. I meet with very great civility from the Consul.

31st-11th. Writ. In the even visited by Mr Purnel, rid out with the Consul in his chaise about the ruins of the walls of the old city Citium south east of the town, the place where Zeno was born;[12] & Cimon the Athenian General died at the siege of it.[13] Went to the town of the Port & very old church of St Lazarus, where I entered into his sepulchre from which the French people carried his body to Marseilles.

Nov. 1st-12th. Mr Luport came to see me & after dinner Mr Charles Ober, nephew of my friend Mr Otto Lewis in Hamburg. Went with the Consul in his chaise to see the foundations of the old city, & returned those two visits, without seeing 'em. Mr Grovesnor came home.

---

9 Theodore of Corsica (1694-1756) was a German adventurer who in 1736 persuaded the Corsicans that he would free them from Genoese rule if they crowned him King. However, he was forced to flee from Corsica later that year.

10 George Barton was English Consul at Cyprus twice, from 1732-33 and again from 1738-39. See Özkul, "The Consuls and their Activities in Cyprus", p. 256.

11 Peter Lupart, whose name was linked with several trading partners in Cyprus, see Wilson (*List of British Consular Officials*, p. 34). Mr Purnel was brother to Consul Purnell at Aleppo and Mr Purnell at Latakia.

12 The Greek thinker Zeno of Citium (c. 334-c. 262 BC) is not to be confused with earlier Greek pre-socratic philosopher, Zeno of Elea.

13 His death took place there in 450 BC.

2d-13th. Walked round on the foundations of the walls of Citium, twixt [between] the Town & the port; writ. Lord Sandwich, Mr Nelthrop [Nelthorpe], Mr Ponsonby, I suppose Lord Duncannons second son & Mr Makai [Mackye] hired a ship & went to Constantinople last summer & returned.[14] One Dr Perry a Physitian, who has been two years in Constantinople, is now towards Jerusalem.[15] At 3 went with the Consul in his chaise to the Salt Lake on the water for which they gather a salt crust; to a litle mosque, where the foster-sister of Mahomets sister was buried: – the Sheke gave us Coffee; – we returned: – play'd at Whisk [whist?]. Consul Barton you may be sure is very young about 24 or 25, when I tell you that in 1730 he was nine months at Monsr. Le Fievres in Southampton to learn French, & came thence immediately abroad; he was so like Adam Speed, that everybody took him for Adam, & Mrs Speed called him her Son.[16] Charles Clutterbuck was his particular friend; – & with him he corresponds, so we have much talk about Southampton, the news of which he knows, & Mr Conduit was his uncle.[17]

3d-14th. I walked out about the ruins, & called on Mr Luport & Mr Purnell, the former is cousin german to Sr Everard Falkener, Ambassador at Constantinople;[18] he presented me with some medals, & the latter with the maimed

---

14 James Nelthorpe (c. 1719-68) was the eldest son of a Justice of the Peace, of Seacroft Hall, Yorkshire. He is recorded as having reached France in 1735 but it is not known how or when he met up with Lord Sandwich. A pastel portrait of him by Liotard in an eastern costume almost identical to that of William Ponsonby (below), though only half length, is reproduced in Neil Jeffares, *Dictionary of Pastellists before 1800*, Unicorn Press, Norwich, 2006, p. 349, the original being in the Holburne Museum, Bath (Ref. A358). Mr Ponsonby (1704-93) is actually Lord Duncannon, having assumed this title in 1724. However, in the sources he is often referred to as Hon. William Ponsonby; and on the death of his father in 1758 he was styled 2nd Earl of Bessborough. John Mackye of Polgowan (1707-97) assumed the name of Ross on his marriage to Jane Ross of Halkhead, 13th Lord Ross. He graduated from Leiden in 1728 and went on his grand tour after a gap of almost a decade. In 1742 he was elected to the Society of Dilettanti and went on to have a distinguished political career. While Lords Sandwich and Duncannon/Ponsonby were founder members of both the Egyptian Society and the Divan Club, Nelthorpe and Mackye were elected members of the latter. For more detailed biographies on all these gentlemen, see See Finnegan, "The Divan Club", pp. 54-55; and for their portraits, see Plates XIII and X, respectively.

15 Charles Perry (1698-1790) whom Pococke was to later meet (see Letter 23) and whose book on the east came out shortly after his own volume (1743). See Introduction.

16 Speed was a "famous family" in Southampton (see Russell, *A History of King Edward VI School Southampton*, p. 185) and Adam was presumably a descendant of the "John Speed jun^r Gent." who had made over to Pococke's father, by deed, "the next presentation of the parsonage of Colmer". See Will of Richard Pococke (Senior).

17 All of this would have been of particular interest to his mother, whose married life was spent in Southampton, and who was associated with the Clutterbucks through her late husband's first marriage.

18 One of the business partnerships with which Peter Lupart's name was linked was Snelling & Fawkener, the latter being Edward Fawkener, the brother of Sir Everard Fawkener. See

head of a statue. This Isle is famous for Hams & pork, but our best butter comes from England. Walked out in the even with the Consul & Mr Grovesnor.

4th-15th. Being rainy staid within all day. Writ; play'd at Whisk in the even.

5th-16th. I read prayers in the Consuls House being Sunday & Messrs Luport & Purnell were here & dined with us. The Guardian of the Terra Santa came & with him Father James Quilk [Quirk] of the county of Galloway in Ireland, whom I met at Jerusalem;[19] drank tea. I sent a box of trumpery [bric-a-brac] from Scanderoon or Alexandretta, in it are two bottles of the best Balm of Gilead or Balsam of Mecca which is a great remedy in many cases, taken a few drops in any liquid, or on a piece of sugar, particularly for the Lungs, & I believe for the Stomach & Cholic, & is a great cooler of the […] & also for the gravel [gallstones], of which you may learn; you must be very choice of it; the larger bottle is very old, & was presented me, & there is another bottle I bought, but not so old, in which when it comes you may send to Ireland, when you have opportunity; half of it to be put in a bottle & given to Captn. Ramsay [Ramsey],[20] the other half to be kept for me there, or used by my Lord or Mr Milles if occasion. I must tell you also that now you must not expect to hear quite so often from me, because in the winter the ships do not go so frequently into Europe.

6th-17th. Rid out in the afternoon with the Consul in his chaise; & after walked out with him & Mr Grovesnor, to see an old church under ground, now a Mosque. We play every night at Whisk.

7th-18th. The language here & in all the Islands in Greek. Monsr. Le Fievre a french Doctor breakfasted with us. I went at 3 to visit the Guardian & Father James of Terra Santa convent with the Consul; then took a walk.

8th-19th. There was a pretty great Earthquake this morn, but being busy in putting some papers in my trunk, I did not perceive it; – but heard a noise which is occasioned by the shaking of the doors & windows, & they say it lasted long, that is perhaps 15 or 20 minutes. I went to see Mr Luport & Mr Purnel; in the even Mr Lupert came to see me. I rid out with the Consul in his chaise; drank tea, plaid at Ombre.

9th-20th I pack'd up to go a Tour round the Isle, being to set out tomorrow. Mr Purnel came to see me. In the even rid out with the consul in his chaise. The guardian of Terra Santa convent & Father James & another Father came to see me, when I was abroad [out]. This Island is exceeding pleasant in the winter. It

---

Wilson, *List of British Consular Officials,* p. 34. Pococke meets the English Ambassador at Constantinople: see Letters 40ff.

19 See Letter 11.

20 Captain John Ramsey, of St. Michael's Parish, Waterford, married Charlotte, daughter of the French exiled apothecary, Jacques Renet, in 1719. See Grace Lee, *Huguenot Settlements in Ireland,* London, 1936, p. 103.

is spring here the corn green & everything pleasant, so that I expect to have a very agreeable Tour. Here are Amianthos stones, like those of the Isle of Anglesey & several painting earths [pigments].[21]

I desire you to present my very kind love to my sister & due respects at Whiteoak & to all friends.

I am Dear Madam your most obedient & dutiful Son, Richard Pococke

LARNICA IN CYPRUS
7/18 December 1738
Pococke to his mother
Letter 23[22]

Honoured Madam,

My last letter to you was from Larnica in the Isle of Cyprus, dated Novr. 10th/21st. The Day I set out on a Tour round the Island. Mr Grovesnor rid near an hour with me, & I went about 18 miles to Famagusta, a strong fortification made a la modern by the Venetians; & bravely defended by Bragadine, who after honourable capitulations, was flea'd [flayed] alive by the Turks here, & his skin stuffed & hanged to a Galley mast in the Port.[23] I went to a Christians house, where I had a room to myself, & was alone & writ. Most people here ride on mules, which they load too with their necessaries, & they pace & are very easy. The soil here is very fruitfull & oranges & Lemons in such abundance that one buys 50 sevil oranges for a Medimn, which is about 3 farthings; & here also they make much fish but great part of the country lies uncultivated.

11th-22d. I went into the Town [Famagusta], walked all round the inside & out; the Fossee is cut into the rock, it is a very good Haven, with a chain across the entrance; the Venetian arms & inscriptions in all places, & a great number of churches going to ruin; three of them have been very fine; St Georges, & another turnd into a mosque, & the old cathedral of the Sophia, a most beautiful gothick building like that of Savum; – but almost all fell down in the earthquake in 35, & buryed 50 Turks at prayers there: – & what remains is a mosque.

21 A type of asbestos stone common in Cyprus (*Lapis asbestus Amianthus Cyprianus*). A specimen of this was the first item from Cyprus to enter the collections of the Ashmolean Museum in Oxford, in the 1770s.

22 BL, Add MS 22998.

23 After 11 months of siege, the Venetian hero Marco Antonio Bragadin (1523-1571) yielded to the Turks, on condition of military honours, but the enemy did not keep its word. His remains are now in the Basilica of St. John & Paul in Venice.

The old House of the Government still remains with the names & arms of the Governors on it, a very litle part of the town inhabited, tho' 2 miles round & every part shatter'd by the earthquake. I came home & dined; & writ.

12th-23d. Went an hour or 3 miles to old Salamis built by Teucer, I traced the walls, saw in it ruins of three churches, an Aqueduct that brings water near 20 miles; a chapel without the walls; & half a mile west the sepulchre of St Barnabas, who was born here, & martyred here under Nero; near it is a convent & large church dedicated to him. Went on about 4 hours & lay [at] a Greek priests house, who was very civil, when he came home from his labour. At Famagasta I met a deposed Greek Patriarch of Constantinople, who intriguing to supplant his successor, was banished hither 3 months ago, dressed in purples lined with furs.

13th-24th. Went about 5 hours west to Galousee [Yialousa], on the north part of the Isle, here we see the coast of Cilicia very plain, which is 30 leagues off. I went to the Priests house, saw the ancient church; dined, & writ; the country I rid through today is very fine, but now become a litle rough, being in a manner only one hill, about 6 miles wide, from the north sea to the south.

14th-25th. The Priests Lady brought me some Pork & bread, but I had taken my Coffee, it is their carnival before the Advent fast. I descended to the narrow plain on the north sea; & went about 5 miles to Carpass, saw a litle church under ground by the sea, the Priests son with us, & his wife carried us to his Father [in] laws house, – where they were very kind like country people in England. I dined on mutton cutlets roasted. I carry my provision with me; there are in this village 7 or more ruined churches, having been probably a retreat for Christians in the primitive times. I went a mile to the sea, to see the ruins of old Carpass & of a large church.

15th-26th. I went out to go to the very East point of the Isle; passed through a ruined village, where were 4 ruined churches; in about 5 hours came to the end, all uninhabited & some bad way through low woods of cyprus &c: & aromatick herbs, particularly very fine sweet marjoram. At this point the Maltese Cruisers often land & make merry; & this is a reason why the country is not inhabited, for to the villages, they often come, take away all they have & make the Turks slaves, & so it is all round the island near the sea. Went to the church of St Andrew, a grot belonging to a convent that was there; & came home to Carpass along a narrow plain on the north side of the Isle, having made it a long days journey.

16th-27th. Went out at noon to return to Galoussee saw by the way the poor convent of Galeusa, with only one papas, caloyer or monk in it, & 3 or 4 boys; came again to the priests house at Galousee.

17th-28th. I went out only with one man above 2 hours off, to see a large sepulchral Grott; – & returned towards the even.

18th-29th. Set out & travelled 6 hours to the mountain, I take to be the Mount Olympus [Khionistra] of Cyprus, & went to the top of it, where are the remains of a very strong castle & where possibly might be that temple of Venus, which no woman could lawfully enter or look upon; & I must tell you about this part of the Island, was a city called Urania, probably sacred to Venus Urania, or the chast[e] Venus; descended to the north side of the mountain near the sea, lay at a Christian village; yesterday at a church we were at, we killed an Asp; – it is a sort of serpent, but some are very large, & their bodies broader than snakes, they answer the description in scripture, of the deaf adder which refuseth to hear the voice of the charmer, being deaf & thence have the name of kouphis in vulgar Greek; they are very venemous rather worse than the viper, & are prepared as they are, in medicine, & by reason of their deafness are easily caught. They have also here a large venomous spider, they call a Tarantula, but it is not the true kind.

19th-30th. Set out & travelled on the north side of the Isle, 5 or 6 hours to a Xtian village.

Novr. 20th-Decr. 1st. I went on westward & came to a convent called Malandria of a few priests: – & soon after to another called Antiphonite, – a good church; – but what is most extraordinary, seven trees of the Lignum Cyprium, called also Lignum Rhodium, because it grows only in Rhodes & here; it is called also Lingnum Rosa from its smell; [This is a species of The oriental Plane tree, I believe that which is call'd the smaller & more deeply divided Plane];[24] & here they call it Xylon effendi, or the wood of our Lord, none of it in the Isle but here, it is like the Plane tree & the Sycamore in its leaves but smaller, & has mast like the plane-tree, which are like the spiked balls you may have seen on each side of some gates: – It is a tree that produces a Turpentine when the vine is out, which has a strong smell, & the leaves & wood have the smell of firr trees; but somewhat different; the monks say, the vine roasted & sucked is good for a fever. I eat here; – they brought olives & honey & a green sort of cheese; ascended up the hills, & came down on the south side into the great plain between Famagosta & Nicosia; – went 2 hours from it, & lay at a Christian village.

21st-2d. Went to a fine village between the hills called Cherkes, said to be the Old Citari; – fine waters that were carried to Salamis, – warm in winter & cold in summer; thence went 3 hours to Nicosia, the capital of Cyprus, fortified weakly by the Venetians. I went to the house of the Consuls agent, where I sup'd & lay.

24 This is a note in Pococke's hand appearing on the reverse of the page.

22d-3d. Went with the Agent to the Mosolems (governors) Droggerman, with a letter from Mr Luport; drank Coffee; – he sent his man with me, to see the fine cathedral, now a mosque, some grand Venetian houses &c: – I went after to the Armenian church; dined; – visited the President of Terra Santa convent, walk'd round the foundation of the old walls; – the present city being fortiyed a la modern by the Venetians, & is about 2 miles round, the old city, 3, tho' ill built & gardens to the houses, all of unburnt brick. Sup'd & packed up.

23d-4th. Set out & went 3 leagues to the convent of St Chrysostimo, north east on the side of the mountains; – saw the handsom old church; – dined with the Superior to whom I had a letter, on fish as big as herrings, & like Anchoves, some fryed, some cold, olives, green cheese preserved in oyl, as all here is, & made of goats milk & honey. 2 of the priests went up to the top of the mountain with me to some poor ruins, called, the hundred chambers. – brought some wine, & the priest pressing me to take a third glass, had the ignorance & prophanity to hold out his 3 fingers & say ἁγία Τριάδα [Holy Trinity]. Went on & crossed over the hills to the north, to a village, where I lay.

24th-5th. Went to a most magnificent ruined convent on the side of the hill, calle[d] Telabaise, from that west to Gerinus [Girne/Kyrenia] the old Cerynia, the town ruinous, a strong fort near it. I went to the house of a French Master of a vessel; saw the church, dined, & went 3 leagues west to a convent at old Lapitho [Lapithos], where I lay.

25th-6th. Saw the ruins of Lapitho – went a league to a poor convent & crossed the hills to the south, saw another indifferent convent, – & went 3 leagues west to a noble convent at Morpho [Morphou], a fine church: – here I lay; the convent entertains you & one gives something.

26th-7th. I went 4 leagues west to the end of the plain, to a ruined city on the sea, now called Aliagora, built by the King of Aepea, at the instance of Solon, who retired here when he was banished; – viewed the old foundations &c: or it may be Arsinoe; & went 3 leagues to Solea, a valley between the hills, probably so called from the old Solea, which might be hereabouts, which was named by the said King in honour of Solon, who perswaded him to leave it, the former name being Aepea, a most delightful spot. Iron mines here [next 2 words scored out]; & the terra umbra green & yellow & other curiosities, & excellent good springs. I went to a small convent where the Bishop of Gerinouis resides, but was not there; [next word scored out].

27th-8th. I went 6 miles along the beautiful vale of Solea to the south end of it, & by woods & cascades to the convent of St Nicholas, a poor place, where we dined; – at convents one meets with dry boyled beans & dry kidney beans, white cabbage dressed like sallad, olives, cheese, Honey & such like, for they never eat flesh, nor anything that comes from it. Went 2 hours N.W. over high hills to the convent of St John in the midst of the hills, like a long farm-house, where I lay – a great compliment to bring Rose-water & pour it on the hands. The superior came & sat with me. I sup'd – & writ.

28th-9th. Set out, went 3 hours all along the sides of the mountains, a good road to the convent of Madonna of Cheque where is a picture of the Virgin Mary with our Saviour in her lap, as a child of 6 or 7 years old, said to be painted by St Luke, & to perform miracles, & is resorted to by the Greeks from all parts, even from Muscovy, as Loretto is by the Romans:[25] – It is on the high mountains & very cold; ten fathers & about sixty other caloyers [Orthodox monks] in it: – the superior met me at the gate, & conducted me to his apartment;[26] – a marmalade a dram & Coffee were brought; I had a sort of collation of pickles, stewed cheese & chana oranges, which are rare here.

29th-10th. Went 6 hours over the mountains to a poor convent, 2 Turks [Turkish] fowlers that came in, went out & shot me two fine wild pidgeons.

30th-11th. Went 5 hours to Baffa the Paphos, where St Paul converted the Governor of this Island, & famous for the worship of Venus; – had a letter to the brother of the late Bishop of Bussa, at whose house I lay.

Decr. 1st-12th. Went with a letter to the Governor, entertained with Coffee; he sent a man with me who took out his hawk, to have caught a hare, if we had seen one. I went to see some sepulchral grotts, – & then the little remains of two temples, & leaving the place, I went 6 hours to a village, where I lay, in 4 hours passing through the ruins of old Paphos, on which a village is built, the other being anciently called New Paphos.

2d-13th. Went 6 hours to Episcopy [Episkopi], called a town, but is only a very large but most pleasant village; here it is said was the ancient Curium [Kourion] built by the Argives, famous for the grove of Apollo. I went to the house of a Christian.

3d-14th. Passing by a ruined monastery in 4 hours I arrived at Limesol, where I first landed, & went to the house of the English Vice Consul Signr Demean.[27]

4th-15th. Went 8 hours to a village, within 4 hours of Larnica, where I lay.

5th-16th. Arrived at Larnica at the Consuls before noon. If it is not too tedious a thing you, you may continue to abridge my letters to send to Waterford. I went with the Consul & Mr Grovesnor to visit Dr Perry an English Physitian

25 Pococke and Milles visited and described the shrine and holy house of Loretto during their tour of Italy. See Finnegan *Letters 1*, pp. 208-12.

26 This is the Panagia Chrysorroyiatissa Monastery whose present building dates from the 1770s. The icon described by Pococke is among the exhibits in the recently established museum.

27 Wilson's *List* does not include any British vice-consul in Limassol until 1799. The person Pococke is referring to remains unidentified.

just arrived from Syria at Mr Luports, he has cursorily visited some of those parts coming from Constantinople, where he had been two years; he is a youngish man & well recommended, & I find we are likely to joyn company for some time; – He was not at home. We took a walk, & in that time, He & Messrs Luport & Purnel came to see me. Play'd at Ombre. Sup'd.

6th-17th. The French Dr breakfasted with us, with whom I had some discourse in my apartment, about some natural curiosities. Dr Perry & Mr Purnel came to see me. They dined here. Father Jacomo & another father came to see me. I walked out with the Consul; – we visited Monsr. Bernard.

7th-18th. You must not expect to hear very soon from me, as I shall have no opportunity of ships. I dined at Mr Luports with Dr Perry & Mr Purnell. Pray present my very kind love to my sister, & due respects at Whiteoak.

I am Dear Madam your most obedient & dutiful Son, Richard Pococke

GRAND CAIRO
20/31 January 1739
Pococke to his mother
Letter 24[28]

Honoured Madam,

My last Letter to you was from Larnica in Cyprus.

Decr. 7th-18th. That day I dined at Mr Luports, & went home & packed up; the Consul went with me in his Chaise to see Dr Perry an English Physician, who has been 3 or 4 months at Jerusalem & Aleppo &c; he was gone to embark; – we met at the Marine; Perry aboard Captn. Tournies, the vessel I went in from Damiata to Joppa; we sailed about midnight, & for 6 days had mostly contrary winds, but sick only the first day.

13th-24th. In the afternoon we landed at Limesol in Cyprus, where I first landed, & had been there since, went to the English Vice Consuls.

14th-25th. Captn. Crispin drove in here, with whom I came from this place to Tripoli, came to see me & some French captains came & sat a while with us. I writ. Captain Crispin dined with me.

15th-26th. Captain Crispin came & we went by invitation aboard his ship, dined, called at our own Ship. Captn Crispin came & sat a while with us.

16th-27th. Captn Crispin dined with us.

28 BL, Add MS 22998.

17th-28th. Writ to the Bishop of Waterford being got only so far as Tripoly in my Account the first time; Captn. Crispin dined with us. Went on board again, but the wind did not favour.

18th-29th. Sailed after dinner & being detained by calms & contrary winds, we did not arrive at Damiata in Egypt [next few words scored out] till the 24th-4th at night.

25th-5th. In the morn went ashore to Signr Hanon Fathers, who put us in lodgings at a Kane. I am come to Egypt to see the Pyramids & some other things, & then embark for Rhodes, where I hope to be before you receive this. On my arrival here I received with much pleasure your letters of May 28th & July 23d, so that none of your letters have miscarried; – tho' I fear mine has from Jerusalem. I sent my Lord [Bishop Milles] 4 letters of my expedition into upper Egypt which I fear came not to hand, having received only my letters from Acre &c: in July. I now received 3 from Jerre, the last dated July 29th. I hope you received the fine flower'd gold Turkish caftan of the Pasha; & also my cloke [cloak], which you mention not, in the box from Cairo; the Nile produces good fish, but the people lazy, [next word scored out] in great plenty at Girge [Girga]; I do not smoak but can sit cross-legg'd well, & eat without a knife & fork. I think I writ from Damiata the beginning of March, which I don't find you received.

26th-6th. Got my things out of the custom house, dined at Signr. Fakers & about 2 set out with Dr Perry in a boat we hired, to Cairo, for about 30s. Consul Butlers Brother of Alexandria who is going to Constantinople came to me just as we were going away.[29] Sailed well, very pleasant; for it is here a glorious country full of people. I find the Dr [Perry] to be a very ingenious man, studied at Leiden; & lived many years in Holland, & has published an account of the Spaw waters of those of Aixe-Chapelle [Aix-La-Chapelle], & is preparing a curious treatise of the causes of all diseases; & I find him to be very agreeable easy company; he is about 40.

27th-7th. We arrived at night at Saramut having passed by Mansera [Mansoura] where Lewis IX was taken prisoner.

28th-8th. Mounted on asses, & in an hour & ½ came to Mehellah [El-Mahalla], a large city on the inside of the Nile, Capital of a Province in Delta; – this is one of the finest countries in the world, level as a Bowling-green, the soil without the least stone, in it a village every mile or half mile, encompassed with

29 Consul Butler may possibly be George Butler, named in a later source (see above), and his brother is the Vice Consul (Samuel Butler), whom Pococke met at Alexandria during his earlier stay in the city. As with Samuel, no biographical details have been found for his brother, though a George Butler is mentioned as being associated with a ship called "The Favorite", in 1764. See Nanninga, B*ronnen tot de Geschiedenis*, p. 600.

groves of Palm trees, all sorts of corn, green, Beans & Lupins in blossom, the former they have now ripe, flax &c: having bore rice Indian wheat & sugar canes & hemp this year before. I went to the house of a Barbary Merchant, to whom I had a letter from Consul Barton;[30] had Coffee, excellent butter, cakes. Walked out to a Christian to whom I had a letter. We sup'd on Pilaw Quarter of roasted veal, fryed Eggs & fine fish &c:

29th-9th. Mounted on asses & went 3 hours East north East to Balbeit, crossing over the middle branch of the Delta; at Balbeit saw the ruins of a most beautiful Temple, all built of red & greyish Granite, the only one of that kind I have seen, but all in ruins; – very fine Hieroglyphicks & figures on the stones, supposed to be the Temple of Isis of ancient Busiris. Came in two hours to Samanuta; – embarked & went on with litle [sic] wind.

30th-10th. Had a good wind, saw several Orange groves covered with fruits.

31st -11th. Arrived at Cairo & went with Dr Perry to the Consuls [Robert Barton], where I met with a most kind reception. Messrs. Congreve & Vernon junr. Sup'd here, the later [latter] lately come, coz. [cousin] of the other here & brother of him in Aleppo.

Jan. 1st-12th. I wish ye all many happy new years. We breakfasted on tea with the Consul & his Lady.[31] Messrs. Congreve & Vernon junr. dined here. Father Georgio Superior here of propaganda fide, who entertained me so civilly at Achmin [Akhmim] came to see me, & many to compliment the Consul on the new year. Mr Upton, Dr Perry & I spent the even [evening] at Messrs Congreve & Vernon many French there, played at Quadrille & were very chearful.

2d-13th. Some rain this day, 2 propaganda Fathers came to see me. We generally play the even at Quadrille, in the Consuls Apartment.

3d-14th. A Cophti father from Achmin came to see me. I went & visited Mr Congreve in the afternoon went with the Dr to see the vice-Prefect of Egypt of the terra Santa – not at home, read till 8 in Dr Shaw, which is come to Mr Upton a subscriber ye first time I saw it.[32]

4th-15th. The Vice Prefect came to visit me. I visited the President &c: of propaganda fide & also Mr Congreve. In the afternoon went with the Consul to see a fine Mosque north of the Town & 2 or 3 houses near, & with him visited his Druggerman. Messrs. Congreve & Vernon spent the Evening here. I writ.

---

30 Presumably George Barton, English Consul to Cyprus, whom he had recently left.

31 As noted by Pococke in Letter 2, Mrs Barton "dresses after the Turkish manner".

32 Thomas Shaw's recently published book, *Travels or Observations*. There is no subscription list printed in Shaw's book, which might have provided a Christian name for Mr Upton. However, it is possibly John Upton, who was licensed to trade in Cairo in 1742 and subscribed to Perry's book, *A View of the Levant*.

5th-16th. Went with Mr Upton to see some parts of the Town. Father Justo the Cophti had been to see me I went to the Fathers of propaganda in some business; & they brought a Cophti father to me in the even to carry me to a church to see the Cophti ceremonies of the Baptism of Christ on the Epiphany: – which is blessing water in a shallow well in the church, & then the people crowd tumultuously, to put in their legs, & wet their face & bodies with it: – it is in memory, they say, that Christ was baptized, that day, which they remember on our Epiphany & keep Epiphany; at Christmas they have musick of two brass plates, which they beat together, & also give about a piece of board an inch & half wide & 6 inches long & a litle hammer with which they beat on it, as the offices continue long, they lean on crutches; I had one given me, being dressed like a Cophti, as in upper Egypt. They perform their devotions with much irreverence talking & laughing, & a pitcher of water often goes about to drink; – they always pull up their shoes in the church, matting being spread & now & then sit down on ye ground.

6th-17th. I went with the Dr to the house of a merchant to see the covering of Mahomets sepulchre &c: – carried from the castle where it was worked, to a mosque against the day it is to be carried out to Mecca & Medina; – it is of several pieces of thick black damask worked in gold, Arabick sentences out of the Alcoran [Koran] out of the Alcoran, about 18 inches long, these pieces were carried open, then came the fine canopy richly embroidered on red & green & made like a pyramid on a square base, this was on the camel & 3 or 4 others like it, finely adorned with Brocades follow'd on led camels; – then followed a small canopy in this shape [sketch of shape like a house] embroidered likewise to be suspended over the place of Abrahams who sacrificed Ishmael as they say at Mount Ararat: – & last of all came the magnificent covering of the door of the Mosque, carried like a picture embroidered in Letters &c: in gold in the richest manner, & was indeed a most glorious sight, as well as the covering on the camel. All the shekes with their flags went in procession with musick & people dancing in the most extravagant mad manner in the world, some naked to their trowzers – & all singing; – & one sett made a noise by breathing; all pressed to touch the things as they passed by, & some threw their sashes to touch 'em. We dined there & came home. Mr Congreve sup'd here.

7th-18th. Father Justo the Cophti Catholick & the Copti Father came to see me. Messrs. Congreve & Vernon dined & sup'd here; rid out with the Consul to the Isle of Rouia to the Mikias to the reservoir of the Aqueduct at Old Cairo, & returned.

8th-19th. Visited the Dr, Messrs. Congreve & Vernon, & then the Vice Prefect & Vice Procurator at terra santa convent. Mr Conduit was Consul Bartons Uncle.

9th-20th. I visited the house of Signr. Bernaki [Bernacci].[33]

10th-21st. Went up to the Castle, dined at Mr Congreves.

11th-22d. Went with the Dr to see the Beye of the Tribute go out for Constantinople preceded by the seven military bodies a very fine sight; some for Parade accoutered with fine bows & arrows, others in coats of mail made of network of wire, & the tribute carried little chests on thirty mules. I sent a small box from Aleppo, but when I pack'd it, I did not think of sending it [...] to England, otherwise I should have put some small thing in it for you; & I believe there is nothing in it, that will much please your taste. I think there is a catalogue with it, & there are many things of the same kind, to put one or two of the best in the collection will be enough. I write to Mr Milles by this opportunity, & there is like to be very few more for 4 or 5 months, & I much fear my letters are gone to Smyrna or Constantinople tho' I directed them to be sent here; – but some time ago I writ for them to be sent to me hither if any are gone there. I am going to see the Pyramids & some other things near Cairo, & I hope before this comes to hand to be in Rhodes. This country is now exceeding pleasant, being the pleasant part of the spring, like May in England, but the mornings cold. I desire you to take care of your health, by exercise & all other means, & pray let not the least care about me or anything hurt your constitution, that we may all meet together in good health. There is much talk of a treaty with the Turk & that France will be sole mediator, for the Emperor must be utterly incapable of carrying on the war; – & a strange rumour that we are to introduce Spanish troops again to Tuscany & that Corsica will be given to Don Carlos. The handle of a knife of Egyptian pebble I designed for Mr Milles I desired you to send, having another for you in the room of it, you need not send him, unless he desires it, because I have bespoke a larger; so that the two may be for a knife & fork & the desert size, of which you may advise him, if he has desired you to send it. If youf do send one, send the shorter. Mr Congreve supp'd here. [Next four lines scored out]. The curiosity of the Wasp – nests & combs are very extraordinary, & I shall be glad to have them in my collection & you will make my acknowledgments, employed all day at home, you may enquire if it would not be better to send my very large box by water to Newbury; – & if so, you may desire Mr Wood to put any large box that may come aboard the Barges; but for small ones I suppose it will be best to send by the wagon. The seed of these countries should not be sown till late, about the time you sow kidney beans & in a dry well exposed ground. I thank God I have my health exceeding well, & if any difference inclined rather to grow fat.

33 This is possibly the Italian merchant who sold Pococke the pair of Egyptian block statues of Osiris and Isis.

12th-23rd. I went out with the Dr & went & saw the new built convent of terra santa; drank Coffee in Mrs Bartons apartment, as I generally do in the afternoon. Mr Congreve sup'd here. [Next line scored out.]

13th-24th. The Dr & I went up to the castle, saw the Beys come out from the Divan & some other things; took Coffee at the Pasha's.

14th-25th. Staid within all day; – had a litle shower in the afternoon.

15th-26th. Captn. Lisle has been waiting 4 years to get the south sea ship that was to go to Vera Cruz but the King of Spain will not grant the schedule as he formerly has done.[34] A small box I sent from Alexandretta or Scanderoon contains in it shells, fossils, minerals, a pint of Balm of Gilead, 3 or 4 books, a collection of simples [samples] some few cornelian stones; – of which when you write you may acquaint Mr Wood. The ship will arrive about April or May. As you see I am a Collector of Curiosities so as you have opportunity, you may desire your friends who live in sea port towns, if they meet with anything curious as shells, coralls, medals, ancient or modern &c: to secure them for me & I will pay the cost. The oyl you found in the box was a very small viol of Balm of Gilead broken. We expect a ship soon from Marseilles, by which I hope to hear from you; a litle shower this afternoon. It is said the Grand Vizier is deposed & likewise the Kiesslar Aga or black Eunuch, who has long govern'd all; if so there must be tumults at Constantinople, & how they may end in regard to the present Grand Signior no one knows. A proverb for you, They ought never to sow who are afraid of sparrows. [Next half line scored out and next two days missing].

18th-29th. At seven went with the Consul to the Turkish Merchants house to see the Grand caravan go out to Mecca. We had sweetmeats & Coffee & then a collation of calves feet boyled, poached Eggs, a great many sorts of sweet puff cakes & Pankakes, puffs of forced meat, preserved plums &c: – & a quarter after 9 the Processon began, with 6 cannon drawn carriages, a great number of camels between, near 1000 finely equipped, many of them carrying provision, water, tents, Litters, kettle-drums & many of them led empty. The Imam Sheke or Chaplain of the caravan rode on a camel Blessing the people with his hand & carrying a Green Banner: – then came the several bodies of soldiers; – then the Beys & last of all the Emir Hadge, or Bey that conducts the Caravan. Then came all the Shekes of Mosques with their Banners, many dancing [next half line scored out] extravagantly & before 'em, & last of all came the camel carrying the covering under an embroidered Pavilion as afore mentioned, with embroidered Trappings, & the like with three others that followed, to do the same duty by turns. The people pressing to touch the cover; & throwing the sash of their Turbans to touch it; – thus ended this extraordinary Procession. We dined

34 See Letter 5, where this matter was mentioned two years earlier.

there on many good dishes difficult to describe: – a sort of sweet Tart; – Ragou very fine, [...] on skuers [skewers] with onions between, boyle'd, Lamb, roasted fowls, soup Balls of rice & forc'd neat rowled [sic] up in cabbage leaves a Ragou, a stew, a sort of custard called cai-milk, rice milk cold, jellies, fruit-rose water – very fine, Coffee & returned home.; a litle rain before 9, & about 4 for a quarter of an hour.

19th-30th. I desire you to present my very kind Love to my Sister, & due respects to Whiteoak & all friends. I rid out with the Consul to the tent of the Bey of the treasure at Adalea [Adalia], saw the house there, & the 60 chests of money for Stambole, chained together; took Coffee. Saw a great mans house & came home; at a distance saw the camp of the Mecca Caravan; & the house of a great man returning. Messrs. Congreve & Vernon spent the even here.

20th-31st. Serene cold weather.

I am Dear Madam your most obedient & dutiful Son, Richard Pococke.

I shall seal this letter with the seal of my name in Arabick as is usual in his country.

Abdel messioh ['Abd al-Masīḥ, literally 'slave of the Messiah' here translated into Latin as Servus Christi]

GRAND CAIRO
5/12 /February 1739
Pococke to his mother
Letter 25[35]

Honoured Madam,

My last Letter to you was from Grand Cairo of Jan. 20th-31st. In which I told you I came to this place to see some things I had left unseen, & hoped to be in Rhodes before that letter came to hand. A Greek Merchant came to see me in the afternoon, I had been acquainted with.

Feb. 1st-21st. We all dined at Mr Congreves. When I send you seeds, that you cannot well raise, – as I have sent some melon seed &c; you had best distribute them; & have a share of the fruit, & they may also keep up the kind; but a litle of every sort I would have preferred, only as a specimen; but not to sow. Our cucumber & melon seeds you may keep, because the older the better.

22d-2d. I went out to buy some things; drank Coffee with some company above.

35 BL, Add MSS 22998.

23d-2d. Went up to the castle reviewed Josephs Hall; – returned & saw some of the best houses of the great people of Cairo on a beautiful Lake in the middle of the city, covered with wild ducks & other fowl. The Houses consist chiefly of 2 or 3 fine saloons all the rest being more for conveniency than ornament. Was visited by the Mehullah Merchant I was entertained by in Delta. I went to see a Greek Merchant, who was not at home; – & after the Fathers of the propaganda fide.

24th-4th. Went with the Droggerman 8 miles out of town to the north to the Birke, a large Lake 12 miles round where the caravan for Meccas is encamped, & thence goes East; the camp is about 3 miles round, it was a fine sight, & diverting to see so many going out on asses, mules, horses & camels, their tents are very fine, they say the Caravan consists of 50,000 people & twice or three times the number from Cairo – about 12000 camels, besides horses, asses & mules & many of the camels carry two. There were also several Litters for the women & great men. I went to the tent of a merchant I was recommended to, & walked about the camp.

25th-5th. Went out to walk about the camp, met Messrs. Vernons, wth whom I joyn'd & went round the camp, they came to my tent & dined; I went about the camp, & they lay in my tent, saw fine fireworks.

26th-6th. Saw corn up here in ear. – We rose early saw the caravan go off towards Mecca & vast numbers towards Cairo. When the Emir Hadge had passed in great pomp; we returned to town & a fine sight it was, to see two such great bodies moving different ways. I drank Caffelet at Mr Congreves, made in this manner; – first you make Coffee of what strength most agreeable, you may try, let it settle & pour it off, & then with that Coffee, make Chocolate, about half the quantity you would boyl in so much water, well sweetened; – it is esteemed more wholesome than either Coffee or Chocolate. In the even I went to visit a Greek Merchant. I find I shall have an opportunity of sending this before I shall be able to fill it, & probably in about 2 or 3 months you may hear from me again. The cold weather is now past, which was a litle severe of late.

27th-7th. I sat a while with the Merchant of Mehalleh, my friend, & in the afternoon I visited Signr. Abraham Delamar a droggerman of the house, & after Messrs. Congreve & Vernons, & drank Coffee.

28th-8th. Messrs. Congreve & Vernon dined & sup'd here; – we all paid a visit to the Venetian Merchant.

29th-9th. Went with the Droggerman to the house of Osman Bey, saw a fine saloon there, did some business with his caia [T. kahya] or steward; – & went to another house to meet a person, we found not at home. This country is everywhere very healthy. Mr Congreve came in the evening & we all went to a Merchants house.

30th-10th. We had colly flower [cauliflower] on the 28th. I walked out into the town with the Druggerman. In the even we went all to a merchants house. When I send several things of the same sort, I would have one or two of the best sort, put in my collection & the rest not shown to anybody whatsoever; so likewise of anything I have a quantity of.

31st-11th. Went with the Druggerman to Old Cairo, to the Consuls house & to the house of a Bey. Messrs. Congreve & Vernon spent the even here.

Feb. 1st-12th. I went to Mr Gabrians [Malim Gabrian], & after dinner to see Messrs. Congreve & Vernon.

I desire you to present my very kind Love to my Sister & due respects to all friends.

I am Dear Madam your most obedient & dutiful Son, Richard Pococke

GRAND CAIRO
2/13 March 1739
Pococke to his mother
Letter 26[36]

Honoured Madam,

My last Letter to you was from Grand Cairo of Feb 1st-12th. That evening I went to Old Cairo, to ye Cashif, or Governor of Fiume [Faiyum], & where I sup'd with him & lay – a very good sort of man: – & as a favour, he gave me a piece of meat in his hand as I was sitting cross-leg'd round with them, & passed me a piece of bread, & to crown all gave me his Pipe: – a high favour of which I took a whiff or two: – I am recommended to him, to go with him two days journey to Fiume.

3d-14th. I went & saw the Cophti churches of Old Cairo, & 4 monasteries south of it; at Old Babylon in one of which they say St Mark preached, & some think is that church of Babylon which St Peter said, saluted them; returned; – writ, dined; – rid out round two Islands & crossed the river Gilch & had a good view of the Pyramids, six miles off; returned, & the cashiffs departure being put off, I returned to the Consuls in Cairo. Tell Mr Milles I hope he corresponds with Dr Hebenstret [Hebenstreit] of Leipsick [Leipzig], which I could wish he would for some certain reasons as he may be useful in procureing [sic] several things one may want that way.[37]

36 BL, Add MSS 22998.

37 Pococke and his cousin had spent many days with the German naturalist Professor Johann Ernst Hebenstreit in November 1736 and had both written to their relatives about

4th-15th. Breakfasted with Mr Upton on chocolate; – Visited the Dr in the afternoon, went with him to Mr Bec's. Mr Congreve sup'd here.

5th-16th. Messrs. Congreve & Vernon dined here. I went to see an old Palace with Mr Upton, & paid a visit to Mr Vernon.

6th-17th. Breakfasted with the Dr, & we went to see the making of Sal Armoniac, & the hatching of chicken in ovens, & took a circuit into the Town. Mr Upton, the Dr & I dined at Messrs. Congreve & Vernons.

7th-18th. I had a summons again to go to Fiume. I prepared, & at 4 went to Old Cairo, but finding the Cashif did not go tomorrow; – I went to see Mr Vernon at Monsr. Condits house, there being much company there to go to the Pyramids; – & so I joined company & lay there, & we were very chearfull.

8th-19th. We set out early & passed through a fine country 6 miles west of the Nile to see the Pyrammids, built on the low chain of hills that part Egypt from Lydia. When I ascended, the heigth so little answered my expectation, that I thought I was not half way up, when I got to the top. The work looks greater at bottom, I think it seems not so high as St Pauls, but the figure deceives, & all the measures are exactly known: A Glorious prospect! The entrance is difficult, being drawn through a narrow hole by the legs; it is hot in winter, cool in summer & great pain to ascend, & a disagreeable stench of large Batts [bats] in the first smaller room; – I went to the Sphynx [sic] the head much worn by time, especially the neck, one just sees the top of the tail & either a tail or a thigh in a sitting posture; – the whole by the nicest culmination I could make seems to be cut out of the rock; – went into some catacombs & round the second Pyramid. We dined together & returned; – & being to go to morrow took leave, came to my lodgings & writ.

9th-20th. Set out early with the Governor of the Province of Fiume, cross'd the Nile in about an hour to the south & travelled S.S.E. We stopt; & a long black woollen cloth being laid ["spread" is inserted above the line] on the ground, with bread raw onions & a salt curdy green cheese. I eat [ate] with the Governor: – was arrived at a village near which we lay in a Grove of Palm trees.

10th-21st. We set out at day-break, & travelled over the low-hills of stone & came to a village, where the Governor was met by many people, & passing over a sandy plain, it was diverting to see the Arabs divide themselves into two parts & ride backward & forward before the Governor with great speed running at one another in single combat with their long pikes; – we came into an

this experience, as it related to the yet unpublished book of Thomas Shaw (see Finnegan, *English Explorers in the East*, pp. 94-96). This request is presumably for Milles to procure some botanical specimens from Hebenstreit's botanical collection, obtained during his recent exploration of Africa, undertaken on behalf of the king.

improved country & arrived at the village of Sennours where we went to the house of the Governor of it: here two black cloths being laid on the ground to make an oblong square, a rampart was made round about 8 inches high of about 2 bushels of bread, in cakes; & 60 dishes being set on 4 in a row, & some, one, on another, & 7 or 8 in the middle of boyled mutton, & a Lamb boyled whole; – we sat down on the ground to supper; – except some good dishes at the upper end, the rest were only the same sort of things repeated, perhaps of 5 or 6 different kinds, – Pilaw being a principal one; at top we had roast fowls & meat & a series of thin flummery very good & much ginger in it, rice mil cold, is a great dish; & meat stewed with onions: – when one set has done, another sits down after them; – & then what is left is given to the poor.

11th-22d. We had a breakfast or collation with bread round, not in such great quantities, but fine white cakes, & on them some made with butter, three or four great dishes of eggs fryed whole, & a great number of plates of olives & honey into which latter they dip their bread. We set out & passed by the very imperfect ruins of two small Pyramids, built of great stones about 115 ft. square; & arrived at Faiume where I am lodged in the Governors house. The two fathers here of Terra Santa came to see me; – they have a litle house or convent here. The President was guardian at Bethlehem when I was there.[38] I dined in the saloon below with the company. I rid out a litle way & went to the silversmiths to get some medals. The Governors steward came & sat with me. I sup'd in my rook, they having sent my supper up at my request, – not being agreeable to set at it in their manner. I writ.

12th-23d. I writ; after dinner the steward came to see me, I walked out about the Town, sat a while with the Governor, who had sent for me. A monk of the convent of St Antonio Abbate, west of the red sea came to me, hearing I had proposed to go to their convent, where [next two words scored out] the monkish life was first founded as the Hermits life was by St Paul Eremite, a days journey south of it. I went & visited the fathers of Terra Santa; who were very civil & gave me two sorts of Good white wine, they make here, there being no wine made in Egypt but here, & in some Christian villages near.

13th-24th. I went out to see the fish market, of the fish of Birkacaon [Birket Qarun] all one sort generally, like broad large carp; – I bought 7 for 7 medimns or 3 farthings, & sent them to the Fathers & that I would come & dine with them; – I went to the silver smiths in a vain search after Antiquities; – dined very cheerfully with the Fathers, took Coffee & then went to the gardens & sat in a Grove of Lemon trees, returned to the convent & sat a while with them, & came home. Writ.

---

38 See Letters 12 and 13.

14th-25th. Tho' I fear it may be disagreeable, yet I must beg of you constantly to abridge my Letters & send them to Waterford, when you have a sheet full, tho' freedom should not be in; for I have not time to write an Account of things to my Ld, but very slowly, I am now no farther than July last, & this may make amends that they [Bishop and Jeremiah Milles] may not take it ill. I began to write to my Lord Bishop of Waterford an Account of Cyprus, which I could best do without my papers, tho' I have from Mount Libanon to Cyprus to give an account of the travels of 4 months.[39] Hail & rain this morning for 2 or 3 hours, which is extraordinary here, & does harm the water of the Nile being sufficient, as they convey it over their lands. Father Januario came to see me.

15th-26th. Father Januarius called on me. I writ to the Bishop of Waterford concerning Cyprus, & after dinner, went & sat with the Fathers: – went to the Governor about my Tour, sup'd with him, the drams went round briskly, but they do not drink after meals.

16th-27th. Set out about 9, with two conductors, we went southward a league to see an obelisk, & thence westward a league to a village & dined with a sheke; – went on a great way through the Groves of Palm trees, & at length came to the canal Joseph from the Nile into Lacus Marcia, & went to the Arab sheke of a village there, to whom I had a letter. I went to the Turkish Governor, who settled affairs for my going to the Lake.

17th-28th. Set out at 4 in the morn South west & passing over the desert, came in about 6 hours to what Lucas calls the Labyrinth , but I suppose was only a Temple built to the crocodiles they worshiped from the releif [relief] of the head of one I saw there; but no rooms without the building, nor many rooms one within another, as he saies, it is probable the Labyrinth might be here, there being great signs of foundations of ancient buildings; – I saw the name of Paul Lucas there, & the only name, nor do I know that any other Frank was ever there.[40] Thence I went to the Lake, the water of which is salt, by reason of the nitre that is in the soil, & is fresh at the end where the Nile comes in. I saw also an Island in it, in which appeared something like an hill or half built Pyramid; – an Isle being mentioned with two Pyrammids in it: – abundance of

---

39 These accounts are reproduced in Chapter 9, below.

40 Paul Lucas (1664-1737) travelled in the east by order of King Louis XIV. His eastern travel writings were based on three separate expeditions during the first two decades of the 18th century. According to Charles Perry's informants, Lucas had been given extraordinary powers and privileges by the Porte to "examine, transcribe, and carry away whatever he pleased", after which he left these precious antiquities (largely inscriptions on slabs) in complete disarray. See Perry, *A View of the Levant*, preface, p. xiv. Perry's negative view of Lucas is further discussed in Finnegan, *English Explorers in the East*, including his refusal to believe that he had ever found or seen the labrynth (p. 171).

fish here, that come out of the Nile, where it is high. We returned to the village about 9 at night, & I lay at the Turkish Governors.

Feb. 18th-March 1st. The Governour accompanied me almost to Faiume where I arrived about noon; – the Fathers came to see me.

19th-2d. Father Januario came to me. I writ all day.

20th-3d. Rid out to see the ruins of old Arsinoe, north of the town & to a village beyond it: – returned. – dined. Father Januario came to see me. I took leave of the Governor & called on the fathers, who set out with me tomorrow for Cairo. I prepared for my journey.

21st-4th. Found the caravan would not depart, so I went to the convent, & thence to the silver smiths in search of Antiquity: came home, – writ. Father Januario came from the President of their convent to invite me to dine on a fine fish of the Nile, which I went to see, then went to the silver smiths & returned – & dined, very chearfully with them; – went home by the silver smiths; Father Januario came to me twice. I writ of Cyprus to my Lord Bishop; the Governors steward came & sat a while with me.

22d-5th. Set out about 8 to return with the caravan, the two Fathers in company; – we went about 5 hours to Tameah, at the entrance of the desert, where I pitched my tent in the Kane, there being no rooms.

23d-6th. We set out about sun rising, & went 9 hours over the deserts, & two hours further, only taking a litle refreshment of a piece of bread in the way. Lay in a Kane in the open air, there not being room to pitch the tent.

24th-7th. The caravan went on, & I went 2 hours south west to Sacara [Saqqara]; went with a letter to the Sheke, who conducted me to his house, we took a collation of Fryed eggs & dates together & curdy cheese, & the sheke & his men went with me to see a Pyramid on the hill, with 6 steps, each about 25 feet high & 11 broad; went to the top of it, the corners being broke so as to ascend. Viewed some other small ruined Pyramids & returned to Sacara at the foot of the hills.

25th-8th. About 8 set out to the South S. West, on the mountains, saw an oblong square solid building, of ten teir [tiers] or large stone, each setting in a foot, seems to be an imperfect Pyrammid; – they say Pharaoh promulged his Laws from it. Saw a ruined small pyramid near, which appears as others, the outer stones being taken out like sandy burroughs. Went 2 miles south to a Pyramid near as large as the famous one of the Giseh [Giza] near Cairo, but not so high being a greater slope. There is an entrance in, 225 to a room built in a peculiar manner, & another within it – in the same manner; & one within that by a hole on high, which has been entered. Went to the top of the Pyramid, & then went to another large one a mile south, in this shape [D], & then a mile east to one ruined, built of large unburnt brick, made of earth, & chopt straw;

– the Pyramid of a middling size & very probably the work of the children of Israel, went to the top of it; returned to Sacara.

26th-9th. Went out near the first Pyramid; – I went to the catacombs of bodies to which one goes down a square well 20 feet deep, & crawling in on the belly, came to the place where the bodies are; – all has been rifled & one sees the bones, skulls, bandage &c: – all is cut out of the rock. Came up & entered the catacombs of embalmed birds in long pots, a more spatious place, but the well 30 feet deep; – brought away some pots of birds, & returned to Sacara.[41]

27th-10th. Set out for Cairo, in about 3 hours came to the river, crossed over, & in about 2 hours came to the Consuls house. The Dr took tea with me. I sat above with the Consul & Lady.

28th-11th. I went out to Old Cairo, saw all the Copti churches & an ancient synagogue; after dinner went out to buy some things. Mr Congreve spent the even here.

March 1st-11th. I went to the Greek convent, after visited the fathers of propaganda, – & in the afternoon Messrs Congreve & Vernons.

I shall set out in a few days for Alexandria to embark for Rhodes.

I desire you to present my very kind Love to my Sister.

I am Dear Madam your most obedient & Dutiful Son, Richard Pococke.

If the seal comes whole it is an intaglio of a Cupid making a sort of a fillet to bind or blind mankind .

---

41 It must be one of these mummified birds (an ibis) that he discussed and had drawn for the Egyptian Society, at a meeting held on 5 February 1742. The minutes for that meeting (written in Pococke's hand) state: "Dr Pococke acquainted the society that he had lately open'd an Ibis with great care, & at the same time shew'd the society some drawings which he had caused to be made of it in several different situations; which were ordered to be copied into the book of the society, & are represented in the following leaf [folio 19]." The text continues with a description of each of the seven views (BL, Add MS 52362, 18-19). This illustration was then embellished, engraved and included as Plate LXX in *Description 1*, with the title "Different Views of Embalmed Birds of Egypt". The remains of this unwrapped mummy is listed as Lot 92, of Day 1, in the catalogue for the posthumous sale of Pococke's antiquities and natural curiosities (Langford, *A Catalogue of a Large and Curious Collection of Ancient Statues, Urns, Mummies, Fossils, Shells, and other Curiosities…*, p. 7).

GRAND CAIRO
26 March/6 April 1739
Pococke to his mother
Letter 27[42]

Honoured Madam,

My last letter I sent you March 2d-13th from Grand Cairo – I went out about a litle business; – sat above with the Consul & Lady [at] Matare [Al-Matariyyah] a 2 leagues, which is the ancient Heliopolis, saw the Obelisk & broken Sphinx there; & came back half a mile to the garden where is the tree & well where it is said the Holy family were; – & in garden the Balsam trees brought by Cleopatra to Egypt;[43] – dined there & called at two or three places coming home. Messrs. Congreve & Vernon spent the even here.

4th-15th. Messieurs Congreve & Vernon & Orbus, the last from Alexandria an Hamburger dined here.[44] I packed up a box for Constantinople. Writ; two french Gentlemen sup'd here.

5th-16th. I went to see Mr Orbus & the Vice Prefetto of Terra Santa; if you have not mentioned to Mr Milles that I designed one of those handles of knives for him of Egyptian pebble, say nothing of it, but keep it; & it will be better to have the blades put in when I come home – after dinner I went out in business; two french gentlemen spent the even here. I writ. I believe I need not desire you to preserve all your curiosities for me & part with none; – such as your sea-feather [crinoid] &c: – the best cuts [prints] I have, particularly the three or four sent from Venice, very fine ones, will keep well in the large Map-book or some other large book of mine.

6th-17th. I went out with the Dr; – we saw the Mikias or measuring pillar of the Nile at Old Cairo; – then to a large mosque in which are near 400 pillars; after to the castle, saw Josephs Well again. Visited a Turk merchant; dined; – went out in business, drank tea with the Dr. – Mr Orbus came to see me.

7th-18th. Went out to the Kane of the people from Sennar & Ethiopia, saw the black slaves there to be sold. The Dr called on me; dined. Signr. Gabrian came to me. Mr Congreve spent the even here.

42 BL, Add MSS 22998.

43 Mark Anthony is said to have taken the precious balsam trees away from the Jews, in Jericho, and given them to Cleopatra.

44 Mr Orbus is described in Letter 2, however, as "a Dutchman, a factor".

8th-19th. Employed all day in settling my affairs. The Consul & Lady going to be at the Droggermans on account of the marriage of his son. The Dr & I spent the even at Mr Uptons room.

9th-20th. Went very early & saw the ceremony of the Jewish marriage in the house, breakfasted there. Spent an hour with Mr Upton.

10th-21st. Breakfasted on chocolate with Mr Upton. Writ all day. Messrs Congreve & Orbus spent the even here.

11th-22nd. About 11 went with the Consul & all the English to the marriage feast of the Drogermans Son; dined & sup'd at the house; much good cheer. What was most particular was Puffs made with brains, & Ragou with preserved Apricots; a stewed dish with young tender Artichoaks – eat leaves & all; Pilaw with currants sugar & grated Cinamon on it; & eat with White wine very good.

12th-23rd. Rid out after dinner to Bulac [Bulaq] with the Dr, we spent the even at Mr Uptons room Mr Orbus & Malim Elias being there.

13th-24th. After dinner visited Messieurs Congreve & Vernon. The Consul came home from the wedding house. Sat with him & his Lady.

14th-25th. I made a visit to one Signr. Agib; & after waited on the Copti Patriarch of Alexandria about some affairs from the society of propagating religion in foreign parts; dined; rid out with Mr Upton to the myrtle garden. Went with the Consul to wait on the Prefetto of propaganda fide. I writ. Mr Congreve spent the even with us. Mr Upton is the person who had an intrigue [affair] with Miss Heathcote, after divorced from Sr William Young, of which you must have heard.

15th-26th. Drank tea with Mr Upton after dinner. I went to a few Suez merchants, & thence to the Greek convent; – sat with Mr Upton & Mr Orbus & then with Consul & Lady.

16th-27th. Writ all day.

17th-28th. Went with Mr Upton & the Dr in the even to see Malim Eias, the broker of the house; & being the Roman Easter, during the Holy days, a table is always spread with cakes & sweet meats for all covers: We eat & took a glass of wine. Mr Congreve spent the even at our house.

18th-29th. We dined at Messrs. Congreve & Vernon.

19th-30th. Called at Mr Congreves, his brother is fellow of All Souls, Chaplain to the Primate & Archdeacon of Armagh; – a young man, but two years in Ireland.[45]

---

45 Charles Walter Congreve (born Stretton, South Staffordshire) was Archdeacon of Armagh from 1738 until his death in 1777. He was buried in Westminster Abbey.

20th-31st. Went a second time to the Pyramids with the Consul &c: dined under out tent. I went into the Pyramid again, & to the top of the 3d Pyramid. Returned. Messieurs Congreve & Orbus supp'd with us.

March 21st-April 1st. Went with the Dr to the Miknias [Mikias] of the Nile, or its measuring Pillar, & to the castle of the Aqueduct; – dined, called at Mr Congreves.

22d-2d. Within all day, the air is mighty sweet now the Oranges are in blossom.

23d-3d. Writ; the Dr called on me. Messrs. Congreve & Orbus spent the even here.

24th-4th. Went out early with the Dr to see a great burial place of the Turks, mostly the Beys, which have fine cupola's built over them. Went up the hill over the castle; & the Nile to the fine Aqueduct from the Nile to the castle of near 300 Arches. Came home. Saw sage & borage flowers good to make tea with, 1 ounce of Balm, 1 ounce of mint, ½ an ounce of the flowers of sage ½ ounce of borrage flowers good to strengthen the nerves & for weak digestion.

25th-5th. Messrs. Congreve, Orbus & Vernon din'd here. I rid out with Mr Upton. I went to the Greek convent & having an opportunity to send this, conclude it. I desire you to send my kind Love to my Sister; & due respects to al friends.

I am Dear Madam your most obedient & Dutiful Son, Richard Pococke.

I expect your Letters from Smyrna writ since July 29th.

MOUNT SINAI
15/26 April 1739
Pococke to his mother
Letter 28[46]

Honoured Madam,

I writ to you from Cairo & sent March 26th April 6th & that afternoon went to see & receive their commands to Mount Sinai. Messrs Congreve & Vernon & Orbus & the Prefetto of propaganda fide – they returned my visit – & in the even I went to Mount Sinai convent in Town, sup'd & lay there.

27th-7th. Mounted a camel to set out with the caravan to Suez, & so to Mount Sinai. When come to the place of Rendevous; – I found they would not

46 BL, Add MSS 22998.

go in two days so returned to Town; dined, went to see Messrs Congreve & Vernon.

28th-8th. Writ & sent 13 letters to the Bishop of Waterford, 5 of them concluding my account of Cyprus, – & being an account of the Government & Customs of Egypt, & part of its natural History; – not having given yet an account of my journey from Balbeck to Cyprus.[47] Set out about 4 to the place where the caravan is & lay there.

29th-9th. I mounted on my camel loaded with Provisions & a skin full of water; sitting like those who ride on panniers, very easy; – but the motion of the camels walk, causes a perpetual motion in the back, which does not tire so much as I expected; – a leathern bottle is hung on one side full of water to drink in this thirsty land; – so I set out with the caravan of corn, going to meet the caravan from Mecca; – we went about 13 hours or 26 miles & lay in the open air; – my litle tent is a shelter when we stop in the heat of the day, but travellers in this country, do not regard lying in the open air.

30th-10th. Went a tedious journey of 16 hours to Adjeroute [El Agrud], supposed to be near the site or Heroopolis. I cover myself from the heat with a sort of veil, or when very hot with my umbrella.

31st-11th. Went about 4 hours to Suez on the Red Sea, thought to be Arsinoe; had a letter to a Greek, but went to the Greek priests house, a jolly young man: this is the port of the Red Sea, where they bring the Coffee &c: – & carry it by land to Cairo & so it is sent to England & other parts; – but a worse Coffee of America has under sold them much, till within these two years. I dined on Sepia [cuttlefish] & fine crabs, different from those of our seas. I went about the litle town; writ; went on the sea shore & picked up shells, which are different from our seas & ye Mediterranean; sup'd on fish & crabs.

April 1st-12th. Sent the baggage to be carried over an arm of the sea; met with a Turkish Captain of a vessel, with whom I had much discourse, they are very civil they had received an account that 17 ships were near coming from Gedda [Jeddah] with Coffee &c: & after I saw four of them; – crossed over about noon, dined on fish by the sea shoar & went 3 hours to the springs of Moses, for hereabouts the children of Israel are supposed to have crossed over the Red Sea: – & as I came before in the road by which they might go out of Egypt, so I shall follow them to Mount Sinai; these waters have a saltish & chalybeate taste, – near them we lay.

2d-13th. Went 3 or 4 hours, pitched my tent, – dined, – & went 3 or 4 hours further passing over the bed of a winter torrent called sedur; – whence it is

47 These accounts are reproduced in Chapter 9, below.

thought the wilderness here was called Shur, all sandy & gravelly, & a few shrubs in some parts, travel in sight of the sea.

3d-14th. Went a litle way to Wardan [Wadi Wardan] & met with good water, & loaded with it; went on for about 8 or 9 hours leaving the sea & came to Corondel, a bed of a torrent where we lay.

4th-15th. Went 7 hours & dined & went on about 4 more.

5th-16th. Travelled about 8 hours.

6th-17th. I went about 4 hours, staid behind at a distance the camels for water, dined; – went on towards Tor [El-Tor]. We are now in Asia & in Arabia Petraea, since we crossed over from Suez.

7th-18th. We travelled about 13 hours southward.

8th-19th. We went about 5 hours & arrived at Tor on the red sea, went to the house of the Greek Priest, to whom I had a letter, & also to the Procurator of the convent on Mount Sinai [St. Catherine's], who was not at home; – but his son brought me a present of shells & fruit, & invited me to sup, which I declined; walked out with the Priest by the sea & found some curious shells; – the boats being abroad with some ships wind bound; – I went into the water to see the woods of madreporae, a sort of coral growing in the water: – I swam over the woods & took up some of the stone trees; I can't say I saw any thorns in them, but found many in my hands & feet, & experienced that the thorns of stone, are worse than thorns of a wood. The Procurators son brought us a dish to add to our supper.

9th-20th. I went out on the sea shore, & again into the Red Sea, thence to the Date grove of the convent of Mount Sinai, where are salt warm springs & a bath; the Greeks say this is Elim, returned to Tor, & set out towards Mount Sinai, went 2 hours the same way we came & then north east about 4.

10th-21st. I went two hours to the narrow vale of Hebron, between high mountains of Granite, met with several springs, & Date or Palm trees. Stopt towards mid-day, – went on till night.

11th-22nd. We arrived at the convent of mount Sinai, at the foot of Mount Horeb; & were drawn up about 40 feet by a windlace [windlass] & received by the superior & members of the convent, conducted to the fine church built by Justinian, & never has come into the hands of the Turks; – they sung a Hymn, [I] was conducted to the marble chest, in which are the reliques of St Catherine; they then led me to my room, where I had a collation of olives, pickled fennel, cheese & Botargo, which is salted roes of mullets. I went to church & after prayers over the Superior showed me the chapel on the spot, where God appeared to Moses in the burning bush, which all enter with bare feet. I writ, & looking in a glass found my beard turned grey since I left Cairo, tho' I had been in no fright, but the Father comforted me, by telling me it was caused by the sun, & that eating & drinking 2 or 3 days in the convent would make it turn again; this convent belongs to the Greeks. The superior by his letters being informed, that I was a Priest, said they had a chapel, where I might celebrtate the Lords Supper if I pleased; – so if I had had two or three English men with me, I could have performed divine service & administered the sacrament; before I went to bed, one of the monks came in, & incensed the picture of the Virgin Mary & our Saviour.

12th-23d. Writ, went to church, dined on dryed plaice, dressed in a soup, sallad, cheese, olives. The monks never eat meat, & in Lent even oyl & olives, only Saturdays & Sundays, & such fish as have no blood in them, such as shell-fish & roes of fish. I spent the afternoon in the fine church of the Transfigurations, standing as it was built by Justinian, & in the chapel of the burning Bush. Writ; the Superior came & sat with me.

13th-24th. Went & saw them celebrate the Lords Supper; dined: went & saw the convent with the Superior.

14th-25th. I went about 7 & saw them celebrate the Lords Supper in the chapel of the Burning Bush; was conducted into the Refectory, where the Superior, I being a Priest, washed my feet; for a Lay brother washes the feet of a Lay Pilgrim. We then dined in the Refectory, had a dish of rice boyled with oyl & onions & dryed shell fish, being a sort of soup, ¼ of a pint of brandy in a porringer to each person, a plate of olives & dryed small beans, sodden in water. I was conducted to the Apartments of the Archbishop of Mount Sinai, who is not here; had a dram & Coffee, & writ my name in a book; – could find only the names of 6 Englishmen, in above 100 years, viz. Hen. Worsely, Sam. Haslewood in 1701, Dr Shaw in 1721[48] & Mr John Coke a Turkey merchant now in London 1722.[49] R. Savage & Raphael Baron in 1704.[50] Went to see some chapels, there being 26 in the convent. Went to the church, saw the ceremony of blessing the 5 loaves in remembrance of the miracle, & after sitting on a bench without the church, the bread was brought to us, & a dram given round, being the only supper, the monks have, this is a ceremony performed on vigils. The Superior came & supped in my room.

15th-26th. Palm Sunday they changed the Litany in procession round the convent celebrated the Lords Supper; – & then showed the left hand & skull of St Catherine. I took Coffee & dined early; & went with a lay-monk & a vassal of the convent, to visit the holy places. We immediately began to ascend Mount

48 Thomas Shaw, author of *Travels or Observations*. See Introduction, above.

49 Mentioned in Letters 4 and 5, above.

50 Pococke was to keep company with a Mr Savage in Ankara (Letters 39 and 40) and met his sister in Constantinople (Letter 40).

Horeb, passed by a spring, St Maries chapel & two gate-ways, & came to a plain spot which separates Mount Sinai proper on the south from Mount Hareb on the north, the whole being Mount Sinai in general; we went in between the several summits of Mount Horeb; – saw several chapels, supposed to be of Hermits, – & came again into the plain spot, where on the foot of Mount Sinai, is the grott of Elijah mentioned in Kings. xix. 9. in this chapel; – & here is one of Elisha adjoying: began to ascend Mount Sinai; – saw in a stone the pretended print of the foot of mahomets camel; – at length came to the summit, where God gave the law to Moses; – it is but small. The Greeks have a chapel on it, & the Latin fathers adjoining. To the latter is a part of the rock higher than any other part of the mount, about 15 feet long & 8 broad, on which probably Moses received the Law: – & under it is that cliff, where they say Moses was, when he saw the back part of the Lord, & seems very probable; – this is the East end or side.[51] On the south west is a mosque, & under it a Grott, where it is said Moses fasted 40 days, which is in the possession of the Mahometans, who sacrifice here & I saw the entrails of the beasts. I well examined this Holy place, where God manifested his Glory in so extraordinary a manner. And from it I salute you all with my due respects, & am

Dear Madam your most obedient & dutiful Son, Richard Pococke

SUEZ ON THE RED SEA
1/12 May 1739
Pococke to his mother
Letter 29[52]

Honoured Madam,

My last Letter to you was dated from Mount Sinai April 15th-26th. I descended from that Mount & lay in the chapel of Elias at the foot of it.

16th-27th. Went West & descended the hill, into the Valley of Rephidim; & took some refreshment at the convent of the 40 martyrs, only a servant there,

51 Plate LII and part of LIII in *Description 1* represent views of Mount Sinai and Mount Horeb. The views in the former are crudely executed, with no dedication, whereas those in the latter are slightly more sophicsticated and are dedicated to "The most Revd. Father in God, Theophilus Lord Archbishop of Cashel, Primate of Ireland and Metropolitan of Munster". Plate LVI of the same series is a Plan of the Convent and Church of Mount Sinai, dedicated to the Archbishop of Canterbury, John Potter, whose daughter Edith was to marry Jeremiah Milles in 1745.

52 BL, Add MSS 22998.

but a garden, in which they have almost all sorts of fruits: – from this place we began to ascend, this very high hill of St Catherine, & it took us up the whole day; to return to the convent of 40 martyrs; – a spring, in the way of which they have a story: – to the top of the hill, they say, the Angels conveyed St Catherines body, who under Maxentius, being tyed to a wheel to die a painful death, the wheel strapt in pieces, & so she was beheaded; – & they show you on the stone some cavities which, they say, are the print of her body; on this hill the stones have on them the figures of trees, & all the mountains are of Granite, mostly the red sort.

17th-28th. We went to the north along the Valley of Rephidim. Saw the chapel & cell of St Onuphrius; & came to the great stone or rock, which Moses struck, & the water gushed out; – there are about 12 Gracks & mouths in the stone on each side; [next line scored out]. The Arabs put herbs into the mouths of the holes, which they give to their camels, as a soverign remedy in all disorders. Came out into a wide valley, which was the desert of Sinai. – Saw the ruined convent of the Apostles. – Went over a hill into another narrow vale, to the west to the convent of St Cosmas & Damianus [Damian]; a vassal only there, but a very pleasant retired place; took refreshment under the shade of the Orange trees & walked about.

18th-29th. Went northward along this vale, saw a cell or two, & crossed over into the desert of Sinai, where they show a place; where they say Korah, Dathan & Abiran were swallowed up; – but this affair seems to have happened at another place; perhaps at Hazeroth; – saw the ruined convent of St Mary & David: & came to the place, where its said the Israelites cast the calves [sic] head, for so they will have it. It is a hole somewhat in the shape of a bullocks head, close to it is an advanced rock, where they say it was placed, & certainly a better place could not be found to place an Idol on: – a litle further is a small hill, where they say, Aaron was consecrated, & nearer the convent a stone, on which was an extraordinary inscription mentioned by Kircher, but defaced by the Arabs, who take of the stone as a remedy, & say they sometimes see a light at it. I went beyond the convent to a hill, where it is said Moses fed Jethro's sheep, when he saw the burning Bush: was drawn up again into the convent, dined on fresh fish brought from the East arm of the Red Sea towards Eziongeber [Ezion-Geber]. The Superior came & sat with me. I went.

19th-30th. Writ all day; dined & sup'd on frish fish. The Superior came & sat with me two or three times.

20th-May 1st. Good Friday. I went to church; – they exposed the reliques; the monks neither eat nor drink from Thursday night till Saturday evening, & then only a piece of bread & a litle cup of Aqua Vitae [spirits], & from the Sunday before Lent at night they do not eat till Wednesday afternoon.

21st-2d. Rose at Midnight to see their devotions on Good Friday & Saturday; – reposed a litle, took Coffee, & went up again to the top of Mount Sinai by another way by which they say Moses went as is probable; returned about 3; the Superior dined with us; dryed Apricots stew'd like pruins & much better. Saw the Library & some other parts of the convent.

22d-3d. Easter Sunday. Rose at midnight to see them celebrate the sacrament; all the Priests & deacons received, as the lay brothers did on Thursday, all went to the Archbps lodgings, where a basket of boyled eggs died red, a plate of bread & another of cheese being on a long table covered with a cloth; – the Superior sat at the upper end, & the priests & deacons down the side: – they first blessed the victuals & then it was distributed to all the company; a dram & Coffee before they rise. About nine dined in the Refectory: – Pilaw & dried fish & wine of the place: – all went into the garden, & then the Priests & Deacons & I with them went in procession to the 26 churches & chapels to pray for our safe journey to Cairo. I sup'd with the Superior, visited some of the chief members of the convent; – & having distributed according to custom a litle money amongst them; several of them brought me presents of curious stones, shells &c: & the Superior came & sat with em.

23d-4th. Having eat some fresh fish, I set out about 9 a clock to return to Cairo; – the Superior & many of the convent coming with me to the window, from which I was let down: – stopt in about 2 hours, to provide some Provender [fodder] for the camels.

24th-5th. We went on, stopt in the heat of the day, – & went on till the even, our travels being through the Desert of Sin, & near Paran, which mount we saw; that night a priest of the convent, a native of Jerusalem, overtook us & we joined company.

25th-6th. We set out an hour before the sun, & went 5 hours, & lay near the Vale Corondel. I sup'd with the Father on salt new cheese & onions shred & oyl & vinegar, which when I tasted I thought good.

26th-7th. Went about 12 hours, stopping 3 in the middle of the day, & lay near the vale Corondel. I sup'd with the Father on salt new cheese & onions shred & oyl & vinegar, which when I tasted I though good.

27th-8th. Travelled till about 2 & then stopt a Corondel a narrow vale, went to see the extraordinary hot Baths of Pharoah near, hotter than those of Albano near Padoa.[53] I oft eat rice with sugar, Brandy & Oranges, & found out an improvement which is butter & sowre milk for fresh sweet milk or flesh you have not; – on that I sup'd this even.

---

53 Pococke and Milles had visited these hot springs during their first grand tour. See Finnegan, *Letters 1*, pp. 225-26.

28th-9th. We travelled all day; – stopping twice.

29th-10th. Came by the fountain of Moses to the Ferry over an arm of the Red sea to Suez, pitched my tent a Sheke Arab came & took a pipe & Coffee with me, & his myrmidon Arabs made a ring round within the Tent. A Tartar I had some small acquaintance with, & some of the Pasha's people, & his friends seeing the Tent, came in a boat to meet me, & I went over with them in the boat: – went to the priests house; – & laid abroad to avoid fleas & to be cool.

30th-11th. Dined on rice-soup made with mutton; – went to see Sede Masson [Sidi Masud or Mansur?], the Turkish Captain, took Coffee with him, & he presented me with some shells. I shall be in a few days in Cairo; – & shall immediately set forward for Candia.

May 1st-12th. Took Coffee with the Tartar, shall go to Cairo in 2 daies journey with the pilgrims from Mecca in 4 or 5 days. So I am now writing to my Lord.

I am Dear Madam your most obedient & dutiful Son, Richard Pococke

P.S. Grand Cairo May 10th O.S.

This day I arrived safe at Cairo, & shall set out very speedily for Candia. Your Letters which must be gone to Constantinople or Smyrna, dated since July, are not yet arrived.

ALEXANDRIA IN EGYPT
2/13 July 1739
Pococke to his mother
Letter 30[54]

Honoured Madam,

My last Letter to you was of May 1st-12th from Suez on the Red sea – with a postscript from Cairo of May 10th, giving an account of my arrival from Mount Sinai.

2d-13th. With the officers of the Custom house I went aboard the ships a league off; they are mostly built in the Indies, & in some things differ from ours ships. The father of the convent came to see me, as he did also – when the pilgrims from Mecca came to Arsinoe 3 hours off.

3d-14th. The Pilgrims seizing on all camels that came, their own being weak; my Arab could not carry me, so lost the opportunity of going to Cairo with them [no entry for next three days].

54 BL, Add MSS 22998.

7th-18th. Spent all the days past in writing. A greek sailor came to me, for a cold in his bowels; & tho' I told him I am no Physician, he would make me give him my advice, which was a Brandy posset & sweat.

8th-19th. Found a caravan would go to Cairo at 10; – about 6, I went to the house of the Governor, who was to go, & had been very civil to me. As soon as it was dark we went, & lay abroad ½ a mile out of the Town.

9th-20th. We travelled & stopt; 4 hours, till 2 a clock in the morn.

10th-21st. At sun rise went on & at noon came within 4 hours of Cairo, to the Birke or Lake where I had seen the Hadgies [Hajjes] for Mecca encamped; – stopt a while & arrived at Cairo – about sun sett.

11th-22d. Messieurs Congreve & Orbus, came to see me in the even.

12th-23d. The Prefetto of propaganda fide & Father Georgio came to see me, & in the even another of that society; with a Neapolitan cavalier of the family of [word missing] who was in the German service, & was advised to go to Mount Sinai for a consumption.

13th-24th. Mr Congreve came to see me, & in the even Mr Orbus & Mr Vernon.

14th-25th. Busy about settling my affairs.

15th-26th. I visited Messrs. Congreve, Vernons & Orbus; & the Prefetto of propaganda fide & the Neapolitan cavalier.

16th-27th. Messrs. Congreve, Vernon & Orbus sup'd here.

17th-28th. I went with the Dr to see the measuring place of the Nile.

18th-29th. Drank tea with Mr Upton in the even. We all go to bed after dinner, which is very refreshing, & enables one to bear the fatigue of the heat for the rest of the day [no entry for next day].

20th-31st. Mr Du Liege dined with us. I rid out in the morning early, when it is very pleasant. Walked in the Jesuits garden. Signr. Gabrian came to see me.

May 21st-June 1st. I rid out. The two officers of the Pasha, from whom I received civilities, came to see me.

22d-2d. Messrs. Congreve & Orbus sup'd here. One Captn. Of [sic] Boyle of Southampton was here whilst I was at Mount Sinai, & wanted much to see [me], remembering me he said at Southampton; but I don't know, or remember anything of the family; tho' his wife is there.[55]

23rd-3rd. [No entry]

24th-4th. I visited Messrs. Congreve & Vernons.

25th-5th. Went out about business.

---

55 Having lived in Southampton while married, and still in Hampshire, Mrs Pococke would have appreciated hearing about matters of local interest.

26th-6th. Went with Mr Upton to the shops. Pray when you write to Mr Milles [Jeremiah] desire him to have another Licence of absence taken out, & registered & attested as a true copy by the Register: if my first was not actually sent or cannot be sent: – for a copy from the original copied into the Registry is of no force, nor can the Register attest it; – desire him not to fail in this, for it is a matter of importance &c:[56] – Mr Congreve came to see me.

27th-7th. I rid out about the Town. Messrs. Congreve, Vernon & Orbus dined here.

28th-8th. Rid out; began to write to my Lord [Bishop Milles] about Mesopotamia, & so design to continue on about Syria. We have had fine cool weather many days past. I am detained here by reason of soldiers landed from Constantinople, at Alexandria who will soon be here; – & it is not good to meet them on the road.

29th -9th. Signr. Bernacci sup'd here.

30th-10th. Went to see Messrs. Congreve & Orbus & then the Prefetto of propaganda fide.

31st-11th. I went & breakfasted with Messrs. Congreve & Vernon.

June 1st-12th. A Prince of Achmim & Malim Soliman my great friend there came to see the Consul & I went to them. I dined by invitation with Messrs. Congreve & Vernons, as did the Dr.

2nd-13th. I called at Mr Congreves – He Messrs. Vernon & Orbus sup'd here.

3d-14th. Went to take leave of the Prefetto of propaganda fide, & after of Signr. Bernacci & the Neapolitan cavalier; dined. – Went to see the Mikias or measuring part of the Nile with Messrs. Congreve & Vernon. Went to a garden; – returned. Mons. Leige [Du Liege] sup'd here. I had visited Malim Elias, Mr Upton being there, sent for me.

4th-15th. Visited the Vice Prefetto of Terra Santa, & after took leave of Messrs. Congreve & Vernons house. Settled my affairs, – & sent away my baggage. Went & took leave of the Prefetto of propaganda fide. The fathers of that person [?] returned my visit & the Neapolitan cavaliere & Signr. Bernacci came to see me; dined; – Messrs. Congreve, Vernon & Orbus came to see me. About 5 I set out to Bulac, & all the English came with me to the Galiote [barge], in which I had taken the cabbin; – we drank a bowl of Punch together, & took leave. A lay Italian Brother of Terra Santa convent, went with me, to whom I gave a place in my cabbin: – about 9 we put off for Rosetto.

5th-16th. Entered into Rosetto branch & went slowly.

---

56 This license was to formalise Pococke's absence from his ecclesiastical duties in the Diocese of Waterford & Lismore, where he had held several lucrative preferments since 1725, without ever having taken up residence there.

6th-17th. Went slowly, – but made about half way.

7th-18th. Made litle way, – but got near two thirds of our voyage. Would you have a royal dish, half fry, chickens or pidgeons or any meat cut small; fry shred onions about a few, then put your meat to boyl in water, enough to cover it till near boyld enough; put to it a few currants, some pounded spice, the juyce of Orange, the yolk of 2 eggs beat, & the fried onions, & a litle white wine, this makes a fine soup & dish; this our great dish now on the Nile; – pidgeons 3 farthings a piece.

8th-19th. Wind rose, & we lay by all day.

9th-20th. Went on slowly all day – forced to lye by from ten a clock; a young Goose dressed [prepared] as above, today.

10th-21st. Arrived at Rosetto about 10 in the morn, went to the house of our Consul. Signr. Fabri of Pisaro, who is married since I saw him, dined;[57] Fra. Ambrosio came to the house. I visited the French Consul [Monsieur Teissan]. Walked out with our Consul, eat mulberries; – took a walk in a pleasant garden.

11th-22d. Some weighty things, that is five boxes will come from Egypt, I have desired may be sent by water to you;[58] – [next line scored out] that nothing may be damaged by land carriage: – unless you should write to Mr Wood to send them by land. I desire you to advise Mr Wood to whom they may be addressed in Newbury to give you notice when they come; & you may send a cart on purpose. A box of Papers markt RP you may open first, for at the top of it you will find a catalogue & an account of all the things. In the afternoon the French Consul with this Droggerman & chancellor came to see me, & after Monsr. Shabere. I writ; – returned the visit of the latter & took a walk with the Consul & some French.

12th-23d. After dinner Monsr. Tisan [Teissin] came to see, me in the even we cross'd the river to the Island, where the French had a Festino, being St Johns Eve, we staid a while & returned.

13th-24th. Writ all day & walk'd in the even; – the country here is now very pleasant, the rice being up & green; – & at Cairo there is none, the corn harvest is past, & nothing green on the ground; – but the country looks black as they pull up the corn by the roots.

14th-25th. The Bishop of Paphos in Cyprus came to the Consul, at whose house I was at Busa or Paphos, – but he was not there.

57 See Letter 4, where Monsieur Fabri was at the time Vice-Consul.

58 These boxes contained items for his collections, including his Egyptian antiquities, such as the statues of Osiris and Isis, possibly his two mummies, and other smaller pieces described and depicted in *Description 1*.

15th-26th. I walk every evening; – the French came to our house & made a sort of a concert. In the even went to see Monsr. Shabere, who showed me his Antiquities: the walks pleasant, fine lanes, & the most agreeable place in Egypt. [Next two lines scored out].

16th-27th. I visited the French Consul in the Evening, who showed me his medals & antiquities: treat, wine & water & then a sherbet; walk'd out with our Consul into an Orange Garden.

17th-28th. Apples here full ripe. At 5, went a league to the castle with our Consul & Lady & family; the French Consul & some french; – took a collation near the river of fruit &c: – walked in the Palm Groves & returned; the rice begins to rise.

18th-29th. In the evening went to visit Monsr. Tisan; – not at home.

19th-30th. Visited the French Consul, & in the even the President of Terra Santa convent. Took a walk with Monsr. Tisan.

June 20th-July 1st. Visited Monsr. Shabere, who was not at home: – dined at the French Consuls by an invitation.

21st-2d. In the even visited Monsr. Jobere, & took a walk with him & heard his litle concert at the Consuls.

23d-3d.[59] Our Consul went with me to take leave of the French Consul, Messrs. Shabere, Legran, Coigny & the President of the Convent. After dinner they all returned my visit, & about 5 the Consul went a mile out of Town with me in the way to Alexandria, & a french gentleman going, many french went with him; – we took some refreshment, before we took leave; & I went in a chaise the Consul of Alexandria [Mr Butler] had sent for me, & a Maltese merchant went with me in it; – We came to the madea or Ferry half way about nine, sup'd in the Kane, & went on about midnight, arrived at Alexandria, about sun rising

24th-4th. & went to the house of Consul Butler, whose Lady [was] with him; – breakfasted on tea with them. In the even rid out with the Consul & Mr Scot round the outside of the walls of the Old city.

25th-5th. In the evening walked out with the Consul & Lady &c:

26th-6th. In the evening took a walk with the Consul & Lady.

27th-7th. Went out with Mrs Butler in a chaise, the Consul & Mr Scot on horseback.

28th-8th. Walked on the Strand in the even; – met Monsr. Chassini [Cassin] the French Droggerman – who since I saw him had been in Paris & at Versailles in his Turkish dress; had much discourse with him.

59 An error is made here in the dates, whereby a day is missed and is not made up.

29th-9th. Went with Mr Scot aboard Captn. Aish, drank tea; – went aboard 2 or 3 other ships – rid out in the even with the Consul.

30th-10th. Breakfasted with Mr Scot, rid out to the. I went to an English house, in business. Walked out with Mr Scot.

July 31st-Aug. 11th. In the even walk'd out with the Consul & Lady.

Aug. 1st-12th. Rid out in the even with Mrs Butler in the chaise to the canal the Consul being with us.

2d-13th. I am embarking aboard an English ship for Canea [Chania] in Candia [Crete].

Pray my kind love to my Sister & due respects to all.

I am Dear Madam your most Dutiful Son, Richard Pococke.

CANEA IN THE ISLE OF CANDIA
16/27 August 1739
Pococke to his mother
Letter 31[60]

Honoured Madam,

My last Letter to you was of July 2nd-13th from Alexandria. That day I thought to embark aboard the Eagle Brigantine, Captn. Skinner for the Isle of Candia [Crete], but the wind did not favour. I went out in business & took a cool tankard with some captains of my acquaintance at an English house; took a walk, – met the two French droggermen; took leave of them.

3d-14th. Captn. Norman sent his boat for me; & Captn. Pulham with Mr Scot in company carried me aboard his own ship, where Captn. Norman came to me.[61] We went off to the Eagle; – the captain saluted me with 5 guns, & as I passed by Captn. Norman was saluted by 5 other guns. These two commanders of ships of 200 & 300 tuns went with me, very kindly loaded me with presents, for my sea store, of wine, beer, cyder & about 7 we set sail. I being sick the first day & not very well for 2 or 3 days.

7th-18th. We were 20 leagues west of Cyprus.

8th-19th. I saw the Carmanial coast, the old Cilicia; about Satalia;[62] these days I was able to read, – made very litle way.

60 BL, Add MSS 22998.

61 This vessel was "bound to Tunis, Algiers, and some other places on the coast of Africa, freighted with Moors on their return from Mecca." See *Description* 2, Part I, p. 236.

62 By this, Pococke means that he he had a view of the modern Antalya Province, located on the Mediterranean coast of south-west Turkey.

9th-20th. [No entry]

10th-21st. Approached towards the Isle: Castle Ross [Kastellorizo], & was well enough to sit up & write. This day we were about south of Perga in Pamphylia [Antalya], where St Paul was; in the evening we were about a small Isle called Castle Ross.

11th-22d. We saw Rhodes at a distance & were south of the ancient Lycia.[63]

12th-23d. Approached nearer Rhodes & saw at a distance the entrance to the part of Rhodes, the only one in the Isle.

13th-24th & 14th-25th. Mostly becalmed East of Rhodes; – having a near sight of it.

15th-26th. A Wind sprung up last night & we saw Scarpanto Isle [Karpathos], the old Carpathusi, whence the Carpathian sea was called, & at noon the East of Candia.

16th-27th. Not being able to keep our course, we came back again to the south of Rhodes.

17th-28th. We came to an Anchor west of Cape Tranquilio,[64] & went ashore to take in water & wood; – we dined by the stream, & went 2 miles to a Greek village [Lahania] & in the even went aboard.

18th-29th. Went ashore & pitched my tent; & the captain & I lay in it.

19th-30th. The Governours people & Greeks came to our tent; we treated them with Coffee & wine, & went with them to the village, visited the Turkish Governor, & then went to an house where they treated us with figs & cold milk, that had been boyled, & the people brought us fowls, sheep, goats, Eggs, figgs & Grapes to buy: – & some of them came with us to our Tent.

20th-31st. The Greeks came to us, with bread we had bespoke & a priest with them & brought us like presents of figs, Lemons, Quinces, peaches & a Melon; we treated them with wine & a dram.

July 21st – Aug. 1st. Dined & went aboard in hopes we might be able to set sail in the even.

22d-2d. The wind still contrary & we staid aboard.

23d-3d. We sailed in the morning, came up near Scarpanto but turned...

24th-4th. Sailed south again & passed Scarpanto, drove by the wind.

25th-5th. So we continued.

26th-6th. The wind slackened & gained a litle to the N.E.

---

63 Though nothing is said to his mother, Pococke notes in his book, "As the plague was at the capital town of Rhodes, we did not think proper to go to it, though the wind was contrary." See *Description 2*, Part 1, p. 237.

64 Spelled "Tranquillo" in the book, *Description 2*, Part 1, p. 237.

27th-7th. We went north east of the East of Candia again. Quince dumpling for dinner – yolk of an egg beat in Tea & Coffee, instead of cream.

28th.-8th. Becalmed.

29th-9th. Becalmed.

30th-10th. Had a litle wind towards night.

31st-11th. A Gentle wind, & we went north towards the East end of Candia; on the 24th-4th about ½ an hour before sunset was an Ecclipse of the sun, by which above 1/3 of it was darkened.

Aug. 1st-12th. But litle wind, had sight of the East of Candia.

2nd-13th – 3rd-14th – 4th-15th Very litle wind, – the last day a litle favourable.

5th-16th – 6th-17th – 7th-18th. These three days very litle good wind.

8th-19th. We passed by the isle of Chabra close to Candia.[65]

9th-20th. We were near Gozo Isle [Gavdos], thought by some to be the Island Claude mentioned in the Acts. We made into a shallow bay to Castle Suatia in Candia & came to an anchor – the captain & I went ashore, where the Greek priest & some people met us & I sent a letter to Canea, 40 miles off to the English Consul, Mr Paradice [Paradise][66] to send me mules to go to Canea, we returned aboard ship.

10th-21st. We went away to the empty Castle Suatia [Frangokastello], called in the maps Sfakia. The Priest & a Turk very civil, met us & presented us with a wild fowl, water melons, which are very wholesome, & Tuberoses, which grow on the shoar in the sand. We took some refreshments & I went with my things two miles to the foot of the mountains, to a village to the Priests house near Suatia; where I am to stay till the mules come: the people came to stare at me, very civil & honest, but not the most polished, being an unfrequented place; all Greek Christians; – the captain came & dined with me on roast fowls three half pence a piece; – they all talk modern Greek, of which I shall get some litle knowledge.

11th-22d. Put my things in order, & writ. The Turk had me to a house & treated me with olives, grapes, wine &c: in the even, The Consuls Janizary came with horses for me.

12th-23d. Set out for Canea; – passed a curious Gulf [Gully] between the mountains, refreshed ourselves at a well, & went on to a village & lay at the Greek priests house.

65 This is one of the uninhabited rocky islands known as Paximadia. According to Pococke (*Description 2*, Part 1, p. 240) there were called "Chabra" by mariners.

66 Peter Paradise (c. 1704-79) became Consul at Salonica in 1741 and again from 1753-54 (see Wilson, *List of British Consular Officials*, p. 22) and it is likely that this is the same man.

13th-24th. We performed this journey of 40 miles & arrived at Canea about 9 a clock, where I went to Consul Paradises house, a very worthy Batchelor about 36;[67] his father & brother were Greeks; but when he was a child his mother was married to an English Gentleman I think of Sr Robert Suttons family, who was an Ambassador at Constantinople,[68] & he was made Consul of Candia, his son-in-law succeeded him, who had been page to Sr Robert Sutton 32 years, when he was Embassador in France. He is a very ingenious many & septi linguis [speaks seven languages], & is most extremely civil & obliging. The Chirurgeon [surgeon] of the English a native of Candia came to see me. I went to repose; – the French Chanceliere[69] & first Droggerman came to see me; – dined; – the 1st English Droggerman & the 2d French Droggerman came.

14th-25th. The Consul breakfasted in my room, & spent most of the morning with me. The Chirurgeon dined with us. A Greek Gentleman came to see me. This is a litle pleasant fortified city, & a fine plain country about, tho' Candia is mostly mountainous.

15th-26th. I went with the Consul to wait on the new Janizer Aga: the day before, the Consul sends a present of cloth & silk for a suit of clothes; – we had preserved Roses, Coffee, Sherbet, & each a present of an handkerchief. I went with the Droggerman & returned my visits. The French Consul,[70] his Drogerman & a Padua Physitian native of Morea came & we played at Quadrille. The Dr Sup'd here.

16th-27th. The Bishop of Canea, attended by his aeconomist or Treasurer & another Priest came to see me; dined; the Bp of Chisamo came to see me. I went with the Consul to wait on the Bishops of Canea & Chisamo & the two Doctors.

I am going towards Constantinople, & you may hear from me again, in a month after this, – but afterwards I may not have opportunity so soon from Constantinople. We had rejoycings here for three days on account of a victory over the Imperialists near Belgrade.[71]

Pray my Kind love to my sister.

I am Dear Madam your most obedient & dutiful Son, Richard Pococke

67 He later married the daughter of Philip Lodvill and his Greek wife, and their son John Paradise (1743-95) was a friend of Dr Samuel Johnson. See Wilson, *List of British Consular Officials*, p. 22.

68 Sir Robert Sutton was English Ambassador to Constantinople from 1700-17, having asked in 1714 to be recalled. See Wood, *A History of the Levant Company*, Appendix 1.

69 André Pinatel, chancellor of the French vice-consulate 1731-1740. Mézin, *Les Consuls de France*, p. 679.

70 Antoine Maltor, in office 1735-1740, who actually vice-consul.

71 A reference to the recent recapture by the Ottomans of Belgrade (which had been seized by Austria in 1717), following a seige of 51 days.

CANEA [CHANIA] IN THE ISLE OF CANDIA [CRETE]
29 September/10 October 1739
Pocoke to his mother
Letter 32[72]

Honoured Madam,

My last letter I sent you from this place was Septr. 2d-13th on the 3d-14th. Set out with the Consul & the Bishop [of] Chisamo to see the west part of the Isle; – went 6 miles by the sea to a shady place on a river & din'd – went 6 or 8 miles to a village, to the house of the Brother in-law of the Bishop, who is his Protopapa. Took Coffee – walked.

4th-15th. The Abbot of Conia [Canea] convent came to see the Consul.[73] In the even we went up the hill to a Grotto, where is a church to St John Eremite; returned we took a walk near the village.

5th-16th. Set out early with the Bishop of Chisamo & the Janizary & went 4 hours & ½ to Magnes, a ruined Port, the ancient Dictamnum [Diktynnaion], on the East side of Cape Spada, the old Promontory Placum [Akra Spatha], dined at a place near; returned in the even.

6th-17th. We all went to Cania convent, a pleasant place on the sea, where we spent the day very agreeably.

7th-18th. I writ in the garden. If when my letters to the Bishop come in any box, the freedom be in, in Ireland, you had better send 2 or 3 every post. I rid out in the even with the Consul, & we paid a visit.

8th-19th. The Bishop of Chisamo, who had been at Canea, returned. We took a walk.

9th-20th. We set out to see some places; – saw a church thought to be an ancient cathedral: – went on south west to Rocca, where we saw a ruin on a high rock: – came near to the sea west of Cape Spada, & dined at the Paternal house of the Bishop of Chisamo, & went south west about 4 miles to Palaio Castro [Palaikastro], the old ruined city of Aptera on a high hill, at the foot of which, the Muses or Sirens had a trial of Musick, & the Sirens being vanquished

72 BL, Add MSS 22998. Mrs Pococke makes an error in the dating of this letter, failing to use the dual dating system at the beginning, then becoming confused towards the end. Her errors are corrected in her son's hand. In addition, she notes that a letter is missing ("A letter wanting" is written at the top of the page). There is a gap of almost three weeks in the journal and this letter refers to a letter of 2/13 September 1739.

73 Mrs Pococke misspells the word "Canea" several times in this letter.

lost their wings.[74] We viewed the Antiquities & came near to the North Sea & lay at a Priests house.

10th-21st. I went a mile with the Bishop of Chisamo to see the ruins of Cysamus, – now called Chisamo, & its ancient port, returned to the Bishops paternal house, dined; – went on to the Bishops brother in Law, where we first were.

11th-22d. Set out for Canea: dined in a shady grove; called at the house of a Turkish Aga & drank Coffee; arrived at Canea.

12th-23d. After dinner went with the Consul to wait on the Bishop of Chisamo; – some french came here.

13th-24th. The Bishop of Canea breakfasted with us. I visited Monsr. Brimont, the Dr dined here, some french gentlemen came in the afternoon.

14th-25th. Some french Gentlemen breakfasted here. In the even some french & the Bishop of Chisamo were here.

15th-26th. The Consuls brother by same mother, Mr Burgoin, arrived from Constantinople, 15 years old & is to go to his native country England.[75] The Bishop of Chisamo dined with us; – the Dr & Chirurgeon dined here; – much company of the french &c: in the afternoon – & some Ladies; the Bishop of Chisamo sup'd here.

16th-27th. The Dr & chirurgeon dined here; – much company of the french &c: in the afternoon, & some Ladies; the Bishop of Chisamo sup'd here.

17th-28th. The Bishop of Chisamo's mother breakfasted here.

18th-29th. Walked round the ramparts & to the Arsenal with the Consul & Brother. Company this afternoon; – the Bishop of Chisamo sup'd here.

19th-30th. Visited the Dr with Mr Burgoin, – & called at Signr. Joannies shop.

20th-Oct. 1st. Went with the Consul the Bishop of Chisamo, Mr Burgoin & the Dr two hours to see a ruined old Town I take to be Cydonia, on a hill, dined on the side of the hill, & returned by a convent of Mount Sinai, where the English bury.

21st-2d. Some French came & staid till 8.

22d-3d. The Bishop of Chisamo sup'd here; this day the first rain.

---

74 Pococke's topographical studies of Crete, in particular his identification of the two ancient sites mentioned above (Diktynnaion and Aptera) have earned him the reputation of inaugurating British exploration of antiquity on a new scale. See Warren, "Early Travellers from Britain and Ireland", p. 4. Plate XXXV of *Description 2*, Part 1, represents Plans of the ruins of Dictamnum, Cydonia and Aptera. He mentions in his book that he procured here "a very antient bas relief", and this is depicted in his Plate IIIVI. Several "basso relievo's" are listed in his sale catalogue of ancient statues and natural curiosities (see L. Flin, *A Catalogue of the Library*) but none fits the description.

75 Pococke meets Mrs Burgoin (mother of Consul Paradise and Mr Burgoin) in Constantinople. See Letter 40, below.

23d-4th. Rain in the morn; – The Pasha's Dr dined here; – I went to see the Venetian Dr with Mr Burgoin. French came to the Consuls house.

24th-5th. The Pasha's Secretary dined here. Some French came in the evening.

25th-6th. Set out to go to the point of land East of Canea [Akrotiri]; – went 13 miles near the end of it a grotto, where I had heard were petrified bones among the stones, hardened Earth;[76] went on towards [missing word] there is a large round Grott & church & descended the hill to a most curious place Catholico,[77] where is a church in a grott, a fine bridge across a vale to two Hermitages in which is a curious Grott with fine petrifications of columns made by the droppings of the water, & goes in about a ¼ of a mile;[78] – returned up to the convent of St John the Hermit on the Hill; – a good convent, where I lay.

26th-7th. Went 2 miles to the fine convent of St Trinity, & thence on to the bay of Suda [Souda] opposite to the Isle & fortress; – came on towards Canea to a village, where we stayed & dined; – came down to the plain of Canea, & saw the 3 convents of St George, [missing word] & Hiemane;[79] & arrived at Canea.

27th-8th. Busy in settling my affairs for my voyage to Scio [Chios]. This comes by France under Mr Maydwells cover & with it 5 letters to my Lord describing this Isle.[80] I went to take leave of the Bishop of Canea, the Dr & several french & went with the Consul to the Bishop of Chisamos house, the Dr came to see me; & the Bishop of Canea.

28th-9th. The captn. fires his gun to go abroad. The first Droggerman came to see me. The French drogerman, the Abbot of Arsani, & my captn. dined here. I have received here very extraordinary civilities from Consul Paradise who is a very worthy man abounding in all amiable civilities; – & has given me letters of recommendation to Scio.

29th-10th. We are setting sail: [....] kindest love to my sister.

I am Dear Madam your most obedient & dutiful Son, Richard Pococke

---

76 In the book he notes that he was told these bones were "of a larger size than ordinary" and that he had actually seen some in the softer part of the rock, though not petrified. See *Description* 2, Part I, p. 264

77 There are various omissions in this entry, where Mrs Pococke had failed to transcribe the placenames and Pococke has later edited the text. However, this has rendered some of the words illegible.

78 These natural curiosities, as well as others such as the Arkoudospilia (Cave of the Bear) are further described in *Description* 2, Part I, pp. 263-64

79 Probably Agios Ioannis o Prodromos.

80 It is not known what became of these letters which, like all subsequent letters to Bishop Milles are not part of the collection BL, Add MS 15779.

SMYRNA
5/16 November 1739
Pococke to his mother
Letter 33[81]

Honoured Madam,

I writ to you last from Canea in the Isle of Candia. Went to see our Droggerman & the Capuchins convent & walk'd about the Town.

30th-11th. About 8 embarked aboard a french ship for Scio, had a tolerable wind.

Oct. 1st-12th. Litle wind all day, – were South of Milo [Milos] & saw most of the southern Isles of the Archipelago.

2d-13th. Had a fine wind passed west of Milo Argentaria, & between Siphanto [Sifnos] & Serphanto [Serifos] & in the night by Scyra [Syros].

3d-14th. Sailed by Delos & between Mycone [Mykonos] & Tine [Tinos]; – saw Andros & Negroponte [Euboea] to the West & Nicaria, Samos & Asia to the East, & about noon Scio.

4th-15th. Early in the morning we came into the Port of Scio: – I went to the house of a Roman Catholick Sciote of Genoese extraction, who is English & Dutch Consul;[82] – who invited me to make it my home, having letters to him. This place is delightfully situated, & in a shallow Bay about 2 leagues long, so improved with gardens & houses, that it all looks like such a city, as perhaps Moscow may be. I walked out in the even, saw a Greek church.

5th-16th. I went in a boat 4 miles north to the School of Homer, who the Sciotes say was born in this Isle; – it is a seat cut round in a rock, & in the middle a square part of the rock left, in which on one side is a broken Alt. [high] Relief of a person sitting, & on each side of him, a lesser person sitting; – & on the other 3 sides a Lyon:[83] near it are the vineyards which produce the best wine, & that of which He [Homer] quaff'd copious draughts of, to make Invention flow; returned half of the way on foot; dined; – went to see some churches; writ; saw the Franciscan convent.

6th-17th. Set out with one of the Town South, to see the Campo [Plain], a most delightfull place, full of gardens [next word scored out] country houses for 2 leagues on the sea, & about a league broad; – we saw a Garden of the

81 BL, Add MSS 22998.

82 Probably Stellio Rafaeli, or one of his sons. The office of Dutch consul in Chios appears to have stayed in this family's hands from 1708 to 1794. Schutte, *Repertorium*, p. 346.

83 This relief is represented in Plate XXXVII, *Description 2*, Part II, with the broken sections "supplied by the fancy [imagination] of the drawer".

Justinian, where we were entertained with conserve of Roses, & then a cup of water brought, which is a genteel treat. We went to the house we were to lye at; dined, – went out & met with an insignificant Greek Inscription, & relief: saw several gardens of orange trees & went to a village & returned to our house.

7th-18th. Went South & saw the convent of St Minas, & thence to the first village of Mastick [mastic], which is made only in this Isle, & drops like a gum from a large shrub, the greatest part is sent to the Grand Signior, for his Ladies to chew; – they eat it also mixed with their bread in the making.[84] I eat, – & then went a league south & saw two convents & returned: dined; – went to sea a nunnery, where men enter & the nuns go out when they please & live with their friends; – & of 100 a great many not professed, have no allowance but their chambers, & live on their fortunes or labours, making Dimothy's [dimities] &c: They sung a Paradesis for us in the church; returned to the village.

8th-19th. Returned north to the hills saw some churches in the way, & dined at a village, & ascended the hills to a large convent, called Neamone [Nea Moni/New Monastery], where there is a fine church, not large, wainscoted [panelled] with jasper of the Island, & another fine marbles; – they showed us the reliques, among them a thumb of St John Baptist, Head of St Timothy, & a bone of St Luke, & a piece of the Cross: – The Abbot entertained us in his room, & came & sup'd in our apartment.

9th-20th. Returned by some fountains, from which the city is supplyed; settled my affairs, writ, & walked out about some affairs.

10th-21st. Set out in a boat to see some parts of the Island, very difficult of access by land on account of the mountains. Came to the north of the Isle to Cardamilla [Kardamyla], a deep Bay, on the West of which are small ruins of the temple of Apollo mentioned by Strabo, the place being called Farum. I pitched my tent on the shoar, & lay under it.

11th-22d. Went 2 miles to the village, returned & saw the ruins of the temple; the wind not being fair, I staid under my house & writ.

12th-23d. I went to see the imperfect ruins of a marble Temple, supposed to be built to Neptune, a league off; returned, dined, & the wind still against us.

13th-24th. Set out about 5 in the morning & went about 3 leagues to St Thelene, where is a Grott & a distillation of water of which the Greeks have strange stories; walked to meet the boat by Melanose, on the west of the Isle; we went on to the south & came to the Bay Volisso, & went 2 miles to the East to the village, over which is a castle said to be built by the famous Belisarius, & some say,

84 In his book, Pococke gives a more detailed account of the mastic villages of Chios (mastichochorea), the privileges they enjoyed on account of the precious commodity they produced, and the the production process itself. *Description* 2, Part II, pp. 4-5.

he was after a prisoner in it;[85] – I went on 2 hours to see a convent, returned, & raining lay at a Priests house.

14th-25th. Came to my Tent on the Shoar. Writ – in the afternoon went to the village & ascended up to the castle. This day we had a sucking pig stewed with butter & wine exceeding good & cost about 6d.

15th-26th. Set out after mignight to go a few leagues to Ipsera Isle [Psara], arrived there about 9 a clock, 'tis 18 miles round, one Town only, & no village. I found some reliefs & insignificant short Greek Inscriptions. Tasted & bought some excellent strong wine; dined, went to the North from the South of the Island, to a litle convent, almost uninhabited, a good soil & produces good wine; all Christians, not a Turk in the Isle; returned; at midnight began to rain.

16th-27th. Removed according to the Oriental Custom, into an open unfurnished chapel, it being rainy weather: & there spent the day in writing: going to the convent yesterday, two plough-men seeing me pass by as they were eating their fish, called to me to set with them & partake; – I complied with their honest simplicity to please them, & took a draught of their wine.[86]

17th-28th. I staid within all day & read & writ. Saw vespers performed in the even for St Luke. The women or children bring a plate or round basket of boyled wheat, covered for ornament with raisins or kernels of Pomegranates, & a litle wax-candle on each, & some cake of bread, this after , all [...] over is distributed among the congregation.

18th-29th. The Greeks had the sacrament celebrated, & many brought a plate of Lent figs & brandy, which were distributed after all was over.

19th -30th. We sailed for Mytilene [Lesbos], & towards the evening came again into the bay of Cardamilla in Scio Island.

20th-31st. Passed the gulph of Smyrna & arrived in the port of Mytilene about ten at night: – we saw 5 large ships which I thought might be our Smyrna fleet, – but were Turks.

Oct. 21-Novr. 1st. Carried a Letter to a Roman Priest of Scio, from the Consul, Chaplain to the French Consul here; – I staid in his house. Waited on the French Consul with another letter. Monsr. Rose came to see me, a French Merchant; – with him the Chaplain &c: – I took a walk in the even. The French Consul first returned my visit.

85 Flavius Belisarius (c. 500-565 AD) was a General of the Byzantine Emperor during the reign of Justinian.

86 The Irish nobleman Lord Charlemont recorded a similar experience on this island, which he visited 10 years later, when a Turkish gentleman seeings him pass by, "immediately sent his servant to desire our company to drink Coffee....[and] received us with the greatest and most unaffected politeness...." See Stanford and Finopoulos, *The Travels of Lord Charlemont*, p. 44.

22d-2d. I set out on a tour round the Island, with the chaplain & the Consuls, going south, – we dined at the hot baths, – & went 8 hours to a village, where we found the English Vice Consul, a naxiote, but bred & lived mostly in Smyrna.

23d-3d. He went with us to Hot Baths near, & we left him & went on, passed through the Gulph of Calone [Kalloni], in a ferry boat, & went & lay at a village.

24th-4th. Went to the ruins of old Eropus [Eresos], where Theophrastus was born, & thence to a convent on a high mountain in the north west end of the Isle.

25th-5th. We went 8 hours to Petra, near the large town of Molivo [Mithymna].

26th-6th. Went to Molivo, old Methymna, situated on the side of an Hill by the sea. Lodged in an house for strangers at the church; walked out in the even to see the Town & antiquities.

27th-7th. Went by two hot baths to a village north of the Island.

28th-8th. Went on to another.

29th-9th. Went to hot Baths where are several Greek Inscriptions & signs of great buildings. – Went on & saw a very fine Aqueduct[87] & arrived again at Mytilene city.

30th-10th. Visited the French Consul; – & went but all the day with Monsr. Rose to visit Antiquities.

31st-11th. Went out again with him, both morning & Evening.

Novr. 1st-12th. Took leave of the French Consul; – & he came to see me. Embarked at ten for Smyrna; – had litle wind till towards the Evening, when the wind springing up we came to Smyrna about 5 in the morning of the –

2d-13th. I sent a note to Mr Philips, but he being returned to England, Mr Masters now in his house, sent the Consuls Droggerman to me, who conducted me to the Revd. Mr Burdetts lodgings in the Consuls house;[88] where I found Mr Masters & the Chancellor Mr Bodington [Boddington],[89] the Consul [Francis Williams] being in Constantinople:[90] – I after went to Mr Masters house where I live; – Mr Lee is here, partner of Mr Lisle in Constantinople, both young

87 This antiquity, engraved by Samuel Wale, is represented in *Description* 2, Part II, Plate XL.

88 Revd Charles Burdett (b. 1700) matriculated at Lincoln College Oxford in 1717, and took a BA (1728), MA (1724) and later at BD & DD (1764). He arrived at Smyrna c. 1724, was there in 1744 and may have remained until 1759. See Laidlaw, *The British in the Levant*, p. 227. Revd Burdett acted as an agent to English and Irish collectors of antiquities, including 2nd Lord Bessborough. See van den Boogert, "Freemasonry in Eighteenth-century Izmir?", p. 110.

89 This is a different Mr Boddington from those whom Pococke met in Acre and Aleppo (Letters 14 and 15).

90 Though no reference is made to his name, the English Consul at Izmir from 1733-41 was Francis Williams. See Wilson, *List of British Consular Officials*, p. 15.

Gentlemen.[91] Mr Burdett came to dinner here. Mr Took & Mr Shaw came to see me, each of them being Masters of the two English houses here.[92] Mr Frem[e] aux a Dutch merchant & his Brother came to see me.[93] This place is very agreeable on account of the company; – there being 4 Consuls, – the Consul of Sweden, being lately come.[94] All the great people here, among whom are to be reckoned our merchants, have fine large houses, with long open Galleries very pleasant on the water side, & they live here in a very hansome manner. We went to supp with Mr Shaw, Mr Burdett was of Lincoln College & about 6 years my senior. Arriving here I received your letter of Septr. 7th. 1738, your last to me being of July 23; all the others being at Salonica, where I expect them. I received also 2 from Jerre of Septr. 2nd. 1738. & June 18th. 1739. I thank you for your design of parting with Jack to me; you may sound him if he would be willing t go to Ireland.[95] I must desire you to abridge my letters & send them to Waterford, for they are of a quite different nature to those I write to that place [i.e. the "accounts" written to Bishop Milles]. The Barometer & Level I hear are at Constantinople, but not Dr Shaws book, which I have seen & do not now want.[96] My Musta [black] chic's are to grow till I arrive in Italy.

---

91 Probably Richard Lee. While he was at this time in partnership with John Lisle (see Letter 2, above) he is recorded as having later (from 1748) been in partnership with Thomas D'Aeth, and was evidently a keen collector of antiquities. See van den Boogert, "Freemasonry in Eighteenth-century Izmir?", pp. 110-11.

92 This was Lethieullier Tooke, born in 1700 to Edmund Tooke (d. 1729), merchant, and his wife Mary (d. 1751). Records suggest that he was formerly consul at Smyrna, but he returned to England and is described in directories, between 1749 and 1753, as a Turkey merchant living in three different addresses in London. He died in 1759. See Wilson, "More Notes", p. 3. A Lethieullier Tooke is also listed as a trader at Smyrna in 1721 and 1743. See David Wilson, *Levant Company: Admissions of Freemen and Grants of Liberty of Trade, 1695-1824*, 2017. Available at: <http://www.levantineheritage.com/pdf/Levant-Co-Members-1695-to-1824-D-Wilson.pdf>. We are later told that Mr Shaw is a brother of Peter Shaw, whom Pococke had met in Aleppo.

93 Either Mrs Pococke or her son had difficulty with this name (which should read "Fremeaux) as almost every entry throughout the remainder of the Aleppo letters has been altered incorrectly (in Pococke's hand) to "Fremo", "Fremos", "Fremon" or "Fremose". The version used in the preface to *Description 2*, p. v, where this gentleman is described in glowing terms, is correct.

94 The Swedish consulate in Smyrna was established in 1736. The first consul was Hendrik Hackson. Leos Müller, *Consuls, Corsairs, and Commerce. The Swedish Consular Service and Long-distance Shipping*, 1720-1815, Uppsala, 2004, p. 71-2.

95 Presumably a servant of his mother's who was a likely candidate for going to work in Ireland for her son. However, on hearing of the death of Bishop Milles, Pococke no longer needs one (see Letter 45).

96 *Travels or Observations*. As noted in an earlier letter, he had seen Mr Upton's copy of this book. Pococke did not possess a copy of the first edition, but his library sale catalogue

3d-14th. I breakfasted with the Gentlemen. The Dutch & English Consuls sent their Droggerman to compliment me on my arrival, & the latter to invite us to dinner tomorrow. After the French Droggermans came with the same compliment. I writ. Mr Burdett dined with us, & he went with me to visit the French, Swedish & Dutch Consuls[97] & Messrs. Fremosi, two French Gentlemen din'd here.

4th-15th. I went to church, dined at the Swedish Consuls, where at a long table, we had 3 rows of dishes, 5 in a row, – mostly in the French manner; – 5 or 6 were removed & others placed, & then a desert of 18 dishes, all very neat & after dinner, we had a concert. I went to wait on Mrs Fremeaux, & then on her mother Mrs Cook [Cooke] widow of an English Consul, but of this country.[98] The Apartments of the Ladies are very Grand, a long room, half of which all down is raised into a sopha [sofa] with cushions all round, & the whole room covered with carpets at the end of the sopha where the Lady sits, is a low square table, spread with silk coverings, under it is a stove of coals not seen, the Ladies putting their feet under the table. The entertainment is Coffee. I went to see a Dutch Gentleman. Mr Burdett & Mr Shaw bred at Eton, brother of him at Aleppo[99] sup'd here. All my letters now go to Constantinople; be pleased to continue to write as usual, till you hear further. Now the peace is made with the Emperor [Charles VI], the correspondence will be more regular, as the roads will be open; – but you must continue to send to Mr Maydwell.

(Flin, *A Catalogue of the Library*, pp. 2 and 8 respectively) lists the 1757 edition and the 1746 "Supplement to Shaw's Travels" as Lots 59 and 280.

97 The Dutch Consul is eventually named by Pococke as "Baron de Hochipied", at the end of Letter 34, shortly before his departure from Smyrna. This is Baron (later Count) Daniël Alexander de Hochepied , who was born in Smyrna in 1689 and was Dutch consul in Izmir from 1720 until his death in 1759. Schutte, *Repertorium*, p. 335. The Swedish consul was probably Anders Johan Magnusson Rydelius, who was born in 1717 and died in Izmir at an unknown date. He was a correspondent of Linneaus and is referred to by him as the Swedish consul in Izmir, in August 1749, in Linneaus's biographical preface to Frederick Hasselquist, *Voyages and Travels in the Levant*, London, 1767, p. iv. Cf. ibidem, pp. 408, 420, 441.

98 Margaret Cooke was daughter of Gertrude Constantia Cooke (who herself was the daughter of former Dutch Consul at Smyrna, Daniel Jan de Hochepied) and John Cooke, former English Consul of Smyrna (1716-21). In 1721 she married the wealthy Dutch merchant James Fremeaux who, in 1738, commissioned Liotard to paint her portrait in eastern costume. Liotard made both a drawing in red-and-black chalk now in the Louvre (Fonds des Dessins et Miniatures, Louvre, RF 1383, Recto) and a pastel on paper glued onto canvas, now in the Musees d'Art et d'Histoire of Geneva (Inv. D 1997-14). See *Jean-Etiènne Liotard, 1702-1789, Masterpieces from the Musees d'Art et d'Histoire of Geneva and Swiss Private Collections*, Somogy Editions d'Art, Paris, 2006, p. 57, where this portrait is reproduced.

99 The brother in Aleppo must have been Jasper Shaw.

I receive great civilities from the Gentlemen here; – being recommended by Mr Lisle at Mr Maydwells instance; which you may mention on proper occasions especially to Mr Wood. You see I have with wonderful fortune got over the greatest & worst part of my journey; & I hope the same good Providence will still protect me. And I am to acquaint you, that there is no plague in any of these parts.

I desire you to present my very kind love to my sister, with my acknowledgments for her Good wishes; & my due respects to Whiteoak & all friends.

I am Dear Madam your most obedient & dutiful Son, Richard Pococke

SMYRNA
1/12 December 1739
Pococke to his mother
Letter 34[100]

Honoured Madam,

My last Letter was from this place of Novr. 5th-16th when we dined with Mr Took. In the afternoon the Swedish & Dutch Consuls came to return my visit. We sup'd at Mr Shaws.

6th-17th. I went to Mr Burdetts room at 10. We all dined at the Dutch Consuls. 2 courses on a long oval table of 15 dishes, mostly after the English manner, & a desert of 21 dishes. We drank tea in Mrs Cooks apartment. The Consul showed me his cabinet of curiosities. We sup'd there, the middle now being part of the desert & five dishes on each side.

7th-18th. Mr Shaw dined here. Mrs Burdett came & we visited a Dutch Gentleman.

8th-19th. I walked out with the Gentlemen of our house. Went to Mr Burdett. We dined at Mr Fremose [Fremaux] by invitation, – 2 courses of 9, & a desert of 12: – sup'd on a course of nine & a desert.

9th-20th. Went to Mr Burdett, we visited the Archbishop of Smyrna, who had sent to welcome me. Showed us a fine MSS of the Pentateuch [Hebrew Bible], with a moment of the year 800; – treated with sweetmeats tea & Coffee; – & another Bishop of the Morea, showed us the Cathedral church. We went to another Greek church; dined. Went to see some antiquities in the Armenian churchyard. I took Coffee with Mr Burdett;- visited Mr Bodington – & after Mr Dobbs & Mr Stephenson, we sup'd at Mr Shaws.

100 BL, Add MSS 22998.

10th-21st. Breakfasted at Mr Burdetts, rid out with him & Mr Bodington to the Circus & saw St Polycarps tomb; – he being torn to pieces by lions there: – went to the castle & thence to some Aqueducts, & came to a place called Homerion, no ruins but a tradition that there was a temple here to Homer;[101] returned, din'd at Mr Bodingtons house. I came home & writ. We spent the even with some Dutch.

11th.-22d. I preached, Messrs Shaw, Burdett & the Dr & two Dutch Gentlemen din'd with us. I went with Mr Burdett to see the Dutch Hospital. We went to a concert at a French house. Mr Shaw sup'd here.

12th-23d. I went with Mr Burdett & we went & finished at the Armenian churchyard. We dined at Mr Tooks & sup'd at Mr Shaws.

13th-24th. Mr Burdett dined here; – he went with me to Mr[s] Cooks, where we met the French Consul & Lady,[102] the Dutch Consul & some other Ladies: first red wine then Cyprus wine, then punch brought to the Gentlemen; after to all, wet sweetmeats & Coffee, then a dish of tea & fine cakes, & last of all dryed sweetmeats & chocolate; – we went & played at the Dutch Consuls. Sup'd at home.

14th-25th. I went to Mr Burdetts , we dined at Mr Fremos without invitations, 5 dishes & one removed, & a small desert. We played in the even – & sup'd. Messrs Masters & Lee &c: being here.

15th-26th. We all dined at Mr Shaws. I went to Mr Burdetts; – they all sup'd at our house.

16th-27th. I shall for the future hardly have time to write to my Ld. Bishop concerning my travels, having brought my account as far as Mytilene. Mr Fremo came to see me in the morn; – & came & play'd in the even; – we went & sup'd at Mr Shaws. When my things come from Egypt let me know what silver

---

101 Pococke regrets in his book that so little is to be seen of the antiquities in this city, "the reason is that the new city is built on the spot of the antient one, and most of the materials have been removed to serve for the modern buildings..." See *Description 2*, Part II, p. 36.

102 Gaspard de Péleran, who had been born in Constantinople, was French Consul at Smyrna from 1731 until his death in 1747 and his wife Marie Boyer, daughter of French merchant Véran Boyer, who, when widowed, married the Swedish Consul at Smyrna and died at that city in 1786. Mézin, *Les Consuls de France*, pp. 481-2. Their portraits by Liotard are in the collection of miniature red-and-black chalk sketches in the Louvre (Fonds des Dessins et Miniatures, Louvre, RF 1365, Recto and RF 1366, Recto respectively) where they are depicted in western costume, with Monsieur Gaspard in a fashionable wig and Madame Gaspard wearing a pointed black bonnet and choker and holding what appears to be a bow.

pieces there are of Ragusa; – a Bishops head & round per te ope divina;[103] – if one or more pieces as big as a crown & one or more as big as an half crown.

17th-28th. Messrs Fremo [brothers] dined with us. Mr Burdett came & spent the even.

18th-29th. I preach'd, went with Mr Burdett & dined at Mr Fremo's. Went to see Mr Vander Sanders [Van der Sanden's] garden; – treated with wine & Coffee. I visited the French Consuls Lady; – we all sup'd at Mr Burdetts.

19th-30th. Dined at Mr Tooks: – went with Mr Burdett to Mr Fremon, who carried us to two fine houses & gardens. Visited the Dutch Consul. Sup'd at Mr Shaws.

Novr. 20th-Decr. 1st.Went out in a boat with Messrs Fremon [the brothers], Burdett, Took, Bodington & the Dr to the North East corner of the Bay, to see some ancient Tombs; dined in our boat: – in return went aboard a Dutch ship; – we all went & played at Mr Burdett's; – sup'd at Mr Tooks.

21st-2d. Mr Burdett & I went & waited on the Dutch Consul, who showed us his medals – we dined there with other Dutch & our nation. Spent the even at Mr Burdetts.

22d-3d. Messrs. Burdett, Bodington & the Dutch Consul rid out to see some few ruins; & we went & dined by invitation at the Dutch Consuls, at Bujaro [Buca], a village a league off, where he has a country house. We after dinner went to see the house of Messrs. Masters & Lee & some others; returned. The English sup'd at our house.

23rd-4th. I rid out with Mr Bodington & the Dr to search after some Antiquities. Our Gentlemen went out to Mr Shaws country house to stay two days. I copied some inscriptions at the Dutch Consuls. I went to Mr Burdetts, & spent the even with Mr Fremo [Fremeaux] & Mr Bodington.

24th-5th. I breakfasted with Mr Burdett, Mr Fremaux & went round the plain north of Smyrna – saw many villages, pleasant country houses; & dined at a Dutch Gentlemans house, returned at night & spent the even at Mr Fremose.

25th-6th. The Dr & I went two hours to Mr Shaws country house: – we went & saw some other country houses, dined, & all came home together. I find

103 This coin was minted in Ragusa in 1735 and depicted St. Blasius, with the inscription around the top. It is impossible to identify this in Pococke's coins and medals sale catalogue, which contains two sections on "Silver Coins and Medals" & "Modern Silver" (Lots 21-34 and 79-93 respectively) since the entries lack detail, being listed as "Foreign coins, in bad silver..." or "Ditto, in good silver...". See Langford, *A Catalogue of a Curious Collection of Greek, Roman, and English Coins and Medals*.

Kinchin has resigned his fellowship.[104] So you may send the seeds by any acquaintance to the Botany Professor if you will without the letter, or as you find best, but pray do not omit sending them & adding such as I have sent before or at the same time & are not in the catalogue.

26th-7th. Mr Burdett came here & we went to see the medals of the French Chancellor, thence to see an old Tomb,[105] & the [word missing] dined at Mr Tooks; – Mr Fremon came & we went to see some old vaults;[106] – & to Greek house. Was at Mr Fremo's till 8 – all sup'd at Mr Shaws.

27th-8th. Went with Mr Burdett, & copyed some inscriptions at the Chanceliere Francois; dined; went with Mr Burdett to the Dutch Consuls. Sup'd at Mr Tooks.

28th-9th. I went with Messrs. Burdett, Bodington & Dr to the hot-baths in the So. West mentioned by Strabo, where we dined, spent the evening at home.

29th.-10th. I went to Mr Burdetts & did some business. Mr Fremose, came here; dined at Mr Shaws. Went & saw the Capuchin & Franciscan convents; – in the latter the Apostolical vicar entertained us with much civility & showed us the Convent: there are also Jesuits here. We went & sat at the Dutch Consuls till 8; most of the nation sup'd at our house. This day I received with much satisfaction your letter of July 19th last past, all your others since July 1738 being at Salonica, except your letter of Septr. 7th. 1738, whence I expect them, which I received when I arrived here. You should have endeavoured to have saved so great a Curiosity, as a calf with two heads, either by stuffing the skin, or in spirits. The two statues that will come, you may if you think proper, be put in the Hall, on such stands as may keep them from being dirted, when the room is washed, or you may lay them by in the boxes as you please.[107]

104 Revd Charles Kinchin (1711-42) came from Hampshire, was educated at Corpus Christi College Oxford (as was Pococke) and was appointed a Fellow of this college in 1731 and Dean from 1736. During the earlier part of this tour Pococke asks his mother to order Dr Shaw's book through Professor Kinchin, as an Oxford colleague. See Finnegan, *Letters 2*, pp. 193-94.

105 Possibly the "very antient sepulchres" mentioned in his book. See *Description 2*, Part II, p. 39.

106 Possibly the remains of vaults briefly described in *Description 2*, Part II, p 36.

107 These are the statues of Osiris & Isis (mentioned again in Letters 35, 38 and 45) which he acquired from an Italian merchant at Cairo, "who could give no account from what part they came but was so obliging as to part with them." See *Description 1*, p. 211. Clearly, they were the pride of his collection, as he includes four engravings of them (front and rear views) as Plates LX-LXIII and dedicates them respectively to the Dukes of Richmond and Montagu and the Earls of Lincoln and Sandwich. The statues remained in his collection for his life-time and are listed in his sale catalogue of antiquities and natural curiosities (Langford, *A Catalogue of a Large and Curious Collection of Ancient Statues, Urns, Mummies, Fossils, Shells, and other Curiosities…*, p. 12) as Lot 102 (Day 2) and described as,

30th-11th. Packed up all the morning. Went to Mr Burdetts & with him dined at Mr Fremose. I went & visited all my friends; – among them the three Consuls. The Swedish Consul came & returned my visit; – the French & Dutch Consuls sent their Drogerman to wish me a good journey. We all supd at Mr Shaws. I was to set out for Ephesus tomorrow but it is deferred to another day. I shall afterwards go towards Constantinople – so you may continue writing thither till you hear to the contrary. Letters from England in 29 days to Constantinople, which I suppose brought me yours of the 19th of July, tho I might have had one of a later date.

Decr. 1st-12th. Busied in settling my affairs. Messrs. Shaw & Burdett sup'd here & the Dr Thus I have spent a month very agreeably in Smyrna having received great civilities, not only from our own nation, but from others, particularly the Dutch & among them in a particular way the Dutch Consul the Baron de Hochipied originally a Fleming, whose Ancestor returned to Holland on account of Religion as did Mr Fremose of whom I shall mention more. The [Dutch] Consul is a Gentleman of great humanity & good nature & very curious in Antiquities which he showed me with much civility, & I was as you see often with him & entertained with much courtesy; – & he has given me letters to places I am going to. Mr Fremo is my Age, married to the [Dutch] Consuls niece, as the Consul was married to his sister, is one of the most valuable men I ever met with, & one finds united in him, the good qualities of all nations, joined with the greatest benevolence of disposition; – he has the sincerity of the English, & exceeds even the French in a desire to please, & with the utmost complaisance, without the too great ceremony of the nation. If I had been of the same nation & recommended in the most particular manner, I could not [have] received greater civilities than I met with from this Gentleman, in carrying me abroad to see every thing that was curious, in entertaining in the most civil manner, several times at his house & in giving & procuring several letters for me to the places I am going to. This Gentleman has great business, a very rich family & has Brothers settled in Holland.

Pray present my very kind love to my sister, & due respects at Whiteoak & to Jerre &c: – who has not favoured me with a letter but once since he was in England. Captn. Bulmer arrived lately at Constantinople from Waterford in fifty days, but I don't find he has brought me any letters.[108] He is coming to this place with the Consul [Francis Williams] & Lady, who have been half a year in

---

"Two large stone statues of *Isis* and *Osiris*". The drawings are reproduced as Plates 18 and 19 in the present volume.

108 See Letter 37, where Pococke notes that this captain has brought a letter from Bishop Milles.

a visit to Sr Everard Falkener at Constantinople. A man of war is gone to Alexandria, & will convey home the ships, with which my things from Egypt will come. No account yet of the arrival of the man of war, they will probably set out at the beginning of the year, & be with you in March or April.

I am Dear Madam your most obedient & dutiful Son, Richard Pococke

EPHESUS
15/26 December 1739
Pococke to his mother
Letter 35[109]

Honoured Madam,

I writ to you last from Smyrna on the 1st-12th of Decr.

2d-13th. I set out for Vourlaw [Vourla/Urla] the old Clazomene [Klazomenai], 12 leagues to the So. West, in the Bay of Smyrna; arrived at the [sea] in the even & went a league to the town, where I lodged with a greek to whom I was recommended.

3d-14th. I writ; – dined after, went & saw the two churches; – near one the Archbishop of Ephesus has an house, spending a month or two here every year. Went to the person I had a letter to. Walked two miles to the seaside where Clazomene stood, & went to the remains of the causeway to the Isle the Clazomenians retired to from the Persians.

4th-15th. Went 3 hours to Severhissar [Seferihisar] a Turkish town, & thence went an hour further to Segigierk [Sığacık] about the spot of Teios [Teos], the city of Anacreon[Greek lyric poet]; a castle there & a fine port; – I lay aboard a Dutch Ship, recommended by Mr Fremose [Fremeaux].

5th-16th. I went ashore, – a Greek merchant walked out with me 2 miles east to a small lake; – returned, dined in a garden, & went aboard ship & writ. The country there is very fine a rich soil, & the fields planted with olives & fig trees & many Elms, up where grow the vine Anacreon so much praises, & which a Poet would say was ungratefully the cause of his Death.

6th-17th. I went in company with the sacred green head of a descendant of Mahomet to view the ruins of Teios, a Theatre,[110] a temple of Bacchus, in each of which I found an imperfect inscription; – a circus – the two latter were very

---

109 BL, Add MSS 22998.

110 This is Represented in Plate XLIV of *Description* 2, Part II. See Letter 5 on the distinction between the various coloured turbans.

fine buildings. I returned, dined in a garden; the captain came here. I went aboard, writ.

7th-18th. I went out with a Turk the same way as yesterday, & reviewed, & went to the Old Port, to the south, returned, dined at a garden; – a fine winter & the mild anemonies are in blossom. I came to the Castle & copyed some imperfect Inscriptions.

8th-19th. I went a league to see some imperfect inscriptions: – I have seen here many of those inscriptions you will see in Chisul [Chishull], which I think you have, which you will see under Teius.[111] I returned, went with the captain &c: & dined in a garden.

9th-20th. At 4 embarked on a Hoy, to go about 40 miles to Scala Nuova [Kuşadası] near Ephesus, arrived there in the evening; – went with letters I had to the house of the successor of St John the Archbishop of Ephesus, who resides here; – I had set before me pickled fish, olives & sallad; – & then was introduced: – He makes a venerable figure, with a red mohair scarf, in other respects as other Greek priests; – he does not rise when one enters. I had Coffee given me, & being late, I lay in his house, on a sopha, where beds are laid for 3 or 4 people with an under sheet & a quilt only over.

10th-21st. The Archbishop made a Reverend figure, sitting on the ground below the sopha by his chimney, & a cloth in his lap; first a dram was given, then sweetmeats & then Coffee & cake, – & I went to my lodging at the kane, & then to some Greek merchants I was recommended to, for this is near them: I dined in my room, writ. Went out to settle about going to Ephesus. Went & waited on the Archbishop, who was in the same posture with some books before him. I took Coffee, came home & writ. The Archbishop had 32 Bishops under him, but now has none, but all the Province is his Diocese. Came to my Lodgings & writ.

The two old Egyptian statues [Isis and Osiris] weigh much, & perhaps it may be better to have them left in Mr Maydwells ware-house, as to which do as you please.

11th-22d. A fine clear frost, my departure to Ephesus put off til tomorrow. I went out about affairs. That you may know how Eastern travellers live in a kane; – they give 3 farthings a day for a room which has only open window places with shutters so the door place is the way the light must enter, which is

111 Edmund Chishull (1671-1733) was educated at Corpus Christi College, Oxford and was Chaplain to the English Factory at Smyrna from 1698-1702 and published two scholarly books of inscription (see above). Pococke refers to his works several times (see also Letters 40 and 44) and the fact that copies of these works do not appear in the sale catalogue of his own library means that the volume referred to was indeed his mother's, rather than his own.

not convenient in winter; – in these rooms is nothing so you send for your fuel & everything; – I have my travelling kitchen furniture of copper Tin'd over, & my food is rice-soup made of Beef: – as it is inconvenient to carry my bed my carpet laid on the stone floor serves instead of that convenience, & a bundle or sometimes my wooden bottle is my pillow. And I pass my time in writing. As to my habit; – now it is winter, I have a blew linnen garment lined with an ordinary fur & over that such a coarse great coat, as the common people here wear, these girded about me; – & today I wear a red cap turned up with black hair, & on the road a red cap of the strangers a sort of grade of the Grand Signr.[112]

12th-23d. I set out for Ephesus 4 hours distant & arrived there & lodged in a very poor house.

13th-24th. I went out to see the Antiquities of this place; – viewed the large building & several other ruins, the Theatre & two or three other things, & returned.[113]

14th-25th. I went up to the castle & saw an Aqueduct near; & then went & viewed the ruins of the famous Temple of Diana, & the vaults under; return'd to the kane.

15th-26th. I went to see the great Mosque, which they falsely say was a church to St John; – thence went to the Cayster [river], & after went & viewed the circus, another large building, & the place of the 7 sleepers, & reviewed some things, & went up to a tower called the prison of St Paul.

I beg the favour of you to present my kind love to my sister & due respects everywhere

I am Dear Madam your most obedient & dutiful Son, Richard Pococke

CORA IN THE ISLE OF SAMOS
7/18 January 1740
Pococke to his mother
Letter 36[114]

Honoured Madam,

I writ to you last from Ephesus on Decr. 15th-26th & that day returned to Scala Nuova.

---

112 This is the outfit depicted in the Liotard portrait, though obviously with a different hat. See McCarthy, "'The Dullest Man That Ever Travelled?'", p. 26.

113 The antiquities are described in *Description* 2, Part II, pp. 46ff and the various buildings are represented in Plates XLIV-L.

114 BL, Add MSS 22998.

16th-27th. I did some business & prepared to go to Samos, & in the even set sail in a small boat, & arrived at midnight at the port of Valeti [Vathi].

17th-28th. In the morning I went up to the Town ½ a mile from the port & found an house to lodge in, the people very civil & excellent white Muscat wine, a Gentman Dr of Corah [Khôra] town came to me.[115]

18th-29th. I rid 6 miles to the east end of the Isle, where there is a very narrow passage between the Isle & Asia Minor, returned in the even.

19th-30th. I went a mile north & then a mile East to the convent Panacea, Ἀγία Ζώνη [Agia Zoni]. I returned, dined, set out for Cora six miles to the south part of the Island, where I went to visit the Drs house.

20th-31st. Staid within all day & writ.

21st-Jan. 1st. Writ all day

22d-2d. After dinner went out & did some litle business; & waited on the Bishop of Samos.

23d-3d. I rid out with the Dr to see two little monasteries, about 2 leagues distance & lay at the village of Baounda.

24th-4th. I went to a village called Milo, & then went to a farm house of St Johns convent of Patmos; & afterwards to see the Pedestals of what remains of the pillars of the famous Temple of Juno.[116] Came home, dined, writ.

25th-5th. Some people came to see us. Staid within all day & writ: we had fine frosty weather for a fortnight, & last week it snowed one night & this day rain.

26th-6th. Fine mild weather; writ all day to my Lord Bishop; this Isle is all Greeks & poor people.

27th-7th. Writ all day.

28th-8th. Rain, writ all day.

29th-9th. I brought down my Account to my Lord Bishop to this Isle in 8 letters, that will probably go with this;[117] & before this arrives, those in my box from Egypt, I suppose will be come to hand, when those things arrive I desire you particularly to mention the receipt of a box of papers; I think no 3.

30th-10th. Spent the day in writing & reading.

31st-11th. Spent the day in the same manner.

Jan. 1st-12th. A Venetian Proverb relating to buying, if you seek for a think & ask for it & seem warm after it, – it is ten pr cent. difference in the purchase.

---

115 Clearly, this gentleman is distinct from "the Dr" frequently mentioned, who must be Dr Charles Perry.

116 Described in *Description* 2, Part II, pp. 27-28 and represented in a plan in Plate XLII.

117 It is not known what happened to these accounts, as they are not included in the same volume as the others (BL, Add MSS 22998).

Alghore, Alrole Dica pr cento di differendza. Have you, will you, is ten pr cent difference.

2d-13th. Went not a tread.

3rd-14th. Staid in all day.

4th-15th. I read & writ all day; fine weather all day.

5th-16th. The Priests this day go about with Holy water to bless the houses & people, & every one puts a piece of mony into the water for the Priests.

6th-17th. I wait here for my letters from Salonica almost all that have been writ since July. 1738. & when I receive them shall go on towards Constantinople.

7th-18th. I prepared letters to send to Smyrna, &c: so take the opportunity to end this, & desire you to present my kind love to my sister & due respects to all friends.

I am Dear Madam your most obedient & dutiful Son,

Richard Pococke

CORA IN THE ISLE OF SAMOS
2/13 February 1740
Pococke to his mother
Letter 37[118]

Honoured Madam,

On the 7th-18th I had the opportunity to send to you from Constantinople.

8th-19th. Staid within all day, & also the following daies.

Of those Dates you will receive from Egypt, pray sow some in the month of May, in the garden & some in a pot, which you may sometimes water, & some in a pot kept without water, – & send some with the other seeds to Oxford; & perhaps you may have Palm trees: – they come up with one leaf, like a Plantane leaf; & when up should be watered a litle; – you may also sow of that large fruit the Dome, of which I have put up two or three for Oxford, & also of the cedar apples, which will without [...] grow, but I supposed they should be opened, & the seeds taken out, – for which purpose choose one of the largest & soundest but not that one which you will find not separated from the bough.

9th-20th to 12th-23d. [No entries]

13th-24th. In the afternoon I walked with the Dr two miles to the village of Mytilene, & went to see some grotts, which were quarries, – & then went to see

118 BL, Add MSS 22998.

the ruins of Old Samos, traced the walls, which were finely built of white marble, & came round by the Port & by several ruins home.

14th-25th. We went to see some grotts, which were quarries, & then went to see the ruins of Old Samos, traced the walls, which were finely built of white marble, & came round by the Port & by several ruins home.

15th-26th. I went to see a Greek who talks Italian, & copyed an Inscription, walked a mile with the Dr to the garden.

16th-27th. I walk'd out with the Dr, & in the evening with him to see a Greek.

17th-28th. [No entry]

18th-29th. Walked out with the Dr to an unfinished convent a mile off, & then to the Gardens, returned, in the afternoon walked about the Town; went to the house of a Greek with the Christian Governour of the Isle; – set before us black olives, curd, when near ripe, the roes of fish salted & dryed, Brandy & wine: – then they came to our house, & staid a while.

19th-30th. The Christian Governour of the Isle came to see me & some other Greeks. In the afternoon walked with the Dr to see the ruins of Old Samos.

20th-31st & 21st-1st. [No entries]

Jan. 22d-Feb. 2d. I walked out with the Dr.

23d-3d. I walked out to the Temple of June, & some other places.

24th-4th to 27th-7th. [No entries]

28th-8th. I walked out & took a further view of the ruins of the old city.

29th-9th. I walked out to the Temple of June. Last week we had bad weather, but now again fine cold weather & northerly wind; here they call me Didaskale knowing that I travel to see antiquities, & as they chanced to be informed, to write a History of them – for that term signifies as much as a scholar or the Learned Man. In Smyrna the Protestant Ladies of Holland, called me Padre Pococke.

30th-10th. I walked out, & my servant killed 3 larks for my supper. Any Letters for the Bishop you find among my papers; read if you please in their order – & send two or three every Post; or in a small box, freedom not in, altogether to Bristol – as you please or find convenient.

31st-11th. I took a pleasant walk to the side of the hill of Old Samos, entered many large Grotts & in one a delightful Hermitage, a large Grott with a litle church built at the end of it [found] cisterns of water – droppings of the rock another church without to the Virgin Mary & litle terraces down the side of the hill, improved into Gardens, & a fine view of the sea: we see abundance of coturni [quail] or red leg'd partridges but cannot come near them, but at ways meet with some game. After supper it being Carnival came in a fiddle, & a man came & two or three others & danced; – the masque was a humorous figure of the devil, with a bottle tyed to his middle, for the wicked one reigns here very much in the Bottle. All the parcels that come sealed among my things are to be opened & then tied up as they were before, only taking out my journals, & put them altogether.

Feb. 1st-12th. I went with the Dr 2 leagues to Vasti [Vathi], dined there; – returned; & received letters from Smyrna, but not the Letters of a year from Salonica.[119] But I had the honour & great pleasure of a most kind Letter from my Lord Bishop, sent from Waterford, by Captn. Bulmer,[120] dated Aug. 31st. 1739. I writ to Smyrna.

Feb. 2d-13th. I forward this to be sent to you & now shall go on towards Constantinople & you will continue writing enclosed to Mr Maydwell. I much wonder I have received no letter since the date of July last, the roads being now open.

Pray let Mr Milles know I expect to hear from him sometimes, tho' he is not in Ireland, & shall write to him, when I receive my Salonica letters.

I desire you to present my very kind love to my sister, & due salutations to everywhere. When my things arrive from Egypt pray mention all the boxes that come to you & their numbers, particularly a box of papers No. 3.

I am Dear Madam your most obedient & dutiful Son, Richard Pococke

GUZELHISSHAR THE
OLD MAGNESION ON MEANDER
28 February/10 March 1740
Pococke to his mother
Letter 38[121]

Honoured Madam,

I writ to you last from Cora in the Isle of Samos on the 2d of Febr. 3d-14th. I walked out to a church.

4th-15th. I embarked about 10 with the Dr for Patmos 40 miles off & arrived there at 6 in the evening.

5th-16th. I went up to the convent called Apocalypsis, half way up the hill on which the Town & great convent stand & here I entered the Grott turned into

119 He eventually received these long-awaited letters on his arrival in Constantinople (see Letter 42).

120 See Letter 34, where Pococke is disappointed at not receiving letters from this captain. This may have been the last letter he receivd from Bishop Milles, who died nine months later.

121 BL, Add MSS 22998.

a church, in which they say St John writ both the Gospel & Revelations. In this convent they teach literal Greek, Logick, Physicks, metaphysicks & Divinity. I attended at one of their Lectures: – went up to the Town, dined, went to see the Great Convent; waited on the Abbot, who treated me with sweetmeats, a dram & Coffee; – I saw all the convent. Met with a Muscovite a great traveller, he went with me to a nunnery of old women, who live by their labour; & then to a Hermitage, & visited the chancellor of the convent, a man of some Learning: I went down & lay in a house at the port.

6th-17th. I walked to see the first church, they say, built here to St John; – met with a Greek that had lived many years in Rome, & another that had lived in Port-Mahon. At ten set sail for Samos where I arrived before night; – & lay in the Watch house; – the Dr bound from Patmos to Smyrna.[122]

7th-18th. Rained much in the morn; – in the afternoon I removed to the Hermitage of St Mary, upon the side of the hill, where I lay.

8th-19th. I walked to the Port & in the afternoon to some inscriptions I had discovered, & had the earth dug away to copy them.

9th-20th. I went to Vathi in order to embark for Scala Nuova, two of my acquaintances came to see me.

10th-21st. Bad weather.

11th-22d. The Greeks being in their drunken carnival, I could not pass over to Scala Nuova today – I walked to the Port.

12th-23d. I had a most pleasant voyage again to the continent to Scala, stopping only an hour by the way at a small Island.

13th-24th. I did some business, visited 2 churches, & set out for Guzelhissar to the East So. East, lay at a Turkish village 9 hours off in a Coffee house: – first part Hills covered with wild olives, Bay, Pine & Laurustin: – the plain on the Meander [Menderes/ Büyük Menderes River] a fine enclosed country; – Almond trees in blossom round the fields.

14th-25th. Went 3 leagues to Guzelhissar the old Magnesia – on the Meander.[123] I met Dr Mamachi a Sciote Physitian settled here; to whom I had a letter from the Dutch Consul [in Smyrna]. He found me a lodging in a Varansara [caravanserai]. This is a large city, mostly Turks, but some Greeks, Armenians & Dr Mamachi came again to see me. I went & visited him, & then delivered a Letter to an officer of the Janizaries, who came to see me & took Coffee with me. I writ & settled some affairs.

122 Here Pococke and Perry parted company but they met again in Athens (see Letter 47).

123 Though Pococke identifies the ancient site of Magnesia with Aydın (and describes it in detail in *Description 2*, Part II, pp. 54-56) it is now thought that this city was located between Ortaklar and Söke, just west of the village of Tekin, in Aydın Province.

15th-26th. Signr. Mamaki came to my room about 10. I set out with the caravan for Melasso [Milas/old Mylasa] to the south, went 4 miles & passed the Meander in a Ferry boat into Caria, went 8 miles further to a village where I lay in a dirty stable, wth the beasts, which had no other title to be called a kane, than that it had 2 or 3 chimneys in it, the rest of the caravan lay without.

16th-27th. Went about 4 miles to Carpousley [Karpuzlu], & found out an old city with a Theatre, & the ruins of a magnificent palace & other buildings, which I take to have been Alabanda.[124] Went up the the top of Mount Latmus [Latmos], where I lay under a rock, making a good fire.

17th-28th. Descended from the mountains into a fine vale, in which stands Melasso [Milas], the old Mylasa, on the other side of it under the hills, I went to a cane [khan] & carried a Letter I had from Dr Mamachi to the Aga, who granted all I desired. I went & examined exactly a most beatufiul Edifice of the Ionick order, built to Augustus & Rome,[125] & some other ruins.

18th-29th. I walked to see an Aqueduct, – reviewed some things I had seen; – went half a mile out of Town to see a building, which some call at Temple; – but seems to have been a canopy raised on Pillars for some statue.[126]

Feb. 19th-March 1st. It rained all the morn, in the afternoon I rid 3 miles to see a ruined town, – which proved not to be very old; – was entertained with Coffee by the Governor of the castle there.

20th-2d. I went out to see some Antiquities, & set out to the north East over the hills to Eskihissar [Eskihisar], the old Stratonicea 6 leagues off, arrived there viewed some Antiquties.[127]

21st-3d. Went to the Aga, saw the Theatre & all the ruins, & about 2 set out again towards Guzelhissar – went 4 hours to Lena, where there is a generous Turk who entertains all comers: – I went to the top of the hill to see the walls of an old town or castle.

22n-4th. Passed over mountains to the north west, & came in about 6 hours to China [Cine], where are ruins of a [next word scored out] castle on a hill, & some litle remains of greater Antiquity; lay in a large Coffee house.

23d-5th. Went 2 hours to Arabihissar [Arab-Hissa], where are great ruins, being, as I suppose, the ancient Lagena, particularly city walls, Theatre & the Temple of Hecate as I conclude: – we went on & came into the road we went

124 He correctly identifies this site, describes at *Description* 2, Part II, pp. 57-58 and the theatre is represented in Plate LIII.

125 The ancient site of Mylasa is described at *Description* 2, Part II, pp. 59-63 and this monument is represented in Plate LV.

126 Plate LVI, *Description* 2, Part II.

127 Described at *Description* 2, Part II, pp, 63-65, but with no illustrations.

in, & at night arrived at Guzelhissar I lay at Dr Mamachies, being disappointed in getting a room.

24th-6th. I removed to the cane I was at before, I settled my affairs; – went with the Dr to the mosolem or Pasha's Deputy, & delivered him the Dutch Consuls letter; who entertained me with conserve of Roses & Coffee, a great honour, the others with me had not & who did not sit down. I writ; rainy afternoon, – as was the day before.

25th-7th. I went out & viewed all the Antiquities of this place, the Armenian & Greek churches. The old city was a fine situation, on the flat top of a low hill, which is almost all round, a sort of hanging perpendicular ground. I came home & dined; rain & snow this afternoon; writ.

26th-8th. I went to see Dr Mamachi, who was not at home; – he afterwards came to me, & we went out to see a piece of Antiquity: – & in the even went to see him, a cold sharp day, I believe I shall not be able for the future to give my Lord Bishop an Account of what I see: – for besides that I send away my Papers by all opportunities, I shall be so taken up in moving from place to place, that I shall not have time. I travel very securely with a servant & Janizary always with me; – the latter a very faithful man who was a slave in Spain, escaped to Gibraltar & was by the English sent into his own country. When you receive my things from Egypt, mention in 2 or 3 letters following that you receive the box of papers. I mentioned to you the 2 statues [Isis & Osiris], in 2 boxes might be left in Mr Maydwells ware house, because they are weighty; – but if they are come may adorn your Hall, or it may be better to keep them in the boxes, whichever you think proper. Tho' if they have not been once exposed, I had rather that all those things should be kept out of sight, except of very particular friends.

27th-9th. Very cold. Dr Mamachi came to see me. I staid within all day & writ: – being the feast of the Turks called Biram [Byram]; – the day the Pilgrims sacrifice on Mount Araret near Mecca, – in memory as they say of Abrahams sacrificing Ishmael; & every one at his door makes a sort of a sacrifice of an ox or a sheep, which he kills himself or if a great person, takes the knife, saies the prayer, & gives the knife to another to kill it. The meat they mostly give to the poor; they are at this time very insolent, if one goes abroad or sprinkle you with rose-water, as a compliment &c: expect money.

28th-10th. I was abroad often to do business & prepare for my departure tomorrow, it is very cold & a sharp wind. I desire you to present my very kind love to my sister & due respects to all friends.

I am Dear Madam your most obedient & dutiful Son, Richard Pococke

ANGORA THE OLD ANCYRA IN GALATIA
12/23 April 1740
Pococke to his mother
Letter 39[128]

Honoured Madam,

My last Letter to you was dated at Guzelhissar the old Magnesia on the Meander on Feb. 28th March 10th.

29th-11th. Dr Mamachi came to see me, & I set out on horse back as I always I travel to the East & in 10 miles came to Sultanhissar [Sultanhisar] near which I went & viewed the ruins of the ancient city Tralles; – then went on 6 miles to a Town called Naslee [Nazilli], went to a Kane, a good Greek of Mytilene came & sat with me. My spare time as usual, I spend in writing of what I had seen; cold frost.

March 1st-12th. Went to the Greek I had a letter to, & then with a letter to the Aga; – hired horses which he procured for me, to go to Geyra [Geyre]. – I went a league to see some ruins: – probably of Nysa, of which some say, St Gregory Nysseus was Bishop, who was brother of Basil.[129]

2d-13th. Set out for Geyra, passed the Meander, went to a great Aga, who was near his house, a hawking; – desired me to go to his house, & he would send a man with me; which he did, & went to a village 2 leagues further & lay at the Aga's house; near this village is a ruined place on a hill called Janitshan,[130] where the famous rebel Soley Ogle was cut off about 10 months ago,[131] I think [...] Antioch on the Meander.

3d-14th. Turned south & came to Geyra, the Ancient Aphrodisium; – went to the Aga's house, an old man who is one of those who entertains all strangers. I went out to see the Antiquities. I spent the 3 following days in viewing the Antiquities of this place,[132] & there being no provisions here, I eat every night with the Aga, of soup made of Barley, & with paste, of a Pilaw of the same, eat with sowre milk, a flummery made sweet with syrup made of Grapes, olives &

128 BL, Add MSS 22998.

129 Gregory of Nyssa, who was Bishop of Nyssa in Cappadocia, was born in Neocaesarea, Pontus (c. 350).

130 The word is partly obliterated by an ink stain.

131 He mentions this also in his book, giving the rebel's name as "Soley Bey Ogle". See *Description 2*, Part II, p. 69.

132 Considering the amount of time spent there, the description in his book is very sparse (*Description 2*, Part II, pp. 69-70), nor are there any plates representing the antiquities.

onions with vinegar & sweetened in the same manner. They have here a very strong well flavoured white wine & the Turks drink heartily of it.

4th-15th to 6th-17th. [No entries]

7th-18th. Set out to return to Naisle, lay at Chislick, where I was very generously entertained by a Turk, a handful of cresse, root & all, laid to every one, pancake with some herbs between, a rich soup, & small grapes like currands preserved in a sweet juice & in the morning boyled milk or caimac [milk fat] with the syrup of Grapes mixed.

8th-19th. I arrived at Nasle, a Greek of Smyrna came to see me.

9th-20th. I went Eastward part of the way to Demizely [Denizli].

10th-21st. Went on the road & lay at a miserable village; – this country being so much damaged by the rebellion of Soley Bey Showeri.

11th-22d. Went on 7 hours to Demizely, a miserable Town, having been destroyed 25 years ago by an Earthquake;[133] took up my lodging in the most private Coffee house.

12th-23d. I went about six miles to Tamboule the old Hierapolis, where are extraordinary hot waters & great ruins of Baths, Theater, Temple, Grand churches &c: return'd ye even.

13th-24th. I & the Janizary walked a league to Eskihissar [Eskihisar], the old Laodicia on the Lycus & one of the seven churches, now entirely destroyed & uninhabited; – except that there are a Circus, Theater & Amphitheatre almost entire; – 2 Temples, an Odeum a church & some other buildings in ruins; returned in the afternoon.[134]

14th-25th. Set out to see the source of the Meander; – travelled not far from the river Lycus for about 3 hours & then the river running between hills, & its course being near south, we went north east over the hills, & at night arrived at a village in a plain near the river where we were accommodated with an uninhabited house: – the Turks immediately brought us fewel [fuel?] & water, & in half an hour two green-heads [Sayyids – descendents of Mohammed] surprised us with a hot supper they brought to us, tho' we were not unprovided, so we sup'd together; & I treated them with Coffee; rainy.

15th-26th. A stormy morn, went over a long plain, crossed the Meander, & lay at a village, where they ofered us of what they had; but being provided, we thanked them for their civility.

---

133 According to his printed account (*Description* 2, Part II, p. 71) this disaster had killed 12,000 people.

134 The antiquities and natural features of these two ancient sites (Hierapolis and Laodicia) are fully described in *Description* 2, Part II, pp.71-78 but there are no plates representing them.

16th-27th. Went 2 hours to Eshekleh a town at the rise of the river Marsyas,[135] which soon falls into the Meander, becoming immediately a large stream, from 8 or 9 great streams which flow out of the foot of the mountain, & make a very delightful scene. I went to the cane, walked out to see the streams & some four Antquities. We had fine crawfish here for about 2d a hundred.

17th-28th. I went to the top of the mountain, over the Town, which they call the Fortress, but nothing to be seen, only in one place, some litle remains of a wall: one of the principal Turks sent to me to show me an inscription in an house, & I went & copied it. I bought a carp 2 feet long, which I suppose weighed 5 pound for about 4 pence. An Effendi of the Law came & sat a while with me, & was inquiring about the age of these antiquities miscalled Celenae, taken by Alexander. This place is, I suppose, Apamia cibota [Apamea Cibotus].[136]

18th-29th. I went up to another part of the mountain, where I saw greater remains of walls: a Turk came & sat with me & informed me about antiquities. When you receive any of my things & find medals in any parcels, let them remain in them, by which I shall know where I bought them. Having passed Ionia, Caria & part of Lydia, we are now in great Phrygia, & but 14 miles from the rise of the Meander & not far from Pisidia, Lycaonia & Sauria & about 50 miles N.W. of Satalia in Pamphylia.[137]

19th-30th. Rain in the morn. I writ to my lord Bishop all day; have feasted these days on excellent carp & crawfish.

20th-31st. Set out early & crossed the mountains to the north, & came into a fine plain; passed by very hot waters used both for Baths & drinking, & in 8 hours or 16 miles came to Sandacleh [Sandikli], that I thought might be Synnada, which probably was in this place.[138]

March 21st-April 1st. Frost, went over the mountains to the north, came into a litle path & then went almost through another chain of mountains, in all about 20 miles & lay at a village where they brought 5 dishes for supper, supplied us with fuel & never expect anything in return; – which is the Hospitality of the East.

22d-2d. We went 2 hours to a large town Carahissar [Karahisar], towards the west end of a large plain, that extends beyond view to the East. A Bohemian young man made a slave at Belgrade, came to me, who turned Turk on his masters promising him a wife. I went into the town to hire horses for Angora

135 The name is given as "Ishecleh" in *Description* 2, Part II, p. 81.

136 This Hellenistic city, founded by Antiochus Soter in the 3rd century BC, superceded the ancient Celaenae and the site is at the modern Dinar.

137 These ancient placenames correspond roughly to the modern province of Antalya in Turkey.

138 The ancient site of Synnada is located in the modern town of Şuhut.

[Ankara], seven days journey; very safe travelling; – going with a caravan, as I shall go from thence to Constantinople.

23d-3d. Went out twice to see the Town & some old stones about it. Writ to my Ld. Bishop. Cold & some hail.

24th-4th. I went to see the Armenian church; – their Bishop came & service was performed. I walked out in the afternoon & saw another Armenian church.

25th-5th. Severe cold wind, frost & snow. Set out with he caravan for Angora & went North East 18 miles between small hills & lay at a Han [khan] in a village, a large stable with bulks or sophas round with chimneys.

26th-6th. I rose early & writ. In caravans we travel 8 or 9 hours without stopping. I have my provision for the road of bread & hard eggs or cheese; & when we come to the Han, have rich soup prepared, sup, go early to sleep, & rise 2 or 3 hours before day; take my Coffee; & write , when others are asleep; a fine frost, we went 18 miles & lay at a Han.

27th-7th. Frost – went 12 miles to a town called Sevrissar [Sivrihisar], 56 miles from Angora; went to a Han there.

28th-8th. I walked out to see the town, writ all day; a frost.

29th-9th. I went to see the Armenian church. A Turk of the caravan came & took Coffee with me; & after two Armenian Priests came & sat with me. I walked out in search of Antiquities.

30th-10th. Set out again, & went only 4 miles – Saw the white Angora goats, of the hair or rather wool of which all camblets [woollen fabrics] here & Prunellas [clerical robes] in our land &c. are made; – the wool hangs down very long, & in curled ringlets very beautifull; & some comes near the ground.

31st-11th. Snow all the morn. We went 8 miles to the river Sacari [Sakayra] the old Sanganus, & the waters being out obliged to stop at a village; bought a carp above 3 feet long for about 8d. The Turks throw away the heads, & skin the fish; no fuel here but dung.

April 1st-12th. Crossed the river on timbers – went 6 miles.

2d-13th. Went 16 miles to a village 12 miles from Angora, where Mr Abbot (to whom I am recommended by Mr Lisle) sent his Broker, Janizary & a servant to meet me, who had prepared an house & brought provisions for me.[139]

3rd-14th. I set out for Angora, & a mile from the Town was met by all the English nation & most of the French & was conducted to a house, where we had a collation, & so being mounted on a fine horse, we all came to town together, & I went to Mr Abbots house, who has in company with him Mr Carew, a young Gentleman of Dorsetshire; – Mr Abbot a gentleman, of great worth,

139 Peter Abbot, English merchant at Ankyra. See Wilson, *List of British Consular Officials*, p. 11.

Learning & Piety, is married to a Greek Lady, has six sons, the eldest at school in England under Mr Maydwells care. Mr Savage cousin german to Mrs Barton of Cairo, the other English house & Mr Franklin his clerk, dined with us. All these Gentlemen that came out to meet me, & some others came to see me.

4th-15th. The Archbishop of Smyrna came to see me, who is the 4th in order of the 12 great Metropolitans under the Patriarch of Constantinople; & also 2 Dutch merchants. In the afternoon Signr. Perry came & sat with me for two hours, a Sciote of Italian extraction.

5th-16th. Mr Carew went with me to return all my visits; they treat with Coffee, sweetmeats & a dram; dined. Some of those Gentlemen came to wish me an happy Easter to the family; – this place is situated on & round a hill, near the mountains to the East, in a fine down country, – but no trees.

6th-17th. I breakfasted in Mr Abbots room, many came to compliment the family on Easter day. I read the service. Mr Savage & Mr Franklin spending the day here. Mr Cary went with me to wait on the Archbishop. The season has been so exceeding cold that there is no plague now in any parts.

7th-18th. Mr Franklin went out with me to view Antiquities. The Archbishop & his mother dined here. We rid out to the Armenian convent a mile off; waited on the suffragan Bishop there; – the Archbishop being on his visitation, saw the church & convent, & returned home.

8th-19th. Went out in the morning & afternoon to see Antiquities. The French dined here, 2 or 3 days past fine mild weather.

9th-20th. Was out all day on the same business. We all dined at Mr Savages.

10th-21st. Spent the day in going to see places. An Armenian church &c: & the castle, visited a relation of Mrs Abbots there.

11th-22d. Writ all day, a litle gentle rain.

12th-23d. Having an opportunity to send this to Constantinople, to go more surely & speedily by the way of Vienna, I lay hold on the occasion; designing to set out shortly for Constantinople, in the direct road, where I shall not stay long, about a month, & then proceed on, in my voyage by land through Greece. And I depend on God Almighty's Providence to protect me, as it has hitherto through all the most dangerous parts of my voyage, which have been hitherto fortunate & happy beyond my expectation, in that I have never been robbed or ill treated, & whenever any attempts have been made by the Magistrates to get money out of me, I have always got the better by resolution & planning one against another being armed with strong Firmans. And what has contributed greatly to the pleasure of my journey, the civilities of all the English, wherever they are settled; – & more especially in this place, from a Gentleman of the greatest worth, who can never be sufficiently esteemed by all his friends. Yesterday I received a Letter from Smyrna, giving me an account that all my packets have arrived that I sent there. My letters indeed arrived too late to reach me & so I shall have the pleasure of receiving all my Letters dated since July was twelve month [i.e. a year ago], except 3 or 4 more I mentioned to you, that I received at Smyrna. As soon as I arrive at Constantinople, where I shall receive them, I shall write to you in answer. I hope before this arrives you will have received my things from Egypt. Pray send the material parts of this to my Lord Bishop, for the conveyance by France being now doubtful, I do not send four Letters I have ready; – but shall send them & perhaps some others, by a ship that will arrive probably in England after Michaelmas. The weather is now come in very fine but I see it has been a severe winter all over Europe. This place is famous for the goats wool, which is sent spun to England; & here they make fine camblet from 2 or 3 threds to 13 which last are very dear, they make also a very fine [...], something like our finest serges, which are dearer than the common camblets. These things are the only trade of this place.

I desire you to present my very kind love to my sister & due respects to all friends. Pray desire Mr Milles to continue to send me an Account of all Bills sent to Mr Hoare on my account from Ireland. I desire you to continue writing to me as usual, & Mr Maydwell will continue forwarding letters for me to Constantinople or Smyrna, till I desire the contrary. If Mr Milles is in London, I desire him to mention it to Mr Maydwell to forward them still to those places or rather to Smyrna if sent by way of France, or if you have occasion to write to him you may mention it so, 'tho my Letters to him were to send my Letters to Constantinople till further order.

I am Dear Madam your most obedient & dutiful Son, Richard Pococke

P.S. April 13th-24th. The messenger is deferr'd a day longer & I performed the morning service to the English.

CONSTANTINOPLE
31 May/11 June 1740
Pococke to his mother
Letter 40[140]

Honoured Madam,

I writ to you last from Angora on April 13th-24th some showers in the afternoon.

14th-25th. I took a walk out to see an inscription; went to visit Monsr. Jonce.

140 BL, Add MSS 22998.

15th-26th. Showers. I went with Mr Cary up to the castle; & saw an Armenian church & some antiquities, visited two people of the country; some French here in the afternoon.

16th-27th. I saw the manner of working the goats wool; we all dined at Mr Savages. I paid a visit. Writ.

17th-28th. Mr Franklin dined here. I went out in the afternoon to see antiquities & began to copy the Will of St Augustus, which you may see in Chisul [Chishull]; hoping to take a more perfect one of it.

18th-29th Showery – in the afternoon visited Signr. Perry & Monsr. Limagot.

19th-30th. Spent the whole day abroad about antiquities. Mr Cary came & dined with me at a Turks house.

20th May 1st. Went to see a relique at an Armenian church, an uncorrupted head of a child. Supposed to be one of the 12 martyred at the same time as St Clemens was; read prayers at home; the English & Mr Doyle [Doyley] a Dutch Gentleman dined here. I writ & walked out.

21st-2d. We went out about antiquities. We all dined at Mr Doyleys.

22d-3d. I went to search for antiquities, paid a visit. In the afternoon went with Mr Abbot & Mr Cary a league to their country house, which they call a vineyard, a very pretty improvement, & produces excellent red wine; pray send me Mr Milles's tour from Trent, through Germany & Flanders, in the two next Letters you write.[141]

23rd-4th. I went out to see some things. We all dined at Mr Savages.

24th-5th. I went out to see the market of the Angora wool & yards; rid out with Messrs Cary & Franklin; – & went to a garden half a mile from the Town, where all the English & French dined with the French Dr Romani.

25th-6th. I walked out in the morning & in the afternoon went to Mr Savage.

26th-7th. Walked out in the even, visited Monsr. Limagot.

27th-8th. I went out to copy a fragment; – read the service to the English, who spent the day here. I took a walk.

28th-9th. I visited the Archbishop of Angora, & all those who had been to see me & some returned my visit.

29th-10th. All came to see me I had visited among them the Archbishop of Angora. I set out accompanied by almost all the Franks, & we made a cavalcade of between 30 & 40; took some refreshment a mile off by a river:side & parted.

141 When recalled to Waterford in 1737, Jeremiah Milles took the following route through Germany: Innsbruck, Salzburg, Munich, Augsburg, Donauwörth, Nuremberg, Würzburg, Frankfurt, Mainz, Koblenz, Cologne, Dusseldorf, Aix-la-Chapelle, Liège, Spa Valenciennes, Lisle, Arras & Dunkirk (see Finnegan, *Letters* 2, p. 275). Pococke wishes to take the same route, which was considered to be the quickest way home. A similar request for this route is made in Letters 45 and 47.

Mr Cary & Mr Franklin going on with me not in the direct road to Stambole [Constantinople], but to the north in hopes of meeting with some antiquities; – we lay 12 miles from Angora.

30th-11th. We went about 20 miles, coming in between hills like Savoy, – saw a chalybeate Hot-Bath in the way, the houses here are built of firr trees, laid one on another & joined at the corners. Saw wild Goosberry trees.

May 1st-12th. Went 4 miles to a hotter & stronger chalybeatt bath; which has been frequented by the franks, & is among other things very good for the dropsy: we dined here & went a league further to a village, where I parted with Mr Cary & Mr Franklin & went on about five miles, still north; – I lay at a wooden village, in a vale beyond the mountains, had snow in passing them. Mr Cary a very pretty good tempered young Gentleman, his fathers sister married to Mr Charles Philips by whom he has an only daughter, a great fortune. Mr Abbot a very friendly valuable man, has read much, & is Octolinguis & writes Portuguese in Hebrew characters as the Jews do.

2d-13th. We are here in Paphlagonia, & setting out came into the great road, from Persia & Tocat to Constantinople & travelling about 16 miles, gaiting half an hour by the way in a more luxurious manner than usual, cold fowl & hung Beef: – & came to the house of a Bey a great Waivode or Governor of this country which belongs to the Waladea [Valide] or Sultaness mother to whom I had a letter. He was not at home, I was put in the rooms allotted for strangers, & when he came home I sent him my Letter & a small present; he was very civil but I found there were no antiquities to be seen.

3d-14th. Returned by another way into the great road & went on west through a pleasant woody country to Ghende, a small wooden town having travelled this day 14 miles, & before noon it began to rain; – I went to a Han, hired 2 horses for Nicomedia, in the way to Constantinople; I travel now in a better manner than formerly, having my own horse & Turkish furniture, wear a large Turkish surtout Habit, & a Green Tarter Carpack or cap like a Turbant, but instead of muslin round it is turned up with a Grey Persian Lambskin. So the Franks go in journey, as well as many Turks, so in the road they giving the Salaum alecum, & I returned it Alecum Saleame. He that hunts two hares looses [loses] one & the other runs away; a proverb.

4th-15th. Set out & travelled 16 miles through an exceeding pleasant country & came to a village in a beautiful vale; – where the man to whom the beasts belong lived. Was accommodated with a good house, & spreading my carpet in a litle grove by a stream, I spent the remainder of the day very agreeably: they make cords here of hemp, not beating it, but taking off the rind with their fingers.

5th-16th. Went 4 miles to Borla [Bolu], which I take to be old Bythineum [Bithynia] afterwards called Claudianopolis [Claudiopolis], where Antinous was born; the old town situated much like old Sarum; but the present on the sides of the hill, & a most beautiful country about it. I went to a stable Han, walked out for 2 or 3 hours; – an Armenian I had a Letter to came to me, & provided me with a room & entertained me at supper.

6th-17th. Travelled 20 miles mostly through pleasant woods of Beach & Hornbeam & variety of trees we have, except Elm & Birch & yew; – as apple, pear, medlar, box, a sort of dwarf Laurel, like the Roman laurel at Oxford, the [...] [next few words scored out], & a dwarf shrub with a pale green flower, like the Lelac [lilac]. Came to the river Hipsiles, & to a village; – where I suppose from some signs of antiquity was Prusa ad Hippilem. Went further & lay in a meadow on the banks of a river; – having laid by under shady walnut trees in the mid-day.

7th-18th. Went 6 miles & laid by 4 or 5 hours; – then went 8 miles further through delightful woods, of very fine tall straight oaks: – lay sub dio [uder the open sky] in a pleasant meadow.

8th-19th. Passed through a wood & came to a litle town in it, at the West end, called Hariadane; – came into the finest plain enclosed country, I ever saw, the fields having large walnut trees in them & hornbeam in a very beautiful manner, we travelled in that plain & passed the Sacari [Sakayra River] 120 paces broad, which I had passed 3 days before I came to Angora, & lay at a village.

9th-20th Went on the plain being divided by fine improved low hills, that make the country still more delightful, nothing I ever saw coming up to it, being a rich soil, not a stone in it, & is the ancient Bythinia. We stopped in a fine mead[ow]. I made Balm tea, with the herb that grows on each side of my carpet; & bought a carp 2ft. 8 inches long for 8d, taken in a stream near a large lake, & the Janizary who went to the Lake brought another somewhat less for about 5d. We went on 8 miles & lay in the open air.

10th-21st. Came six miles by sunrising to Ismit [Izmit] the old Nicomedia, built on the side of the hills & between the foot of them, & the sea & most beautiful situation: here all people & Goods from the East & south east embark for Constantinople. I walked round the town & after to a Greek church, & in the even to the Armenian church.

11th-22nd. I walked out to the Greek church, where I saw the tomb of St Pantalion. Set out about 10 towards Constantinople; – I went out of the road, to a mineral water of alum, & going near the sea, lay at a cane 10 miles from Ismit.

12th-23d. We set out & went by Gebseh a small town, stopt here to graze our cattle; an uneven country partly corn & partly pasturage, & but few trees, tho' naturally running into wood. Passed by Pandeh [Pandik], the old Pandichio, within 10 miles of Scutari [Üsküdar] & went a mile further & lay abroad.

13th-24th. I set out at sun rise, passed through a litle town called Corsol, passed near Chalcedon [Kadikoy] & arrived as Scutari, a large fine situated town opposite to Constantinople.[142] I went to a Han, writ a Letter to Mr John Lisle, who came over to me with his Janizary & Droggerman & showed me the way to his house from Asia into Europe. His Brother Richard is his partner. Mr Levett came to see me:[143] all the English & one or two more, sup'd at his house.

14th-25th. Breakfasted together & Captn. Crispin I came with from Tripoly to Cyprus.[144] Spent the morn in settling my affairs. I had the pleasure on my arrival here of receiving your two Letters of Decr. 30th 1739 & Feb. 17th. 39/40, which I shall defer answering till I receive the great cargo of letters, which came too late for me at Smyrna, & which I expect here every day, only let you know that the small bottle of Balm of Gilean, which is a white one, is to be sent to Ireland: – half of it to my Lord & half to Captn. Ramsey, & it would be better to put it into two bottles, in a white thick flint bottle, & the cork tied down with a bladder.[145] Mrs Lypeat will tell you the virtues of it. That in the black bottle is also Balm of Gilead, which keep by you, & use if occasion, but not part with any unless to a very good particular friend, on an urgent occasion. I received also Mr Milles's of Feb. 11th & 21st, for which I thank him, & shall answer all together when I receive the next. The beauty of this town in its view every way, & in the prospect it affords at a distance is beyond anything I could imagine, & certainly is the most beautiful situation, that & Scutari being on rising grounds; – many litle gardens to the houses with cypress & other high trees in them add greatly to the beauty. There is no plague here, & my Barometer is at temperate air; – having been a very cold winter everywhere. In a month I shall leave this city. Messrs. Barker, Jennings & Deth [D'Aeth] came to visit me in the afternoon.[146] Mr Lisle went with me to wait on the Marquis de Villeneus the French Ambassador, who has had the honour to make peace between the Turks the Emperor & the Muscovite. After we went to see the Swedish Ministers, Baron Hopkins & Mr Casclou, the later only I saw, he is a very pretty young Gentleman

142 Pococke's map of the Thracian Boshphorus is reproduced here as Plate 22.

143 Francis Levett (1700-64) was the chief representative of the Levant Company in Constantinople from 1737-50 and sat for Jean- Étienne Liotard in Turkish costume at least twice. The miniature sketch for one, in the Louvre (Fonds des Dessins et Miniatures, Louvre, RF 1377, Recto) is described in the inventory as, "Portrait de M. Levett, Négociant Anglais, en Costume Tartare".

144 See Letter 22.

145 Similar instructions were issued in Letter 22.

146 Benjamin (?) Barker and Thomas D'Aeth are recorded as being the Radcliffe agents in Constantinople during the mid 1730s. See Davis, *Aleppo and Devonshire Square*, pp. 201-2. A James Jennings was a Treasurer of the Levant Company from 1735-36. See Wilson, *List of British Consular Officials*, p. 11.

talks English & was in Egypt. They are building a fine palace. I then went to see Mrs Burgoin, Consul Paradises mother & his two sisters Miss Burgoins:[147] & after to see Miss Savages, – sisters of Mr Savage in Angora, born here, of an English family. Archdeacon Pain came to Town from the Ambassadors country house, & came to see me.[148] All the English sup'd here. Mr Barker & Mr [Thomas] Deth [D'Aeth][149] are one house, Mr Levett another, Mr Jennings a 4th, & Mr Pemberton a 5th.

15th-26th. I went with Mr Lisle in his chair, to visit the Ambassador in the country but was turned back by the rain. The Archdeacon came to see me. I writ some Letters to Smyrna & Angora. We all dined at Mr Barkers. I went with Mr Pain to visit the Venetian Ambassador Eritzo [Erizo?]. We all sup'd at Mr Barkers.

16th-27th. Mr Pemberton came to see me, who had formerly a great house of trade here. We all supt at Mr Levetts.

17th-28th. In the afternoon I went north west to a village called Belgrade [Belgrad Forest][150] with Mr Lisle where the English, Swedes & Dutch have their country houses, to wait on the Ambassador [Sir Everard Fawkener], who is a person of great sense & a most condescending behaviour in every respect, a very fine Gentleman & universally esteemed & beloved.[151] In the evening was a great Assembly, of the Dutch Ambassador & the Swedish minister &c: & a concert. The Ambassadors house being full, I lay at Mr Levetts.

18th-29th. Mr Falkener the nephew & Mr Lisle breakfasted. I waited on the Dutch Ambassador & Swedish minister; – & walk'd out with our Ambassador: – the above minister & much company dined with him. Archdeacon Pain returned the 12 miles as it is called, with me to town, tho' it is not above 8. We all sup'd at Mr Barkers.

19th-30th. I went to the English palace to Archdeacon Pain; – then visited the Muscovite minister, & after Baron Hopkins the Swedish minister & Mr Pemberton. [Word scored out] I dined with the French Ambassador, met Baron Wallis there of Ireland, nephew to General Wallis,[152] one of the 6 German

147 Pococke had already met their half-brother, aged 15, in Chania. See Letter 32.

148 Revd Thomas Payne, whose name is spelled variously in the letters, was Chaplain to the English Factory at Constantinople in 1718, 1725 and 1733-36. See Wilson, *List of British Consular Officials*, p. 12.

149 Christine Laidlaw, *The British in the Levant*, 209.

150 This village was named after the capture of Belgrade (the Serbian village in Hungary) in 1521 by Suleiman I.

151 For a brief account of this man, see Introduction.

152 George Olivier, Count of Wallis (1671-1743), the young man's uncle, was a field master of Irish descent who had lost the Battle of Grocka or Krotzka (1739) fought between the Austrians and the Ottoman Empire, which had led to the Peace of Belgrade.

hostages for the surrender of Belgrade [Serbia] & Mr Lesley son of General Lesley in the Muscovit service, a brave youth about 20, taken by the Cham of Tartary & treated well by him in his house & it is said was in love with his daughter & she with him. His father killed in the action. We went on the water, up the river a league to Sadatat a country seraglio of the Grand Signr. Saw it within & the womens apartment a canal to it 1600 paces long & 37 broad. We saw the Grand Vizier come & go into his Barge: [We] had with us the Swedish Chaplain; the English sup'd here.

20th-31st. I went at two to wait on the French Ambassador, or to go with him to Audience of the Grand Signior, saw all, but the Grand Signior, the numbers of person being provided before: at 3 the Ambassador preceded by some Turkish officers & by his D[roggerman]s & domesticks, was carried in a very fine chair down to the water side, all the nation following him; we crossed over in boats to Constantinople, where we were all mounted on the Grand Signiors horses to go to the Seraglio, stopped to let the Grand Vizier pass clothed in white Sattin, we then went on, mounted to the Gate of the 2d court of the Seraglio, Janizaries marching all the way on each side of the Ambassador, & with them other officers both the courts were lined, waiting at the 2d gate the Janizaries went in, & the Ambassador going in on foot: – the Janizaries, the signal being given ran to take their Pilaw, set for them on the grass -plat in dishes & then retired to one side & eat it: – this inner court divided into grass plats is planted with trees. We went into the Divan where the Grand Vizier sat in the middle, they spread a gold flowered handkerchief in his lap; – the Ambassador sat on a stool, & in the Divan the two cadelishiers of Romely & Anatole & the Keeper of the Seal & Treasurer: – Petitions were read & signed by the Grand Vizier, the Petioners being severally brought before him: the Grand Signior, not seen, is at a Lattice window over the Vizier. All being finished a stool was put to the 5 persons in the Divan, & covered with a dish as big as a table; – on which was placed a great number of dishes, one by one, before 6 a clock. The Ambassador eat with the Vizier & 5 Gentlemen with the Keeper of the Seal, & 5 with the Treasurer, after all washed & were perfumed & incensed: – Then the Firman was brought in of the Grand Seignr., being the order to introduce the Ambassador to have the treaty renewed with the Port: we all went out were clothed with such customs, as I sent you from Egypt; & after a while the Grand Vizier went in state to the room of Audience of the Grand Signior; – & then the Ambassador & 12 other persons were introduced by a chious on each side of them; I saw them thus led in. The Droggerman of the Port speaks what the Ambassador has to say, & then the kings letter, if any, is given, & they soon retire. We returned as we came; – waiting for the Grand Vizier & other great officers to pass. I waited on the Ambassador, who invited me to dine. Saw some drawings at the room of Count Guicciardi a modonese & one of the hostages. Archdeacon Pain came to me; writ.

May 21st – June 1st. Set out early with Archdeacon Pain & Mr Graltow a Swede, & went to St Sophia & walked all about it for near 3 hours; – went over to Chalcedon saw the church in which the council was held: – then saw some sepulchres of sultans, & then went into the fine mosque of Sultan Achmet; saw the wild beasts, & the burnt Pillar, & dined at the Venetian Ambassadors. Visited a French Gentleman, saw his Library & medals: visited Mr Lister & then Mrs Burgoin.

22nd-2nd. We rowed along the Ports & east side of Constantinople, & walked along the south west side, the later [latter] 4 miles in all 12 round. Pera & Galata &c: excepted out. Saw a Greek church & a building called Constantines palace, but not so old; dined at the Dutch Ambassadors, where the English Ambassador also dined, visited [Antonio] Pisani[153] the Droggerman & then went to Archdeacon Paynes, where Mr Gonnet the Dutch Chaplain, Messrs Lister & Pemberton came. Sup'd at Mr Barkers.

23d-3d. We went out with Col. Rusigni a Venetian; – saw the Pedestal & base of the Historical pillar, which is all that remains; – saw Marcians pillar: – the finest mosque of all called Solimonea, exceedingly magnificent; walked about in it, tho' the women & many were come to hear the sermon. Saw the madhouse, & tho' the only one for Turks, but [only] 14 in it. We dined on chops at an Armenians, saw the Armenian Press & a most magnificent cistern under ground, came home by the Atmedan, sup'd at Mr Listers.

24th-4th. Went out & found nobody at home. Archdeacon Payne came to see me. Rainy. – I writ. Visited Mr Ips a German merchant: – went to the Palace with Mr Payn, all sup'd here.

25th-5th. I writ all day. The Dutch chaplain came to see me. We sup'd at Mr Barkers.

26th-6th. Breakfasted at the Palace with Mr Mackenzy [Dr McKenzie].[154] I hope you will be looking out for a servant for me.[155] I went & saw a beautiful seraglio of the Grand Signiors.

---

153 On the Pisani family, see Frank Castiglione, "'Levantine' Dragomans in Nineteenth Century Istanbul: The Pisanis, the British, and Issues of Subjecthood", *Osmanlı Araştırmaları* 44 (2014), pp. 169-195.

154 Dr Murdoch McKenzie was the Levant Company physician to the British community in Constantinople. See van den Boogert, "Freemasonry in Eighteenth-century Izmir?", p. 117. Some of his letters from Istanbul to John Clephane were published in *Philosophical Transactions of the Royal Society* vol. 47 (1752): LXIII: "Extracts of Several Letters of Mordach Mackenzie, M.D., Concerning the Plague at Constantinople".

155 His mother's most recent letter to him must have confirmed that Jack (see Letter 33) was not taking up his offer of employment in Ireland.

27th-7th. I went to see Archdn. Payn; dined at home. Visited Mr Abbot & the Bishop of a Roman, went to Mr Payns; – all sup'd at our house.

28th-8th. Went to Archdn. Payn & to Mr Jennings; din'd at Mr Barkers.

29th.-9th. Visited Col. Rusigni & Mr Ips; – went to Mr Payns, – visted Mr Pemberton.

30th-10th. I went over to Asia, to see the city from several places, entered the mosque Waladea & went up the minaret.

31st-11th. I purpose to stay a few days with the Ambassador at Belgrade, & shall leave the place soon. Pray my very kind love to my sister.

I am Dear Madam your most obedient & dutiful Son, Richard Pococke

CONSTANTINOPLE
19/30 June 1740
Pococke to his mother
Letter 41[156]

Honoured Madam,

I writ to you last from Constantinople on May 31st June 11th – the day before I had been over into Asia, a litle north of Scutari, & had a fine view of Constantinople from two hills; – we thence went behind Scutari near to Chalcedon, & crossing away to a seraglio of the Grand Signiors south of Scutari, we went to Scutari, & dined & then walked out to the south towards Chalcedon & saw some of the most beautiful views in the world; & went round on the outside of the Grand Signiors seraglio of Scutari over the sea at the Boshorus [Bosphorus]: – we walked about Scutari, & went & saw the finest mosque & ascended the minaret, & had a most beautiful view of Scutari, which is like a Theater situated round hills which make a semicircle; – & in the area between, all the litle yards & gardens, having trees planted in them, it looks like a city in a wood, which is the great beauty also of Constantinople, & of the places about it: we came to Mr Payns room & drank tea, & I came home & writ – we all sup'd at Mr Levetts.

June 1st-12th. I went to see the Dutch Chaplain – & he afterwards came to me, & to his lodging, I went & met Mr Payn; – we called on Col. Rusigni – & went in a boat about halfway towards the Black Sea, to the house of Count Bonneval, so famous all over Europe, who was first an officer of France, – then a General in the Emperors service, & is now a Pasha about 70 years old, & a fine

156 BL, Add MSS 22998.

old Gentleman of great parts & Learning, particularly in History, having a most surprising memory, is most agreeable & entertaining in conversation:[157] – we dined with him; – having sent beforehand to desire to be permitted to visit him; & it was intimated that we should come & dine: – our discourse was in his mother Tongue the French; & I left him highy satisfied, – & went near to the Black Sea to Boyucderry,[158] where the French & Venetian Ambassadors have houses, – whence we went about 4 miles west to a village called Belgrade, situated in a wood to the house of Sr Everard Falkner he English Ambassador.[159] The Dutch & Swedish Ambassadors have houses here & some Franks. His Excellency [Sir Everard Fawkener] who was rid out, came home, a Gentleman of great good sence, politeness & affability, & most deservedly esteem'd & beloved by all, & a great medallist.

2d-12th.[160] Walked out & called at the Swedes, & at Mr Levetts. Prayers were read at the Ambassadors. Walked out with the Ambassador & Sweeds & much company for 3 hours in the woods: the Swedes sup'd with us.

2d-13th. I rid out with Mr Payn to see the magnificent Aqueducts a league off, returned to dinner. Walked out in the even, – the Swedes sup'd with us.

3d.-14th. Visited the Swedes; – walked out with A. Payn & a Jesuit. In the afternoon we rid out 5 miles with the Ambassador, to the Black Sea, which is not so salt as the other seas, we returned by what the Franks call Ovids Tower.

4th-15th. Spent part of the morn in conversation with the Ambassador about antiquities medals &c: walked out with Mr A. Payn; – after dinner set out for the city with Dr Mackenzy the Ambassadors Physican: – we passed by the Okmeidan [Okmeydanı], a fine down where the Turks go out to shoot with bows, being over the port opposite to Constantinople & from thence there is a fine prospect of the city: – we came to the Ambassadors house & the Dr came with me, & we all sup'd at Mr Lisles.

157 Count Claude Alexandre du Bonneval (1675-1747) was a French army officer who converted to Islam under the name Ahmed Basha. He sat for a portrait for Liotard in Turkish dress, the miniature sketch for which is in the Louvre (Fonds des Dessins et Miniatures, Louvre, RF 1387, Recto).

158 Büyükdere, now a part of the Sarıyar district of the Province of Istanbul, is located on the European shore of the Bosporus. In the 18th century it was a popular retreat among European residents of the Ottoman capital.

159 This village was known for its great beauty and tranquillity. Lady Mary Wortley-Montagu had spent time there twenty years earlier, when her husband Edward was English Ambassador to Constantinople, and described it as "perfectly [answering] the description of the Elysian fields". Quoted in Wood, *A History of the Levant Company*, p. 243.

160 Pococke or his mother gets muddled in the dating at this point, which it is impossible to correct.

5th-16th. I went to the Palace, & we dined & sup'd at Mr Barkers. I paid some visits in the afternoon.

6th-17th. I went with Dr Mackenzy & dined at the Swedish Envoys. I visited one Mr Herbert a gentleman of Ireland who is a merchant here of small things; – & then visited Mrs Parmentier, – & after the Dutch chaplain.

7th-18th. Two days ago there were rejoicings for the surrendry of Belgrade [Serbia]; in the afternoon, I set out with the Dr for Belgrade; it began to rain yesterday, having been a very hot day. We arrived at the Ambassadors house, where the English that come out here sup'd.

8th-19th. Breakfasted with his Excellency; write; walked out with the Ambassador; dined. We all waited on the Ambassador 4 miles to Boiyucderry, to see the French Ambassador; – where we had a collation in the garden & returned.

9th-20th. I set out with A.D.n [Archdeacon] Payn & Dr Mackenzy to the west about 15 miles along by the Euxine Sea; – stopt at a village off by the sea: – which I take to have been Philea; returned to the village.

10th-21st. We returned to Belgrade, lying by at a village during the heat of the day & came to Belgrade at night.

11th-22d. Spent part of the morn with the Ambassadors; Count Guicciardi & Baron Wallis, two of the German hostages, dined here. Walked out with the Ambassador & Swedish minister & joined the Dutch Ambassador, & returned.

12th-23d. I came with our Ambassador to town. Went & waited on the Ambassador; came home. Settled some affairs. We all dined at Mr Barkers. Father Bernardino came to see me, – whom I had met in upper Egypt. We all sup'd at Mr Barkers. Miss Sally Lisle is to be marryed to a Turkey merchant; I hear great Balls & entertainments were given this winter, to the expence of 500 a night: – it was proposed to Lord Romney to give one: – who answered "that the many objects of misery pointed out to him, how he ought to dispose of his money, that he did not doubt but that others had enough to divert themselves & to succour the distressed; – but as he had not, he thought it his duty & it was his pleasure to dispose of his superfluities, in assisting his Fellow creatures." Lord Shanon was so charmed with his virtue & good sense, that it is said, he immediately offer'd him his daughter, & with a good fortune.[161]

13th-24th. The English Ambassador & nation dined here. I visited Consul Paradises mother. We all sup'd at Mr Levetts.

14th-25th. The Ambassador & nation dined at Mr Levetts. I waited on the Ambassador to his palace, who showed me his medals, a very fine collection;

161 This was the Irish nobleman and politician Henry Boyle, 1st Earl of Shannon (1682-1764).

his excellency presented me with three fine large Greek medals; all sup'd at our house.

15th-26th. The Ambassador & Count Solm dined at our house. I took a walk round Galata.

16th-27th. I went down to Galata with Mr Lisle to his warehouse; – we also went to see the convent of the Capuchins, Dominicans & Jesuits: – & went up to the top of the high round tower of Galata; dined at home. Rid out 5 hours with Mr Lisle to see some fine prospects of the Boshorus; returned; & all sup'd at Mr Barkers.

17th-28th. We went to wait on the Ambassasdor to accompany his Excellency to complement the new Vizier; – the old one being put out of his office: the Procession began to the water side on foot two & two; – first six Janizaries, then 18 servants in the white habit of the country, – then the maitre D'Hotel followed by 10 servants in blew Liveries, richy laced with silver lace, 4 Heydukes & gentlemen of his Excellency in Laced cloths, the Droggerman, his Excellency followed by the Gentlemen & merchants of this nation, passing the water to Constantinople. The English shipping fired, as they did at the return, on the other side of the water. His Excellency mounted on horseback & was preceded by his master of the horse, & before the Livery servants were led 2 or 3 of his Excellencies fine horses in Turkish furniture & as many in our furniture. We came to the Visiers Palace, & were first led into a room, where the Droggerman of the Port a Christian Prince of Moldavia, received his Excellency, & sat a while, & then were led into the room of Audience, the Vizier after a while came, preceeded by several officers; He sat on the sopha & the Ambassador faceing him on a stool, when he come in an officer cryed out Alecum Salam! which is a welcome, & then a great many persons cryed out, Long live the Grand Signior! the ambassadory the Droggerman of the port complimented the Vizier, & mutual compliments passed, an embroidered Handkerchief was laid in their laps; & sweetmeats brought, first to the Vizier & then to the Ambassador; & afterwards Coffee, when we withdrew; & very soon after the Ambassador, as I was informed, being incensed, came all & we returned as we came. The presents made the Vizier & other charges, cost the company three purses of money, each of which is about £700.[162] As soon as we came to the Palace, the Ambassador presented me with the Viziers Present to him; which was a crimson gause [gauze] handkerchief, with a large border of silver, which you must wear, when it comes home; & a muslin pocket handkerchief, an oblong square, worked at each end with gold flowers. I went with Mr Lisle by invitation to a great entertainment of a Dutch house, where was Mr Barkers

162 A vast fortune, considering Pococke's substantial income of £300 per annum.

family, all the Dutch, & some French to the number of about 40 Gentlemen & Ladies. After dinner was dancing I withdrew at night; most of the company staid late. I sup'd at Mr Levetts.

18th-29th. I do think it a litle hard, that my last letters from England were of the middle of February, now that the way is open by Vienna, & letters have arrived from England of the 24th of April – & news of the third of May. The Dutch Chaplain dined with us: – I went & spent the afternoon at his Lodgings with Mr Leotar [Liotard];[163] the English sup'd at our house. An opportunity now offers of sending this by Vienna. I propose very soon to leave this place, but would have my letters sent hither till I desire the contrary. My Lord Granby, eldest son of the Duke of Rutland is daily expected from Belgrade [Serbia];[164] – & the Emperors Embassador will probably be here in August. I cannot omit mentioning to you the friendly civilities I receive from Mr Lisle, & the favours of his Excellency Sr Everard Falkener is pleased to show me, which together with the civilities I receive from others, make this place very agreeable: – there being not the least infection here: & the beautiful prospects it affords, are a continual delight to the eye. Before I depart I shall spend a few more days in the country at his Excellencies. There are here all the summer refreshing north winds which render the climate very temperate. So that if there were a good Government here & true Religion settled; – & proper measures taken to prevent & put a stop to the infection of the plague; – it would be one of the finest places in the world to live in.

I desire you to present my very kind love to my sister, & due respects at Whiteoak & all friends.

I am Dear Madam your most obedient & dutiful Son, Richard Pococke

163 This is the first of the three known meetings Pococke had with the Swiss portrait painter.

164 Lieutenant-General John Manners, Marquess of Granby (1721-70) was the eldest son of the 3rd Duke of Rutland. He became a member of the London Divan Club, along with his tutor, Mr Hewitt (see Letter 44, below) but his considerable military demands took him away from London and as a result he did not attend regularly at its meetings. See Finnegan, "The Divan Club", p. 49.

CONSTANTINOPLE
22 June/3 July 1740
Pococke to his mother
Letter 42[165]

Honoured Madam,

My last Letter I sent you was from Constantinople on June 19th-30th, on which day I went to see a sort of mahometan devotion in a mosque of singing & dancing in a circle in a most extravagant manner. We sup'd at Mr Barkers.

June 20th-July 1st. I went to see the Grand Signior go to the mosque; – he is a very handsome man & has a good face. I came home; paid visits at two convents to Fathers I had met in other parts. Visited Consul Paradices mother. The Gentlemen sup'd with us.

21st-2d. This day I had the pleasure to receive my great packet of Letters that has been at Salonica & Smyrna; – among them, what gave me the greatest satisfaction your four letters of Oct. 26th. & Decr. 24th. 1738, March 8th. 1738/39 & May 17. 1739/40 so that I suppose the two letters I received here before of Decr. & Feb. were those between the two Letters I received from you in Smyrna of May & Decr. 1739, which I have not had by me.[166] The letters you observe smelling of brimstone, are smoaked with sulphur against any infection. Mr Pullinger was of Petersfield. You are generally to take Eastern beggars for cheats. I still wear my Whiskers, 'tho sometimes I wear a wig. It would be very well, if you could get anybody to take that Bond & security of [off] you, & to give you the money, that is to transfer the Bond to another person.

I very much rejoice at my Coz. Jermy's [Jeremiah's] good fortune, in being taken more immediately under the patronage of that noble family, which has been so great a Benefactor to us all:[167] & I do heartily congratulate him on it; & I do not in the least doubt, but he will behave in such a manner as to deferre any favour that may be showed him.

The English sup'd at hour house.

22d -3rd. I set out at 5 with Mr Lisle , to go a new way to Belgrade, by the Aqueduct & arrived there in 4 hours & ½. I writ. We had service; & all the

165 BL, Add MSS 22998.

166 The dates show that Mrs Pococke wrote to her son approximately every two months; and the paragraphs to follow answer her queries and comment on her news.

167 It is not clear which member of the Herbert family Pococke refers to: either Herbert, 9th Earl of Pembroke, 6th Earl of Montgomery (1693-1750) or Hon. Robert Sawyer Herbert (1693-1769), who inherited Highclere Castle under the will of his maternal grandfather, Sir Robert Sawyer (*d.* 1692).

English dined here. I propose & in about a week set forward in my journey; in a ship going to Marseilles I send this, tho' another [was] sent by Vienna on the 19th. I desire you to present my kind love to my sister; respects to Whiteoak & service to all friends.

I am Dear Madam your most obedient & dutiful Son, Richard Pococke

CONSTANTINOPLE
7/18 July 1740
Pococke to his mother
Letter 43[168]

Honoured Madam,

I writ to you last by Marseilles from Constantinople on June 22d-July 3d, having sent another to you a few days before by Vienna; that day I went with Mr Lisle to Ambassador Falkeners house at Belgrade. We had service: & all the English that were out dined at his Excellencies; in the afternoon we walked out with the Ambassador.

23d-4th. I rid out with Mr Falkener the Ambassadors nephew, Mr Payn, Mr Gonnet[169] & the Jesuit to the mouth of the Boshorus at the black sea 3 hours off; & saw on a rocky Isle, one of the Janinian rocks, what is called Pompeys Pillar. We dined there, returne'd, found the Muscovite ministers in a visit to the Ambassador.

24th-5th. I writ, walked out with A.deacon Payn; – after dinner rid out with the Dr.

25th-6th. Went out with Mr Pain to Boycuderry: – We crossed over into Asia, & went up a hill[170] to see what is called Nimrods tomb 44 feet long & 4 broad. We descended to the summer house of the Grand Signior in a vale called Jacat, had a very pleasant walk to the water; – & returned to Boyucderry, where we all dined at the French Ambassadors, being in invitation to the English Ambassador; after dinner went with A.deacon Payn to wait on the Venetian Ambassador, came to Constantinople in a boat with the Superior of the Jesuits, & paid a visit in the way to the Minister of the King of Naples. The English sup'd with us. When you abridge this, put in this Paragraph, in 1738 Lord Sandwich, Messrs.

168 BL, Add MSS 22998.

169 Jean Gonnet, Minister to the Dutch ambassador in Istanbul from 1734 to 1743. Schutte, *Repertorium*, 317.

170 This is perhaps what is now called Yuşa Hazretleri Tepesi (Joshua's hill). Nimrod's tomb is generally believed to be located in southeast Turkey or in Iraq.

Ponsonby Nelthorp & Mackey came by sea from Leghorn to Constantinople & returned thither again; & in 1739 Lord Sandwich embarked again at Leghorn, & went to Egypt, but soon returned [next word scored out], arriving there about the month of July.[171]

26th-7th. Writ & settled affairs, dined at Mr Barkers. Went over to Chalcedon; walked about the site of that ancient city, & thence went to Scutari & crossed over. Sup'd at Mr Barkers.

27th-8th. I went with the Janizary & my servant over to Constantinople; saw an old church near the seven towers, now a mosque, 14 fine verd antique pillars in it; – went to the seven towers, to a Greek & Armenian church, to 2 churches now mosques & to Marcians pillar; returned. Archdeacon Payn came to see me; – I went with him to take leave of the Muscovite & Swedish ministers & the Roman Bishop of Santorini. We all sup'd at Mr Levetts.

28th-9th. The Bishop of Santorini came to see me. Afterwards Mr Payn came & we went to wait on the Patriarch of Constantinople who was not well. We then visited the Patriarch of Jerusalem, who talks the old Greek. Saw the handsom Library, we went to the Turkish printing house, where they are printing again, after 4 years intermission. We returned; dined all at our house, & some Dutch. Set out with Mr Pain for Belgrade, overtook Mr Listers family at a collation, staid with them; – & went on a roundabout way & arrived at his Excellencies.

29th-10th. We walked, & after dinner returned with Mr Pain to Constantinople, we sup'd at Mr Barkers.

30th-11th. Went with Mr Lisle over [to] Constantinople to see the flower market chiefly of ordinary carnations, some in pots & some cutt; returned. I made some visits & writ. We sup'd at Mr Levetts.

July 1st-12th. I set out with Mr Jennings, Mr Dath [D'Aeth] & Mr Leotar to go to the Princes Islands;[172] we arrived at the furthest of them called Princes Isle [Büyükada], 30 miles off, & went to a house borrowed of a French merchant, where we dined, & then rid to three convents on the Isle, which is about 5 miles round.

2d-13th. We went to the Isle of Halthi Χαλχη [Heybeliada], half a mile west of it about 4 miles in compass to a litle town there is on it, as that also on the other Island: – this is the pleasanter Isle, & we went & saw the three pleasant convents on it, where we lay.

171 This is contrary to the news conveyed in Letter 22.

172 As suggested in Chapter 2, though there is no reference to it in the correspondence, it was possibly during this trip that Liotard drew the miniature sketch for Pococke's famous portrait (Fonds des Dessins et Miniatures, Louvre, RF 1379, Recto) depicted on the cover of this volume.

3d-14th. We set out on our return, west of this is the Isle Antigone [Burgazada], with a litle town & one convent; – & west of that Prote [Kınalıada], with only a convent on it. We went ashore at the Farar point East of Chalcedon] in Asia, to see a ruinous summer house of the Grand Seigniors; & arrived safe at Mr Lisles; dined at Mr Barkers, & went to wait on Sr Everard Falkener, who was in town, came home, writ & settled some affairs. Visited the Father I met in upper Egypt [Father Bernardino]. Sup'd at Mr Barkers.

4th-15th. I went over to Constantinople & visited the Patriach; – & after Ibrahim Effendi, who manages the Turkish press, which is now going on again, with a History of the Turks:[173] – I went to the English Palace, & then visited Mr Pemberton.

5th-16th. I went over to Constantinople & saw churches & mosques, in the afternoon paid some visits of conge [leave taking/farewell].

6th-17th. I went with Mr Lisle in his chaise to Belgrade, to take leave of his Excellency Sr Everard Falkener, & all there waited on the Dutch Ambassador.

7th-18th. Packed up & settled affairs for my departure, this day, for Adrianople [Edirne], thence I shall come back to the Dardanelles, see some places that way, return to 'em again, & travel over Greece by Land.

My kindest love to my sister & respects to all friends.

I am Dear Madam your most obedient & dutiful Son, Richard Pocock

BRUSA
17/28 August 1740
Pococke to his mother
Letter 44[174]

Honoured Madam,

On my departure from Constantinople on 7th-18th of I writ to you by way of Vienna; – that morning I was visited by the Egyptian Father, Signr. Pisani, Mr Payn, Dr Mackenzy, Mr Leotar & Mr Herbert the 4 latter dined with us; – & about 3 I took leave; – crossed the water to the West side of Constantinople &

173 This is *Ibrahim Müteferrika* (1674–1745), the first Muslim to direct a printing press with movable Arabic type. The work referred to is the *Tarih-i Çelebizade* (1741), the first publication after the press had been closed for some years. Maurits H. van den Boogert, "The Sultan's Answer to the Medici Press? Ibrahim Müteferrika's Printing House in Istanbul", in Alastair Hamilton et al. (eds), *The Republic of Letters and the Levant*, Leiden, 2005, 265-292.

174 BL, Add MSS 22998.

mounted to go to Adrianople, we went 5 miles to the litle bridge & staid 2 hours, about 9 set out & went about 6 miles, & lay in the fields.

8th-19th. In a mile we came to the great bridge, & 12 miles further to Selivree [Silivri] the road being all the way, by, or near the seaside; at this ancient Selgimbra [Selymbria], I saw the old walls from whence probably went the wall across the Thracian Bosphorus, I think to Salmidissus on the black sea, built by the Emperor Anastasius. I saw also the Greek & Armenian churches, which are both old, we set out at 6 & went 10 miles & lay abroad.

9th-20th. Set out & went 16 miles & lay in the meadows.

10th-21st. Went 4 miles to Bourgass, where I spent the day in a stable cane, in eating, sleeping & writing; – till 3 a clock when we passed by Baba 8 miles further a small country town, – & went 8 miles further & lay in the fields.

11th-22nd. Went 4 miles & passed through another litle town called Hapsa, & in 8 miles further arrived at Adrianople. Went to a cane. Mr Damerate,[175] who was bread at Worcester [College] in Oxford, to be made a Priest or a Druggerman being a Smyrniote & is a merchant here came to me, & pressed me to go to his house, having received letters that I was coming with this caravan, & that I had letters to it, so I dined & moved to his house.

12th-23d. The French Consul came to see me, I did not go out all day.

13th-24th. I went with Mr Damerate, saw a very fine mosque & several parts of the town, & visited the French Consul, Mr Palmoniere[176] dined here a merchant. I walked out with Mr Damerate [d'Amirat?] to the rivers, called on the French Consul.

14th-25th. After dinner I visited Mr Palmoniere; – & then rid out with Mr Damerate a league off to a summer house of the Grand Signiors, from which hither is a very fine view.

15th-26th. We went to the French Consuls house, & then to the skirts of the town, to see the entrance of the Imperial Embassador. After dinner we went again to the French Consuls to wait on Count Ulefel [Leo Graf Uhlfeldt] the Emperors Embassador extraordinary. After that I waited on the Marquise of Granby, the Duke of Rutlands eldest son, who came from Belgrade with the Ambassador & Mr Huit [Hewett], a person towards 50 who has the care of his

175 In the records of Oxford's short-lived 'Greek College" the name of this alumnus is given as Michael Damiral. E.D. Tappe, 'The Greek College at Oxford, 1699-1705', in *Oxoniensia* 19 (1954), 92–111, 105.

176 Neither Palmonière, nor the French consulate in Bursa is listed in Mézin, *Les Consuls de France*.

lordship;[177] – my Lord is a pretty affable gentleman without any pride or state, – was of Eton & is not I suppose 20 years old.

16th-27th. Mr Hewet came here & writ Letters. I spent most of the afternoon with Lord Granby. Visited the French Consul & Mr Palmoniere.

17th-28th. I set out to the south & went 12 miles to Demotica [Didymóteicho], the old Dyma, I suppose, having with me a Janizary, an Emir or Prince of the family of mahomed, with his Green Turbant; in the even I viewed the remains of the old town on the hill & saw the house in which Charles XII of Sweden lived, & from which he went when he returned home.[178]

18th-29th. We crossed the Maritchen [Maritsa], the old Hebrus, in a boat & went 6 miles to a town called Ouzun Cupri [Uzunköprü], where we lay by, & after went 12 miles to Hieropoli, where we lay.

19th-30th. We went 8 miles & dined at a village – & then went 8 more to Rodosto [Tekirdağ] on the Propontis or White Sea, being the Old Bisanthe after called Resiston & Radestum [Raidestos]. I went to a cane.

20th-31st. I went & saw a Greek & Armenian church; – & in the afternoon embarked for Gallipoli – sailed all night.

21st-Aug. 1st. We put in on the Asia side at Limen much whereabouts Parium was in Mysia, & the ancient kingdom of Troy, to drop some passengers: – I went ashore & took Coffee, & we sailed again & arrived at Gallipoli – I went to a cane, & the Jew Droggerman of the French nation came to offer his service, & to let me know the French Consul of the Dardenels [Dardenelles] was here, but not at home. I went & saw a Greek church, where I met a Father I had seen at Mount Sinai. I then went to the church, where the suffragan Bishop of the Archbishop of Heraclea resides, who invited me to come into his Apartments, & entertained me with Coffee, & showed me the church with much civility & invited me to come again, if I staid or returned to this place – I dined, – walked to the old castle & port & then carried Letters I had to the French Consul Monsr. Le Fabre.

22d-2d. The French Dr came to invite me to dine with him, the French Consul being to dine with him also; – I went according to invitation; – & drank tea with the Consul. It is my good fortune to have a very cool summer, to travel in as it continues to be at present, & now the summer is almost past. The Consul called on me & we dined at the Doctors. I went with the Consul to his house, & staid a while & then came home. In the even walked round the town. When

177 William Hewitt (b. 1693) was tutor to Lord Granby during his trip to Constantinople and on his return became a member of the London Divan Club. He attended the meetings regularly, serving once as Reis Effendi and four times as Vizier. See Finnegan"The Divan Club", pp. 56-57.

178 He was exiled to this place from 1713-14.

I am fixed in any place I wear a furr Bosnack cap, green at top, like the Hungarian Cap, which commands respect.

23d-3d. The French Droggerman came to see me. I went to the port, & at noon according to invitation to dine with ye French Consul, at the Doctors, came home walked to the Port & to the Consuls.

24th-4th. Embarked on a Greek vessel, with the Consul, for the Dardanelles; arrived there on Asia side in 4 hours; – went to the Consuls house, where the English Droggerman came, – & I gave him a letter from Mr Lisle; & he would have me take up my lodging at his house, – so I dined, & went to his house, took a walk.

25th-5th. Rid 2 hours to see an old castle, dined in a garden, returned. Visited the Consul, he walked with me to the castle.

26th-6th. Saw the Greek church & manufacture of earthern ware, like Delft; found the Consul at our house, who invited me to dine; – I went with him, & after dinner, he went over with me to the castle in Europe, supposed to be, Sestus, as the other to be Abydus. We went to a garden, & then to a Coffee house & returned, I took leave of him & came home.

27th-7th. Set out with the Droggerman to the other castles 12 miles off, at the southern mouth of the Hellespont: – I passed over the Scamander, which runs north of the castle, & went to a cane there, went in the even a mile to Jenichacun [Yenişehir], the old Sigeum,[179] & copyed the famous inscription you may see in Chisul [Chishull].[180]

28th-8th. I set out to the East to go to the silver mines near Eskistambole [Eski Stambul] or New Troy;[181] went 4 miles along the southern banks of the

179 It is on basis of his writings in this area (the Trojan Plain) that Pococke earned his name as a pioneer in the field of Homeric topography, though only the barest details are given in the letters. Yenişehir, which means "new city" is the most prevalent placename for this site since Pococke's visit in 1740 and given the archaeological evidence that has subsequently come to light, he was correct in identifying it as Sigeum (or Sigeion). See Cook, *The Troad: an Archaeological and Topographical Study*, pp. 154ff.

180 This is a reference to the Sigeum stele, a bas relief with inscriptions (CIG i 8) that was coveted by many early travellers (including Robert Wood) and was finally removed by troops in 1799 at the instigation of Lord Elgin. See Cook, J. M. Cook, *The Troad: an Archaeological and Topographical Study*, pp. 26 and 155.

181 Eski Stambul is the modern name for the ancient site of Alexandria Troas (founded as Antigonia at the end of the 4th century BC) leading some early travellers (such as Robert Wood) to believe it was the site Troy or Ilium. See Luce, "Robert Wood and Homer", p. 85, who faults Wood for doing so and for "not taking more seriously the work of another eminent traveller, the Rev. Richard Pococke, which was certainly known to him and his companions". Pococke hired two janizaries that morning to accompany him to this area, "the road being very dangerous". He gives a detailed account of the site (see *Description 2*, Part II, pp. 108-10) and a plan of the site, showing buildings, is represented in Plate LVII.

Scamander, & then leaving it, & going east so. East, we came after travelling 12 miles in all, to the mines, & went to the Aga's house, to whom I had a letter. The mines are near a hill called Chigrek,[182] on which are ruins of an ancient city, I take to be Scephis [Scepsis], which was thought to be the Principality of Aeneas;[183] – & here Aristotles Library lay concealed for many years. I viewed the manner of melting the silver, lead & copper ore; – there are also Iron & alum mines, now not worked.

29th-9th. Went to see a ruined fortress on a high hill, & then the mines, which are worked in like Rabbit-burrows, without ladders: – the alum mines are dug in like chalk-pits; returned & writ. Walked again to the furnaces, the Ware house & the Greek church, for the miners are vagabond sort of Greeks.[184]

30th-10th. We set out at 3 in the morning & went eastward to the Scamander, travelled on the banks of it between the hills, & crossing over the sides to the north west came to the plain we passed through before, & to another river which must be the Simois & falls into the Scamander in this plain & on the high ground hereabouts between the two rivers, it is supposed old Troy was, tho' the site of it was not known in Strabo's time.[185] We continued on towards the Dardanel castles & stopped 6 miles from them at Telmesh, & dined, & then went on & arrived at the Droggermans house at the Dardanels.

31st-11th. Set out to the north & went 12 miles to Lampsae, the old Lampsacam [Lampascus] given to Themistocles by the Persians to furnish him with wine.[186] Went a mile further to a village, where there is a cane. The Aga a good old man came & took Coffee with me. I sent my servant over to Gallipoli in some business, & his brother came to me.

Aug. 1st-12th. Spent the day in writing: & at night embarked for the Islands about Marmora in the Propontis; – fine weather but contrary wind; so that it was noon on the

2d-13th. [No entry]

3d-14th. before we arrived at Afsiah [Avşa] a small Isle, with two villages in it, south of the Isle of Marmora. Saw the churches & eat some oysters: & went

182 At *Description* 2, Part II, p.107, he calls this "long rocky mountain" Chigur.

183 He notes the location of the mines (*Description* 2, Part II, p. 107) as being close to the village of "Eskijupee" – known even in the late 19th century as "Eski Skisepje". This is now, as he correctly supposed, identified with the ancient site of Scepsis.

184 Pococke had a particular interest in mines and mining and together with Jeremiah Milles visited a number of mines in Germany during the earlier part of his voyage, 1736-37. He also sent back a miner's lamp as a souvenir. See Finnegan, *Letters* 2, pp. 72-74, 110, 148 and 209.

185 He further identifies this site with the city of old Troy at *Description* 2, Part II, p. 108.

186 He notes in his book, that he was prevented from staying at this place because of the plague (*Description* 2, Part II, p. 111), a matter not divulged to his mother.

a league to the small Isle Cutalei [Çatal], west of it; – where I went to a room and a church; – saw their churches, & walked to the top of a hill. They have only one town of 70 houses all Greeks; – in the other Island they are partly Turks.

4th-15th. Went 2 leagues to the north, to the Isle of Marmora, & a priest of the Greek church that went over with me, carried me to his house, very pleasantly situated: – a young Turk of condition came & took a glass of wine with my man, while I slept & sent to his vine yard for some grapes for me. I went out to see the churches; & about 4, crossed over 4 leagues So.So. East to Alonia,[187] where we were led to the Episcopal house; & the Bishop came to me & received me with great civility, entertained me at supper & lodged me in his house, tho' I produced no letter to him.

5th-16th. The Bishop came to me & gave me a letter to the Vicar General of the Archbishop of Cyzicum [Cyzicus], & I set out for Artacui 18 miles off, where the Archbishop of Cyzicum has his cathedral & Palace, a league from the ruins of Cyzicus;[188] – had a tedious journey, the wind being contrary, but fine weather; arrived about 3 in the afternoon. The Vicar General invited me to a part of his house, permitting me to provide my own Provisions; – when I came Coffee was served. He brought his fasting provision & sup'd with me, – a brave [sturdy] old Gentleman. I walked up to the top of a fortified hill near, that probably belonged to the city of Cyzicum; return'd.

6th-17th. Went to a litle isle near & to see some churches. The Vicar General had invited me to dine with him, on fish, which I did. In the even walked out with a Greek priest to a fine fountain, & then to the vineyard of my Landlord & eat muscat grapes, that being a great place for excellent dry white wine, like Listan, the muscat of the red kind only for eating. The vicar general brought his fish, & supped with me.

7th-18th. Set out about 6 & went an hour East, to the ancient famous city Cyzicum, – saw all the very imperfect ruins of it, went to a village & took some refreshment, & went on 4 miles further to Panormo a country town; went to a Han, tho' I had a Letter to an Aga, I did not think fit to deliver it; walked out to see some parts of the town.

8th-19th. Went 12 miles east to Mehullitch, stopping at a village in the way, dined at the town & went 4 miles to the Port on the river Ryndacus [Rhyndacus/ Mustafakemalpasha], the [...] between Bythinia [Bithynia] & Mysia, the

---

187 An island "eighteen miles round" with five villages and a town of the same name. For a more detailed account of this place, see *Description 2*, Part II, p., 113.

188 This place is located in the Kapıdağ Peninsula or the Peninsula or Cyzicus, and was an island until joined to the mainland by an isthmus.

port 4 miles from the sea: – met with a French Factor here, from Smyrna Monsr. Le Harte, supped with him at the custom house.

9th-20th. Designed to embark here for Montagna near Brusa [Bursa], to wait for a caravan. We walked out.

10th-21st. I rid, – & conversed with the French Gentleman.

11th-22d. I spent the day in writing to My Lord Bishop.[189] Walked out with Monsr. L'Harty.

12th-23d. Writ all day to my Lord Bishop of Waterford.Walked out in the even to see a fine Antiquity.

13th-24th. I set out at 10 to go eastward to Brusa; – we travelled 4 hours, passing by Lupat, & along near the great Lake Abelleonty [Apolyont Gölü] & stop'd; – went 2 hours more, & slept till mid night & then went on to Bursa, stopt half an hour by the way & arrived at Brusa about 7 in the morn, & went to a Han: dined – went out & saw the three Greek churches, at one the priest had me into his garden, treated me with fruit, bisket & wine of the place; a nosegay &c: – but they are not above accepting a small gratuity. I after went up to the castle to a church, where Orchan is buryed, who took the city. Went to a fine stream of water, where the Turks sit; – for this city being at the foot of Mount Olympus [Uludağ], abounds in fine springs that rise in all parts, & the water runs through the streets; a most beautiful country round, with mulberry trees for the silk, planted thick like nurseries: – returned. A Corfu Dr 83 years old in the same Han, sent that he would be glad to see me, he is off going abroad; – I sent my servant to see who it was; – & he sent me a present of cherry-brandy & Lemons, – & I went to see him.

14th-25th. [No entry].

15th-26th. I walked a mile west to see the hot waters, which are so hot one cannot put ones hand into them & about them are five or six baths. I went & sat in the court of Sultan Morats [Murad's] mosque, before the door of it; – took Coffee &c: – & then went & saw the Archbishops garden; in a room of his house he has a small octagon fountain full of fish. The Archbishop not in town, & visited the Dr; went in the even to the house of the master of the Han, without the town, an Armenian, supt & lay there.

16th-27th. Early in the morning set out with men of the Governor to ascend Mount Olympus, at the foot of which this city [Bursa] is built; – the first half way very steep covered with trees: – we stopped & eat at an encampment of herdsmen, – & went the other half, which except towards the top is covered

189 Pococke has not yet heard the "melancholy news" that his uncle died, on 13 May 1740, three months earlier. See Letter 45.

with Pine & Pear & [...] from what I had seen [next half line scored out];[190] the whole ascent is about 6 hours, probably about 10 miles: – always covered with snow, we ascended above the clouds, for above it was clear, & below the clouds intercepted our view; returned to a stream, where we dined on small trouts with red spots on them, as I supposed & common. We returned down by night, & lay at the same house.

17th-28th. I saw the Armenian church, & the Archbishop sitting in a chair before an altar close to it, with his crozier in his hand, & some [...] standing up. I went & saw the outside of a mosque built of marble: returned to the Han. Writ, & settled my affairs in order to go for Nice [Nicaea] & thence Westward. I send this by Smyrna & Marseilles under cover to Mr Maydwell.

Pray my very kind love to my sister, – due respects to Whiteoak.

I am Dear Madam your most obedient & dutiful Son, Richard Pococke

SALONICA
19/30 September 1740
Pococke to his mother
Letter 45[191]

Honoured Madam,

My last Letter to you as of 18th-29th of August from Brusa, having writ another by Smyrna & Marseilles the day before the latter going by Vienna, – in both I acquainted you that I was coming home by the Morea & Italy as fast as I can, on Receiving the melancholy news of the Death of the Bishop of Waterford, by a Letter from Mr Milles by Captn. Hayes on board the *Stamboline* in the road of Tenedos. I am here in the house of Consul Horsewel [Horowell],[192] who was at Waterford when Mr Milles had the smallpox & is exceeding civil. I set out in two days for Larissa, thence to Athens & so to Patros [Patras]. I desire you in your three next to send me Mr Milles's rout [route] through Germany, – the

190 This page of the manuscript is particularly difficult, with many words being very faded and others rendered illegible through Pococke's later editing.

191 BL, Add MS 22998. At the top of this letter Pococke has written: "I am afraid one lost" and "A letter wanting, before this", which explains why we are not told of his journey from Bursa to Salonica. However, the reader may follow this journey in his printed account, at *Description* 2, Part II, p, Part II, pp. 122ff and view the plans and monuments in Plates LX and LXI.

192 The name is verified in Wilson, *List of British Consular Officials*, p. 61 and Despina Vlami, "Entrepreneurship and Relational Capital in a Levantine Context....", *The Historical Review/La Revue Historique*, Institute for Neohellenic Research, vol. 6, 2009, p. 136.

first recommend [addressed] to Mr Allen in Naples;[193] & the two next writ in three weeks of one another to Leghorn & then to Geneva, as Mr Milles will give you the directions. I desire you in your 2 or 3 next letters to mention the particulars you received from Egypt, & particularly box No. 3 of papers, medals &c. for I have no account of the arrival of these things in England; – the statues if not already done, expose not to publick view. I shall not I believe want a servant. I rid out with Mr Lee round the city walls, five miles in compass, but half the space within only built on. I waited on the Dutch Consul;[194] – I writ to [you] 19th-30th – yesterday by Constantinople & Vienna, the day before by Smyrna & Marseilles, this crosses to Venice, to let you know that I am well in health & proceeding on my journey very successfully.

Pray my very kind love to my sister

I am Dear Madam your most obedient & dutiful Son, Richard Pococke

THEBES IN BAOLIA
27 September/8 October 1740
Pococke to his mother
Letter 46[195]

Honoured Madam,

My last Letter to you was of the 19th-30th Sept from Salonica. That morn, I walk'd out with Mr Lee; saw a piece of Antiquity, & a mosque; which was the fine old church of St Demetrius in which are many fine antique verd Pillars, & all cased with marble. We then went up the hill within the walls to a monastery, returned, paid a visit & came home; dined; Mr Bobet,[196] an English merchant came to see me. The Dutch Consul came to take leave of me; I packed up; – waited on that Consul; returned. The superior of the Jesuits came to see me. Mr Lee went with me to the boat, aboard which I embarked to go to Caritza [Karitsa] in order to proceed to Larissa,[197] & we sailed about midnight.

193 This is Edward Allen, a banker who was Consul at Naples from 1726-48. See Ingamells, *A Dictionary of British Travellers*, p. 15.

194 The Dutch consul in Salonica was Frederik Willem van Frijbergen, in office from 1739 until his death in 1745. Schutte, *Repertorium*, p. 324.

195 BL, Add MSS 22998.

196 Possibly Henry Bobbitt, an English merchant in Izmir, who died prior to 1759. <http://www.levantineheritage.com/pdf/Biography-of-William-Barker-Levant-Company-Merchant-Marjorie-Rear.pdf>.

197 As explained in the book (where it is spelled "Claritza") this is because travelling from Salonica to Larissa by land was too dangerous. *Description* 2, Part II, p. 152.

Septr. 20th-Oct. 1st. The next day in the even arrived at Caritza a port in Thessaly under Mount Pelion.

21st-2d. I went a mile to a convent on the side of the hill, & thence to a village near, & hired mules to go to Larissa, returned to the Port, & set out at noon for Larissa; – had Mount Olympus to the west & Pelion to the East, on the foot of the latter by the river Peneus [Pineiós], we went, having passed the fields of Tempe, near the mouth of it & after 8 miles arrived at Baba, where we lay.

22d-3d. I went through a fine country, seing Mount Ossa to the East, & arrived at Larissa, – went to a cane – & out to see the Town, & took a walk round part of it.

23d-4th. Set out the first time with Post horses, having disengaged myself of most of my baggage, which is almost necessary in this country. Went 2 miles to Catania, through the plain of Pharsalia, famous for the decisive battle between Caesar & Pompey: – & Catania I believe is the old Pharsalus. I dined there & took other horses & went 36 miles to Zeitoun I believe the old Lamia. [Here was a terrible earthquake & half the kane fell down in which I lay.][198]

24th-5th. Rained hard so set not out till noon. Went 20 miles in the country of the Locri Epicni medii [a Greek tribe], by Thermopylae famous for the rout of the Persians: – lay among the mountains called Octa, at a Christian village.

25th-6th. Crossed the mountains into Boeotia, & passed near Turcocori, mentioned by Spon & Wheeler;[199] – & passing another vale came to Livadia the old Lebadia, [next half line scored out]; dined; went to see the Town, which is near Mount Parnassus & Helicon, which I saw, & set out & went 24 miles to Thebes. Lay at a cane.

26th-7th. [No entry]

27th-8th. Went to see the churches, & some litle ruins, this city being destroyed by Alexander the Great. Met with a Greek Priest of [next word missing] dined; was disappointed in getting horses, – so went out again to see other places & to see the Greek Archbishop. Sup'd & lay at the Priests house.

I am Dear Madam your most obedient & dutiful Son, Richard Pococke

198 Added later in Pococke's hand, and graphically described at *Description* 2, Part II, p.155.

199 See Introduction, note 30.

ATHENS
29 September/10 October 1740
Pococke to his mother
Letter 47[200]

Honoured Madam,

I set out from Thebes on the 28th-9th & passing Phyle to which Thrasibyllus fled, when he was expelled by the thirty Tyrants, & stopping a while, arrived in the even at Athens, about 30 miles from Thebes: went to the English Consuls house a rich Greek; – where I met Dr Perry my fellow traveller from Cyprus to Egypt; – who had been here 3 days, & being pressed to return, sets out tomorrow for Patras, where I shall follow him in 3 or 4 days by the way of Corinth.

29th-10th. I waited on the Wainode having sent a present before, showed him the Grand Signrs. Firman, a letter from the Grand Vizier & another from Janizer Aga; – was treated with Coffee. The Consul then went with me to wait on the Archbishop, who talks Italian, & treated me with sweetmeats, a dram & Coffee.

I have writ three letters to Mr Milles & pray let him know that it is my opinion, it is best to be still in everything, unless where absolutely necessary to move, till the Will is proved; only to have everything prepared against my return in Drs commons for the probat[e] of the Will.[201] This goes by Dr Perry by way of France or Italy. I enjoy my health very well; – & in a very short time shall embark for Christendome, & as soon as I arrive, you shall hear from me direct your first letter recommended to Mr Allen in Naples & 2 or 3 next to Mr Blackwell in Leghorn. Mention the receipt of my things from Egypt particularly a box of papers & a litle box with medals, & send me Mr Milles's rout home by Germany. I go through Italy by the Lake Como & by the Vale & not by the Alps to Swiss [Switzerland] & Geneva. From Mr Milles you will know how to direct to me in Geneva. Pray my very kind love to my sister, & due respects to Whiteoak.

I am Dear Madam your most obedient & dutiful Son, Richard Pococke

200 BL, Add MSS 22998.

201 Bishop Milles's Last Will & Testament, dated 18 April 1738, appoints both his nephews as his executors but Jeremiah as his heir. It appears from the context that Milles was in a hurry to give up his Irish connections and return to England. Clearly he heeded his cousin's cautious words, as he gave up his last position in the diocese (when replaced as Vicar of Kilrossanty) in 1749. See Rennison, *Succession List*, p. 175. On his return to England, his first appointment was as Rector of Saltwood with Hythe, Kent, in 1744.

CORINTH
6/17 October 1740
Pocoke to his mother
Letter 48[202]

Honoured Madam,

My last Letters to you were from Thebes & Athens by Dr Perry whom I met at the lat[t]er place, & by the Post from Italy or France, wherever he lands; he departed on

Septr- 28th-Octr. 9th – That afternoon I went to visit the Temple of Augustus [Rome & Augustus] & Jupiter Olympius [Olympian Zeus], & some churches to find inscriptions.

29th-10th. I went to the Temple of Theseus [Hephaisteion], to Areopagus the Musaeum & the temple in a Grott [Panagia Hrysospiliotissa or the Chapel of Our Lady of the Cavern][203] & what they call Demosthenes Lanthorn [Choragic Monument of Lysicrates], at the Capuchins; dined, – went in the afternoon to the bed of the river Elissius, now dry, – to the Stadium, to the temple of Ceres & to Adrians [Hadrian's] palace.

30th-11th. I went up to the Acropolis which is now a castle, where is the famous Temple of Minerva [Parthenon], which I viewed, & also the temple of Erecthea [Erechtheion] & another they call Lycurgus's Arsenal; dined, – went to several churches to fine Inscriptions, – & to the Archbishops.

Octr. 1st-12th. Went again up to the castle & the Consul with me, & review'd all & the Theater of Bacchus [Odeon of Herodes Atticus, or Herodeon] below; dined; – went to some churches.

2d-13th. I rid out with the Consul to Salpodia & to two Pillars at the foot of Mount Ancismus [Lycabettus], near Mount Hymettus, which are remains of Adrians [Hadrian's] Aqueduct [Dexameni]; came home, dined. I went & reviewed Adrians Palace & the Theater of Bacchus without.

3d-14th I went to the Temple in a Grott; & after dinner, to the Temple of Theseus; came home, prepared for my departure.

4th-15th Set out before day; – went by the port Piraeum [Piraeus] & in 8 miles arrived at Lessina [Elefsina] the old Eleusis, where I viewed the small ruins of the Temple of Ceres & Proserpine [Temple of Artemis Agrotera in Agrae, or Panagia stin Petra] & other antiquities & went 8 miles further to Megara: – saw the Antiquities & lay there.

202 BL, Add MSS 22998.

203 Pococke's engraving of this building is reproduced in the present volume as Plate 21.

5th-16th. I set out two hours before day & crossed the mountain to the Isthmus of the Peloponesus [Peloponnese]: – saw the place where they began to cut a channel through it; & remains of the wall that was built across it; – & in the afternoon arrived at Corinth, now a small village. Saw some old pillars of a Temple & ruins of another Building.

I am Dear Madam your most obedient & dutiful Son, Richard Pocöcke

MESSINA
18/29 November 1740
Pococke to his mother
Letter 49[204]

Honoured Madam,

On 6th-17th of October I left Corinth & went westward, near the Gulph of Lepanto [Corinthian Gulf], about 30 miles & lay at a cane in the country.

7th-18th. Went 12 miles to a litle town on the Gulph called Vortitza, slept half an hour & went on about 20 miles to Patras, where I went to our Consuls Mr Paul, being recommended to him by Mr Lisle, & his very particular friend Archdeacon Payn, so I met with an extraordinary civil reception, the Consul being a most obliging wel-bred Gentleman; has his Brother with him, a young man, tho' himself advanced in years & both Batchelors;[205] & here I met Dr Perry again, & so take my letter of him to forward it myself, when I shall arrive in Christendom.

8th-19th. I went out with Mr Paul & Dr Perry on horse back to St Andrews convent; saw the Tomb of St Andrew & the cell, where they say he lived; – & at some distance the place where they say he was crucified. After took a circuit & saw another convent; declining supper, I spent the even in the Drs room, – the Consul & Brother being with us part of the time.

9th-20th. Walked out with Mr Paul to see the Town; – & in the even took a walk with Mr Paul & the Dr – The Consul came & sat with me in my room, & afer we went to ye Drs.

10th-21st. Rid out in the afternoon with Mr Paul & the Dr; saw some churches & a Corinth [corrected to "currints"] Garden. – The tree is exactly like the

204 BL, Add MSS 22998.

205 The former may be either James or Daniel Paul, both of whom served as English Consuls in Greece during the period, the former at Zante and the latter at Patras/Morea. See Wilson, *List of British Consular Officials*, pp. 27 and 28 respectively.

vine, & the fruit grows like Grapes, a small red fruit, as big as a small pea; – many here, but not altogether so good as those of Zanth [Zante]. Spent the even in the Drs room.

11th-22d. Walked out in the even with Mr Paul & the Dr.

12th-23d. Performed divine service in the consuls house twice.

13th-24th. Writ all day, & 14th-25th & so to the 17th-28th. Went out with Mr Paul to some churches & spent the rest of the day in writing, as also the 18th-29th.

19th-30th. Read morning & Evening Service.

20th-31st. Prepared everything for my departure, on board a Genoiese vessel & in the evening went on board, & about midnight sailed to be set down at Messina.

Octr. 21st-Novr. 1st. Had a moderate wind seeing Ithaca & Timona & got between Cephalenia [Cephalonia] & Zanthe.

22nd-2d. The wind being very contrary, we went in the morn into the port of Cephalenia, called Agostolie [Argostoli].

23rd-3rd. We went ashore to the chief city of the isle called Argostoli: – went to the health-office but the General being here with two men of war, & having no Lazaretto [quarantine station], I was not permitted to go to a private house, with one of the office as a guard, which at other times is permitted. So I went aboard the ship.

Be pleased to write to me once, after the receipt of this, recommended to Mr Blackwell in Leghorn, which may come to me to Milan or Geneva; & after write to Geneva. We were detained by contrary winds till –

Novr. 7th-18th. When the wind coming to North & being very fine weather, we set sail.

8th-19th & 9th-20th. [No entries]

10th-21st. Came under Calabria, & saw Sicily & Mount Aetna.

11th-22d. Wind being contrary came to an Anchor at the road called Fossa, olim [once] St Joanni in Calabria, a league from Reggio & about 2 from Messina.

12th-23d. We turned all day & got only as far as Reggio.

13th-24th. Before noon we arrived at Messina in Sicily, a very fine harbour & Quay near half a mile long; – with the houses all built finely in a regular manner, & does really exceed the Quay at Waterford, being the finest I have seen. We went there to produce our Patents of health, at the health-office. I sent to Messrs. Allen, Crawley & Porteen [Porten], an English house here;[206] – another

206 The two Allen brothers were Edward (English Consul at Naples, see above) and William, who took over the latter's business and consular duties in 1748. James Porten was a

being Messrs. Chamberlayne & Cook, Mr Chamberlayne is Consul;[207] – others house is Tatnam & Hopgood: – I had letters particularly recommending me to the first house; – they soon sent their Brother; – & after Messrs. Crawley & Porteen came, Mr Allen being in Naples: – & I went to the Quay, & standing in the boat talked to them; for it is noli me tangere ["touch me not"] till the Quarantine is over.

14th-25th. Before noon, we were all conducted to the health-office, where Mr Crawley met me, & the whole crew were visited by the Physicians – they shore their groynes & armpits, – but the Physician told me I had a very healthy look & there was no need of visiting me: – so we went aboard again, – & soon afer the Guardian called on me to go to the Lazaretto into Quarantain; – I went with my baggage, where the Vice-Consul met me. It is a building round a court, in an Island within the Port: – where I have a good room, & the Brother buys my provisions, & my servant, dresses them here; we can converse with persons at a distance so as not to touch. At this place I hear of the Death of the Emperor,[208] & so in all likelyhood a great war will come on; – but be assured I will take care to come home in a safe road.

I settled my affairs. This is a country abounding all sorts of good provisions, & towards night I was locked into my apartments, being a kitchen below & a lodging room above. I see the Bishop of Ossory [Charles Este] is made Bishop of Waterford, & that Charles Knollys is Earl of Banbury.[209] I was glad to hear the King was arrived.[210]

---

brother-in-law of the Allen brothers who died in 1747, leaving a great debt to to Edward Allen unpaid. For further details on their careers, see Ingamells, *A Dictionary of British Travellers*, pp. 15-16 and 783. Charles Crawley was an eminent merchant in Messina. Two years later, in 1742, he would be appointed British consul there. *The British Chronologist*, vol. 2, London, 1775, p. 275.

207 William Chamberlayne was the "new" British consul around 1725. He was a nephew of Thomas Chamberlayne, the previous consul. Salvatore Bottari, «Consoli Inglesi nella Sicilia del Settecento: Funzioni Institutionali e Attività Informali", in Arnaud Bartolomei et al. (eds), *De l'Utilité Commerciale des Consuls. L'Institution Consulaire et les Marchands dans le Monde Méditerranéen (XVIIe-XXe siècle)*, Rome and Madrid, 2017, p. 182.

208 Charles VI, whom Pococke and Milles had met earlier in their travels (see Finnegan, *Letters 2*) died on 20 October 1740, at the Favorita Palace, Vienna. As with Pococke, it is thought that the Holy Roman Emperor died from mushroom poisoning.

209 Charles Este (1695-1745) had been Bishop of Ossory since 1736 and succeeded Pococke's uncle as Bishop of Waterford & Lismore. On the death of the 4th Earl of Banbury in 1740, his son Sir Charles Knollys (1703-71), who was Vicar of Burford in Oxfordshire, inherited the title.

210 King George II, whom Pococke & Milles had also seen while travelling in Europe I (see Finnegan, *Letters 2*) had spent the summer of 1740 in his native Hanover.

15th-26th. Messrs. Allen, Crowley & Porteen sent me a present of a [...] fruit, raisins, Lent figs & 12 bottles of wine. I find I am to stay in Quarrantain in all 22 days, so shall come out the 5th of December.

16th-27th. Mr Porteen came to see me & in the even Consul Chamberlayne. The Quarantine is 3 weeks but I hope to get out a litle sooner, & to go to Naples from Regio by land. We have had much rain.

17th-28th. Is fine weather. Mr Porteen & Mr Hopgood & the Vice Consul came to see me. I hope to be in Naples on Xtmas day n.s. (new style).

I am Dear Madam your most obedient & dutiful Son, Richard Pococke

This day the Post goes out; if anything of importance write both to Leghorn & Geneva.. Pray direct [...] you may know from Mr Hoare.

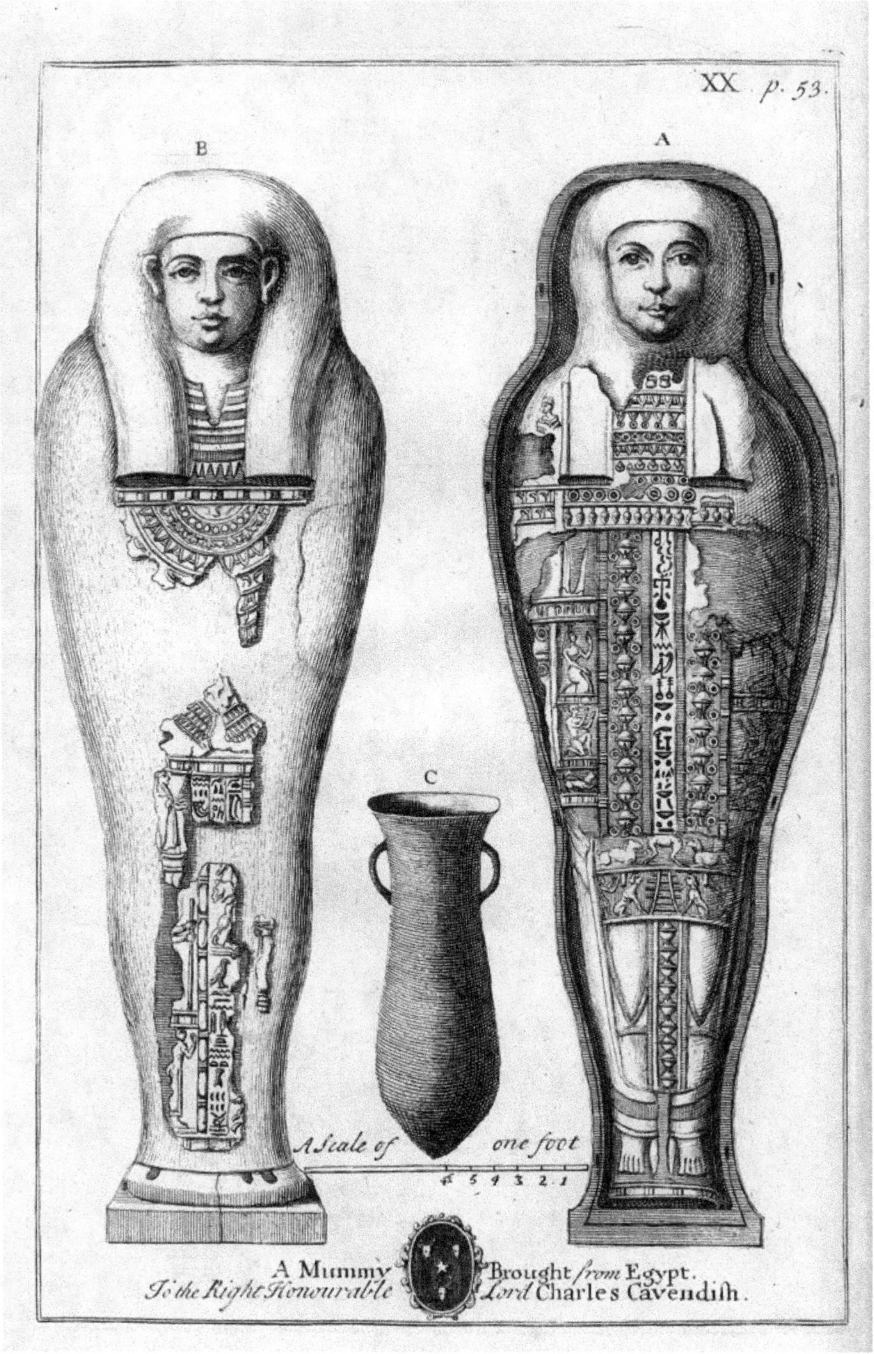

PLATE 16 Egyptian Mummy Brought from Egypt, *Description 1*, Plate XX
PHOTO DAVID KANE

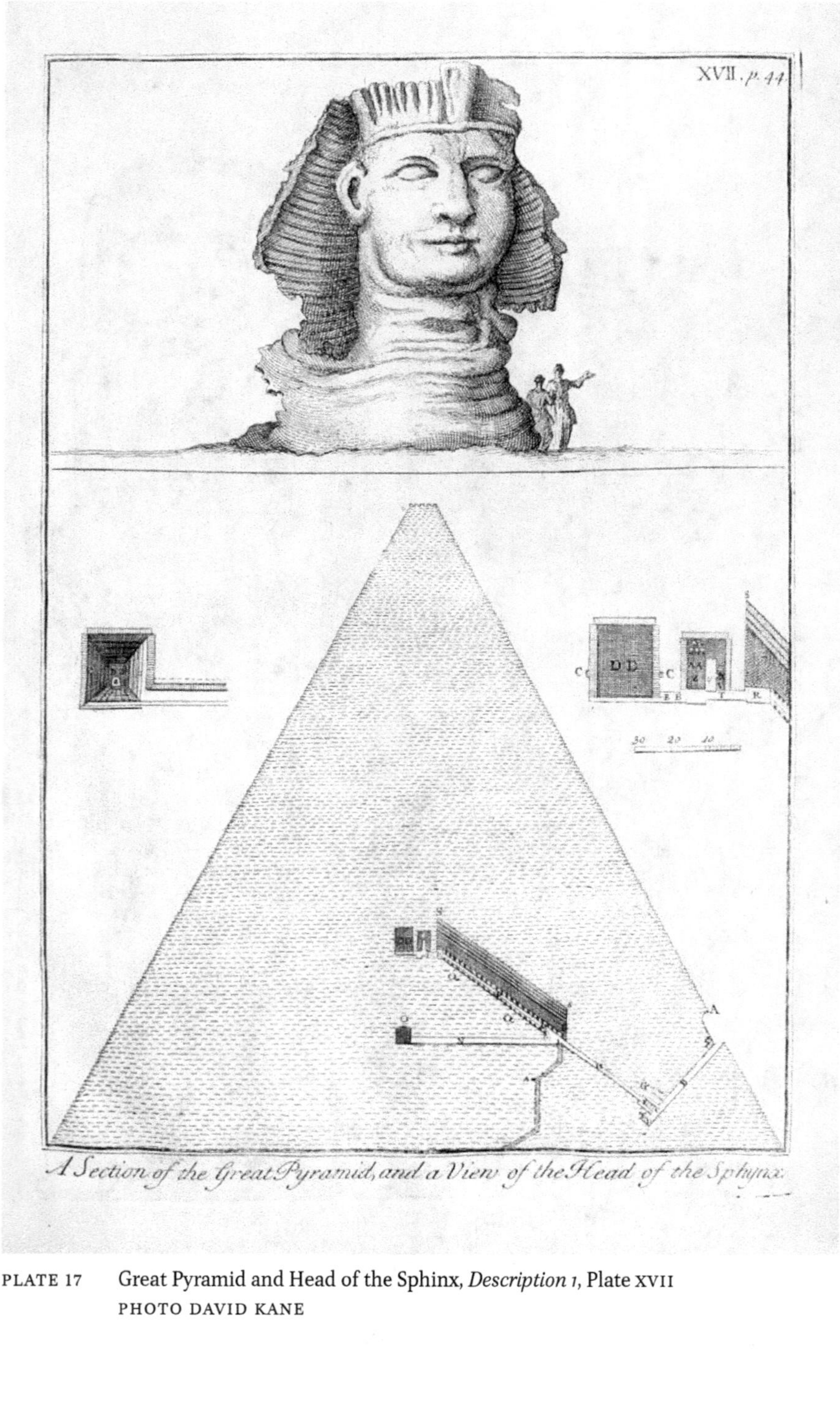

PLATE 17 Great Pyramid and Head of the Sphinx, *Description 1*, Plate XVII
PHOTO DAVID KANE

PLATE 18 Pococke's Statue of Isis, *Description 1*, Plate LX
PHOTO DAVID KANE

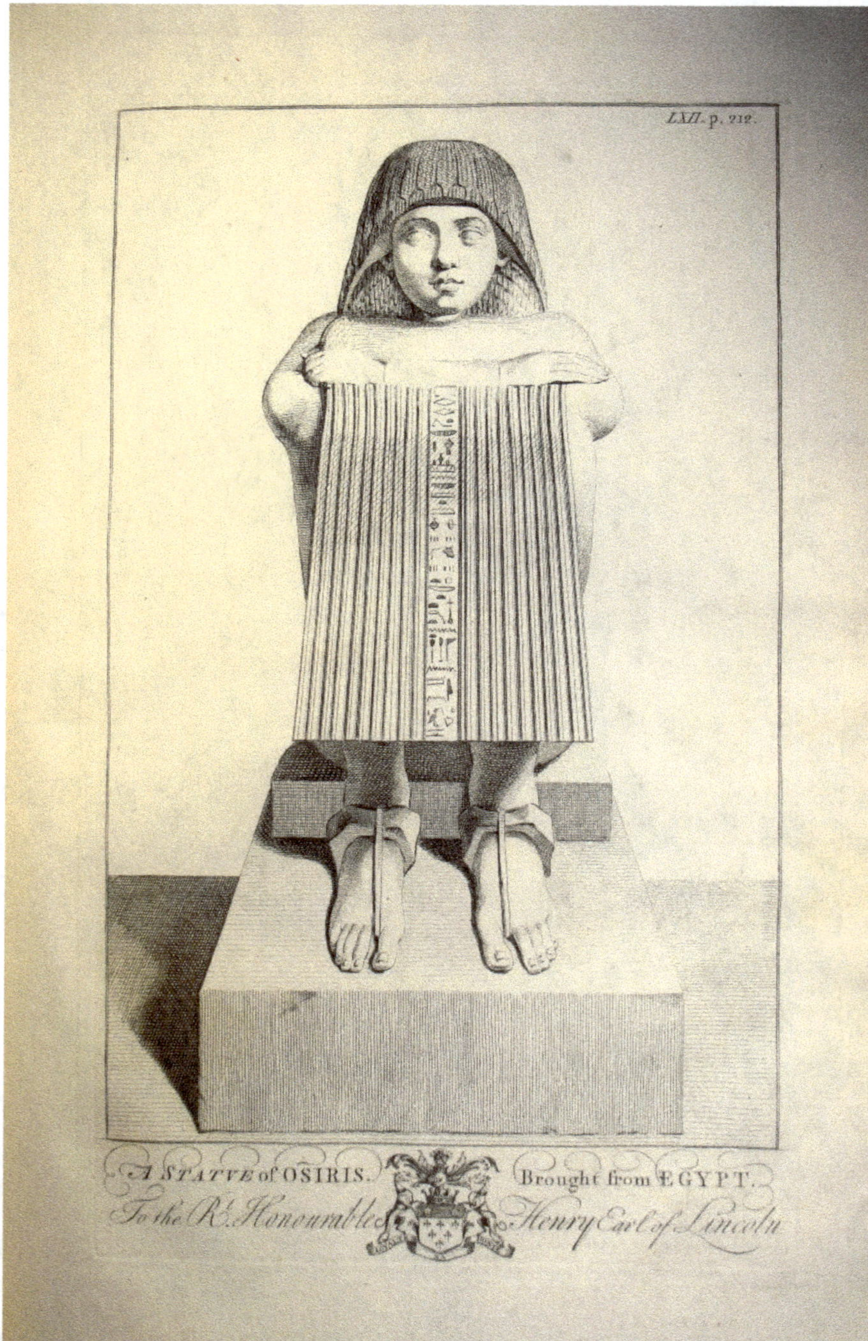

PLATE 19 Pococke's Statue of Osiris, *Description 1*, Plate LXII
PHOTO DAVID KANE

PLATE 20 Tower of Andronicus, Athens, *Description* 2, Part II, Plate LXXV
PHOTO DAVID KANE

PLATE 21 The Chapel of Our Lady of the Cavern, Athens, *Description* 2, Part II, Plate LXX
PHOTO DAVID KANE

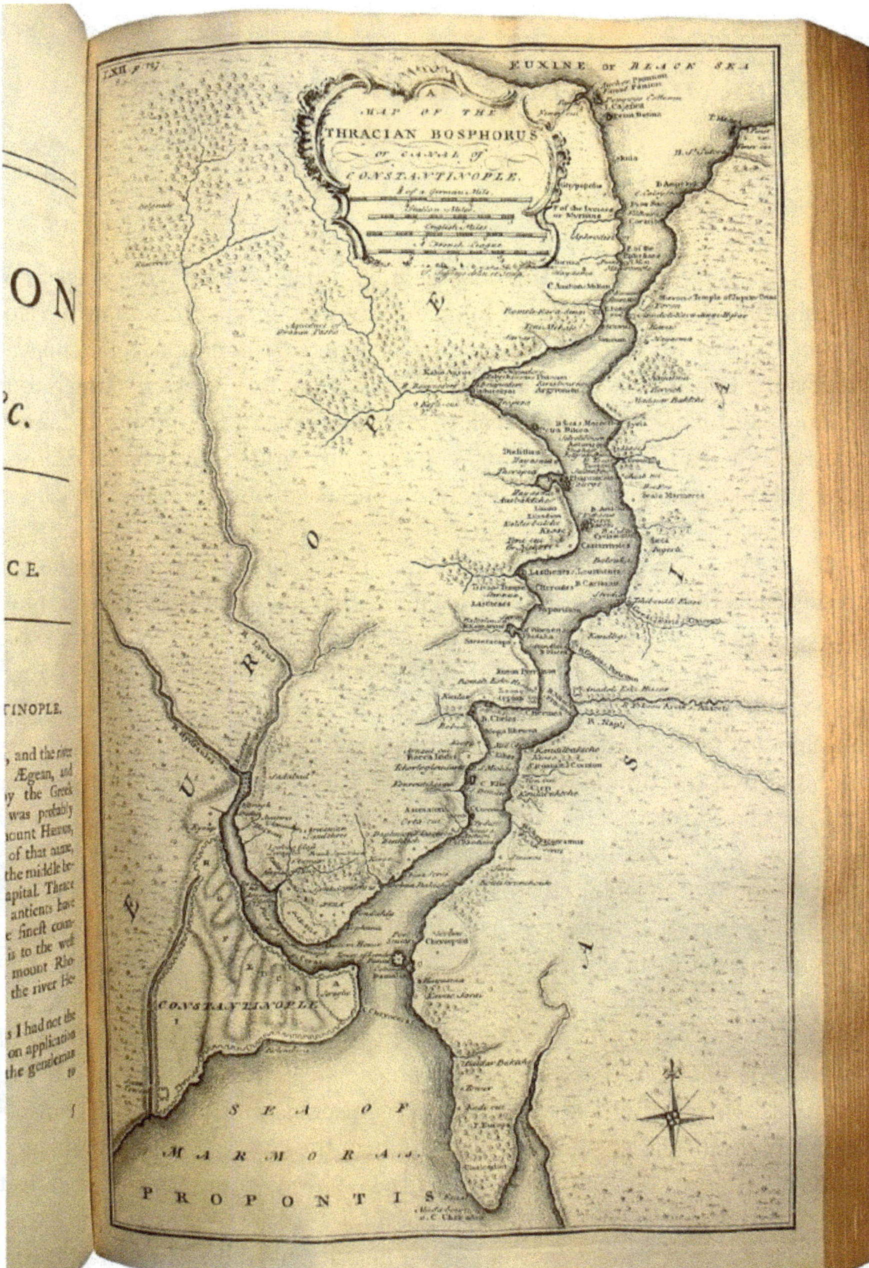

PLATE 22 A Map of the Thracian Bosphorus, *Description* 2, Part II, Plate LXII
PHOTO DAVID KANE

CHAPTER 9

Accounts from Cyprus and the Second Tour of Egypt

The 15 letters reproduced in this chapter are the final texts from the Register of Letters from Pococke to Bishop Milles and cover a period of four months (November 1738 to March 1739). First are the six letters (Accounts 20-25 inclusive) describing Cyprus, where Pococke spent more than five weeks making a tour of the island. The reader will recall that he was slow in composing these accounts, as he had been so busy, and consequently wrote them several months after this tour, and without the aid of his notes (see Letter 26 from Pococke to his mother).

In Account 26, which is a cover note accompanying the final batch of letters in this collection, Pococke informs his uncle that he has already sent two letters relating to Cyprus and that he is enclosing five more. However, as there are only six accounts in this collection, rather than seven, either he made a mistake about the number, or more likely, one is missing. If this is the case, the missing text probably relates to the early part of his stay in Cyprus, as there is nothing from the time of his arrival on the island (28 October/8 November 1738) until a fortnight later, when he writes from Famagusta (Letter 20, dated 11/22 November). The five accounts reproduced here comprise accounts of "the nine kingdoms into which Cyprus [is] divided", with particular reference to the geography, climate, economy, and natural history of the island. The documents also provide details of the many churches and convents visited by Pococke during this two-month period.

Following this is a collection of eight letters from Pococke's second tour of Egypt (Accounts 27-34), which were among a batch of 13 letters from Pococke to the bishop, together with a cover note mentioned above (Letter 26) outlining his plans for the final part of his journey in the east. This correspondence spans a period of two months, from 25 December 1738 to 7 March 1739, but sadly, eight of the promised letters are missing.

A significant difference between these final letters and earlier ones in this collection is the fact that they show no evidence of editing by the recipient, Bishop Milles, which is hardly surprising since he was very ill in the last months of his life and the letters probably did not arrive until well after his death. Consequently, not only are the sentences unduly long, but, in most cases (apart

 | DOI:10.1163/9789004440050_010

from Letters 27 and 28) the text is not divided into paragraphs – errors religiously corrected by the bishop in the earlier examples.

FAMAGUSTA IN CYPRUS
11/22 November 1738
Pococke to Bishop Milles
Account 20[1]

My Lord

It is not very difficult to make a conjecture, concerning the situation of the nine kingdoms into which Cyprus was divided, as there are remains of about nine principal cities which seem to have been the capitals of them, and as the countries about each of them are very naturally bounded by mountains, of which there are two ridges that run from East to West, the Norther[n] most begins at the Eastern point of Cyprus, and runs along about three quarters of the Island, and end at a bay which probably was the Northern part of the kingdom of Ægea having to the West the bold point called antiently Caleinusa. The Eastern part of these mountains takes up almost all that narrow tract, which seems to have been called the Olympial promontory, and probably the highest part of these mountains was called Mount Olympus, on which was a temple to Venus Urania or the chaste, because there was a city here of that name destroyed by Diogenes Poliorcetes, and also because it was not lawfull for any women to enter this temple, or so much as to look at it. All this promontory seems to have been the Kingdom of Carpasia, of the capital of which of the same name I saw some small ruins; especially the remains of a wall not half a mile round on the North sea with a peer built from it, and some signs of a tower at the end of the peer; this seems to have been only a castle for the defence of the port. East of it is a very good church remaining in the Greek stile [style] which belonged to a monastery called Stinphilose and I was told that the present name [of] this place has [was] of Salaruina as given it by some religious who began to improve this place not a great many years ago, but were obliged to leave it on account of the Maltese [privateers?],[2] the antient name has been transferred to a village North of it called Carpus which is the name of

1 BL, Add MS 15779.

2 The word in the text is unclear, but a corresponding passage in *Description 2*, Part 1, p. 219 refers to Maltese privateers in this context.

the whole country. About that village are a great number of small ruined churches, or chapels that might formerly be built of the houses of wealthy families who might retire thither. It was here on the Carpanan shoars that Diogenes Poliorcetes carried his army. The East point of the Island is called Cape St Andria from a church and little Monastery cut out of the rock dedicated to that saint, which some years ago was inhabited by a monk or two. Opposite to the North East corner are the Isles called Clides by the antients, the largest is not a mile round; authors differ about the number of them, those who name only two, took notice only of the two longest, there are two more that appear only as rocks, the furthest not being a mile from the land, a fifth which has some herbage on it, and may be the second in size is so very near to the land, that probably it has been separated from it since those authors writ, at this corner is a Grott cut out of the rock which seems to have been a sepulchre and some signs of a large enclosure round it, and higher, are several sort of oblong square monuments of hewn stone and that only just seen above the ground, and appear to have had covers over them, which doubtless are sepulchres of very great antiquity; and one which is grander than the rest made me conclude that they might be the sepulchres of the antient Kings of this part of Cyprus, that structure has but two teers [tiers] of stone above ground and one foot nine inches thick, within it at the distance of two feet six inches is another, and at the same distance within that a third which has a ledge cut round within as to let in a cover as the others before mentioned, where doubtless was the entrance to the burial place, all the walls are of the thickness of one stone, it is possible the two outer walls might be built up higher, and have entrances through them to the sepulchre, however it was, they are very particular works, and such as I have not seen before. On a little mount which is a rock of marble of different colours, that makes out into the sea, are the signs of a foundation of a building, which might be a very good situation for a light house tho' there are some remains on a little point very near it that have more appearance of such a building. All this country East of Carpass for about twelve miles is almost uninhabited, except that there are a few Turkish herdsmen on the North side where there is a fine narrow plain, this occasioned by the constant depredations of the privateers from Christianity as all over the Island so more particularly in these parts. About a league West of Carpass is a small Greek convent called Eleusa, and about two leagues West of that are several ruins on the sea of buildings of large hewn stone, which seems to have been a city and a port, they now call it Selenia, and have a tradition that it was formerly the residence of some King. Two leagues further to the West is Ialousa, probably the bay it is near is the antient Sinus Isicus placed hereabouts by the old Geographers, for there is a place now called Ialousa about the Sinus Isicus in Asia, whether this part had its name from a colony from those parts or from the notion that Cyprus was separated from the continent, every one may conjecture as he pleases. From all this shoar one sees the coasts of Caramania the antient Cilicia, which is about twenty leagues distant, and from the East point I saw very plainly mount Cassius near Antioch, and the mountain of Rosssus now called Cape Hog which are between Seleucia Piena and Alexandrette, and are about twenty leagues from Cyprus. About two leagues North East of Ialousa near a place called Aima is a large Grott cut into a mountain very difficult to access, and two leagues East near a village called Galliporno is a large Grott cut into the cliffs of the mountain, and not easily enter'd, it is a gallery with four compartments on each side in most of which are holes cut down for sepulchers, which are now filled up, on the hill above it are some small ruins of an antient place which might be Urania taken by Diogenes Poliorcetes; and I saw near the Grott a great number of sepulchres cut into the rock, many being in the manner of our graves, which seem to have had stone covers over them. Towards the West end of the promontory the mountains are very high, and the foot of them stretches out in the North sea, where there is no passage on the North side of them and these on the North Westside I suppose to the bounds of the Kingdom of Carpassia; a low mountain by which there is a narrow pass over the sea near the West end of the promontory having been probably the bounds to the South West. On the heigth of those mountains is a very strong castle, almost entire wch as two or three others on this ridge of mountains is called in the modern Greek Εκατοενασπίτε (or the hundred and one chambers).

From this I descended to the North side of the mountain where there is a very narrow strip of land not above a mile wide but extending Westward for near thirty miles, near to that bay where these mountains end, this I take to have been the Kingdom of Lapithia, of the supposed capital of which the same name, I shall speak in its proper place. In this part are several Greek Convents, the East end being a very indifferent country is thinly inhabited, the first monastery I came to is called Malandrina in it is a well adorned church, and South of it up the hill is the small Convent of Gera Abbate, and a league East of it on the side of the mountain is the Convent of Antiphonete with a church painted in their manner, and a Cupola over it, what this monastery is most famous for is the lignum Cyprium, of which there are seven trees being the only trees of the kind in the Island, and there are of them also in Rhodes, an account of which may be seen in Pomet and I shall mention them in another place.[3] Four or five leagues West of this place is a poor ruined port called Gerines it is the

3 Possibly *A Compleat History of Druggs*, London, 1712, by Pierre Pomet, Chief Druggist to King Louis XIV.

antient Cervina, the ruined walls are about half a mile round and seems to be on the same foundation as the antient walls, because on the Westside there is a fine fossee cut out of the rock, and the old town might extend further East beyond the present fort, which is a Tetragon about a quarter of a mile round, what appears is raised high, but they say it is only a sort of a counterscarp to the inner works which are in the same form but stronger, and yet the Venetian Governour when the Turks were marching towards it after they had taken Nicosia shamefully sent to surrender it before they had set down before the place. West of the town are a great number of sepulchral Grotts, and I saw some pillars standing, and remains of the foundations of an antient building which might be the temple of Venus and Isis, in the town is a Greek church, and two or three in ruins the [word missing] residing in a Convent at Solea there being not above five or six Christian families in the place. The cheif trade here is between Selefki [Silifke] in Caramania which is the antient Seleucia in Cilicia carried on by two small French vessels who carry thither Rice and Coffee that is brought from Egypt, and bring back Storax [styrax balsam], and some other goods, and many passengers they also sometimes go over to Salalia. Klefki is thirty leagues distant and the nearest part of the coast is twenty leagues. South South East of Gerines near two leagues on the side of the mountain in the way to Nicosia, is a very magnificent Convent uninhabited but almost entire, it is called Telabaiese, it consists of a very beautifull Cloyster, on one side of it is a most magnificent refectory, on another side, a fine room up one flight of stairs which might be a Library, and under it two very handsome rooms one of which might be a common refectory, and the other a room to receive strangers, on a third side is a church of a more antient and clumsy building, all the rest being a very fine Gothick architecture, and in the Cloyster they have made a Cistern of a beautifull Sarcophagus of white marble adorned with Bulls heads, Cupids and Festoons of exquisite worksmanship. About three leagues from Gerises to the West and East at a rich Convent called Aicropede are great ruins of the antient Samtha. I saw several ruins cut out of the rock, and one entire room above ground in the same manner. Over the sea are remains of some towers and walls which show that it is not a great many ages ago that this place was inhabited, the old name is translated with a little difference to a village near called Lapta, where there are some sources of very fine water. Between this place and Gerises on a high point of the mountain is another Castle called the hundred and one chambers, and a league West of Lapta is the poor Convent of Aia Parastege.

I am my Lord

NICOSIA IN CYPRUS
21 November/2 December 1738
Pococke to Bishop Milles
Account 21[4]

My Lord

The other ridge of mountains which run along the South side of Cyprus, begin at the point East of Larnica which is called Cape Pyla, and toward the West become very broad as well as high, and stretch away to the North West corner of the Isle. Between this and the Northern range of mountains is a large plain seven or eight miles wide, and between thirty and forty long; as it is one of the best parts of Cyprus and most secure from the privateers, so it is most inhabited by Turks, the Christians living more on the mountains and near the sea, being exempted from that slavery which falls to the lot of the Turks when they come into the hands of these privateers. This plain seems to have been the antient Kingdom of Salamis founded by Teucer the Capital of which was at the East end on the sea, where there are still great heaps of ruins, and the traces of the antient walls, the old city being not less than four or five miles in circumference, with its port at the South end, where the small river Fedius empties itself into the sea. Antient Geographers mention two Islands off from Salamis, which are not now seen; on tracing the ground I imagined the sea might have left these Islands, and I saw near the port some rising grounds with sort of channels round which might formerly be filled with the sea and make those islands. The port which was a bason probably made by art is almost entirely filled up, it appears that there was a more modern city here than that large antient one which might be founded by Teucer, and of the new town there are great remains of the foundations of walls, being about half as big as the old city, on that side of it which is next to the part, there are the ruins in it of a large church, and a small one, and North of this town are some very thick walls which probably are ruins also of a church, near a square plain spot that might be either a piazza, or a bason for water. On the North side of the new town just within the gate are several Grey granite pillars lying on the ground, and two or three Corinthian Capitals of Grey marble cut in a very beautifull and particular manner which seems to have belonged to a temple there. All this is now called old Famagosta, being about four miles from the city of Famagosta, there are the remains of an Aqueduct to this city, what arches I saw of it are Gothick and an inscription on it in Greek, which makes mention of an Archbishop, the antient

4 BL, Add MS 15779.

aqueduct being probably repaired when the new city before mentioned might be built after the establishment of Christianity in these parts. One sees some of the arches all along the plain running to the mountains to the North West, on the side of which mountain the water was conveyed from a plentifull source I saw at Chertes [Cherkes] thought by some to have had its name from the old Cytherea which must have been further to the South, the Fables placing Citari in the road from Salamis to Tremitus now Nicosia. Chertes is six or seven leagues to the West North West in a valley between two hills finely improved with mulberry gardens for the silk worms. A considerable way up between the hills are those plentifull sources of water which supplied this Aqueduct, I found the water luke warm and they say in summer it is very cold, and I have been assured there are many sources in Stiria and Carinthia that are in the same manner hot in winter ad excessively cold in summer. But to return to Salamis, West of the city is a small ruined church, and a little further is a small church built and arched over with very large stones it is now half under ground and is dedicated to St Catherine who was daughter of King Costa who they say was founder of the present Famagosta and that from him it has its name. In this church there is a well, and on one side is a small chapel, built of three stones, the upper one being cut in such a manner that the top of the chapel ends in an angle like our sorts. I think they say this saint was buried here, and it is certain all round the Chapel for about three feet up the stone looks fresh as is if there had been a tomb. A mile to the West is a monastery and a large church which seems to have been a fine building dedicated to St Barnabas, the church has been ruined and rebuilt, for a few feet East of it are the foundations of the East end of the old church in three semicircles, about half a furlong East of this church is a descent by several steps to a sepulchral Grott cut in the rock with a nich for a body in three sides of it, here they say the body of St Barnabas was deposited, that Saint a native of this island having suffered martyrdom at Salamis in the time of Nero, at the entrance of the Grott is a well of saltish water, and over the Grott is built a small Chapel which dos not seem to be of any great antiquity, as this Grott is probably the very place where the body of the Saint was laid, we may conclude that in the earliest times they never buried the bodies of the greatest Saints in their Churches, or built Churches over them. The city of Famagosta is about two miles round, and finely fortified by the Venetians, it is an oblong square the Bastions are all semicircular, and where it is any way commanded there are high walls within called [word missing] to plant Canon to command the beseigers, a rising ground runs along from North to South on which they took the advantage to build the rampart to the West, which makes it exceeding strong this way, a fossee been cut here into the rock, as it is on two other sides, and on the West fossee are covered ways to sally out, this high ground which is the strength of the West side exposes the others, for it was from this that the Turkish General battered the South gate, which leads to Amathus, which is the only gate towards the land, and it is probable from this high ground on the North side they planted their batteries against the North East corner of the sea, where there is a strong castle fortified also within; there is a gate to the port, which is well shelter'd by several rocks and the entrance to it which is at the North West corner is defended by a chain, drawn across to the Castle. It was here the brave unfortunate Bragadines skin being stuffed was hung up at the yard of a Galley after he had been most inhumanly flead [flayed] alive by the false Turks after he had bravely defended the city.[5] On the ramparts are the names of several of the Venetian Governours of Cyprus, and by the gate is a very large statue of a Lyon and a lesser, which were probably set up on some pillars in the principal parts of the city after the Venetian manner. The antient Piazza seems to have been very handsome, the house of the Governor with a Portico before it, and some grand buildings within runs on one side, on the other is the West end of the church of St Sophia a most beautifull Gothick building now converted into a Mosque, and about three years ago two thirds of it was thrown down by an earthquake and the greatest part of the city of Famagosta. On a black stone which might be on the pedestal of a statue is the following inscription

ΝΕ ΡΟΥΑΝ ΤΡΑΙΑΝΟΝ ΚΑΙΣΑΡΑ
ΝΤΕΡ ΜΑ ΝΙΚΟΝ ΥΙΟΝ ΘΕΟΥ
ΣΕΒΑΣΤΟ ΥΗΠΟΛΙΣ ΛΓ[6]

Near the North West corner of the church are two pillars which probably had on them the Venetian ensigns, near these is such another Sarcophagus as I saw at Telabaiese of white marble adorned with Lyons heads and festoons held by Cupids. It is surprizing to see what a great number of Churches there are in this city, one of the finest St Georges was thrown down by the earthquake, another large one I think St Catherines is now the principal Mosque. There is very little trade at this place, no Christian is suffer'd to live within the walls unless it be in confinement, in which condition I saw a Greek Patriarch of Constantinople who being deposed, and found intreiguing [plotting] to supplant his successor was banished to this place a few months ago, nor do they suffer a Christian to go in or out of the city otherwise than a foot. Of the space within the walls not

5 See Letter 23, above.

6 In the original, the first two letters of the third line are dotted to indicate that they were not visible/legible and that Pococke has guessed them.

above half of it is built and not a quarter of it inhabited, the other buildings being all in ruins. They have very good water brought three or four miles off by a low Aqueduct which is built with arches only in two or three parts. Arsinoe is mentioned by the antients at this end of the isle as well as at the other but the situation if it is not known, they mention also the Grove of Idalium sacred to Apollo where this was is unknown the most probable place is Ormedia a very pleasant vale and finely watered, being about three hours West South West of Famagosta, and to the North West of Cape Pyla. The South East point of the Isle beyond Cape Pyla is now known by the name of Cape Grega, this is thought to be the promontory Pedalium mentioned as opposite to Mount Olympus on which promontory was a small temple dedicated to Venus, what seems to make it more probable is that this promontory is mentioned to appear like a table, and from the West at a good distance it shows it self like a square high spot and in figure like a table. Towards the West end of this plain, is the city of Nicosia supposed to be the old Tremitus which is the Capital of Cyprus where the Morotern or Governour resides. It is a fortified city consisting of very large ramparts but without a fossee and consequently is much exposed, and tho it has a spacious appearance being lined with the hewn stone of the old walls yet it is a very weak fortification, the circumference of it is about two miles, the walls of the antient city built with semicircular towers may be traced all round, and it seems not to have been much less then [than] four miles round. There are in the city still remaining several magnificent houses, mostly I beleive of the times of the Kings of Cyprus, some of them are repaired according to modern architecture by the Venetians, and to a house which they say belonged to the Venetian General is a most beautifull Corinthian door case. The Cathedral church now a Mosque, is a large building and exceeds that of Famagosta in the front as much as it falls short of it in other respects there is another Church dedicated to the Holy Cross, and one of the Augustinians which are both Mosques, and the Greeks have several new built churches in the City, and the Latin Fathers of the Terra Santa have a small Convent, there are very few Armenians here who nevertheless have possession of an antient church. There is a great Manufactory of Cotton stuffs particularly of very fine Dimities, and also half Sattins of a coarse sort. They have here the best water in Cyprus brought by an Aqueduct without arches some miles off from the Mountains.

I am my Lord

BAFFA IN CYPRUS
1/12 December 1738
Pococke to Bishop Milles
Account 22[7]

My Lord

Two leagues North North East of Nicosia on the other side of the mountain, is the rich Convent of St Chrysostum, belonging to the Greek Convent of the Holy Sepulchre at Jerusalem, over it is a third place called the hundred and one chambers which consists of several buildings one over another towards the top of the mountain, the highest is very difficult of access and might easily be defended. They say a Queen of Cyprus who had the Leprosy could be easy only up in this airy region, they say that St John Chrysostum advised her to build the Convent below and following his counsel she was cured of her Leprosy, others add that she bathed also in a water there which is still resorted to by persons in that distemper who find benefit in bathing in it. The monastery has been a very large building tho' much of it has been ruined, it has two churches one ruinous is dedicated to St Helena, the other covered with a Dome and painted all over within is dedicated to St John Chrysostom before it is a handsome portico from which there are three doors with fine marble door cases that seem not to be very old, behind the middle folding doors were formerly deposited two Scepters the figures of which are painted on the walls, and at the bottom is a hole where the Crown was kept, all the account they can give is that they belonged to some Queen and that they were taken away by some Pasha of Cyprus, as it is not unlikely that they were here, it is most probable that the Regalia of Cyprus were deposited in this place. This Convent is near the road which leads to Gerines by the famous Convent of Telabaiese before mentioned. This place is bounded to the West by some low hills which stretch from the end of the Northern mountains, to the Southern ones of these hills is a small Convent like a farm house called Panteleemon with a handsome new church to it, on the other side opens a large bay, and here seems to have been the Kingdom of Ægea where the famous Solon took refuge when he was banished out of Greece, it is said he advised the King of Ægea to leave the city of Ægea among the mountains and to inhabit a plainer country. I was told that up in the mountains there is a place now called Ege where there is a church, and beyond the plain that is South of the bay is a most delightfull narrow valley that winds between Southern hills from seven or eight miles, which is called Solea, and it

7 BL, Add MS 15779.

is probable that here was that city which the King of Ægea built and called Solea as history informs us from Solon. Beyond the end of this vale a little advanced in the hills is the small Convent St Nicolas, from the mountains runs a great quantity of water which supplies all the vale, and at the entrance into it forms a great number of beautifull Cascades which together with the wood and fertility of the soil renders it a very delightfull retreat, in this country also is a very small Convent belonging to the Bishop of Gerines where that Prelate resides. There is a great number of churches in this vale which have a very mean appearance, but they are old arched churches some of them built with little cupolas, to preserve which they have built very ordinary walls round them, and raised a roof on them covered with tiles. About two leagues East of this country, is a Convent at a village called Morpho it appears to have been a very grand design of two Cloysters left unfinished, having between a very magnificent church built of hewn stone, and dedicated to St Martha whose sepulchre is here, this Saint is had in great veneration in Cyprus, and they have some legend concerning her riding on a Lyon in which manner she is always painted; the building is not of modern architecture and not being very old I conclude it might be built just before the Venetians had possession of the Isle, being the benefaction of some great family of Cyprus. At the north West corner of the bay where the southern hills come to the sea are the ruins of a very considerable city, on the West it was bounded by those hills, on the North and East by the sea, a wall being drawn from the hills to the sea some remains of which are still seen, and of a Mole made by art within the land, for the shipping to lye securely in: the cheif part of the ruins of this place are a little way up the side of the hills to the West, where I saw the remains of a wall which made a small part of a circle with some signs of an opening like a window, but I could not judge whether this was the East end of a church ending in a semicircle, or an antient temple or theater. Lower on the plain are there peices of wall remaining eight feet thick I could see it had been turned [?] in four arches each fifteen feet wide and the peers between ten feet wide strengthened by Corinthian pilasters the Capital of which are very well executed; this seems to have been a portico to some grand building; the front being to the North and within in every peer is a nich about eight feet high and four feet broad which were doubtless made for Statues. This place is now called Aliogora [Aligora] that is the Sea Mart, and seems to be the antient Arsinoe which is mentioned as near Solea, and some modern writers have placed at Lefcare there being a village now called Lefca about a mile North of it. This place was famous for the Groves of Jupiter. West of this and Solea come the Southern mountains which here extend to the Sea on the North side making the promontory of Acamas, and taking up the whole breadth of the Island, towards the North West corner of the

Island, here is some improveable land between the mountains and the sea inhabited by Turks, and a small bay with an Island in it called the antient Stiria on this bay are some ruins called now Agama which are the remains of the City Acamus built by some Athenians. Part of these mountains and this small spot on the Sea might make a fifth small Kingdom. On the highest part of the mountains is a Convent called Panaia Cheque or the Madonna of Cheque, where they have a miraculous picture of the Blessed Virgin and our Saviour painted as they say by St Luke and brought from Constantinople by a King of Cyprus they call Isage which must be Isaac King of Cyprus son of the last Greek Emperour of the East.[8] This place is as much resorted to by the Greeks as Loretto is by the Latins,[9] and they come to it even from Muscovy. Passing over these mountains one comes into a plain in the South West of them which may be about three miles wide, and about fifteen long in which stood new Paphos and the port of old Paphos, and probably made a sixth Kingdom of which Paphos might be the Capital. From the North West corner of the Island the shoar runs to the East South East, the most Northern part for three or four leagues is mountainous, and about three leagues East South East of those mountains are the ruins of that city which was antiently called new Paphos, on a rising rocky ground in a narrow plain, below the great plain and West of some low rocky cliffs which very antiently might be washed by the sea, and are now full of sepulchral Grotts which doubtless served for the people of this City. West of the City is a sort of a point of land to the South East of which in the angle made by this promontory was the old port defended by peers built out into the sea, some remains of which are still to be seen, the City seems to [have] been East and North of the port, and one may discover the large fossee cut out of the rock, and North of the town, where they might dig their stones for building, they have cut into the rock several large high rooms, and many small ones, one tho' now the side of it is broke away seems to have served for a large Cistern there being a hole in the top to draw up the water, and a pair of stairs down into it cut out of the rock on one side, it is probable this was filled in the winter by an Aqueduct from the mountains of which there are some remains near the town by which means this city might in the summer be supplied with good water, of which there is a great scarcity in this Island. To the North of the port, on a ground raised by art, are some signs of an antient temple by the manner in which the Grey Granite pillars lye, and by the disposition of the ground I judged there was a Colonade

8 Isaac Komnenos or Comnenus (c. 1155-1195/1196), who ruled the island until it was conquered by Richard the Lionheart.

9 See Finnegan *Letters 1*, 208ff, for an account by Pococke's cousin (Jeremiah Milles) of what they saw in Loreto during their earlier travels.

round it and a portico to the West with a double Colonade, the pillars are about two inches diameter. Half a furlong East of this are the foundations of a smaller building of hewn stone, near the corner of the port which might either be a temple or some other publick building. As much farther to the East, are the remains of a large church which probably was the Cathedral, and seems to have been built on the foundations of a great temple for near it are standing some very large finely polished pillars of grey Granite about three feet diameter. It is needless to mention that both these temples were without doubt dedicated to Venus, for whose worship this City in particular was famous. Near the Cistern before mentioned, is a church cut out of the rock, and under ground dedicated to the seven sleepers, and in the town are ruins of several churches, and also of houses uninhabited mostly. This City is famous in Sacred History for being honoured with the presence of St Paul who converted Sergius Governour of the Island to Christianity and restored the blind man to his sight.

I am my Lord

COLOSSE IN CYPRUS
3/14 December 1738
Pococke to Bishop Milles
Account 23[10]

My Lord

Cythera is mentioned as four Stadia West of new Paphos, this must have been on the point which is West of this City, there being ruins all the way to that head of land and the distance answers well. About a mile North is a high rocky ground near the shoar, cut out into sepulchral Grotts many of them are plain rooms and some large; five or six I saw which might be for families of a superior rank, were cut out into a court with a Colonade round of two Dorick pillars in front and three on each side with an entablature over cut also in the rock and some of the pillars are fluted, on three sides of these courts is a room with door places cut in the rock in a very beautifull manner. Half a mile East of this is the new town of Baffa where the Governor resides, new Paphos being now called old Baffa, and inhabited only by a few Christians, and a small Garrison in a castle on the fort. At new Paphos was a great meeting once a year for the worship of Venus from whence they went sixty Stadia in procession to the temple of Venus at the port of old Paphos, where the Fables say that Goddess being

10 BL, Add MS 15779.

born of the froth of the sea came ashore on a shell, I could learn nothing justly where this port was or the Temple but as there are about three small bays to the East South East of new Paphos, I conjectured it might be the most Eastern bay which is nearest Concleh, where one sees some small ruins of old Paphos, on a high ground, the port being mentioned only as a mile from the City, and the nearest the sea comes to this place is rather a greater distance. Old Paphos is at the beginning of the Mountain from this plain which a little beyond it take up the whole breadth of the Island, being low towards the sea in comparison of those to the North, at the sea they end in high white cliffs, and where they make out in a point the farthest to the South they are known to mariners by the name of Capo Bianco. Over these hills we travelled to the East and came to two beautifull villages that are contiguous, one is called Episcopi, and the other Colosse, which being finely watered, are most beautifully improved with Mulberry trees, and a great number of Oranges and Lemmons, at the South end of Colosse is an antient Preceptory of the Knights of St John of Jerusalem in ruins. Some high Arches which conveyed water to a reservoir to it from the hill are still standing, and there is a stone in the Preceptory with this inscription on it:

HOC OPUS. CONSTVXLT. VENELIS.
MILLES DNS ER
L DOVICVS. DE.
MAGNACO. LINGVÆ. ALVERNE MAGNE
PRECENTORIE. FINICE QVE. ET. NO.
AC. DE CAMBARIAN. PRECETPTOR. 1453

It is the opinion of some that the antient city of Curium was here, but I could not see the least sign of any ruins, and on the hill to the West I observed the foundations of a thick wall that seemed to have encompassed some antient city, which probably was the old Curium, and it is not unlikely that the Grove sacred to Apollo near Curium was where Episcopi now is which is a place abounding with water. They think also that the low promontory called Cape Gatto was Promontorium Curias, whence they threw any one into the sea who dared to touch the altar of Apollo, but as this is very low land, it is more probable that it was from some point to the West of Curium where there are high cliffs and might possibly be what is now called Capo bianco. To the East of Episcopi is a small river which is the Lycus of the antients. The promontory called Cape Gatto is South of Episcopi, the Northern part of it is a Morass on the West side and a large salt lake on the East, which is filled by the winter rains, and is almost dry in Summer. The North part of this promontory is a barren rocky soil, and on it is a ruinous uninhabited Convent called St Nicolas, and it is said the Monks of this Convent keeping Cats to hunt and kill the Serpents of which there are great numbers there, gave the name to the promontory of Capo Gatto: and they add withal that they left off their diversion and returned home at the ringing of a bell. At the West corner of a bay made by this land and a point East of it, is a pleasant town like a village called Limeroc, where they have built a Castle and a platform to defend themselves against the Maltese. The Greeks have two churches here, one of which is a very handsome new built fabric. From Episcopi two leagues distant is a narrow plain, which extends about two leagues East of Limeroc, at the end of it on a hill and down the side of it to the sea, are ruins of the City of Amathus, probably the head of the eighth Kingdom, as Curium might be of a seventh, there are considerable remains of the walls to the West on the hill, which seems to have been a sort of a large Castle, and probably was the original antient town which might extend itself to the East where are great heaps of ruins, and a handsome Church in ruins also, perhaps built on the very place where might be the Temple to Venus and Adonis, and where the feasts of the latter were annually celebrated. It is said King Richard of England not being permitted to take in water on the Island when he was going to the Holy land, at his return took Isaac King of Cyprus prisoner here, and sent him in silver chains to Tripoly in Syria. East North East of this place about seven leagues is a mountain called Oros Staveros and by the Franks Monte Croce, by some Geographers this is called Olympus and compared to a breast, the reason of the present name is a Convent at top of it dedicated to the Holy Cross. From these mountains rises a rivulet which I think is called Gidy, it runs into the sea at a head of land called Gidy point. Some where here was the river Tetius, and probably it was this very river, and that from it the river or at least the point has derived its name. To the East of it are those large salt lakes from whence every year they collect from the top of the water great quantities of Salt, the rain fills the lakes and the soil being full of Nitre communicates that quality to the water, and when the water is sunk in Summer they take off the Cakes of Salt, when there is too much water occasioned by extraordinary rains it will not be salt enough to consolidate into Cakes, and for this reason the Venetians had drains to carry off the water, which are now neglected. The West of these Lakes is a small Convent with one Dervise only in it, they have a sepulchre here had in great veneration, by the Mahometans, it being the place where the foster sister of Fatiman Mahomets sister is buried. These Salt Lakes extends almost to Larnica the City where the Franks reside, and make it the most unhealthy place in the Island. This place is situated a small mile from the sea, where at the Port is a small town called the Marine, the Port is naturally well sheltered but the ships lye off at some distance, and the boats come ashore on an open beach and are drawn up to land.

This place tho' so unhealthy being within six leagues of Nicosia where the Government resides, is the reason why the Franks have settled here. At the Marine town is a large antient church dedicated to St Lazarus, where they show the sepulchre of that Saint a small Grott cut out of the rock, for they say being put in a boat at Joppa and committed to the mercy of the sea he was drove hither, and became Bishop of this place, and that his body was stolen away by the French and carried to Marseelies [sic], whereas the French say he was drove to Marseiles. In the church yard belonging to this church the English bury and have handsome monuments with Latin inscriptions on them, and one has a Greek Epitaph inscribed on his tomb. Between Larnica and the Marine are the ruins of the antient City of Citium probably the Capital the ninth Kingdom, it is famous for the birth of the great Philosopher Zeno, and on account of the death of the renowned Athenian General Cimon who expired at the seige of it. Ptolemy the son of Lagus destroyed this city, and removed its inhabitants to new Paphos. The City was about three miles round and it seems as if in very antient times the Sea had washed the South walls of it tho' now it is a quarter of a mile distant. To the East is Multangular bason which served for the security of the shipping and was defended, and was defended by a strong Castle as appears by the foundations of it there were no entrances into the City from the West, and the walls round seem to have been very strong, in the foundations are found many stones with inscriptions on them in an unknown character which I suppose is the antient Phenician, and if the city was ever rebuilt after it was destroyed by Ptolemy these stones might then be built into the walls. A great number of antient Sepulchers are found in and about the City of Larnica, some I saw built of hewn stone, one with the stones laid along at top like large beams, and others laid over them like a floor another ending at top in an angle and both of excellent workmanship and finished in the most perfect manner. In this town the Fathers of the Terra Santa have a large Convent, and the Capuchins have a small one, and the Greeks four or five very good Churches. And the Republick of Ragusa have a Consul here, as well as the French and English.

I am my Lord

LIMESOLE IN CYPRUS
4/15 December 1738
Pococke to Bishop Milles
Account 24[11]

My Lord

The climate of Cyprus is not so temperate as many other parts that are in the same Latitude, the cold winds flowing from the high mountains of Celicia in winter make it very cold, especially the Northern parts of it, and the high hills also of the Island some of them covered with snow, make fires very necessary throughout the whole Isle during the cold season, which are little used in any other parts of the Levant, the clouds also breaking on these hills fall down in heavy rains often from many days together, inso much that I have heard it has been known to rain for forty days incessantly. The same mountains of Cyprus, and the shallow soil mostly on a white freestone make it excessively hot in summer, and the Island very unhealthy especially to strangers, who often fall into fevers here which if they do not carry them off continue for a considerable time, the disorder lurking in the blood and occasion frequent relapses. The soil of Cyprus is for the most part rocky, there are in it some entire hills of Talc or Gyrse, some running in plates and others in shoots like Chrystal, and this latter they use in many parts especially at Lar[n]ica as stones to build walls, they have also in the mountains near that city a sort of a marble paving stone, that cuts like chalk with a common saw, and much of it is found in the old walls of Citium which being in very large peices seems to have been laid to bind the walls, and somewhere near Nicosia they have a yellowish marble which being burnt I was assured would produce some small quantity of sulphur. At a mountain towards Solea is found the Asbestus stone or Amienthus it is something on [of] a blackish green, but running in veins in such a manner that the steeple of it is not above half an inch long, it is much to be doubted if they could spin it to a thread, but by some experiments tried with it I have reason to think an incombustible paper might easily [be] made of it, such as they make of that of Muscovy. Near Baffa is a hill that produces a stone that is called the Baffa diamond it is a very hard stone and seems rather to exceed the Bristol and Kerry stones. Cyprus has been also very famous for its minerals and for many sorts of pretious stones probably found in the mines when they were worked. In going round the Island I saw only two Iron mines which are not now worked, because in Cyprus they want hands to cultivate the ground, nor is it agreable to the

11 BL, Add MS 15779.

inclination of the people to work these mines because they would be ill paid by the officers of the Grand Signior. One of these Iron Mines is about half a days journey East North East of Baffa, the other is at Solea where a large hill seems to be composed of this Ore, which is very fine and light being porous and crumbling, and of a red colour; they have here also the several sorts of earth used by painters called Terra, Umbra, Verde, Rossa and Ialla [?], and I was assured that not long ago was found by a traveller a very fine Azure earth which is uncommon and either is not much known or in so small quantities, that it is not exported. The antients mention three rivers in Cyprus, the Lycus, the Fetius, and Pedius, tho' at best they deserve only the name of rivolets, and I suppose the water seldom fails to run in these, tho' it is generally said there are no rivers in Cyprus, and certain it is they have no fresh water fish except small Crabs which are in most of the rivers in Asia. All round the Island are beds of winter torrents that run from the mountains after rain, but during the summer months when there is never any rain in these Southern parts they are entirely dry, excepting some few springs which have been rarely known to fail, the water that is drawn out of wells is almost all brackish, occasioned by that Nitre in the soil which produces the Salt in the Lakes before mentioned, and at Lar[n]ica they send above a league for all the good water they drink the water of the Isle seems to depend on the rain, and when clouds have been wanting to fall down in rain, the springs have failed, which had been fed by those vapours lying on the mountains; and historians give an account, that not having rained for thirty six years, the Island was abandoned in the time of Constantine for want of water. The Cypress trees in this Island gave name to it, and there are a great number of them on the Eastern promontory, and in the Northern part of it, and all over the Isle is a tree much resembling it in every thing but its seed which is like the Juniper, it is called in Greek Avorados in German Segen Pitum, but grows rather like a large shrub than a tree; they have also the Juniper the Pine in great numbers on the mountains, with which they make tarr. A tree wch is in the Holy land called in Arab Kaikoub, and in modern Greek Andraugra, they have also the Caroub called in Greek Keraka which is supposed to be the Locust tree, the fruit of which in this Isle exceeds any other, growing like a flat bean, and is exported both to Syria and Egypt, most of the trees in the Island are evergreen, but what it is most famous for is the tree called Xyton Effende the wood of our Lord, by the natives and by naturalists Lignum Cyprinum, and Lignum Rhodium because it grows in these two Island, it is also called the Rose wood by reason of the smell of the wood, some say it is in other parts of the Levant, and also in the Isle of Martinica, it grows like the Platanus or Plane tree and bears a seed or mast like that only the leaf is rather smaller, as well as the fruit, the leaves being rub'd have a fine white Turpentine, especially when any incisions are made in the bark, it is of this I suppose they extract a very fine perfumed oyl; both the oyl and the wood they say have the virtue to fortify the heart and brain, the common people here cut off the bark and wood together toast it in the fire and sucking it look in it as a specific remedy in a fever, and seem to think that it has a miraculous operation. They make here Labdanum or Ladenum of a very small shrub called Ladany and by Botanists Cistas Ledon or Cistus Ladanitera it is a sort of a Balsamick Aromatick herb, it is said that the Goats feeding on it in the month of May, a juice sticks to their beards, and makes a sort of a cake which being taken off they purify it and make the Labdanum, this is in some measure true, but this sort requires much labour to clean it and is not perfectly sweet, so in Cyprus they use the same method as in the other Islands with an instrument they call Stavera because it is like a cross; it exactly resembles a Crossbow and to it they tye peices of yarn about three feet long which in the month of May they draw over the leaves, and the fine substance sticking to the yarn, they lay it in the sun, and when it is hot draw it from the yarn, the common people to make it weigh mix it up with sand, which is what the Druggist[s] call Labdenum in sortis and so it is commonly sold, but being purified from the sand it comes to be of the nature of warm bees wax, which is what they call Liquid Labdenum, it is esteemed a remedy both taken inwardly and appli[e]d outwardly for many disorders, and the smoak of it is good for the Eyes; but what it is most used for is against the infection of the plague, by carrying it in the hand and smelling to it. The Island produces also Cotton, a root called Fuy being a sort of Madder, Coloquintida and Lemmons and Civil [Seville?] Oranges in great abundance, and they have great plenty of red wine, the rich Cyprus wine which is so esteemed in all parts it being very clear and produced only about Limesole being much of the nature of Canary wine, but rather heavier and more luscious, and in some few places they make a good red wine. They have very few horses in Cyprus, for the plough they use their cows which they do not milk, looking on it as cruel to milk and work the same beast, but perhaps may have rather regard to the young that are to be nourished by it; this loss is made up by their Goats, which are the most beautifully spotted beasts I have seen of that kind, and for them a great part of the soil of Cyprus is more fit than for large cattle, these they milk in a cruel manner by pressing the duggs, and of the milk they make that Cheese which is so famous all over the Levant and the only good to be met with in these parts, they make them very small and thick much in the shape of the weights of the antients, they keep them in oyl otherwise when new they would breed a worm, and when old would soon grow dry. The Turks have such an aversion to swine agreable to their law, that the Christians dare not keep them where they have not as much power as they have in Cyprus, but from this place the Christians in all parts are supplied with excellent hams, which they cure in a particular manner by salting them, pouring the rich wine on them and then pressing them very dry then hang them up. For burden and the saddle they use Mules which are very good, and also Asses: they have hardly any other wild beasts, but foxes, and hares, and wild Goats, and very little variety of game, the cheif are a very beautifull partridge called in Italian Coturni which I beleive is the same of the red partridge in France, and a beautifull bird called in Italian Francoline in Greek Astochinara they are almost as large and as beautifull as a Pheasant, and are in greater plenty in Syria. There are a surprising number of Snakes here but most of them harmless except a small one, and another that is generally esteemed to be the Asp and here supplies the place of the viper, and is said to have its vertues, it is called Kouphi by reason that it is deaf, they grow near two inches thick, and bigger in proportion than Snakes the head being rather small with regard to their body, and it is positively affirmed that they have been known to swallow a hare whole, their bite is exceeding venemous, but the curing of it depends on the place where they bit, the poison is in bags on their teeth as in the viper, and it has been cured by medicines, and by the serpent stone. I have been informed there is an Asp in Italy which is not deaf, it is possible the Psalmist might speak of this creature, when he makes mention of the deaf adder which refuseth to hear the voice of the charmer. They have an exceeding large broad Spider somewhat resembling a small crab, the Franks call it the Tarantula, which is only in Apulia and those parts of Italy; but there is a brown house Lizard called a Taranta, and if it walks over any part of the body it causes a very great itching which continues for some time with much pain. I do not find that they have Scorpions, which are so common in Syria, but the locusts when they come ravage the country in a most terrible manner, destroy a whole country of corn where they alight, and eat the greens of the Mulberry trees on which their silk depends. The Cyprioles are the most subtle and artfull people of all the Levant, and as they have not more veracity then their neighbours, so their words are not to be depended upon, as they make use of all means that way to deceive. The women are little superior to their ancestors with regard to their virtue, and as they go unveiled so they further expose themselves in a manner that in these parts in looked on as very indecent, they seem to retain something of the heathen custom of going in procession to the Sea annually in remembrance of Venus's coming out of the Sea, which it is said was attended with some other circumstances; for every Whitsunday they go out to the Sea side. They retain here the barbarous custom of the other Easterns of treating their wife as servants, for they wait on them at table, and never sit down with them unless in such families as are civilized by much conversation with the Franks, for here having been under the Greek Emperors and the

Venetians, they have come very much into the European customs: they make us[e] of chairs and tables, and lye on oblong square tables, probably to be more free from Noxious animals in the Summer, and from the damp occasioned by great rains in the winter. They have also carriages with two wheels drawn by two Oxen. The dress of the women is much like the Irish a jacket and petticoat and the head bound round with a handkercheif.[12] The dress of the countrymen is a great pair of trowsers and a very short jacket, those who value themselves on being some what above the common rank dress as the Turks, but wear a red cap turned up to the top with a furr which is the proper Greek dress and used by those of the Islands in whatever parts of the Levant they are.

I am my Lord

LARNICA IN CYPRUS
7/18 December 1738
Pococke to Bishop Milles
Account 25[13]

My Lord

Cyprus on account of its situation and the cheapness of all sorts of provisions, is the place where almost all ships touch in their voyages in these parts, and by this way a correspondence is carried on mostly to and from Syria, both between all the places of the Levant and Christendom, so that furnishing ships with provisions is one of the principal branches of the trade of the Island, tho' it is contrary to their laws yet they sometimes export corn to Christendom. They send their Cottons to Holland, England, Venice, and Leghorn, and wool to Italy and France. A root called Fuah in Arab, in Greek Lizare, and in Latin Rubra Finctorum [Tinctorum] they send to Alexandretta, and by Aleppo to Diarbeck [Diyarbakr] and Persia to dye red, serving only for Cottons for which it is also used here, it is the root of a herb, and is called by the English Madder, but is doubted whether it is the Madder [known] so well in Holland. Another thing they export for dying woollen stuffs red, is what is falsely called by the English vermilion which is known to be made of Cinnaber [Cinnabar] but this is nothing else but the produce of the seed of Alkermes called in Latin Ilex Coc[c]ifera, in the seed is a small hole, out of which comes a very fine powder which is

12 This analogy to the Irish dress is not given in Pococke's book, but would have been of interest to his uncle, the Bishop of Waterford.

13 BL, Add MS 15779.

called the powder of it, of which the famous Syrup of Alkermes is made, and the seed after serves for dying, and both are exported to Venice and Marseilles. Coloquintida is cultivated here and esteemed better than that of Egypt, which being larger does not dry so well, it grows like the Calabash, it is used in medicine, the pulp of it which is round, the seed of it is sent mostly to England and I have heard that being a strong bitter they use instead of hops to put into beer, the seed is sent to Germany being much used there in embalming bodies. In Egypt they fill the shell with milk and let it stand some time and take it as an emetick. They make a great quantity of yellow red and black leather they send to Constantinople much, and the[y] export yearly near a hundred thousand pound weight of raw silk to London and Marseilles, of those fine plain Cotton dimities made at Nicosia. In a word it is surprising, being a hard weighty silk much esteemed to make gold and silver laces sold by weight, and also for sewing [half a line missing][14] thing to see Cyprus maintain its own people in so great plenty and export so many things abroad, when one consider[s] the small extent of the Island and that half of it at least is mountainous, and much of it near the sea lyes uncultivated by reason of the Corsairs, and the Island is not well peopled, eighty thousand souls being the most that are computed in it, whereas historians say in Trajans time the Jews massacred here in one day two hundred and forty thousand persons, since which time they have never permitted any Jews to live in the Island. Two thirds of the inhabitants are Christians there being twelve thousand that pay the tributes exclusive of the women and children. The Christians are mostly Greeks, there are indeed near Nicosia some few villages of Maronites; and in the city of Nicosia a small number of Armenians who are very poor and have an Archbishop, and a Convent in the country, the Mahometan men marry very much to Christian women, and keep the fasts with their wifes, and many of them are thought not to be ill inclined to Christianity, nevertheless the Turks are so jealous of the Christian power here that they will not suffer them to buy any black slaves or others that are Mahometans, which former are brought hither frequently from Egypt and sold to the Turks. The Greeks have one Archbishop of Nicosia and three Bishops, that is of Larnica, Gerines, and Baffa. The Greeks have every where their churches but cannot repair any that are ruined without a license, they all end in a semicircle and are arched over, the larger sort have their naves and have generally as many small cupolas in the middle one as there are arches turn'd on

14 The scribe seems to have confused two lines of the original script here, as can be seen comparing this letter with the almost identical text of *A Description of the East*, Volume 2, Book 1, 233. The sentence should read: "In a word it is a surprising thing to see Cyprus…" and the clause "being a hard weight silk…" belongs to the previous sentence.

pillars in the length of the church, or at least one in the middle; but those cupolas are but of a mean architecture, the basis of them being generally a square rounded at the corners, it is said formerly they had a custom to hoist flags at the West end of their churches on Sundays and Holy days, and I saw in some of them stones at top with holes in them to set in the flag staff. There are a great number of Monasteries in the Island, they are to be looked on only as religious societies that go out to labour on the lands that belong to them with the Superior to oversee them, this is their employment all day, and half the night is spent in performing their long services, they may be also esteemed as places of education, where the youth who labour by day learn to read and chaunt [chant] their offices at night, the lay servants who are distinguished only by a cap, answer to the lay brothers in the Roman church, but they never take the vow, and may leave the Convent and marry, and in these respects the Eastern churches pretty much agree. There is no nunnery in Cyprus, and I met with one only of the Greeks throughout all Syria being very uncommon in these parts, except among the Maronites of Mount Libanon, they take only the vow of chastity and obedience, and I was a witness in many places of the inconvenience of allowing property in a monsastick life, every Monk generally finds himself in clothing and pays his tribute out of his own purse, which is made of the charity cheifly of those who come to the Convents whether out of devotion or to retire or divert themselves. When a Convent is well situated, the Turks often come and stay in it and put the Convent to some expence and never make any return, they are also like ruins [inns] where every one turns in, but the Christians always extended their charity, and what the Monk leaves when he dies falls to the share of the Bishop. The preists here are as ignorant as the other preists of the Eastern Churches, and the Greek is their Mother tongue, yet they do not so much as understand the antient Greek of the New Testament tho' the modern Greek differs from it very much [little] if they were used to pronounce the antient Greek as they do the modern; indeed in Cyprus the Greek is more corrupted than in many other Islands as they have taken some words from the Venetians whilst they were among them, it is notwithstanding a sweet language but they speak it exceeding fast.

Till within this thirty years Cyprus was governed by a Pasha, but now it is governed by a more inferior officer called a Mosolem, the late Grand Signor gave this Island as a dowry to his daughter that was married to the Grand Vizier Ibrahim Pasha and since that the Grand Vizier has always had the Island he legally makes of it about seventy five purses a year, each about seventy pound sterling but then he has only a share of the Harach and Nozole I shall after mention,[15] and I have been informed that the whole Island brings in five hundred purses a year, there are besides fines for offences, and upon account of any death that appears not to be natural in which case the village pays one purse. All the land of the Island belongs to the Grand Signor which he has sold to the present possessors and their male heirs without paying rent, and in default of male heirs the Grand Signor sells it again. The title of the land which doubtless belonged to the church is granted to two sorts of military bodies, one of them are called Zains of which here are eighteen heads, who having the titles of Suck lands are obliged to send so many men to war. The others are called Timariots,[16] under the latter name, lands are granted all over the Turkish Empire, on condition that they furnish so many men to the war, there is also a pole tax called the Nozool of about six dollars a year paid by all that are not obliged to go to the war both Christians and Turks, and the Christians pay a tribute called the Harach which is universal over the Turkish Empire it is from ten to fifteen dollars a head, there is also a small duty of twenty two Timeens or forty four Medeens a head which is about three shillings English, paid yearly to the village wherever one was born. The salt and customs belong to the Janizaries who are about a thousand and have an Aga sent to govern them generally once a year from Constantinople. The Cyprioles having their lands at so easy a rate, any one would imagine they must live very easily but the Mosolem is almost continually harassing the Christians, who often leave the Island and go to the coast of Cilicia and as often return again out of that natural love every one has for his own country, many of them notwithstanding are in all the sea part towns of Syria which very much thins the Island. Cyprus is now divided into sixteen Cadelists, each having its Aga or Governour and Cadi or minister of justice, being the following towns and a district round them. Cherties [Cherkes], Nicosia, Gerines, Morfo, Lefca, Solea, Baffa, Arsinoe, Aitimo, Chrusafou, Limesole, Episcopi, Larnica, Messarias, Famagusta and Carpass.

I am my Lord

The first letter from Egypt was written the day after his arrival, on Christmas Day. The letters, as mentioned in Pococke's cover note, comprise "an account of the government customs and part of the natural history of Egypt", and according to the address, the first five were composed while he was travelling in Damietta (Letter 27), El-Mahalla (Letter 28), Cairo (Letter 29), Faiyum (Letter 30) and Saqqara (Letter 31). Letters 32-34 were written while he was in Cairo.

15 Spelled variously as Nozool (below) and in other sources (including his book) Nozoul.

16 Pococke refers to these in his book as tythes. See *Description* 2, Book 1, p. 234.

GRAND CAIRO
7/18 March 1739[17]
Pococke to Bishop Milles
Account 26[18]

My Lord

I send to your Lordship this day thirteen letters dated from different parts, many of them from whence I had not an opportunity of sending them, I have sent two before relating to Cyprus, and now five more, the rest being an account of the government customs and part of the natural history of Egypt. Some particular reasons determined to write to your Lordship on these subjects rather than at present to continue on my account of Syria which I had brought down to Balbeck. Your Lordship will be so good as to excuse many things in them under the many inconveniences I write and especially my bad writing for I had not time to write over a second time.

I shall set out shortly for Candia and thence shall proceed with the utmost expedition to Constantinople, some accidents [circumstances/incidents] with regard to seeing some places, having made my stay in Egypt much longer than I designed.

I writ to your Lordship four letters from Egypt giving a general account of my journey into upper Egypt which I fear miscarried because they were not come to hand, when those from the Holy land were arrived.

I have received no letters from England or Ireland dated since July, so conclude they have been sent to Smyrna or Constantinople by mistake, to which places I writ for them to be sent here about two months ago, and hope to receive them before my departure from this country. My humblest salutes attend the Chantor [Jeremiah Milles]. Begging your Lordships blessing I am with the greatest respect and submission

My Lord

17 This letter, though dated March 1739, appears in the manuscript as the first letter in this section.

18 BL, Add MS 15779.

DAMIATA IN EGYPT
25 December/5 January 1738/39
Pococke to Bishop Milles
Letter 27[19]

My Lord

Egypt is divided into two parts, the upper, and the lower; the whole is comprehended under the Arabick name of Messa, but upper Egypt is particularly distinguished by the name of Said; the whole is divided into provinces governed either by Sangjacks or Beys those antient governours under whose standard all the military men of the province were obliged to rank themselves, whenever they pleased to call them, or by their Castafs [Cashifs] during their absence. Lower Egypt is divided into nine provinces, two of them are in Delta that is Mehullah and Menoufick [Menoufieh], on the East of the Nile are Balbeis, Mansoura and Cherke Atsich. On the West are Baheira, Gizeh which extends to the mountains of Faiume and allways belongs to the Testerdar or Lord high Treasurer of Egypt, who governs it by his Cashif, the other two are Faiume, and Benesuef. Upper Egypt was formerly divided into twenty four provinces, but now the greater part of it is under these five Arab Shekes. On the West the Sheke of Aboutig who has also some country on the East, the Sheke of Pardice near Girge who has a very small territory there, and a larger about Cous and Luxerim; the Sheke of Furchout, whose territory extends on the West near as far as the Cataracts and has also a country on the East beyond that of the Sheke of Paradice. On the East is the Emir of Aksuim, who has a large country also on the West. A Sheke who resides at Elbanaut and has a small territory about Kepht, and is the brother of the Sheke of Furchout, and then the country before mentioned of the Sheke of Pardice, besides these Shekes on the West are the Cashifs of Minio and Mansalouth, over their respective provinces, and the province of Girge under the Sangjack and his Caimacam. The country on the East being mostly in the possession of Arabs that are not under any regular government, there is but one province under its cashif as well as I could be informed which is that of Sciout, and the Cashif of Ibrahim both on the East, and West above the first Cataract.

All Egypt on the part of the Grand Signor is governed by a Pasha who having in reality but very little power, his business seems cheifly to consist in the following particulars, to communicate to the Divan of Beys, and to the Divans of the several military Adjack or bodies, the orders of the Grand Signor, and to see

19 BL, Add MS 15779.

that they be executed by the proper officers, if the farms of the Grand Signor, the fines paid when any life drops on the lands sold by the Grand Signor during life, as are all the lands of Egypt, to be well informed of it, and to raise money by selling those; to keep up as good an interest as possible with the persons of greatest, with the Sheke Belcet and the leading men of the military bodies, to watch their designs, and if he finds them prejudicial to the interest of the Port, to foment divisions among them, and if he cannot obtain his ends, to make the best party he can, and by the best means he can cut off those he finds too aspiring, tho' it will certainly end in his own deposition; about which he not be very solicitous, as his person is always held sacred, and as his removal will be a sure step to a more profitable employment. As to the first, the orders of the Grand Signior his business is not very great, because they will not suffer any any [sic] such are sent. If he has a good interest with the Sheke Belcet and the leading men of the military bodies, and they attempt nothing against the Porte, and the Porte leaves every thing without innovation the office of the Pasha is very easy, if he is to raise divisions he or his Caia ought to be men of parts to manage intreigue [intrigue] and employ fit instruments, and if it is necessary to cut off some leading men, money must be well and plentifully applied to engage persons least suspected to act with the utmost secrecy and treachery.

If some might conjecture at the original of a Sheke, a Besset or head of the city either that he answers to such an officer of the Arabs during the Marmaluke government, or one may imagine that the people at some time or other might require the Port to nominate such a Bey as might be agreable to them to be their head and take care of their interests, a person himself always of the greatest interest, and is in reality Lord over all the land of Egypt, and will act with the greatest prudence, when he makes himself esteemed and revered by the other leading men, keeps up a very good correspondence with the Porte, and prevents any innovations that may be attempted contrary to the interest of the Porte, and takes care that they send no orders that may encroach on the liberties the country enjoys, and if any tumults happen against the orders of the Porte, to endeavour to palliate them as well as possible at Constantinople that no more may be heard of it, and in general to take care that nothing be done either at home or abroad that relates to his country without being communicated to him or without his advice.

As upper Egypt is under such powerfull Arabian Shekes, so it is necessary to send a Sanjacke governour there to collect the tributes due from them and from the Cashifs under him, this governour resides at Girge, with his officers almost in as much state as a Pasha has his Divan, and detachments from the military bodies reside there, he is named yearly by the Divan at Cairo.

Historians give us an account that Sultan Selim [I, r. 1512-1520] utterly destroyed the Marmalukes [Mamluks] when he conquered Egypt, it is probable he might leave them the same form of government as they had before, but probably it was only a shadow of government, and that he made Beys of his own creation in the country; it is possible these Beys might come in length of time to be succeeded by their slaves they had advanced to offices, and so the Beye at length might all have been slaves, and thus it might approach nearer to the Marmaluke government, notwithstanding the Beys seem not to have had great power, but the military bodies particularly the Janizares and Azots growing powerfull might begin to attempt some innovations in the government, for in a list of Pashas we find everything went on very quiet till the year 1602 when a Pasha was massacred, and twenty eight years after the military bodies deposed a Pasha which is the only instance of the kind from the time of Sultan Selim to the year 1673, tho' it is said of late years they have often been obliged to descend by the soldiery. It is probably the Porte perceiving the military bodies grew too powerfull were well pleased to sell the villages to the leading men among them, which formerly they would not purchase that they might not subject themselves to the Beys, to whom now they pay court, the Beys as governours better, for the power being in the hands of the Beys, if they grow dangerous it is more easy to cut them off, without creating any disturbances among the people, whereas any injury offered to a leading man among the military bodies, might stir up the resentment of the whole body, and be of dangerous consequence. The present succession of the Beys as they are not entirely attached to the Porte, is better than it was hereditary for the Hasnedan or treasurer because of a Bey or some other great officer or Cashif, that was slave to the deceased Bey, marrying the widow is obliged to part with a great part of the estate to make him self friends, to secure his succession, which prevents very great estates centering in one person. But the Porte seems to have been much mistaken in their measures to secure the subjection of Egypt, one great means towards it would be to have constantly changed the military bodies every year, and not let them settle and have a succession of natives of the country, and then to have had Beys always Turks and creatures of their own, and cheifly to have strictly prohibited the importation of slaves into Egypt, they being the strength of the present government against the Port, for a great man having given his slaves liberty they are made Caimacams then Cashifs and then begin to purchase other slaves and at length to give them liberty and al are strictly dependant on the great master.

The Pasha has a Caia [T. *kahya*, steward] a Bey pro tempore by his office who is his prime minister and generally holds the Divan, the Pasha like the Grand Signior sitting behind a lattice in a room at the back of the Divan, and one of the great officers that always attends him when he goes out is the Tergjuman Aga [Chief Interpreter] who is not only an interpreter but more especially acts as master of the ceremonies, he has the Chouses Shateri Bostangees and a horse guard of Turks on whom he cheifly depends both for his safety if there was any danger and also in sending all his dispatches.

Next in rank after the Pasha is the Emir Hadge [*Amīr al-ḥājj*], or Prince of the Pilgrims that go to Mecca, named yearly from Constantinople, and generally continues in three years and sometimes six but if they carry the Caravan seven years the Grand Signior presents them with a golden collar their persons are esteemed sacred, they cannot be publickly cut off, he is sovereign prince of all the estates that belong to Mecca, and sends his sardars to govern them.

The next office is the Tefterdar [*Defterdar*] or Lord high Treasurer of the tribute paid out of the lands to the grand Signior, he is a Bey as well as the Emir Hadge, named for a year likewise by the Grand Signior, but is generally continued in the office many years, being the Bey generally of greatest interest after the Shekeel Belcet, and in his absence is commonly made Caimacam to govern when the Pasha is deposed by the soldiery or recalled.

Cairo is under the guard of the Janizaries, old Cairo is guarded by a Bey who resides there and is changed every month. So likewise is the country North of Cairo about Adalia, and the Asobs have the charge of the country around the City.

An officer patroles about the City more especially by night – who is called the Walla, answering to the Turkish officer called Soubahsa [*subaşı*, the chief of police], he takes up all persons he finds committing any disorders, or that cannot give an account of themselves, and often has their heads cut off on the spot, if so be they are not under the protection of the Janizaries or of any of the military bodies. As he is the terror of rogues so for presents made to him, he is often their protector, and without those presents they are sure to be cut off, and to him the great men send for any villains that have render'd themselves obnoxious to them, and they are sure to have them. Another officer is the Metefsib[20] who has the care of all weights and measures and to see that every thing is made justly according to them. Under the Cashifs in their provinces is a Caimacam in every great village, who may have eight or ten or more small villages under him, each of which have in them a Sheke el Bellet either a native Egyptian, or an Arab where the Arabs are settled, and these Caimacams as well as the Cashifs have to manage with the Sheke Arabs, who in many parts have

20 Muhtesib, the market inspector and the principal enforcer of *ihtisab* – the controls governing social, economic and religious life. See the entry by S. Aynural, in G. Ágoston and B. Masters, *Encyclopedia of the Ottoman Empire*, New York, 2009, pp. 264-5.

the real power, and the Cashif governs by making a leading Sheke Arab his friend by presents and so forth, who perhaps reside with him in the same town. All the annual officers are appointed by the Divan the last day of August, and the cheif business of all these Governors besides keeping the country in order is to get in the money for the Grand Signior, but more especially whatever they can for themselves.

I am my Lord

MEHULLAH IN EGYPT
29 December/9 January 1738/39
Pococke to Bishop Milles
Account 28[21]

My Lord

There are five bodies of Spahis on horse in Egypt, the two principal area the Chaous and the Muteferrika, who were originally the guards of the Sultans of Egypt, and their leaders were his two Vizers that always accompanied him, the Chousler Caiasi on his right hand, and the Muteferrika bashee on his left, and they now always go out with the Pasha. The body of Chaouses seem originally to have been the guard of which the Sultan used to send persons to execute his orders, for the Grand Signior has a bod of Chaouse for that purpose, but they are not reckoned among the Spahis.

The Muleferrika are in themselves of the greatest dignity, which that words implys, being generally persons of some distinction, and the Grand Signior has a guard of this name that are not in the body of the Spahis's; these are here sent to garrison Castles, and at present are in the Castles of Adjeroule, and Yembo in the road to Mecca, it is not improbable that they began to send them into these garrisons when the Circassian Marmalukes or slaves had drove out the first Marmaluke succession of kings and Turcoman slaves called Mahariah, who originally were sent to guard the fortresses on the sea coast, as the Circassions were sent to the inland garrisons, and were also he guards of the Seraglio of the Sultans. These Carcassions seem to have been the body of Spahis called Cercasi, that these two are the Jjomelu and Tufersi [*tüfekçi*, musketeer],[22] these

21 BL, Add MS 15779.

22 He spells these as Giomelu and Tufecsi in *Description 1*, p. 166. The first part of this chapter, entitled "Of the Military Bodies in Egypt", seems to have been based almost verbatim on the letter.

three last have at their heads, each of them an Aga, and when they have passed through that office, they are called Ikiars, or seniors, and compose the council of their respective Divans in the house of their Aga, but these bodies are of little interest, when the Divans of the Janizaries or Azabs would have any thing done, they send a Chaous to the Pasha, to have an order for it, which must not be denied, the order is brought to the Caia in charge, who gives it to the Aga to put in execution.

The slaves of the officers of the military bodies, when they give them their liberty by ordering them to let their beards grow, become Janizaries and advanced, and so it is realy a Marmaluke government thro' every part, but the slaves are by no means a despicable people, they are the fairest and most promising Christian children taken for the tribute brought here, sold and become Mahometans, are well clothed and fed, taught to ride, throw the dart, and shoot with great dexterity, and every one of them has a servant to take care of his horse, to wait on him and attend him on foot by his horse whenever they go out, and to say the truth they are in the hands of very kind masters, and are as observant of them, for of them they are to expect their liberty their advancement, and every thing, so that a slave behaves himself as one that is to become a governour of towns and provinces, and if he has more ambitious views as one that may come to succeed his master, and this is observed in the strength of the present government against the Grand Signior. In Constantinople it is the slaves of the grand Signior that are advanced, what has transferred the power in a great measure, from the two military bodies of infantry to the Beys as observed in the leaders of those bodies purchasing lands of the Grand Signior, which obliges them to be submissive to the Beys that they may not ruin their villages, whereas formerly the bodies were rich had a treasurer, and an estate mostly in Cairo as they have still, but the revenues of it the Divan divide among themselves, and when the publick body was rich, each particularly was poor, and no one would purchase any village that he might not be subject to the Beys, to whose houses they would not then go, and this was what secured their power, whereas now they have lost that influence and liberty they enjoyed by submitting themselves. At present indeed they have a considerable share of power by the policy of the present leading man, who, at the same time they pay court to him, and the other Beys thinks proper to do nothing without the advice and approbation of the leading Caia among the Janizaries, in order to establish himself.

The revenues of the Grand Signior in Egypt consist of three branches, that which arises [sic]from the lands, the customs and the Pole tax on Christians and Jews. The immense riches center in the Grand Signior, for the little officers oppress the people, the great officers squeeze them, and the Pashas all the people under him, and the Pasha becomes a prey to the great people of the Porte if not to the Grand Signior himself, and the Grand Signior at last seizes the riches of the great officers about him.

All the villages in Egypt pay a certain yearly rent to the Grand Signior, which is fixed and this is the Hasha or Treasurer that is sent every year to Constantinople, how easy the rest is may be concluded from the sum which is raised which amounts only to six thousand purses each of twenty five thousand Medins, which is about eighty pound Sterling, out of this twelve thousand soldiers are paid, and corn flower, Oyl &c sent yearly to Mecca, which reduces the treasure to twelve hundred purses, out of which also they deduct from fifty to two hundred purses according to the pretences they can make, which are mostly with regard to the conveyance of the water of the Nile to their lands as in opening and cleaning some great common Canals, and repairing some walls that relate for keeping up or turning the water. This treasure used to be sent by sea, but once being taken by the Maltese, they have since gone by land taking the other treasures in the way at Damascus and other places, under the conduct of a Bey and a detachment from each of the military bodies, they return by sea under the convoy of four men of war, and have the privilege of bringing what merchandise they please custom free. It seems if Egypt was formerly divided into Timars or Knights fee that is land granted for life on condition of furnishing so many men for the war, who were obliged to rank themselves under the Sangiaks or Bameers [?] of their province, from whence came the title of Sangiaks for governours of a province, but I cannot now find any thing of this, and it may be that the extraordinary advance the country gives to the three thousand soldiers sent once in three years at the Gran Signiors command, is in lieu of this. These lands are sold during life at that fixed rent and the money arising from those sales, is commonly given to the Pasha, who pays to the Porte about eight hundred purses a year, but much roguery is practised in this, for a great man buys them in the name of a slave, and has two or more slaves of that name, and often pretends when one dies that the village was bought in the name of the other. There are lands that belong to the Mecca, the revenue of which is received by the Keiler Aga or black Eunuch, who sends a deputy here to manage the revenues of those estates which are mostly paid in corn rent to supply the country of Mecca. And as to the Grand Signiors rents if the Nile dos [does] not rise to sixteen pikes, when the [word missing] at Cairo is cut they are not sent because the consequence of it would be a want of every thing throughout the land of Egypt and occasion a famine. The second branch of the Grand Signiors revenues are the customs of Egypt. The Ajack of the Janizaries farm most of the customs of the Grand Signior, and last year the Grand Signior sending a person to take it out of their hands they managed so that it had no effect. The customs of Damiata are let for four hundred purses and yield six hundred. The customs of Alexandria for two hundred and fifty purses and produce about four hundred to the Janizaries. The customs of Bulac the port of Cairo and of things landed from upper Egypt at Old Cairo are likewise set to the Janizaries, for the former they give purses and make of it [several words missing] for the latter [several words missing] and it yeilds them.

All these customs are given to the Pasha to support his grandeur, and those of Suez which may be worth eight hundred purses, being about forty shillings on every bale that is two Funducllee [?] sequins he keeps in his own hands, as likewise of the merchandizes brought from Mecca which is about half a Guinea on every camel. The customs of upper Egypt are given to the Bey of upper Egypt; the customs arise from a duty of ten per Cent, on all goods imported, and seven more when they come to Cairo, but the English and French pay only three per Cent instead of seventeen, which is by treatys with the Porte. The Swedes also have lately obtained an order for the same purpose. The Venetians, and Dutch by reason of some debts contracted here, do not send Consuls, and have lost their privilege, and all persons pay three per Cent for goods exported, all the customs area second time farmed to Jews. The other branch of the Grand Signiors revenue in Egypt is the Pole tax on the Christians and Jews, called the Harach this till within a few years was in the hands of the Janizaries who gave eighty purses a year for it, but Harach Aga being sent from Constantinople, by applying a great sum of money to the leading man got possession of the Harach, and it is said he makes of it for the Grand Signior eight hundred purses, before this the Christians paid but a trifle; but now they pay according to their substance, two dollars and a half, of five dollars, or seven dollars and a half, each dollar being about half a crown, for which a certain number of papers are sent yearly from Constantinople as to other parts, which must be either returned, or the money answering to the sums contained in them, which papers are given to those who pay the tribute, and are their acquittance; the persons that pay are only men after they arrive at the age of sixteen, the Jews say their number is but four thousand in Cairo and if so not above a thousand pay, for the number that pay tribute is commonly computed at a fourth or fifth of the whole, and suppassing these there are eighty three thousand Christians that pay to make up the number of purses, the whole is the most that there are, and those cheifly in great towns and in upper Egypt.

As to the trade of Egypt that which is within itself is in supplying the lower parts from above with corn and Dates, and the upper part from Delta with Rice and linnens, and from Cairo all sorts of things imported into Egypt. As upper Egypt has not communications with the red sea, or any other parts, that can supply them with those things before the way was found to the East Indies by the Cape of Good hope, Egypt had a great trade by landing all Indian and

Persian goods at Cosseir on the red sea, whence they were carried over Europe by the Venetians which was the great riches of that state, which as ever since declined, bringing them to Kept four days by land, and then carrying them to Alexandria from the Nile by a canal at Force a few leagues above Rosetto whence this easy method without going to sea as they do now from Rosetto to Alexandria, was interrupted by the depredations of the Arabs, and the Canal was choaked up, it was last year cleaned, but the trade still continues by seas, as more secure, some attempts have been made to carry corn to Corsier to be exported, and to bring Coffea thence,[23] and some other goods most wanted in upper Egypt, but the Arabs plundered the Caravans and obliged them to desist, and the people of Cairo could not be much concerned at it because a trade there might much prejudice the City; as might take its old course to Alexandria, the navigation North of Cossier to Suez being very dangerous, and when they go South and have not very fair winds they coast along and come to Anchor every night, whereas from Cossier they would have only to cross over the red sea to the South East to Moca, but Suez is now the port of Egypt on the red sea, where they bring Coffea from Moca and many other goods, both of Persia and India, but the richest come by the Caravan from Mecca as the most secure way, and these are sent to Constantinople and to all parts of Turky, the rich goods by the Caravan of the Hasha or treasurer who pay no custom going or coming, but India goods are much dearer here than they are in England. Coffea and Rice are prohibited to be exported out of Egypt into any parts out of Turkey, but presents make all those things easy, notwithstanding the West India Coffea is much inferior to this it has within this two years ruined all trade of Coffea to Europe, as it sells at a much lower price, but the market being risen they have exported it these two last years. Many sorts of India and Persian drugs are exported to Europe, and of the growth of this country Senna Cassia and Coloquintida and a red dye called Saffranoun and Indigo made of a herb called Nil. The import is English, French and Venetian Cloth, some drugs and dyes, many silks from Leghorn and Venice, pewter from England, tin lead and marble blocks from Leghorn, sorts of small wares from France, Venice, and Constantinople, from the latter Furrs all sorts of copper vases and plates which are much used being tinned over, and from Salonica they bring all their Iron in Turkish ships, that it may not be carried out of Turky, which is strictly prohibited, and Carpets from Asia Minor; with the parts of Barbary is a great trade in red caps, and many sorts of woollen things used for coverings.

I am my Lord

23 Pococke consistently spelled coffee in this way in his letters, but his mother always corrected his spelling to "coffee" when transcribing his correspondence.

GRAND CAIRO
1/12 January 1739
Pococke to Bishop Milles
Account 29[24]

My Lord

Justice is administered in Egypt in the same manner as in other parts a Cadeliskreri [Cadiliskier] like a Lord high Chancellor is sent yearly from Constantinople to Grand Cairo, to whom they may appeal from the Cadis and many causes of importance in law go immediately before him, he has his deputy called Nagib, and his house is the place of justice, these are in Grand Cairo also eight Cadis in different parts. In every ward is an officer called Kaban who is something like a publick notary, for by him all obligations to be valid ought to be drawn, here is likewise a publick weigh Master by whom every thing ought to be weighed. The City is divided into as many parts almost as streets, which divisions have gates to them, which are shut up at night, and so every part is a gaurd of two Janizaries. To Alexandria, Rosetto Damiata and Girch, the Cadi is sent yearly from Constantinople, and to other towns the Cadiliskier send them from this place, for the law is much studied here, and they speak the best Arabick of all the Easterns, except those of Micca [Mecca], whose expressions are stronger and they speak with greater guttural force which is the greatest perfection in the pronunciation; they have a saying that the Law cuts the sword but the sword cant cut the Law; for the Grand Signior himself cannot take off a person of the Law, but if any great man in that profession has render'd himself obnoxious, he sends him a horses tail, by which he is made a Pasha or General, and then he can send him a bowstring, but those executions of no kind are even ordered in Egypt least the people should take part with the offender, who is commanded to some other place to wait his punishment. As the Mahometan faith is divided into speculative and practical, they study the latter as it relates to their morals and their Laws, and they go through much the same course of study to be officers of their religion and of their law, only the most able men apply to the latter, and more particularly direct their studies to the knowledge of their laws; the religious persons who have the care of their Mosques here are called Shekes, and Turkish they are called Imaurus, which is also an Arab word, signifying as well as Sheke a Head; they have more or less to one mosque, according to its size and revenues, one is head over the rest, who tho' he answers to a parish preist, yet his office is rather that of a Clerk and his inferiors are the sextons of churches, it is the business of all of them to open the Mosque to cry

24 BL, Add MS 15779.

out to prayers, and of the head to begin their short prayers at the head of the congregation who stand rank and file in great order and make all their motions together, even these I have been told sometimes have been advanced to be Cadis, and Cadis are sometimes when unfit for business made Shekes of Mosques if they desire it, that is if they have not saved fortunes, for here the Cadis that are put in by the Cadelisker remain in their office many years. The relations of Mahomet called in Arab Sherif or Noble, by the Turks Emir or prince, have the priviledge of being exempt from appearing before any judge but their own head, sent also yearly from Constantinople being himself a relation of Mahomet and is called Naikibe elesharaf [*Naqib al-ashraf*]; but here they are so little esteemed that they often do not wear their green sash or Turban but a white one with a green string tied round; and if they do any injury to any one of the military bodies they take off their Turbant and kiss it, and then subject them to punishment as well as any others.

As the military bodies especially the Janizaries, have so great a share in the government it will be proper to give some account of them. There are five bodies of Spaki or horse who are little considered, but the two bodies of foot the Janizaries and Azabs have a great influence in all affairs. The Janizaries or Jenitcheri (which word signifies the new Band) are composed of a certain number of Companies called Odas or Chambers, over each of which is a head called Obadashee, these officers in processions march with Couks or high stiff Turbants, and a sheild slung behind them: from this office those that are thought fit to be advanced are put into an office called Boab Odabashee, who is something in the nature of a high Constable, and walks every day with many attendants about the principal parts of the City, to keep order and to see that all things are regular even to the dress. This office is for three months; they are after advanced to be Serach (a servant that holds the stirrup and goes on messages) of the Caia of the Janizaries in charge for a year, and then to the same office to the Aga of the Janizaries, he may after be chose into the number of Chouses or messengers of the Council of the Janizaries, and is called Cachouse or little Chouse in which office likewise he remains a year, as he dos in another called Alloy Chouse that is the Chouse of the Ceremonies who has the care and direction of all processions, from this office he may be removed to be Petelma which is somewhat in nature of a Procurator to the body, having the care of their effects, and when any one dies, under the protection of this body he seals up their houses, to secure the tenth part, which is due to them out of their effects: after this he comes to be Bas Chouses or head Chous, and enters into the Divan or Council of the Janizaries, has that title for a [word missing][25] and

25 The amount of time he remains in this position is not stated in the section of the book devoted to the role of the Janissaries (*Description* 1, pp. 167-8).

then remains in the Divan with the body of the Chouses, he must then either go to the ward, or with the Caravan to Mecca, or with the Treasure to Constantinople, and then he is made Waugh Caiasi or Caia in charge for a year, who is the judge in all affairs that relate to the body, from that office he is made Bas Caia or head Caia who presides in the Divan and when he has passed through this office he is in the body of the Caias and take[s] his place as the youngest below the rest. The office of Caia is properly a deputy or steward one that acts for a person or body. The Chouses are like pursuivants at arms, and being always sent by the body, approach nearer to the nature of Ambassadors or Emours,[26] and their persons are held very secret, and they are people of great authority and yet those in office I have mentioned go about the city go always clothed in black on Asses and have a particular broad Turbant, except the Bas Chouse who mounts a horse. The Janizer Aga or General of the Janizaries is chose by the Divan of Janizaries out of what body they please, but generally from the horse and that body of them called the Mutteferrika, he has no seat in the Divan, but is superior to any one person of it, out of the Divan, and the Baslaia holds his stirrup when he mounts, he executes all orders of the Divan that belong to his office, as in dangerous times he Patroles once a day about the City, and publishes any orders they think fit should be known, his office is during pleasure. In Constantinople they have Serbajies over every chamber, but there a Serbajie is only a honorary thing, like a brevet Colonel. When any detachment is sent to war, or on any other affair they are under the command of a Sardar taken from the Odabashees whose office is at an end at his return, he has his deputy called a Jemac, and two Sabederiks or Secretaries this body thus detached is called a Bouluke, but the whole body of Janizaries in general, and their Divan is called, the Ajack of the Janizaries, both these and the Azabs have their Divan at the Castle, each at a gate of it called the Porte of the Janizaries and the Azabs.

Azab signifies an unmarried person, and was a new recruite of young single men added to the Janizaries and became a distinct body; they have been great rivals with the Janizaries in Egypt, and sometime the Azabs have got the better, as in the tumult in 1715, but now the Janizaries have the upper hand, notwithstanding some of the Caias of the Azabs are men of great interest their institutions and offices are the same as those described of the Janizaries, only from Odabashees, they are made Serbajies, and from that office are made Caias, and come into the Divan, when as if they go through those other offices of Seraches, and Chouses they never are advanced to be Caias, nor have they office of Boab Odabashee which regards the government of the City, with which they

26 In his book, he uses the word "envoys". *Description 1*, p. 168.

have nothing to do, it is on the contrary among the Janizaries, when any one is made Serbasie, it is laying him aside and he is no farther advanced.

These two bodies are the great protectors of the people and by their Waugh Caia all things regarding those under their protection are judged, and almost all the people are under the protection of one or other of them, and if they apprehend they are wronged by one body they can fly to the protection of the other which often causes great embroils. They make themselves independent of the Pasha and every body have their Bas Chous, and if occasion requires, their Bas Caia always in the Divan of the Pasha, to oppose any thing they think proper, and when orders are sent to their Divan from the Grand Signior, by means of the Pasha, if they are displeasing to them they return them unexecuted. And these two bodies execute a power of turning out the Pasha by sending a Bas Chouse from each body who turning up the corner of his Carpet signifies to him they are no longer content with this government and then he descends an honourable prisoner to a house prepared for him and they name a Caimacam out of the Beys to govern, there is this difference between the tumults here, and those of Constantinople, that the latter are commonly begun by some resolute follows among the Janizaries, whereas here the mob is generally raised by some great man who envies one that is a rival to him, for as long as the Cairoles are poor and weakened by former divisions, they are quiet, but when they grow rich and great, they envy one another and so fall into divisions, which is only avoided by a prudent person that has the sway makes all the other great men his freinds, and adjusts all difference between them. The Janizaries in Constantinople have no Divan, but the Janizer Aga enters the great Divan, and from the Grand Vizier he receives his orders, and the body of Azabs they have not. The Janizaries also here have different offices from those, by which they raise here to the highest degree. Every Janizary here has a great power which is usurped, and when they find any roguery among the common people they give them the bastanado without any further ceremony,[27] and there is no remedy for those who have no money and when they are going to war they are Lords of the property of every one, insomuch that a stop is put to all trade, the shops are shut, and theres no security, but in keeping out of the way; for of all the soldiery in the Grand Signiors dominions those of Cairo are most insolent and injurious. Egypt is obliged to furnish the Grand Signior with three thousand soldiers every three years he demands them that is twelve hundred Janizaries, nine hundred Azabs, and nine hundred Spakis.

I am my Lord

27 Bastinado is a punishment or torture in which the soles of the feet are beaten with a stick.

FAIUME IN EGYPT
12/23 February 1739
Pocockе to Bishop Milles
Account 30[28]

My Lord

The manufactories of Egypt are mostly spent among themselves, they consist cheifly of three branches, Linnen, Woollen and Silk, the woollen is unknaped carpets used mostly for the seats of Divans or Sophas, and for coats for children, all made with broad stripes of different colours, and little other variety, these are made at Benesuef towards upper Egypt, the silk brought to Damiata from Syria, they make with it large hanchercheifs for womens veils, and a very rich sort of hanchercheif worked with Gold and in flowers, used likewise on several occasions by the Ladies, and sometimes they make Cushions, and covering for divans in their manner which are very costly; in Cairo they make great variety of Sattins many in imitation of those of India, but none of them very good. The Delta and other parts of Egypt produce a great quantity of flax, they do not spin with a wheel but letting the spindle hang down draw out the threads from the distaff. Egypt is not now remakeable for its fine linnen, which seems to be owing to the little care they have for it, because among the better sort they wear a sort of Muslin which is much proper for so hot a climate, what they make to wear is exceeding cheap, and becomes white, it is made cheifly at Rosetto where they also manufacture many striped linnens used cheifly about beds as a defence against gnats at night they also make of to wear both at Cairo and Faiume, and at the latter great quantities of sacking brought to Caiio; at Inbbabe opposite to Cairo and the villages about they make a coarse strong linen used for sheets, but the very best linen used for sheets, but the very best linnen is made is about Delta and Damiata, especially the latter, it is for napkins and towels, and long narrow clothes thrown round the dish at eating to be used by the guests, it is a plain well wove linen tho' not fine, but being worked with a striped silk border sells dear. The Turks have a genius for merchandise so that most arts that require any ingenuity are here generally in the hands [of] Christians, particularly the Silver Smiths trade in all parts which is a great trade here by reason of the ornaments of the women and of the trappings of horses; but they use no plate in their houses nor can the Mahometans wear a gold ring, but the women are very costly in gold bracelets and other ornaments, because throughout Turkey a woman by the law after her husbands death unless

28 BL, Add MS 15779.

particular donations are made has nothing but her apparel which is often sold with her jewels and Gold ornaments to maintain the poor widow. They turn here very curious turned lattices before their windows which when made with the utmost art are very costly, and before their Mosque windows they have of this sort made of iron, and brass in the most perfect manner being all of round barrs let into one another so as to make the window in many small squares, and embellished with very proper ornaments; but these seem to be remains of their works, under the Mamaluke government when they were very magnificent in their archjtecture: the Egyptian pebles they work here, and polish in great perfection for handles of knives and snuff boxes, I know not whether they can do it so well, or so cheap in any other parts, it is done in the same manner as they work jewels and is in the hands of one Jew. They make also red leather at Cairo but the best is made a Alexandria, which does not come up to the perfection of the Morocco leather that is of a brighter red, for all arts they are reckoned much inferior here to what they are in Constantinople, which makes every thing esteemed that comes from that place.

As to the state of religion in Egypt, the Copti is that of the natives of the country there are many Greeks in Cairo and Damiata, but very few in Alexandria and Rosetto, and in the other parts of Egypt only a few merchants in some of the principal town, there are a very few Armenians in Cairo, who have a church there given them by the Coptis in lieu of a chapel given to them by the Armenians in the church of the holy sepulchre at Jerusalem. The Christian religion would be at a very low ebb if the people did not find it convenient to have Copti stewards of their estates, who are well acquainted with all affairs are very dextrous of keeping accounts which they do in sort of Copti characters, and language, understood by no others, and it is probable these people area more under their command, and they might have them more in their power in case of any infidelity, than any others. These in every village are sort of Lords and are protectors of the Christians in it. The Coptis of all the Easterns are the most irreverent and careless in their devotions, the night before Sundays and festivals they spend in their churches, and the holy day in sauntering about and sitting under their walls in winter, and under shady trees in summer, seem to think that their whole religion consists in repeating their long services, tho' without the least devotion and in strictly observing their numerous fasts, except the Convents of the deserts of St Macarius, and St Antoninus, and one at Esne, the Convents are inhabited by one or two married preists , but the Patriarch is a man never married and taken out of one of those Convents, they are all exceeding ignorant both preist and people, the former perform the service in the Copti language by rote, which they do not understand, but they have books writ with particular rites of the Alexandrine church. Strabo observes of the Egyptians an extraordinary custom in these words [two lines of Greek script referring to male and female circumcision][29] all which they strictly observe at about ten years of age, but neither is a religious rite, and they give some reasons for it. The Mahometans likewise, originally natives of Egypt, and consequently their ancestors Christians, observe both these customs. The Coptics bear an implacable hatred towards the Greeks even since the famous affair of pressing them to receive the Council of Chalcedon, and when the Greeks get the upper hand it is said they treated them with the utmost impudence and in the most injurious manner. The Jews have one particular thing here, for fear they should drink wine offere'd to Idols in the time of Paganism they had a custom, to have all the wine they drank made by their own people, and sealed up to be sent them, and this custom they still observe in all the Eastern parts. They have thirty six Synagogues in Cairo, and one in old Cairo, in which they say was the prophet Jeremy. Among them is a particular sect who live by themselves, and have a separate Synagogue, and as the other Jews area remarkable for their eyes, so are these for their large noses,[30] they are the antient Essenes, and have now the name of Charaims, from Mikra which is the name by which they call the five books of Moses, which the observe to the letter not receiving any written traditions, it is said that the others would join with them but not having observed the exact rules of the law with regard to divorces they think that they live in adultery.

The Mahometan inhabitants of Egypt are either original natives in the villages, called Filaws, or Arabs, of two sorts, those from the East, mostly on the Eastern side of the river, and those of the West, called Mugrabi, or Western people, who have come from the parts of Barbary have different manners and customs, from the others and are rather worse. Many of the Arabs are settled in the villages, and are honest people, especially in upper Egypt hose who live under tents are called Bedoni living mostly by cattle they graze. The natives of Egypt are now a slothfull people, and delight in sitting still and hearing tales, and indeed seem always to have been made more for the quite life, than for any active scenes; and this idle manner of living is probably one great reason of the fruitfulness of their invention with regard to their antient religion, and of their making so many extravagant fables, out of which the Greeks might like some of the most beautifull, as a foundation for their religion, and poetry, and so that passed to the Romans: they are also malicious and envious to the last degree,

29 The script is so difficult to read/transcribe and differs so much from the version given in *Description 1*, p. 178, that it is best to omit it from here.

30 In his book (*Description 1*, p. 177), Pococke modifies this comment to, "so they observe these are for their large noses".

which keeps them from uniting, and setting up for themselves, and tho' they are very ignorant yet they have a natural cunning and articice, as well as falseness, and this makes them always suspicious of travellers, that they want to find treasures, and that if they have been on the place and carry away something they can by Magick art draw away the money they think may be there being so ignorant, which notion of theirs often occasions a traveller much trouble and sometimes prevents his seeing every thing as he would they have however learned from the Arabs hospitality, and some thing of that strict virtue of standing by those that are under their protection. The people of the country are mostly employed in tilling the ground, which is attended with much labour, and in bringing water to it, which is often very difficult. The Arabs love plunder, and that roving sort of life, this passion leads them to the business of attending cattle seems most suited to their genius, they have good horses and manage them and their pikes with much address, those on foot use poles with which they fence off the spear with great art. Another sort of people are those they call Turks in distinction from other Mahometans, being those who are sent by the Grand Signior, the slaves and the governing part that rise from them, and all their descendants, these are most covetous of money most desirous of power and withall most subtle and engenious in carrying on any affair to attain their ends; the most surprizing things having been managed and brought about by them in Cairo, with the utmost policy and secrecy, and these distinguish themselves from the others by what is strictly either the Turkish or Marmaluke dress of which more particularly in another place.

I am my Lord

SACCARA IN EGYPT
23 February/6 March 1739
Pococke to Bishop Milles
Account 31[31]

My Lord

The education in Egypt is seldom more than to read and write which the Coptis generally learning, and their manner of keeping accounts; but the Arabs and native Mahamotans very rarely can read: the best education is among their slaves, who understand Arabick and Turkish, and often write both, and go through their exercises constantly to ride, shoot and throw the dart well, being

31 BL, Add MS 15779.

esteemed great perfections. The beleif of predestination is very strongly rooted in them, especially in those who are properly Turks, which often inspires them with very great courage and quiet[s] them in a very extraordinary manner, when they are thrown from the hight of power, into the most miserable condition, in which circumstances they say it is the will of god, and bless God; and indeed they behave rather better in adversity than prosperity, tho' when they are in stations, they carry themselves with much becoming gravity; and when they do a favour it is with a very gracious countenance, but the love of money is so rooted in them, that nothing is to be done without brevity, and it is an affront to come into the presence of a great man without a present or having signified to him that something is designed. The same notion of predestination makes them use no precautions against the plague, but even go and help to carry body [bodies] dead of it, which is reckoned a very great charity, tho' it is said there is no infection in a dead body. They think the greatest villaines are expiated when once they wash their hands, and feet, which is their preparation to go to prayers, which all the polite people constantly do, and they have so much sense that their religion is in fashion among them, their prayers are very short, and five times a day, which they perform all at once always praying on a carpet, or some cloth; and in the most publick places where ever they are. Resignation to the will of God, and perhaps little affection for their relations, makes them little lament and soon forget the loss of them unless it be of their children, for they are very fond to have children, and are very affectionate towards them. Their words pass for nothing either in relations, promises, or professions of freindship. The use of Laudanum so much in vogue formerly, is succeeded by drinking cheifly strong waters, which they take plentifully at their meals, tho' a great many will not drink and those use heating things to chear them; the Arabs indeed do not drink, or very rarely, and the common people pound the leaves of green hemp make a ball of it an swallow it down to make them cheerfull. They have a great notion of the Magick art, have books about it, and think there is great virtue in Talismans and charms, but particularly are strongly possessed with an opinion of the evil Eye, and when a child is commended, except you should give it some blessing, if they are not very well assured of your good will, the use of charms against the evil eye, and particularly when they think any ill success attends them on account of an evil eye, they throw salt into the fire. The meanest Mahometan thinks himself above any Christian, and when there is no dependency, but themselves on a rank, and seldom preserve any tolerable good manners, especially the Turks; the Arabs and people of the country being civil enough, and then if it is their way, by coming and sitting about you, tho' they are troublesome by being too observing, curious, and inquisitive, the Turks also will be often very civil either to get presents or to find out your designs, and inclinations in which they are very artfull; but where there is any dependency among one another they observe a great decorum, all rising up when a superior comes in, and in a regular meeting in Divan, the inferior takes up the papouches of the superior set them by him and kisses his hand, and after receives the same regards from his inferior, and a superior of great dignity holds the stirrup t a superior when he mounts, as a two tailed Vizier holds the stirrup to one that has three tails. The way of saluting as you approach or pass, is by stretching out the right hand and bringing it to the breast, and a little inclining the head to the side. The Arab salutation is by joyning hands and often bowing the head to the side of the head of the person saluted, asking him how he does, if he is well, and bidding him peace for several times; when they take any thing either from the hands of a superior, or that is sent from a superior, they kiss it and as the highest respect put it to their foreheads: and if you demand any thing that relates to a service or protection of you, when they promise it they put their hands up to their turbants, as much as to say be it on their heads, or that they will carry you on their heads, which they sometimes say, for nothing can be imagined finer than the Arab manner of expressing civility or friendship, and if the news only is told of any ones death, they always say may your head be safe, and on some particular occasions to a great man in times of confusion when any of his enemies are taken or dead, they say may all your enemies be as he is. Among the Coptis in publick company a son dos not sit before his father, or any superior relation without leave being ordered to sit more than once, and there is great reason why superiors should keep inferiors thus at a distance, and exact so much respect of them especially with regard to civil governours, for the people [who] are of slavish minds must be kept under and if any liberty is given, they immediately assume too much. The manner of eating is thus, a little stool is set on the divan on a coloured cloth laid on the ground, round it they throw a long cloth, to be put in the laps of the guests, and to every one a napkin is given to wahs, which-he likewise lays before him, and wipes with it when he washes after dinner, on the stool is put a dish of Copper tinned over, about six feet diameter, which is as a table, round it is put bread small dishes of pickles salads and the like, and then they bring two or three dishes it may be two or three times, if theydo not drink wine, they seldom drink while eating, they either sit cross legg'd or kneel and eat with the right hand, without knives or forks tearing the meat with the hand, and in his hand the master of the house often throws peices to the guests, immediately after Coffea is brought; this ragouse, pieces of meat cut small and roasted etc. all is taken out, and eat by inferior servants, not by the slaves who have a dinner laid for them in another room. With the Arabs and people of the country either a round skin is laid on the ground for a small company, or large woollen clothes for a great one all down the room, and about ten dishes repeated six or seven times over laid all round at a great feast, and whole sheep lambs boyled and roasted in the middle, when company has done, another sits even [?] to the meanest till all is consumed, for they never lay by any thing, and an Arab Prince often dines in the street before his door, and calls to all that pass, even beggars, in the usual expression, Bismeleah i.e. in the name of God, who come and sit down, and when they have done give their Hamdelliliah i.e. God be praised: for the Arabs are great levellers, put every body on a footing with them, and it is by this their generosity, and hospitality, that they maintain their interest, it is diverting to see them eat, take a handful of Pilaw toss it in their hands to make it into a ball and then cram in into their mouths, but the middling people among them and the Coptis live but poorly, I have often sat down with them only to bread raw onions and a seed pounded and put in oyl which they call serich made of a herb called huisim to eat with their bread, which his made as often as they eat in thin cakes broke to pieces and mixed up with a sort of treacle made of the sugar cane, which cane when it is green, is a great desert, sucking the sweet juice out of it, they also eat a sower milk covered with seeds, their great meal is generally at night taking a light collation in the morning of fried eggs, cheese, and if it is a great table, olives and honey, and this is also usual with the Turks especially if they go out to do business or for diversion to stay out most part of the day otherwise they dine at noon, and sup early in the evening: when they have no company they commonly go into the Haram, or womens apartment, and eat with one of the wives in her separate apartment, who either prepares the dinner, or inspects it and directs, tho' they may be great persons; having their offices adjoining to their rooms, and a great man who has four wifes, has five kitchens one for each of them managed by their several slaves or servants; and one great one for publick entertainments, for the master when he dines out of the Harem, and for the slaves and servants. At a Turkish visit is immediately brought Coffea, Sherbet, if it is a visit of ceremony then according to the dignity of the person, incense and rose water to perfume, which is a genteel way of dismissing the company. Friends who visit, especially women, stay a night or two or more carrying their beds with them tho' in the same town, and Coffea or a sweet water boyled on Cinnamon are brought at least once in an hour. If one goes to the house of an Arab or to his tent bread is immediately made, sower milk and cucumbers in it brought when in season, fried Eggs and oyl to dip your bread in, cheese like curds and such like, and they do not take it well if you do not stay and eat, and they think it such a favour to come to their places, and put your self as it were under their protection, that where there have been any enmities or if one goes to the others house and eats with him all is forgot, and I have seen them sometimes show a resentment by refusing to take Coffea or any thing offered, like the Janizaries when they mutiny will not eat the Grand Signiors Pilaw: and if pressed to eat, and comply, it is a sign the resentment is past. They generally rise early at break of day, and often the common people at least go then to the Mosque, thence to Coffea houses, and having taken their collations go rather later to their shops, and shut them about four in the afternoon. The great people either visit or are visited. In Cairo on Sundays, Tuesdays and Thursdays they go to the Pashas Divan and these are general days of business. Fridays they stay at home and to their Mosques at noon; and tho' with them it is their day of devotion yet they never abstain from their business, the three other days are Bemich days so called from the garment of that name, which is not a habit of ceremony, they then go out early in the morning with their slaves to publick places out of town, where they have sort of open summer houses, and see their slaves ride, shoot, and throw the dart, and regale themselves with their pipe and Coffea. Few people ever read in books, but have generally company with them, and as they have time to think much that is one very great cause of their refined policy in many respects. When at home from twelve to four they stay in the womens apartment, anf rom the time they go to sup till the next morn, and when they are there it must be business of very extraordinary importance, if they are called out, especially if they are great men. The month of fasting of the Mahmetans called Ramedam is very severe, for from twelve a clock at night till sun set, the next day they must not eat or drink or smoak, and it is for the most part pretty strictly observed, but the great people turn night into day, and when it happens in the summer it is very hard on the working people who cannot drink, being only allowed to wash their mouths, for this month in thirty three years happens at all times of the year, the Turkish month being Lunar, they begin it the day they can see the Moon, whereas the Jews begin it the day the Moon makes, which is a day before the Turks. The Coptis begin the year the 29th of August which is the first of their month Thoth. The Copti month is thirty days, and every year they have five intercalary days, and every fourth year six. Their Ara begins three hundred and two years after Christ from the martyrdom of the saints in Egypt under Diocletian.

I am my Lord

GRAND CAIRO
26 February/9 March 1739
Pococke to Bishop Milles
Account 32[32]

My Lord

The most simple dress in Egypt resembles probably the primitive manner of clothing, for it is only a long shirt with wide sleeves sometimes tied about the middle, and many children in the country go naked all the year round, as most of them do in the summer. The common people wear over this a brown woollen shirt, and those of better condition have a cloth coat over and then along blew shirt, and a dress of ceremony over this instead of blew is a white shirt which they put on upon festival days, and to pay great visits in upper Egypt, but lower [below] it is a shirt or garment made like it of black woollen, which is sometimes by the more genteel left open before, and then is properly what they call a Ferijee [cape], and some of the first condition have them of cloth and furred, the Arabs, and natives wearing their clothes with large sleeves, like the dress of ceremony of the Turks called the Ferijee, made like a night gown, the other dress of the Turks being with streight [sic], under all most of them wear a pair of linnen drawers, and when their vests are open before, after the Turkish manner, it is an odd sight to see the shirt hang down under all which they do not but into the drawers after the Turkish manner. The most simple Turkish dress is such drawers over which the better sort wear a pair of red close drawers down to the ankles called Shackshires[33] to which are sown stockings of yellow leather, called Mess which come no higher than the ankle, and under them they wear socks of yellow leather called Fenlick. But the Christians of the country wear red leather at least their papouches or slippers and the Jews wear the blue leather those who wear not the Mess, wear the Fenlick, and paupouses, the leggs [sic] appearing bare up to the drawers, the Arabs and Egyptians wear shoes of red leather like a slipper with hind quarters called Merkoub and the Janizaries wear of the same colour, and little different, called Jeminee, and the Turks and Christians in the city within doors out of frugality wear a wooden sort of pattern something like the wooden sandal of the monks, some of which are made very fine and it is to be observed that the dress of the feet, and of the head very much distinguish persons in these countrys, and they are

32 BL, Add MS 15779.

33 This word, which describes Turkish breeches, can also be spelled "Chackshires", according to N. Bailey, *The Universal Etymological Dictionary*, vol. 2, London, 1731, p.

fined if they go contrary to the custom. Frank Christians only have the liberty of wearing as the Turks yellow paupouches. The Turkish garments of the body are first a sort of short Cassock without sleeves of dimitie or linen it is often not open before, but tied n the side, this is called a Zipone, over it is a short Cassock with sleeves called an Antery in Turkish and Bombos in Arab, over that is a garment like a long Cassock called Doliman in Arab, and a Caftan in Turkish, these two last are either of died linnen or striped Damascus cotton stuffs, or of striped or pain silks are Sattins; and in summer the fashion is to wear the finest white Calicoes of the latter, over this is a close sleeved garment like the Greek gown, and is called a Venish which is the common dress, over that is the Ferijee mentioned the dress of ceremony, they have another sort cut differently in the sleeve and is not so high a dress and is called a Cherikee these two last garments are either of Cloth, Mohair, of Silk according to the season, the latter indeed is never of silk nor is the Benish ever wore here of silk [word/s missing] a girdle goes round the Caftan which is called a Couishak [cummerbund] in Turkish and Arab into which they stick a knife. The people of Egypt wear a blue cloth about their necks with wch they cover their heads against the cold and Sun, this is called a Foursa in Arab and Pestamel in Turkish, and some of them are very large, and it is almost a general custom among the Arabs and Mahomatan natives of the country to wear a large blanket either white or brown, and in summer a blue and white cotton sheet which the Christians constantly use in the country putting one corner before over the left shoulder they bring it behind and under the right arm and so over their bodys, throwing it behind, over the left shoulder so the right arm is left bare for action; when it is hot, and on horseback they let it fall down on the saddle round them, and about Faiume I particularly observed that young people especially and the poorer sort had nothing on whatsover [sic] but this blanket, and it is probable the young man was clothed in this manner who followed our Saviour when he was taken, having a linnen cloth cast about his naked body, who when the young men laid hold of him, left the linnen clothe, and fled from them naked. The head dress is either that high stiff cap called a Caouk with a long peice of white Muslin wound round it, called a Sheash in Arab and in Turkish sarik, both which together is what we call the Turbant [turban] or in the place of the Caoak they wear a red cap of wool which comes close to the head, this is wore by the Arabs and native Egyptians, except merchants and great stewards of Arab Princes, and Copti preists who wear the Caouk, this cap is called Tarboush in Arab and Fesse in Turkish, under it all wear a little white skull cap called in Arab Takich in Turkish Kellespoush, I was told at Damascus, that one whose family had not worn a Caouk nor Messes on the feet cannot begin to wear them there, unless by virtue of some office or employ, the shash that goes round, is the great distinction. None but the relations of Mahomet can wear green, the white is proper to Mahometans, but in Syria where there are villages of Christians they wear white and any one may wear white, if they put a coloured string to distinguish, as the Samaritans do, and the Jews of Saphet in Syria having red and blue strings sown along down their shash, but this is not much used. The Christian and Jewish in all parts are either blue, or blue and white striped, the Franks in Cairo did wear white then they wore Carpachs, or red caps turned up with furr as some do now but they generally wear the red cape, and a red muslin shash, in all other parts they wear hat and wig, as the English Consul does when he goes out in publick. The French Consul and some of his officers go in Frank habit, and so have the English Consuls formerly as they do at Aleppo, and Cairo is the only place I have heard of in Turkey, where none but the consuls ordinarily rid on horses. The common people of the country either wear no shash, or one of coarse white Cotton or of yellow, or red woollen stuff. The dress of the women according to the manner of Constantinople is not much unlike that of the men, only most of their under garments are of silk as well as their drawers, and all but the outer vest are shorter than the mens, and their sleeves are made to hang down, a sort of gause shirt coming down near to the ground under all; their heads are dressed with an embroidered hackerchief [sic], and the hair plated [plaited] round, having one a white woolen [sic] skull cap, the ordinary women wear a large linnen or cotton blue garment like a surplice, and before their faces hangs a sort of a bib, which is joined to their head dress by a tape over the nose, the space between being only for the eyes, the others wear this garment of silk, have a large black veil that comes all over, and something of a gause that covers the face, it being reckoned a great indecency to show the whole face, generally the mouth and one eye, if they do not cover the whole. Women ride on asses in Cairo with very short stirrups, which it is a dishonour for the men to use when they mount asses, they have a prophecy that Cairo will be taken by a woman on horse back which makes them strictly forbid women mounting on horses, the time when the women go out is mostly on Frydays to the burial places, to adorn with flowers and boughs the sepulcher of their relations, to hang a lamp over them, and pour water on their graves, and place water in vases near; when they ride they commonly wear a short light yellow sort of a boot, and Ladies of distinction, have many female slaves follow them on asses who have not those large veils that come over the body, but have only a covering over the forehead and lower part of the face, leaving a space for the eyes; and their close garment has a cape that hangs down behind, and they make a much better figure than their Mistresses, they also go out of the Bagnios which are reserved for them at a certain time every day, this is their rendevouz to hear news, and for their discourse together, which I have been told is very far from being edifying. There are women who go bare faced about the streets dancing, singing, and playing on some instruments, these may not be supposed to be very vertuous [sic], and yet they say they are so in some respect, in which them might be most suspected, there are not withstanding common women who are in a manner licensed and pay a tribute on that account. The vulgar women paint with blue their lips and the tip of their Chin, and those of better condition paint their nails yellow and their feet with a dye made of a seed called Hennah, and they black also their eyelids with black lead and so do many even of the men among the Turks, imagining it good for sore eyes. Coffea houses are not so resorted to by the best company, only by people of middling rank, they have Musick at certain hours of the day in some of them and in others a man tells some history with a good grace, probably something lake the Arabian tales, which is a method they take to bring company: to the Coffea house they often send for your ordinaries,[34] and those who have nothing to do, pass whole days in them. Both Turks and Egyptians are very frugal in the manner of living, the latter seldom eat meat, and the expence of the great men in their tables is but small considering the number of attendants they have in which they are very extravagant, it is common for them to have fifty or sixty slaves and double the number of other servants besides dependants, they are at great expences in clothing their slaves, and in keeping a great number of horses, from fifty to two hundred, they are excellently well managed do not seem to know their power or force always standing with their fore legs tied, they feed them with grass, and corn, and they lye on their dung without straw: they commonly stand abroad especially in Upper Egypt all the year round, they walk finely never trot but gallop swiftly, turn suddenly, and stop in a moment in full speed; but they never gallop but for diversion or run away. The Mahometans have a certain veneration for fools and mad people, as thinking them acted by a divine spirit and look on them as sort of saints, they call them here Shekes, there are some women of them, but they are mostly men who go about are received in all houses and at all tables, the people kiss their hands, and pay them great regards, and I have seen some of these who go about their towns entirely naked and in Cairo they have a large Mosque with buildings adjoining, and great revenues to maintain such persons. As these are recommended by their want of reason, so the Derviches [Dervishes] are by their want of money, poverty being esteemed by a Turk as a great degree of perfection in every one but himself: there are hardly any of these in Egypt having but one Convent

---

34 This passage is not very clear but is clarified by the corresponding passage in the book: "They often send to the shops for their provisions to be brought to the coffee-houses …" See *Description 1*, p. 193.

near old Cairo but not of the [missing word] order. The Egyptians are but an ill looking people and tho' many fair enough when young, they grow swarthy by the Sun, they are also a dirty slovenly generation, especially the Coptis, whose dress at best appears but ill on them, and as table linnen is very rarely or every used by the Arabs Egyptian or Coptis, I have seen the latter after washing their hands when they have eat, wipe them on the great sleeve of their shirts. The true Marmaluke dress is the short garment called Anterry, put into their great red trowsers, which are tied round the leg at each ankle, the foot being left bare, and they wear the sort of shoes wore by the Arabs when they ride and at other times the Fenlick and papouches, in other respects they dress like the Turks, and this is the dress of the salves, and likewise of many of the great men, when they are not in dress of ceremony.

Whatever is tolerable at Cairo in Architecture is of the times of the Marmalukes of which one also sees great remains in Damascus and Aleppo, being very solid buildings of hewn stone inside and out, the windows are oblong squares, and perfectly plain, as well as the doors, except that the latter which are at the inside of the wall have a sort of Gothick carved work from the top sloping out to the outside of the wall, which has a grand appearance, over all openings the stones are either narrower at bottom than at top which gives them the strength of an arch, or being indented on the sides they are made to fit into one another, so as they cannot give way, and in most of t heir Mosques are small well proportioned Cupolas; to some of them are exceeding fine nicharets [minarets][35] with several stories of Galaries [sic], lessening till the nicharet ends in a point. These are also about Cairo some very grand gates, with a semi circular tower on each side. Their houses consist of one or more large saloons with a sopha [sofa] at each end and square sort of Cupola in the middle, wainscoted about six feet high with panels of marble having round them a sort of Mosaick work of different sorts of marble fit [arranged] in different figures, and in this last manner the middle part between the sophas are paved, as for the other rooms they are generally small for convenience as they live and often lay, in these having their beds brought on the sophas, and in these receive all their company.

I am my Lord

35 This word is spelled correctly in the book, so it is likely that the person who transcribed the account did not recognise the word. See *Description 1*, p. 194.

GRAND CAIRO
1/12 March 1739
Pococke to Bishop Milles
Account 33[36]

My Lord

The climate in Egypt is very hot in summer by reason of its situation between two ranges of mountains, and also on account of the sandy soil.[37] In the sun towards the middle of the day, it is hot even in winter, but the nights and mornings are very cold, occasioned by the Nitre in the air, which is penetrating; and on that account it is very dangerous to catch cold, especially in the head which often occasions a humour to fall n the eyes, which is thought to be the cause of the great number of blind, and of those who have sore eyes, they think it is also bad for the eyes to suffer the dew which falls very plentifully by night to light on them, which are the reasons why they carefully wrap up the head, and cover their eyes at night, the coldest time here is the beginning of February. Near the sea they have great rains sometimes from November to March, but up higher about Cairo they have seldom any rain but in December, January, and February, and these light showers for a quarter or half an hour. In upper Egypt they have sometimes a little sprinkling and in eight years it had been known to rain twice very hard for about half an hour., the West and North East are the winds that bring the rain. Thunder they have in the summer, but without rain, and at a distance, from the North East, but never have any earthquakes that do mischief, and those that rarely happen are hardly perceivable. There are two sorts of wind, that blow mostly in Egypt, the South or a few points from it, and the West wind, the South wind is called Merisi it is a very hot wind,[38] and sometimes blows some points from the East of [to] the South, when it is exceeding hot, at others some points from the West to the South, when it is rather windy and not so hot, it blows also very hard sometimes from the South East, and when these winds are high, it raises the sand in such a manner, that it darkens the sun, and one cannot see a quarter of a mile distant, and the dust enters into chambers the closest shut, into the verry beds and Scrutores [closets], and to give an instance of the Romantick manner of talking, among the Easterns, they say the dust will enter into an entire egg, through the shell, the wind is often so

36 BL, Add MS 15779.

37 The opening sentence of this letter is identical to that of Chapter 6 of the book (see *Description 1*, p. 195) and much of the content is very similarly worded.

38 Spelled "Merisi" in the book. *Description 1*, p. 195.

excessively hot that the air is like an oven, and people are forced to retire into lower rooms and shut themselves close up; for the best fence against it is to keep the rooms close shut, this wind generally begins about the middle of March and continues till May, it is commonly called by the Europeans the Hamsine [Hamsin] wind, because that season of the year when it blows is so called by the Arabs. The North wind is called Meltem, and begins to blow in June, before the Nile rises, this is a refreshing wind, and makes the excessive heats of summer supportable, it brings with it health, and the happiness of Egypt, in being the cause in more respect than one of the overflow of the Nile, and continues blowing till November, and without this wind, they could not navigate up the Nile, all that time when its current is so very rapid. It is this wind which puts a stop to the plague, as it generally begins in Egypt if it breeds of it self in March when the warm weather comes in, and rages, and is very mortal during the hot winds: but they have the plague but rarely in Egypt, unless brought by infection. It is supposed that in temperate warm weather this distemper breeds, and that excessive cold and heat stop it, so that they have not here nor in Constantinople in winter, nor in Egypt in Summer of what it is here, hence some imagine it is bred and infects by some small insect, that the great head and cold destroy, and that dies with the body, which they say does not infect when dead, and also because several things are known not to communicate the infections as water, iron, wood or any thing made of the date tree, they use the ropes made of a part of this tree. The Franks shut up, but others go about as at other times, tho' some great men observing the Franks always escape retired in the last plague, notwithstanding their notion of predestination. The compute the number of the people at Cairo to be two Millions, tho' they may exceed half a million, and say that seven thousand have died in one day, which they have computed by the money paid for the bier, tho' it seems incredible, the most mortal plague is that which comes from upper Egypt being worse than any infection from Constantinople. The air of Cairo is not reckoned very wholsome [sic] at night especially, for then fires are made throughout the City, the great [smoak] hangs as a cloud at night and the smoak over the City when there is no wind and thickens the air. At spring and fall people are much liable to fluxes, but in the Summer they are not dangerous, and in all distempers Soups made of meat are not reckoned proper: the people of the country are much liable to broken[39] bellys which are thought to be caused by heats and colds and also by straining their voices so excessively in singing, those who cry off from the Mosque being most commonly afflicted with this distemper. A fright causes persons here to break out into blotches like

39 The word "ruptured" is used in *Description 1*, p. 196.

a foul disease, to prevent which bleeding is thought a proper remedy. The most unhealthy time is when perspiration is checked by the cold, for every thing here is very nourishing, but when they have the benefit of a free perspiration it carries off all superfluous humours and the great quantities of water. Sherbets &c that are drunk do not harm; but if a person does not sweat he burns as in a fever, in hot weather, in which case the bagnios are a great releif [sic] which carry off or at least diminish a great many disorders in the blood, and prevent their appearing, about May most persons break out in a rash which continues during the heats, and it is the fashion a stick called Manarously[40] which is made at the end like a file, and with this it is no shame to skub [rub] themselves. The soil of Egypt except what additions it has received from the overflow of the Nile is naturally sandy, and any one that sees the hills on which the Pyramids near Cairo are built full of petrified shells, and the stones of which the Pyramids are built particularly one large one and a smaller unfinished at Sacara built of stones that are as it were composed of oyster shells cemented together, and also some low hills of the mountains going South South West to Faiume, and some close to the Nile on the East which are nothing but heaps of large Oyster shells, some remaining in their first state, and others petrified, where there was sufficient moisture, I say one who is an eye witness of this will find it difficult to account for it, unless he supposes the sea remained here, some ages after the flood, in a sort of gulph like the red sea, as Herodotus says the Egyptian historians relate it did; and that Egypt below the Thabais is the gift of the Nile: and tho' in many points, especially about Faiume, there is a rich black soil, tho' the Nile does not rise it may be within five or six feet of the surface of the ground, yet that soil might be brought there by the stream of the Nile, as it is now for several leagues out into the sea, as they find by their soundings; and tho' the sea might cover those hills, coming out of the Nile ran, which mush have been confined between the Eastern and Western hills: in some of the Canals with high banks worn by the violence of the stream about Faiume, I saw the black soil near ten feet deep; and in some about four feed deep, and then two feet of a sandy layer, and black again for four or five feet to the bottom of the Canal. Whether that sandy layer might every be occasioned by any extraordinary movements of the sands by very violent winds, I leave any one to conjecture, as he thinks probable. The soil of Egypt is full of Nitre or salt which occasions those nitrous vapours that make the night air so cold and dangerous, where the Nitre is very plentifull, and there are small lakes made by the rains, as the waters dry they leave a cake of salt on the surface which is collected for use; and the beast often eat the very earth when they find it mixed with the

40 In *Description 1*, p. 197, this is spelled "Maharoshy".

salt, it is this and the rich quality of the earth which is the sediment of the water of the Nile that makes Egypt so fertile, and sometimes they temper the soil by bringing sand to it, but the sandy soil, tho' they bring water to it will not produce a crop [that] will answer the expence, but does very well for trees, for about a mile from the mountain, or more the country is commonly sandy, and along at the edge of this desert they generally have villages at a proper distance, with plantations about them, as well as at some little distance from the river, and if the hills are above four or five miles from the river they have villages along in the middle which are built on raised grounds here the Nile overflows. As to the water of Egypt it is very much to be doubted whether they have any, that does not depend on the Nile, that of Josephs well which is mentioned by some as the only spring in Egypt certainly does, for it rises and falls with the Nile, and has some small degree of saltness by producing through the nitrous soil: toward the Malorea [Mataria] they find water for a league along the edge of the desert digging from four or five to ten feet, it is very fine clean water without the least saltness in it; and some have said it is lighter than the water of the Nile, and it is indeed difficult to determine whether this may be some source from the Eastern low hills, or whether the soil may not be salt here, but rather purify the water, and so may be derived from the Nile, or from the lake a league North called the Birki,[41] where it is certain the water is not salt, tho', it is in most lakes in Egypt, which seems to be a proof that the soil is not salt there, it is also to be observed, that the ground here is lower than it is between that and the Nile which may answer the objection that nearer the Nile they must dig lower down in the earth to find water, in case any one is not satisfied that it may be derived from the lake. All along the sea coast towards Alexandria they find a water by digging wells which is so little brackish [salty] that it may be drank, as Caesar found it when he was besieging Alexandria, according to the account of Hirtius Pansa;[42] which water may pass through the earth from the Canal that leads to Alexandria which is but a few miles distant from the shoar. There is indeed one water which must be a spring, it rises in several parts among the rocks and even drops from them at the ruins of a Convent of [Dermadoud][43] to which is near two leagues North East of [Akmim] in Upper Egypt, the water runs from the basons in the rock, and forms the pools, and here grow several trees especially the Palm, which are properly the plantations

41 Possibly the Birk-El-Hadjee (Pilgrim's Lake) referred to in *Description 1*, p. 291.

42 This name is an invention that combines the names of the two Roman consuls for the year 43BC. See Jan Felix Gaertner and Bianca C. Hausberg, *Caesar and the Bellum Alexandrinum, An Analysis of Style, Narrative Technique, and the Reception of Greek Histiography*, Vandenhoeck & Ruprecht, Göttingen, 2013, p. 170.

43 Possibly Dair Madfid.

of the Monks. In Upper Egypt especially about Erne, they dig large wells a mile or two from the river, and draws [sic] up the water from them, and in all parts they have wells of brackish water, and such water can no where be wanting within a league or two of the Nile, it being the nature of a sandy soil that the water penetrates into it as high as the superficies of the waters if not higher, both by a communication of humidity as in Sugar, and also by an attraction of the vapours upwards: and where the Nile overflows not only there the earth must be full of water, but also in the neighbouring parts, the water lying on the ground for three or four month, which accounts for the Niles being the bountifull giver of water thro'out the land of Egypt. This river is truly the great curiosity of the country, and would be still more wonderull in the eyes of the world, if the Niger which has its sources near had not all the same appearances. The Antients [sic] knew the cause of this extraordinary overflow, the North winds which begins [sic] to blow about the latter end of May, drive the clouds formed by the vapours of the Mediterranean South wards to the high mountains of Ethiopa which stopping their course they condense and fall down in violent rains, it is said at that time, the men by their reason, but the very wild beasts by a sort of instinct leave those mountains. This wind which is the cause of the rise of the Nile by driving the clouds against those hills, is also the cause of it in another respect, as it drives in the water from the Sea, and keeps back the waters of the river in such a manner as to raise the waters above: the increase of its rise every day must be greatest during the time it is confined within its banks. By accounts I have by me of its rise for three years, I find it rose the six first days, every day from two inches to five inches, for the twelve next daies [sic] from five to ten, and so continues rising much in the same manner, but rather abating in its rise every day, till towards the time, when it arrives at the heigth of sixteen pikes, when the Calige or Canal at Cairo is cut; afterwards tho' it goes on rising six weeks longer, yet it does not rise so much every day, but from there to five inches; for spreading over the land and entering into the Canals, tho' the quantity of water that descends may be much greater than before, yet the rise is not so great, for after the Canal at Cairo is opened, the others are opened at fixed times, those which water the lowest grounds being cut open to it. From these Canals when full the country is overflowed, and not from the great body of the Nile where the banks are high, and then Canals are carried along the highest parts of the country; that from them the water may have a fall to all other parts when the Nile rises, and they draw the water out of the great Canals into small channels to convey it to all parts.

I am my Lord

GRAND CAIRO
7/18 March 1739
Pococke to Bishop Milles
Account 34[44]

The Egyptians especially the Coptis were very fond of an opinion that the Nile always began to rise on the same day, it does indeed begin in generally [sic] about the eighteenth or nineteenth of June. They have a notion also of a great dew falling the night before the day that they perceive it to rise, and that this dew which they call Nokta purifies the air, and causes the water of the Nile to ferment, and turn green, which it certainly does, as soon as the Nile begins to rise, and continues so for about thirty daies, but never less than twenty, nor does it exceed forty; then the waters are very unwholsom [sic] and purging, and at Cairo they drink of water preserved in Cisterns under the houses and Mosques; and this might originally be a reason why they let not the water into the Canals, which would fill the little lakes that then are to every village, and afterwards spoil the good water that might come into them. It is supposed that the sources of the Nile beginning to flow plentifully, the water at first that green filth which may be about the lakes at its rise, or at the rise of those small rivers that flow into it, near its principal source, for tho' there is so little water in the Nile when at lowest, that there is hardly any current, in many parts of it, yet it cannot be supposed, that the waters any where in the bed of the Nile should stagnate, as to become green. After this the water becomes red, and very turbid [clouded], and then it begins to be wholsom, and is drank by the common people; but most people have large Jarrs which they rub within with pounded almonds which causes the water to ferment, and settle in four or five hours, the water continues reddish till the rapidity of the stream begins to abate in December and January, and the river becomes lower and lower almost to the time when it begins to rise, the waters being yellowish and colour the Sea several leagues out, in March I found it about nine pikes high, in January about four feet higher and at lowest it is five or six pikes, which settles in the measuring place called the Mika being about five feet deep. The precise day the Coptis would fix the beginning of its rise is the twelfth of their month Keah which is the fifth of June Old Style, and this being their Festival of St Michael they make a miracle of it. Certain it is, about the time or rather about St John the plague begins to stop when it happens to be here, and becomes less mortal, tho' it seems rather to be owing to the change of the wind, which happens before and then they begin to find the effects of it. This time the Nile is at sixteen pikes

44 BL, Add MS 15779.

from the twenty fifth of July to the eighteenth of August, the sooner the better prospects of a high Nile it has been so late as the first and even the nineteenth of September, but they have been afflicted with plague and famine, the Nile not rising to its proper heigth. Eighteen pikes is but an indifferent Nile; twenty is middling, twenty two is a good Nile, beyond which it seldom rises, at it is said if it rises above twenty four pikes it is an inundation and is of bad consequence, as the water does not retire in time to sow the corn, but I cannot find any certain account when this has happened. As many parts of Egypt are not overflown, one would imagine when the Nile does not rise high they might supply the want of water by labour, and raising water as in Upper Egypt, but it is said if it is a low Nile the water retires too soon and the earth must be sown in a proper time after before it drys into hard cakes, in which case: the hot weather still continuing breeds or preserves a worm that eats the corn, as well as burns up the young corn, and tho' it is not suffered to overflow at all, yet the Nile being low and the ground entirely dry and parched the case is very different from what it is in upper Egypt, where if the Nile is of a proper heigth, tho' it does not overflow, yet it fills the earth to the heigth of the surface of the water, and when the Nile does not rise to sixteen pikes, whilst they expect it to rise higher, it begins to fall, and the ground is to be laboured out of season during the hot weather, and the water is flowing from them to increase the immense labour of watering, by raising it higher to a dry thirsty land that will drink it up, and if they sow too soon the same inconveniences would follow as above. As they have dikes to keep the water out of the Canals, till the proper time comes to let it in, so they have contrivances to keep it in some Canals, after the Nile is fallen; as well as to keep it up in certain lakes, when the Nile grows low, and from [there] let it out at pleasure on lands that are higher than the channels of the Canals. Of these methods to hinder the water from flowing in, or going out when it is in, Strabo takes notice, the expression he makes use of for confining the water is ταμιενωσιν and a place near Faiume where the water is confined in a lake, and let out at pleasure into different Canals is now called Tameiah, which is a plain proof, that there had been such a work there in antient times.[45] Towards the mouths of the Nile the banks are low and the water overflows the land soon, and there it has its vent into the sea, so that the water does not rise at Rosetta and below Damiata above three or four pikes, and I was assured it does not rise above four or five at Assouan, just below the Cataracts, the reason of which may be that the Nile below is very broad, and that the banks are not as in other places perpendicular, but a sloping, so that the water is not confined but it spreads up the bank, to which on the West side the low hills come with a

45 Pococke does not mention this fact in his book.

gentle descent. As the Grand Signior has not a title to his rents, till the Canal is opened at Cairo, by breaking down the bank that is made before it which is not to be done till the Nile rises to sixteen pikes; yet when the Nile once did not rise so high, and the Pasha caused the Canal to be opened, the people notwithstanding would not pay the tribute. The Nile has rose [sic] out of season, and particularly that it did so in the time of Cleopatra, when Egypt was taken by the Romans; but however nothing happened the year following, but a very plentifull rising of the Nile which is the blessing of Egypt. The time when the Nile is at highest is about the middle of September. They have different methods of raising the water; where the Nile is not much below the surface of the ground, as at Rosetto, and Damiata, they make a hole and put in a wheel made with boxes round its circumference, which receives the water, and as the wheel goes round the boxes empty at top into a trough made to catch the water, where the water is too deep to be raised in their way, they put a cord round the wheel, which reaches down to the water, to it they tye earthen jars which fill with water as it goes round and emptys [sic] themselves at top in the same manner, being turned with oxen. Where the banks are high the most common way is to make a bason in the sides of them, and fixing a pole in an axle on two forks they tye a pole to that at the end of which is a leather bucket next [to] the river, and a stone being tied on the other end, two men draw down the bucket into the water, and the weights brings [sic] it up the men directing it, and turnign the water into the bason; from this it runs into another bason, whence it is raised in like manner, and I have seen five one over another, in the upper parts of Egypt, which is a great deal of labour;[46] another was mostly used in raising water from the Canals, is a string tied to each end of a basket, with one man to each string, who let the basket into the water and drawing the strings tight, raise it three or four feet to a little Canal, they empty it into, that conveys the water over the land, and so continue raising the water. The waters of the Nile are esteemed to be very wholsome and nourishing and are drunk very plentifully without any inconvenience, in hot weather. I know not whether there are any sort of fish in the Nile that are in the rivers in Europe except Eels and Mullet, which last [the latter of which] and some other fish in it come from the sea at proper seasons. There is a sort of shell fish in the river but in Bahr Joseph [Bahr Yussef] the Canal near Faiume which runs into the Lake Caroun is a large Muscle [Mussel] having within a mother of pearl of a reddish cast. They have some stories of sea horses, near the Sea of Damiata, and that they have come ashore and have destroyed feilds [sic] of corn but I very much doubt either of one or the other[r], though Herodotus mentions these fish and describes them

46 This process is represented in *Description 1*, Plate VIII.

particularly, and says they were worshiped [sic] in a part of Egypt. Of the fish mostly esteemed are the Bucti,[47] which is very plentifull, and is some thing like a large Roche [Roach]; the Sebuga, which is full of bones, and eats [tastes] much like a Herring: the Rai [Ray?] which is what is called the Cefalo in Italian, and comes from the sea, but what is still more esteemed is a fish called in Arabick [word missing] which the French call the Variole because it is very much like the Carp, they they [sic] are sometimes found of two hundred weight, tho' this may exceed, yet I have seen them very large. But the most delicate fish is the Kesher[48] which are caught only towards Upper Egypt, the skin appears like the Salmon, it has a sort of a long narrow snout with so small a mouth, that from that as well as its inside, one may conjecture it lives by sucking a juice either from weeks, or out of the ground, as the herring is supposed to do and the woodcock kind among the birds. When I was in Upper Egypt they told me there was a large fish, called Latous [Nile perch?] which must be the Latus that was worshiped by the Egyptians from which the City called Latopolis had its name, they told me this fish at Cairo was called Cusir which I suppose must be this Kesher which is the name of the above mentioned particular sort of fish at Faiume: tho' at Cairo I am told they call the Variole Kesher, and they informed me if I mistake not that the Latous grew to two hundred weight, so that whether of the two is the Latus of the antients may be a doubt. There is a small fish called Gurgur in Upper Egypt and Shalk[49] in Cairo it is at most about a foot long, its head is a strange bone like the Sturgeons, it has a fin on its back, and on each side under the gills, each fin is armed with a strong sharp bone, they have a notion that this fish enters the Crocodile and kills him, this probably may be what Pliny falsely calls the Dolphin, which he says has a sharp point on his back with which getting under the Crocodiles belly he wounds him. It is a general observation that the Crocodile has no tongue, and Herodotus affirms it, but he has a tongue which is tied all along by a fleshy substance of the breadth of the tongue to the lower jaw, which may serve to turn his meat; opposite to two long teeth at the end of his lower jaw are two holes in the upper jaw into which these teeth go when he shuts his mouth. I experienced that the Crocodiles are very quick sighted, for making a circuit to come directly behind them to shoot at them, I always observ'd they began to move gently into the water as soon as I came in sight of them, and behind their eyes is a sort of chanel on the head to each eye by which objects are conveyed to them; their Eggs are not bigger but rather longer, than a goose Egg, they make a hole about two

47 The spelling is Bulsi in *Description 1*, p. 202.

48 Depicted in *Description 1*, Plate LXX.

49 Spelled Shalh in *Description 1*, p. 202.

feet deep in the sand, in which they lay their Eggs, and cover them over often going to the place, and take care of their young which immediately run into the water, they lay fifty Eggs, which are twenty five or thirty daies in hatching: they [Egyptian people] search for the Eggs to destroy them with an iron pike. I could learn no account in upper Egypt of the Ichneumon [wasp] destroying the Eggs or entering by the mouth of the Crocodile into his bowels and killing him, and it seems improbable that he should do this and not be stifled, the animal they have here which is shown for it in Europe is something of the make of a Stote [Stoat], it is not improbable they may destroy their Eggs. The Crocodile keeps close to the water with his head to the river on Sandy islands with sloping banks, and if disturbed, walk in gently to the river and disappear by degrees, tho' its said they can run fast. Herodotus says he [the crocodile] eats nothing during four months in the winter and Pliny that they lye hid [in] that space in some cave; but I saw them in great abundance all the month of January, and was assured they never go above thirty of forty paces from the river, and that by night, tho' it is probable they are mostly out of water by day to sun themselves, in the winter as I observed, and it is probable that they keep in the water by day in summer when the Sun is hot: the people say they cant take a man swimming in the water but standing by the river at one jump out of the water seize him with their fore claws, and so likewise cattle, and that if it is too far make a spring and beat down the prey with their tail; tho' I believe the most common way of killing them is by shooting them so that the ball may take place about their under parts where their skin I soft and not cover'd with scales as their back are, yet they give an account of catching them something like that which Herodotus mentions they make some animal cry at some distance from the river and when the Crocodiles [sic] comes fix a barbed spear into him, to which a rope is tied let him go into the water to spend himself and then drawing him out with a pole into his mouth and jumping on his back tye up his mouth, they frequent places mostly where there are shallow island and for that reason there are very few below [Akmim] and in these lower parts the currents may be more strong which they avoid as well as places where the Nile runs among rocks, as it does at the Cataracts.

I am my Lord

CHAPTER 10

Conclusion

1 Return Journey

Following his initial letter from Messina, Pococke sent two further notes to his mother (dated 25 November/6 December and 2/13 December 1740), the latter enclosing "two sheets, that were designed for the Bishop of Waterford". These papers give an account of Cephalonia, where he had been delayed by the weather, and were written before he heard news of his uncle's death. From these notes and the subsequent letter from Naples (30 December/10 January) we learn that Pococke continued to socialise with his compatriots in Messina, including Dr Perry, and that on the memorable day of 26 December/7 January, "I did cut off my whiskers, which I had left to grow about 3 years & 3 months, & habited myself almost entirely in the European dress, which I had disused for nearly the same time." The restrictions on his movements were gradually relaxed and he spent the next week or so making short trips around Messina until released from his period of quarantine and ready to depart from Messina on 17/28 January 1741.

From here he proceeded to Naples, where he arrived in 10 days, having missed the opportunity to visit this city during his earlier travels.[1] Here he collected a trunk that he had left in Venice, picked up a box containing his cousin's manuscripts, notified his mother about the recent movements of Lady Wortley Montagu and Lady Walpole, both of whom were travelling in Italy, and thanked her "for the valuable present of [his] Great Grandmothers embroidered purse", which he considered to be a "valuable piece of Antiquity." Two more letters were sent from Naples, dated 11/22 January and 20/31 January 1741, the former of which (Letter 52) gives the following curious account of his visit to Mount Vesuvius, on 3/14 January. "…going forty paces, [we] descended about 20 feet & going 54 paces more, descended 15 feet, & going 30 paces more, descended 6 feet [next 2 words scored out], & going 6 paces more descended 9 feet; – then going 10 paces more, descended 8 feet, & going 15 more came to the last hole, which I guessed might be about 50 feet deep; – down which nobody had been, & no one would venture to go, & I believe would be difficult without

1 See Finnegan, *Letters 2*, pp. 73, note 21 and 125. However, it appears from the correspondence for the eastern voyage that his cousin had been more anxious to visit this city than him. See Letter 10, above.

 | DOI:10.1163/9789004440050_011

ropes. Returning the sweat fell down in great drops…." Not satisfied with this, he made a second visit to the volcano (as recorded in his letter from Rome, dated 30 January/10 February 1741) but this time did not descend into it, merely looking down to the bottom and admiring from it the "fine view of Naples & Campania."

The remainder of Pococke's route home can be ascertained by the postmarks of his subsequent letters to his mother, which are from the following places: Rome (30 January/10 February 1741) where, among other British visitors, he met the literary scholar and anecdotist Revd Joseph Spence;[2] Florence (21 February/4 March 1741); Milan (lost but dated 9/20 March 1741); Innsbruck (21 March/1 April 1741); Munich (11/22 April 1741) and Kassels (2/13 May 1741). He then proceeded to Geneva (29 May/9 June and 15/26 June 1741), where he records that on 7/18 June he "called on Mr Dassier the famous medallist, & went with him to see the Brothers of Mr Leotar the painter at Constantinople".[3] Two days later he famously ascended the Glacier at Chamonix,[4] for which he resumed in fun his oriental dress, and the account of which is reproduced in the following extract from Letter 61.

8th-19th. I set out with seven other English viz. Lord Hadington, his brother Mr Bailey, Messrs. Windham, Aldsworth near Twyford, Price, Chetwind & Stillingfleet to go to the Glacieres or mountains of Ice, to the south east in Savoy.[5] We all took the name of Arab chiefs & officers belonging to them, & went 16 miles to Bonneville; – where we walked & diverted ourselves; – our baggage not being come up.

9th-20th. We set out, passed between the mountains by la Clase to Salanche [Sallanches], having very extraordinary views of fine plains & sides of hills finely improved: high rocky mountains, & fine Cascades: We encamped. – I dressed myself privately in the Arab dress & surprized the company, & were

2 Joseph Spence Professor of Poetry at Oxford who between 1730 and 1741 made three tours of the continent, spending five and a half years in Italy, France and Holland. His meetings with Pococke and Lord Sandwich are referred to above and it is likely that he would have been known to Mrs Pococke, not only from his Oxford and church connections, but also because he was from Hampshire.

3 He may well have been making enquiries about or conducting some transaction relating to the portrait he had commissioned by Jean- Étienne Liotard.

4 Though Pococke does not mention this in *Description*, details of this trip were published by William Windham, in 1744 (see Bibliography).

5 The company were as follows: the Scottish nobleman Thomas Hamilton, 7th Earl of Haddington (1721-94); his brother George Baillie-Hamilton; William Windham (1717-61); Richard Neville-Aldworth (1717-93); Price (unidentified); William Richard Chetwynd, 3rd Viscount Chetwynd (1685?-1770); and Benjamin Stillingfleet (1702-71), who was tutor to Windham.

all exceeding cheerful. I hope you will not think it much, that I take one weeks diversion in such good company after my long travels & fatigues: – We went on about 8 miles & lay on straw; after having spent the evening very agreeably; – our journey being all alonge the Arve[ron].

10th -21st. We set out & came in between the mountains, the tops of which are covered with snow & Ice, but fine narrow vales between, we arrived at Chamogny [Chamonix] where we encamped, & after drinking tea, we began our great Tour up the mountains, & after climbing above 4 hours, we came to a deep channel, in which is a bed of Ice; that may be 40 feet deep, looking like a congealed river, & a most extraordinary & wonderful thing it is; – the reeks above rising in a most beautiful manner, tis much higher, as we had already come. We returned down after about 8 hours travail: – lay at a polite Priests house, within six miles of Pays di Valors. This is the highest mount after Teneriff.

11th-22nd. We set out on our return, dined, where we had lain two nights before, & went on to Salench, walked about the town & lay there.

12th-23d. A litle rainy, – we went on to Cluse [Cluses]; dined, & came to Bonneville, where we had lain the night before we took our walk.

13th-24th. It being proposed last night to walk up a fine high mountain called the Mole; – we set out this morn & ascended it inabout 4 hours, & had a glorious view from it. We drank on the top of it, to the health of Admiral Vernon,[6] & to our own under the name of the Eight English men, which we had given ourselves, & descended, having been out above 8 hours: – we drank tea & set out to the So. West, by La Rocke [La Roche-sur-Foron] to Anesy, 20 miles & arrived about 10 at night.

14th-25th. We walked on the fine Avenues on the Lake, on which the Town stands most pleasantly; – saw the churches & the castle, & in one church the Tomb of St Frances di Sales. The Bishop of Geneva resides here; & here all the monks & nuns came from Geneva after the Reformation: – we took a collation & set out at 11, dined at a village & arrived at Geneva, after a most pleasant Tour of 7 days."

6 Edward Vernon (1684-1757) visited the east in 1704, aged 19, when, having been appointed Third Lieutentant on the *Lennox,* he was sent on a convoy to Smyrna with the Levant trade. His naval career continued until he was dismissed by the king in 1746.Vernon became one of the more senior members of the Divan Club, having been proposed by Lord Sandwich, but attended the meetings infrequently on account of difficulties he was encountering in naval matters. See Finnegan, "The Divan Club", pp. 53-54.

This memorable event, involving a group of British gentlemen behaving like members of a club,[7] the use of Arab names and titles, and the donning of oriental clothes, must have been the inspiration for Pococke's establishment of the two London clubs (the Egyptian Society and the later Divan Club) only six months later.[8] Following this brief interlude, he continued his homeward journey through Basel (7/18 July 1741), Liège (9/26 August and 19/30 August 1741) and Dover (29 August 1741). The final letter from the collection describes his calm passage from Ypres and his landing at Dover at sunrise on Saturday 29 August. It also informs his mother of his intended movements across the south of England before their promised reunion in Newtown: "I set out at noon, in my chaise, & shall arrive about Monday noon at the Bell-Savage London: & as I must wait on the Bishop of Waterford [Bishop Este], I would willingly slip down to Dulwich; so I shall not be in Newtown till Wednesday night; or possibly it may be Thursday morning, as I doubt [...] I go through in a day...". Added to his customary signature at the end of the last letter from his travels (Dover, 29 August 1741) are the words "Dei Gloria". It is not clear whether these are the words of Pococke or his mother. Whatever the case, a great debt of gratitude is owed to this extraordinary woman for "insisting" on her son's sending him a daily journal in the first place, for carefully managing his papers while he was abroad, for painstakingly transcribing into three volumes the whole of his voluminous collection of travel correspondence, and for contributing to the preservation of these important manuscripts.

Tribute must also be paid to his uncle, Bishop Milles. Not only did he facilite Pococke's eastern voyage, by allowing him to be absent from his church duties for so many years without detriment to his career, but he was also indirectly responsible for much of the more formal writing from this tour. Pococke felt obliged, through their relationship of patronage, to send the bishop regular accounts of his travels, and the bishop reciprocated by making copies that could be preserved for posterity.

2 Pococke's Reputation

A Description of the East was greatly acclaimed from the time of its publication and, though its scholarly failings have often been noted alongside its merits, it has continued to be regarded as a most worthy and pioneering work. The 18th-century historian Edward Gibbon, for example, described the book as being of "superior learning and dignity", though he objected that its author too often

7 The "Common Room", mentioned by Pococke in his journal entry of 5/16 June 1741, was a club of like-minded men formed by William Windham during his tour of Italy.

8 On the Ottoman titles of their officers, see Finnegan, "The Divan Club".

confounded what he had seen with what he had read;[9] and Pococke has been described more recently as: "an able scholar, but not a great one .... a hardened and determined traveller, an intelligent observer, and a clear writer...."[10] Though, as we have seen, the book had its share of forthright contemporary critics, it was a very popular work and during the latter part of the 18th century was translated into French, German and Dutch, as well as being included in various travel collections or compendiums.[11] Not surprisingly, original sets, of which only 1,250 copies were produced, are valuable, particularly when all the plates and folding maps are intact.

The appearance of an anonymous and obscene satirical poem (*Meekness or Ambition: the Hypocrite Detected*)[12] suggests that towards the end of his term as Bishop of Ossory, Pococke was unpopular with his immediate colleagues. There is only one known original copy of this pamphlet,[13] which suggests it may have been suppressed at the time of publication, and a recent analysis of the contents reveals that the scandalous subject matter contained within it was loosely based on fact.[14]

The pamphlet consists of seven pages of poetical text in various meters and a further 12 pages of "Remarks and Variations", or explanatory notes, signed by fictional scholiasts such as "Scriblerus", "Theobold" and "Jansius". The subject is a poetical dialogue between the "Hero", Bishop ("Dick") Pococke and his alleged lover Susan, who can be identified as the wife of Revd William Cockburne ("Billy Cock"),[15] with whom he regularly dallies in the Robing Room of

9 Cited by Elizabeth Baigent, in "Pococke, Richard (1704-1765)", *Oxford Dictionary of National Biography*, Oxford, 2004. Available at: <https://doi.org/10.1093/ref:odnb/22432>.

10 Bartlett, "Richard Pococke in Lebanon, 1738", pp. 31-32.

11 For a full list of these publications, see Finnegan, *English Explorers in the East*, Appendix, pp. 293-4.

12 Anon., *Meekness or Ambition: the Hypocrite Detected. In a Dialogue between R–h–d, an I-r-h B-sh-p, and S-s-n, a Favourite Ch-b-m-d. On Occasion of his L–s–p's Being Resolutely Oppos'd and Ignominiously Defeated in an Encroachment He Had Long Meditated and Lately Attempted upon the Perogative of a Certain D–n, and the Rights and Privileges of his Ch–t-r, London*, no date (probably false imprint and should be Dublin, 1765).

13 This is in the BL (General Reference Collection 11631.i.14) and all subsequent copies available in libraries are based on this, as can be seen from the annotations. A Gale ECCO Print Edition was published in 2010: ISBN-10: 117025523X.

14 See R. Finnegan, "Meekness and Ambition, or, the Hypocrite Detected: A Satirical Poem about Richard Pococke, Bishop of Ossory", *Old Kilkenny Review. The Journal of the Kilkenny Archaeological Society*, vol. 67, 2015, pp. 72-84.

15 Revd Cockburne (1739-76) was the son of a Kilkenny surgeon and became Vicar General of Ossory in 1759. In 1763 he married Susanna Judith Savage (d. 1790), youngest daughter of Robert Savage of Rock-Savage, Cheshire (see Leslie, *Ossory Clergy and Parishes*, p. 107). The notes to the poem (Anon., *Meekness or Ambition*, p. 11) state that Bishop Pococke, her

his Palace.[16] It concerns Pococke's scheme for vengeance following his recent humiliating defeat by the Dean and Chapter of St Canice's Cathedral, a body to which, he had (in reality) long been opposed.[17] An analysis of the cathedral records ("Chapter Books", 1672-1758, and 1758-1850) allowed the present editor to identify the thinly disguised names, as well as to verify the diocesan disagreements referred to in the poem.[18] This suggests that the author was either a member of the Dean and Chapter itself or was closely associated with this body. Though the pamphlet is not dated, it appears from the context that it was published in Dublin,[19] in 1765, shortly before Pococke vacated the See of Ossory, and therefore just before his death.[20] Furthermore, the poem, which is primarily an attack on Pococke's character and career, scorns the fame that he derived from his eastern travels, and questions its fitness for him to acquire episcopal status.[21]

Regarding other contemporary views of Pococke, the English dramatist Richard Cumberland, once stated: "That celebrated oriental traveller and author was a man of mild manners and primitive simplicity: having given the world a full detail of his researches in Egypt, he seemed to hold himself excused from saying anything more about it and observed an obdurate simplicity."[22] While he may have given this impression in some circles, to like-minded

---

"present K-p-r [Keeper]", had "given universal Offence, as well as Astonishment to the Publick" by providing Cockburne with a dowry.

16 In the poem this building is called "D-s-t's [Dorset's] *Bower*" (Anon., *Meekness or Ambition*, p. 3) and is further explained, at Anon., *Meekness or Ambition*, p. 9, as, "A Pleasure'house (so called) in his L'sh'p's Garden, wherein the scandalous Jokers of K-k- y [Kilkenny] tell us, that Susan and R-h-d, every Summer Evening – In dalliance sweet, in am'rous Sport and Play, With Lip to Lip, beguile the tedious Day." For a full account of the Robing Room, see *Cóilín Ó Drisceoil, Excavation & Archaeological Building Survey at the Robing Room, Bishop's Palace, Church Lane, Kilkenny 2011 (11E157) Preliminary Report,* Heritage Council, Kilkenny, 2011, especially pp. 41ff.

17 The Dean and Chapter are the ecclesiastical governing body of a cathedral, and consist of the Dean, who is the chief, and his Canons or Prebenderies.

18 Finnegan, "Meekness and Ambition", pp. 72-84.

19 The imprint of "London (E. Curle)" is regarded as false.

20 The last section, entitled "The Author's last Words, or Farewell to his Hero", urges him to, "Be meek and gentle, without Affectation, Enjoy, content, your present Lot and Station, and drop your little Schemes for a Translation" (Anon., *Meekness or Ambition,* p. 20).

21 When Pococke's declares, "I'll teach the Gentlemen... that I did not visit *Grand Cairo,* and wash my Hands in the *Nile,* for nothing", his enemies responds: "By the L-d, he should have 'wash'e his dirty *Palms,* while his *Hands* were in.' – But – perhaps, he foresaw, that, on his Return, the Fame of his great Travels wou'd (among as great Blockeads as himself) raise him to a M-t-e [Mitre]; in which View, he might have considered the washing them, as unnecessary Labour." Anon., *Meekness or Ambition,* p. 12.

22 *Memoirs of Richard Cumberland written by Himself...,* London, 1806-7, p. 121.

individuals his extensive knowledge of foreign parts and antiquity, his writings and his collections would have been the basis for much scholarly discourse. Ironically, again it is the more satirical record that has has survived, in an account given by Mrs Delany. When anticipating her imminent visit to his Dublin residence in 1752, she declared: "Next Tuesday we are to ... go in the afternoon to an Egyptian drum (I suppose) at Dr Pocock's. Instead of spreading his table with cards, I hope he designs to cover them with drawers and curiosities, and instead of the tittle-tattle of the town to give us some philosophical lectures! It would be pleasant enough to see the surprise of the smart beaux and belles when they observe such an entertainment prepared for them; and instead of the rooms being decorated with china, japan, Indian paper, and looking glasses, to observe nothing but Egyptian deities on pedestals, tables covered with precious fragments such as toes and fingers, lumps of stone that have neither shape nor beauty of colour, Turkish robes hanging on pegs, travelling kitchen utensils, etc."[23]

Whatever impression Pococke may have given to the world, one thing that can be said for certain is that his passion for travel remained with him for the rest of his life. Not only did he perform journeys around England, Ireland and Scotland, the valuable correspondence from which was published in the late 19th century,[24] but it is clear from the contents of his library sale catalogue that he continued to collect foreign travel books and related literature for many years after his grand tours. As with many serious scholar-travellers of the period, he collected numerous prints of drawings, portraits and views from his travels, among which was a set of prints listed in the catalogue (Lot 127) as "Dalton's Antiquities of Greece, 23 Plates". This was the first issue of *Antiquities and Views in Greece*,[25] with 23 plates engraved after original drawings made by the artist during a tour of the east with Lord Charlemont (1749-50).[26] The volume includes views of the Temples of Theseus, Hercules, the Erectheion and the Tower of the Winds; and though, as already observed, we do not know where Pococke obtained the drawings of these famous Athenian monuments for his own book, it must have been a pleasing thought that his engravings, as

23 Quoted in Katherine Cahill *Mrs Delany's Menus, Medcines and Manners*, Dublin, 2005, p. 335.

24 Kemp (1887), Cartwright (1888-89) and Stokes (1891), see the *Bibliography*

25 The full title is Antiquities and Views in Greece and Egypt, with the Manners and Customs ofthe Inhabitants from Drawings Made on the Spot AD 1749, London, 1751-52.

26 See Stanford and Finopoulos, *The Travels of Lord Charlemont*, pp. 232-33, for a discussion of Charlemont's draughtsman and their quarrel.

acknowleged by some modern scholars (particularly Irish ones, who claim him as one of their own), were superior to those later presented by Dalton.[27]

---

27 W.B. Stanford notes that Dalton was "far from being a fine draughtsman" and cites architectural historian Maurice Craig, who expressed a similar view. See "The Manuscripts of Lord Charlemont's Eastern Travels", *Proceedings of the Royal Irish Academy,* Section C, vol. 80, no. 5, 1980, p. 82 and note 13. For a full discussion of this subject, see Chapters 8 and 9 of Finnegan, *English Explorers in the East.*

APPENDIX

List of Grand Tour Letters and Accounts Reproduced in This Volume

Letters from Pocoke to his mother transcribed in this volume:[1]

Letter 0	LEGHORN 16/17 September 1737
Letter 1	ALEXANDRIA IN EGYPT 30 September/11 October 1737
Letter 2	ALEXANDRIA IN EGYPT 5/16 October 1737
Letter 3	ALEXANDRIA IN EGYPT 23 October/3 November 1737
Letter 4	GRAND CAIRO 11/22 November 1737
Letter 5	GRAND CAIRO 15/26 November 1737
Letter 6	GRAND CAIRO 5/16 December 1737
Letter 7	ACKMIM IN UPPER EGYPT 21 Dec./1 January 1737/38
Letter 8	FURCHUT IN UPPER EGYPT 3/14 January 1738

1 Apart from Letter 0, which is Letter 27 from the collection BL, Add MS 22997 (letters from the second tour), all letters are from the collection BL, Add MS 22998. The original spellings of placenames are given in this list to match the headings of the letters and accounts in this volume.

 | DOI:10.1163/9789004440050_012

Letter 9	GRAND CAIRO 28 February/11 March 1738
Letter 10	DAMIATA IN EGYPT 8/19 March 1738
Letter 11	JERUSALEM 26 March/6 April 1738
Letter 12	BETHLEHEM 8/19 April 1738
Letter 13	JOPPA 2/13 May 1738
Letter 14	ST JOHN D'ACRE 16/27 May 1738
Letter 15	SIDON 30 May/10 June 1738
Letter 16	TRIPOLI 11/22 June 1738
Letter 17	DAMASCUS 1/12 July 1738
Letter 18	ALEPPO 31 July/22 August 1738
Letter 19	ALEPPO 12/23 August 1738
Letter 20	ALEPPO 18/29 September 1738
Letter 21	LATICHEA THE OLD LAODICIA 9/20 October 1738
Letter 22	LARNICA IN CYPRUS 10/21 November 1738

Letter 23 LARNICA IN CYPRUS
7/18 December 1738

Letter 24 GRAND CAIRO
20/31 January 1739

Letter 25 GRAND CAIRO
1/12 February 1739

Letter 26 GRAND CAIRO
2/13 March 1739

Letter 27 GRAND CAIRO
26 March/6 April 1739

Letter 28 MOUNT SINAI
12/26 April 1739

Letter 29 SUEZ ON THE RED SEA
1/12 May 1739

Letter 30 ALEXANDRIA IN EGYPT
2/13 July 1739

Letter 31 CANEA IN THE ISLE OF CANDIA
16/27 August 1739

Letter 32 CANEA IN THE ISLE OF CANDIA
29 Sept. 10 Oct. 1739

Letter 33 SMYRNA
5/16 November 1739

Letter 34 SMYRNA
1/12 December 1739

Letter 35 EPHESUS
15/26 December 1739

Letter 36 CORA IN THE ISLE OF SAMOS
7/18 January 1740

Letter 37	CORA IN THE ISLE OF SAMOS 2/13 February 1740
Letter 38	GUZELHISSHAR THE OLD MAGNESION ON MEANDER 28 February/10 March 1740
Letter 39	ANGORA THE OLD ANCYRA IN GALATIA 12/23 April 1740
Letter 40	CONSTANTINOPLE 31 May/11 June 1740
Letter 41	CONSTANTINOPLE 19/30 June 1740
Letter 42	CONSTANTINOPLE 22 June/3 July 1740
Letter 43	CONSTANTINOPLE 7/18 July 1740
Letter 44	BRUSA 17/28 August 1740
Letter 45	SALONICA 19/30 September 1740
Letter 46	THEBES IN BAOLIA 27 Sept./8 October 1740
Letter 47	ATHENS 29 Sept. /10 Oct. 1740
Letter 48	CORINTH 6/17 October 1740
Letter 49	MESSINA 18/29 November 1740

Accounts from Pococke to Bishop Milles transcribed in this volume:[2]

Account 0	VENICE 19/30 August 1737
Account 1	LEGHORN 6/17 September 1737
Account 2	ALEXANDRIA IN EGYPT 30 September/11 October 1737
Account 3	ALEXANDRIA IN EGYT 30 September/11 October 1737
Account 4	ALEXANDRIA IN EGYPT 6/17 October 1737
Account 5	GRAND CAIRO 1/12 March 1738
Account 6	DAMIATA IN EGYPT 8/19 March 1738
Account 7	JERUSALEM 2/13 April 1738
Account 8	BETHELEM 6/17 April 1738
Account 9	JOPPA 23 April/4 May 1738
Account 10	ST JOHN D'ACRE 6/17 May 1738

2 Apart from Account 0, which is from the collection BL, Add MS 15774 (letters from Jeremiah Milles to Bishop Milles), all letters are from the collection BL, Add MS 15779. The original spellings of placenames are given in this list to match the headings of the letters and accounts in this volume.

Account 11	NAZARETH 11/22 May 1738
Account 12	TIBERIUS 13/24 May 1738
Account 13	TYRE 28 May/8 June 1738
Account 14	SIDON 31 May/11 June 1738
Account 15	BEYREUT IN SYRIA 1/12 June 1738
Account 16	TRIPOLY IN SYRIA 9/20 June 1738
Account 17	CANNOBINE ON MOUNT LIBANON 13/24 June 1738
Account 18	BAALBECK IN SYRIA 9/20 June 1738
Account 19	ALEPPO 31 July/2 August 1738
Account 20	FAMAGUSTA IN CYPRUS 11/22 November 1738
Account 21	NICOSIA IN CYPRUS 21 November/2 Dec. 1738
Account 22	BAFFA IN CYPRUS 1/12 December 1738
Account 23	COLOSSE IN CYPRUS 3/14 December 1738

Account 24	LIMESOLE IN CYPRUS 4/15 December 1738
Account 25	LARNICA IN CYPRUS 7/18 December 1738
Account 26	GRAND CAIRO 7/18 March 1739
Account 27	DAMIATA IN EGYPT 25 December/5 Jan. 1738/39
Account 28	MEHULLAH IN EGYPT 29 December/9 Jan. 1738/39
Account 29	GRAND CAIRO 1/12 January 1739
Account 30	FAIUME IN EGYPT 12/23 February 1739
Account 31	SACCARA IN EGYPT 23 February/6 March 1739
Account 32	GRAND CAIRO 26 February/9 March 1739
Account 33	GRAND CAIRO 1/12 March 1739
Account 34	GRAND CAIRO 7/18 March 1739

Bibliography

(A) Manuscripts

Pococke's Travel Correspondence (British Library, herafter: BL)

Jeremiah Milles, "Notes in a Journey from London to Holyhead, 6-13 May [1740?]" (BL, Add MS 15776)

Richard Pococke, Letters to Mrs Elizabeth Pococke, copies (BL, Add MS 22997)

Richard Pococke, Letters to Mrs Elizabeth Pococke, copies (BL, Add MS 22998)

Richard Pococke, Register of Letters from R. Pococke to Bishop of Waterford, copies (BL, Add MS 15779)

Richard Pococke, "A Voyage From Leghorn to [Alexandria]" (BL, Add MS 22995)

Other Manuscripts in Public Archives

"Al-Koran", the Minute Book of the Divan Club (National Maritime Museum, London, Ref. SAN/V/113)

"A Voyage Round the Mediterranean in the years 1738 & 1739" (National Maritime Museum, Ref. SAN/F/50)

"Chapter Book", 1672-1758 (St Canice's Cathedral, Kilkenny)

"Chapter Book", 1758-1850 (St Canice's Cathedral, Kilkenny)

Election Citation for Richard Pococke, 1741 (Royal Society, London, EC/1741/15)

"Geographical and Antiquarian Notes Relating to Greece and Asia Minor" (BL, Add MS 4824)

Jeremiah Milles, "Sketch Book of Antiquities in Germany" (BL, Add MS 14263)

Last Will & Testament of Lister Bigge (The National Archives of the UK [hereafter: TNA], Prob 11/759)

Last Will & Testament of Thomas Milles (TNA, Prob 11/717)

Last Will & Testament of Mrs Elizabeth Pococke (TNA, Prob 11/858)

Last Will & Testament of Elizabeth Pococke (Spinster) (TNA, Prob 11/999)

Last Will & Testament of Richard Pococke (National Library of Ireland, D. 27, 161)

Last Will & Testament of Richard Pococke (Senior) (Hampshire Archives and Local Studies, Ref. 1710B/42)

Last Will & Testament of Paggen Shaw of Tottenham, Middlesex (TNA, Prob 11/960/131)

Last Will & Testament of Rowland Sherman (TNA, Prob 11/771)

Letter from the Duke of Devonshire to the Right Honorable Henry Fox Esq., Dublin Castle, 14 February 1756 (TNA Kew, 5P 63/414)

Letter from Thomas Johnston to David Latouche & Sons, 16 September 1765 (Derbyshire Record Office, D3155/C3725)

Letter from Thomas Milles to Mrs Pococke, Waterford, 9 January 1724 (Waterford Museum of Treasures)

Letter from Richard Pococke to Dean Charles Lyttelton, 25 June 1756 (Warwickshire County Record Office, Ref. 125B/798)

Letters from Richard Pococke to Sanderson Miller, 8 and 17 February 1757 (Warwick County Record Office, 125B/801 and 802)

Letter from Richard Pococke to Sanderson Miller, 19 February 1757 (Warwickshire County Record Office, CR125/B802)

Letter from Richard Pococke to William Stukeley, 4 March 1745 (Oxford, Bodleian Library, MS. Eng. misc. c. 114, fol. 147r)

Letter from Sir Robert Wilmot to Mr Fox, 25 February 1763 (Derbyshire Record Office, D3155/C3207).

Minute Book of the Egyptian Society (BL, Add MS 52362)

Minute Book of the Florists' Club with Names of Members, from Its Foundation in 1746 to Its End 1766 (Royal Irish Academy, 24 E 37)

Minute Book of the Physico-Historical Society [of Ireland] 14 April 1744-22 March 1752 (Royal Irish Academy, 24 E 28)

Minutes of Waterford Corporation (Waterford City Archives, LA1/1A/07)

Pembroke Estate Papers (National Archives of Ireland, 2/5/32)

"Portrait de l'Archéologue et Théologien Richard Pococke" (Fonds des Dessins et Miniatures, Louvre, RF 1379, Recto)

"Portrait de Gaspard de Péleran Consul de France à Smyrne" (Fonds des Dessins et Miniatures, Louvre, RF 1365, Recto)

"Portrait de Madame James Fremeaux Née Margaret Cooke, Smyrne" (Fonds des Dessins et Miniatures, Louvre, RF 1383, Recto)

"Portrait de Madame Gaspard de Péleran, Épouse du Consul de France à Smyrne" (Fonds des Dessins et Miniatures, Louvre, RF 1366 Recto)

"Portrait de M. Levett, Négociant Anglais, en Costume Tartare" (Fonds des Dessins et Miniatures, Louvre, RF 1377, Recto)

"Portrait du Comte de Bonneval, Dit Ahmet Pacha" (Fonds des Dessins et Miniatures, Louvre, RF 1387, Recto)

"The Chapter Book of Lismore, 1663-1829" (Cotton Library, St Carthage's Cathedral, Lismore)

(B) Printed Books and Articles

Ágoston, G. and Masters, B., *Encyclopedia of the Ottoman Empire*, Facts on File, New York, 2009

Annis, M., "The First Egyptian Society in London (1741-1743)", *Bulletin de L'Institut Francais D'Archaeologie Orientale* 50, 1952, pp. 99-105

[Anon., probably Mrs Elizabeth Pococke] *An Account of the Life and Conversation of the Reverend and Worthy Mr. Isaac Milles, Late Rector of High Cleer in Hampshire*, London, 1721

[Anon.] *Meekness or Ambition: the Hypocrite Detected. In a Dialogue between R–h–d, an I-r-h B-sh-p, and S-s-n, a Favourite Ch-b-m-d. On Occasion of his L–s–p's Being Resolutely Oppos'd and Ignominiously Defeated in an Encroachment He Had Long Meditated and Lately Attempted upon the Perogative of a Certain D–n, and the Rights and Privileges of his Ch–t-r,* London, no date (probably false imprint and should be Dublin, 1765)

Aynural, S., in G. Ágoston and B. Masters, *Encyclopedia of the Ottoman Empire,* New York, 2009

Baedecker, K., *Palestine and Syria: Handbook for Travellers*, London, 1876

Baigent, E., "Pococke, Richard (1704-1765)", in *Dictionary of National Biography*, Oxford University Press, Oxford, 2004. Available at: <https://doi.org/10.1093/ref:odnb/22432>

Bailey, N., T*he Universal Etymological Dictionary*, vol. 2, London, 1731

Bard, K, *Encyclopaedia of the Archaeology of Ancient Egypt*, London, 1999

Barnard T., *A New Anatomy of Ireland: The Irish Protestants, 1649-1770*, Yale University Press, New Haven, 2003

Bartlett, J. R., "Richard Pococke in Lebanon, 1738", *Archaeology & History in Lebanon,* Issue 16, Autumn 2002, pp. 17-33

Beaufort, D. A., *Memoir of a Map of Ireland; Illustrating the Topography of that Kingdom, and Containing a Short Account of its Present State, Civil and Ecclesiastical, with a Complete Index to the Map*, London, 1792

Bence-Jones, M., *A Guide to Irish Country Houses*, London, 2nd revised edition, 1990

Benezit, E., *Dictionary of British Graphic Artists and Illustrators*, vol. 1, Oxford, 2012

Black, J., *The British Abroad: the Grand Tour in the Eighteenth Century*, Sutton Publishing, Stroud, 2003

Boogert, M. H., van den, *The Capitulations and the Ottoman Legal System,* Leiden 2005.

Boogert, M. H., van den, "The Sultan's Answer to the Medici Press? Ibrahim Müteferrika's Printing House in Istanbul", in Alastair Hamilton et al. (eds), *The Republic of Letters and the Levant*, Leiden, 2005, 265-292

Boogert, M.H., van den, "Freemasonry in Eighteenth-century Izmir? A Critical Analysis of Alexander Drummond's Travels", in Maurits H. van den Boogert (ed.), *Ottoman Izmir. Studies in Honour of Sander de Groot*, Nederlands Instituut voor het Nabije Oosten, Leiden, 2007, pp. 110-21.

Bottari, S., "Consoli Inglesi nella Sicilia del Settecento: Funzioni Institutionali e Attività Informali", in Arnaud Bartolomei et al. (eds), *De l'Utilité Commerciale des Consuls.*

L'Institution Consulaire et les Marchands dans le Monde Méditerranéen (*XVIIe-XXe siècle*), Rome and Madrid, 2017

Bradley, R., *The Country Housewife and Lady's Director in the Management of a House, and the Delights and Profits of a Farm*, London, 1732.

Brewer, J. N., T*he Beauties of Ireland Being Original Deliniations, Topographical, Historical, and Biographical, of Each County*, vol. 2, London, 1826

Brunton, D. A., "Isaac Milles (1638-1720)", in *Oxford Dictionary of National Biography*, Oxford University Press, 2004. Available at: <https://doi.org/10.1093/ref:odnb/18749>

Buckler, J. C., *Sixty Views of Endowed Schools: From Original Drawings by J. C. Buckler With Letter-Press Descriptions*, London, 1827

Cahill, K., *Mrs Delany's Menus, Medcines and Manners*, Dublin, 2005

C*alendar of Treasury Books and Papers*, vol. 2, 1733-34, Her Majesty's State Paper Office, London, 1898. Available at: <https://www.british-history.ac.uk/cal-treasury-books-papers/vol2/pp406-411>

Cartwright, J.J. (ed.), *The Travels through England of Dr Richard Pococke, Successively Bishop of Meath and of Ossory during 1750, 1751, and Later Years*, Camden Society, London, 1888-89

Casey, C. and Rowan, A., *The Buildings of Ireland: North Leinster*, Yale University Press, London, 1993

Castiglione, F., "'Levantine' Dragomans in Nineteenth Century Istanbul: The Pisanis, the British, and Issues of Subjecthood", Osmanlı Araştırmaları 44 (2014), pp. 169-195

Chandler, R., *Ionian Antiquities, or, Ruins of Magnificent and Famous Buildings in Ionia*, Society of Dilettanti, London, 1769

Chandler, R., *Inscriptiones Antiquae Pleraeque Nondum Editae*, in Asia Minore et Graecia, Clarendon Press, Oxford, 1774

Chandler, R., *Travels in Asia Minor: or An Account of a Tour Made at the Expense of the Society of Dilettanti*, Clarendon Press, Oxford, 1775

Chandler, R., *Travels in Greece, or An Account of a Tour Made at the Expense of the Society of Dilettanti*, Clarendon Press, Oxford, 1776

Chaney, E., *The Evolution of the Grand Tour*, Frank Cass, London, 1998

Chetwood, W. R., *A Tour through Ireland In Several Entertaining Letters Wherein the Present State of that Kingdom is Consider'd and the Most Noted Cities, Towns, Seats, Rivers, Buildings &c. are Described...*, London, 1748

Chishull, E., *Inscriptio Sigea Antiquissima Boustrophedon Exarata/Commentario eam Historico, Grammatico, Critico Illustravi*, London, 1721.

Chishull, E., *Antiquitates Asiaticae Christianam Aeram Antecedentes: Ex Primaris Monumentis Graecis...*, London, 1728.

Clarke, W., *Repertorium Bibliographicum; or Some Account of the Most Celebrated British Libraries*, London, 1819

Clayton, Revd R., *A Journal from Grand Cairo to Mount Sinai and Back Again, Translated from a Manuscript, Written by the Prefetto of Egypt in Company with the Missionaries of de Propaganda Fide at Grand Cairo, to Which are Added some Remarks on the Origin of Hieroglyphics and the Mythology of the Ancient Heathens, by the Right Reverend Robert Lord Bishop of Clogher. Dedicated to the Society of Antiquaries, London*, London, 1753

Cohausen, J. H., *Hermippus Redivivus, or, The Sage's Triumph over Old Age and the Grave*, London, 1742.

Cook, J.M., *The Troad: an Archaeological and Topographical Study*, Oxford University Press, Oxford, 1973, Special Edition, 1999

Cooke, J. (ed.), A *Voyage Performed by the Late Earl of Sandwich round the Mediterranean in the Years 1738 and 1739, Written by Himself*, London, 1799

Cruttwell, C., *The New Universal Gazetteer*, vol. 3, London 1798

Cumberland, R., *Memoirs of Richard Cumberland written by Himself ...*, London, 1806-7

Dalton, R., *Antiquities and Views in Greece and Egypt, with the Manners and Customs of the Inhabitants from Drawings Made on the Spot AD 1749*, London, 1751-52.

Davis, R., *Aleppo and Devonshire Square: English Traders in the Levant in the Eighteenth Century*, Macmillan, London, 1967

Dehérain, H. "L'Orientalisme Français en Égypte au XVIIIe Siècle", *Journal des Savants*, vol. 6, 1931, pp. 261-272

Dibdin, T. F., *The Library Companion, or the Young Man's Guide, and the Old Man's Comfort, in the Choice of a Library*, London, 1824

Doble C. E. and Rannie, D. W., *Remarks and Collections of Thomas Hearne. Nine Volumes, Covering the Years 1705-1714*, Oxford Historical Society, 1885-98, vol. 1

Ellis, D., *Newtown at the Millenium*, Newtown Parish Council, 2000

Finnegan, R., "The Classical Taste of William Ponsonby, 2nd Earl of Bessborough", *Irish Architectural and Decorative Arts. The Journal of the Irish Georgian Society*, vol. 8, 2005, pp. 12-43

Finnegan, R., "The Library of William Ponsonby, 2nd Earl of Bessborough, 1704-93", *Hermathena, A Trinity College Dublin Review* (In Honour of George Huxley), no. 181, Winter, 2006, pp. 149-87

Finnegan, R., *Letters from Abroad: The Grand Tour Correspondence of Richard Pococke & Jeremiah Milles*, 3 vols, Pococke Press, Piltown, 2011-2013

Finnegan, R., "The Divan Club: an 18th-Century Club for Travellers to the Ottoman Empire", originally published in *Electronic Journal of Oriental Studies*, 9, 2006, pp. 1-86. Available at: <http://www.pocockepress.com/DivanClub.pdf>

Finnegan, R., "Bishop Pococke's Improvements to St. Canice's Cathedral, Kilkenny", *Irish Architectural and Decorative Studies: The Journal of the Irish Georgian Society*, vol. 11, 2008, pp. 12-55

Finnegan, R., "Manuscripts, Mummies & Minute Books: the Case of Pococke & Sandwich in the East", in L. Mulvin (ed.), *The New Griffon. 13, A Culture of Translation:*

British and Irish Scholarship in the Gennadius Library (1740-1840), Athens, 2012, pp. 47-59

Finnegan, R. "Thomas Milles Bishop of Waterford and Lismore", *Decies, Journal of the Waterford Archaeological & Historical Society*, 69, 2013, pp. 53-80.

Finnegan, R., "Meekness and Ambition, or, the Hypocrite Detected: A Satirical Poem about Richard Pococke, Bishop of Ossory", *Old Kilkenny Review. The Journal of the Kilkenny Archaeological Society*, vol. 67, 2015, pp. 72-84.

Finnegan, R., "The Travels and Curious Collections of Richard Pococke, Bishop of Meath", *Journal of the History of Collections*, vol. 27, no. 1, 2015, pp. 33-48

Finnegan, R., *English Explorers in the East (1738-1745). The Travels of Thomas Shaw, Charles Perry and Richard Pococke*, Leiden, 2019

Finnegan, R., "Richard Pococke's Herbarium and Manuscript Index", *Journal of the History of Collections*, 16 April 2020. Available at: <https://doi.org/10.1093/jhc/fhaa004>

Finnegan, R., "Letters from Vienna: Richard Pococke en Route to Egypt", in Mladen Tomorad (ed.) *Egypt and Austria XII: Current Research. Proceedings of the Conference Held at the Faculty of Croatian Studies, University of Zagreb (September 17th-22nd, 2018)*, Oxford-Zagreb: Archaopress-Faculty of Croatian Studies, 2020, 91-96

Flin, L., *A Catalogue of the Library of the Late Right Revd. Dr Richard Pococke, Lord Bishop of Meath Deceased*, Dublin, 1766

Foster, J., *Alumni Oxonienses, 1715-1886 (L-Z)* (1888), Oxford University Press, Oxford

Gaertner, J.F. and Hausberg, B.C., *Caesar and the Bellum Alexandrinum, An Analysis of Style, Narrative Technique, and the Reception of Greek Histiography*, Vandenhoeck & Ruprecht, Göttingen, 2013

Gardiner, R.B., *Registers of Wadham College*, vol. 1, London, 1889-95

Gough Nichols, J. *The Topographer and Genealogist*, vol. 1, London, 1846

Guillet de Saint-Georges, G., *Athenes Ancienne et Nouvelle, et l'Estat Present de l'Empire des Turcs, Contenant la Vie du Sultan Mahomet IV...*, Paris, 1675

Hall, A., (ed.), *Autobiography and Correspondence of Mary Granville, Mrs. Delany*, vol. III, London, 1861; digitally printed by Cambridge University Press, 2003

Hasselquist, F., *Voyages and Travels in the Levant*, London, 1767

Healy, J., *History of the Diocese of Meath*, vol. 2, Dublin, 1908

Hegarty, M., "Dr. Richard Pococke's Travels in Ireland, England and Wales", *Old Kilkenny Review: Journal of Kilkenny Archaeological Society*, 1987, pp. 388-98

Huxley, D. (ed.), *Cretan Quests: British Explorers, Excavators and Historians*, British School at Athens, 2000

Huxley, G., *Homer and the Travellers: A Lecture on Some Antiquarian & Topographical Books in the Gennadius Library of the American School of Classical Studies at Athens*, Athens, 1988

Ingamells, J. (ed.), *A Dictionary of British Travellers in Italy, 1701-1800*, The Paul Mellon Centre for Studies in British Art, Yale University Press, New Haven and London, 1977

Ireland, A., "A Gentle Luxury: Collectors and Collecting in Eighteenth Century Ireland", in J.V. Luce *et al.* (eds), *The Lure of Greece: Irish Involvement in Greek Culture Literature History and Politics*, Hinds Publishing, Dublin, 2007, pp. 55-69

Ireland, A., "Richard Pococke (1704-65), Antiquarian", *Peritia*, vol. 20, 2008, pp. 353-78

Jean-Etienne Liotard, 1702-1789, Masterpieces from the Musees d'Art et d'Histoire of Geneva and Swiss Private Collections, Somogy Editions d'Art, Paris, 2006

Jeffares, N., *Dictionary of Pastellists before 1800,* Unicorn Press, Norwich, 2006

Kemp, D.W. (ed.), *Tours in Scotland 1747, 1750, 1760 by Richard Pococke Bishop of Meath,* Edinburgh, Publications of the Scottish History Society, vol. 1, University Press, Edinburgh, 1887; Reprinted Maryland, 2003

Klaver, Jan Marten Ivo, "Richard Pococke and the Natural Curiosities of the East", in Jan Loop and Jill Kraye (eds), *Scholarship between Europe and the Levant. Essays in Honour of Alastair Hamilton*, Leiden, 2020, pp. 260-275

Klima, S. (ed.), *Joseph Spence, Letters from the Grand Tour*, McGill-Queen's University Press, Montreal and London, 1975

Knolles, *Generall Historie of the Turkes*, London, 1603

Korte, J., *Reisen ach dem Gelobten Lande, Aegypten, Syrien, und Mesopotamien,* Altona, 1740; with Three Supplements, Halle, 1746

Laidlaw, C. *The British in the Levant: Trade and Perceptions of the Ottoman Empire in the Eighteenth Century*, Taurus Academic Studies, 2010

Lane, E. W., *An Arabic-English Lexicon, Book I*, London, 1885

Langford, A., *A Catalogue of a Curious Collection of Greek, Roman, and English Coins and Medals, of the Right Reverend Dr Pococke, Lord Bishop of Meath, Collected by his Lordship, during his Travels*, London, May 1766

Langford, A., *A Catalogue of a Large and Curious Collection of Ancient Statues, Urns, Mummies, Fossils, Shells, and Other Curiosities, of the Right Reverend Dr Pococke, Lord Bishop of Meath, Deceased; Collected by His Lordship, during his Travels*, London, 1766

Lee, G., *Huguenot Settlements in Ireland,* London, 1936

Le Grys Norgate, G., Revised by J. C. Walton "Thomas Milles (1671-1740)", *Oxford Dictionary of National Biography*, Oxford University Press, 2004. Available at: <https://doi.org/10.1093/ref:odnb/18754>

Lennon, C., *Irish Historic Towns Atlas (IHTA), no. 19, Dublin, Part II, 1610 to 1756*, Royal Irish Academy, Dublin, 2008

Leslie, J. B., *Ossory Clergy and Parishes: Being an Account of the Clergy of the Church of Ireland in the Diocese of Ossory, from the Earliest Period, with Historical Notices of the Several Parishes, Churches, &c.*, Enniskillen, 1933

Lowndes, W. T., *The Bibliographer's Manual of English Literature … by William Thomas Lowndes*, vol. 2, London, 1834

Lucas, P., *Voyage du Sieur Paul Lucas, Fait en MDCCXIV, &c. Par Ordre de Louis XIV, Dans la Turquie, l'Asie, Sourie, Palestine, Haute & Basse Egypte, &c.,* Rouen, 1724.

Luce, J. V., "Robert Wood and Homer", in J. V. Luce *et al.* (eds), *The Lure of Greece: Irish Involvement in Greek Culture Literature History and Politics,* Hinds Publishing, Dublin, 2007, pp. 71-87

Luce, J.V., Morris. C., and Souyoudzoglou-Haywood C. (eds), *The Lure of Greece: Irish Involvement in Greek Culture Literature History and Politics,* Hinds Publishing, Dublin, 2007

Masters, B., *Christian and Jews in the Ottoman Arab World: The Roots of Sectarianism, Cambridge Studies in Islamic Civilization*, Cambridge University Press, Cambridge, 2001

Matar, N., "Islam in Britain, 1689-1750", *Journal of British Studies*, vol. 47, no. 2, 2008, pp. 284-300

Maundrell, H., *A Journey from Aleppo to Jerusalem at Easter A.D. 167,* Oxford, 1703

Maundrell, H., *A Journey to the Banks of the Euphrates at Beer, and to the Country of Mesopotamia*, Oxford, 1699

McCarthy, Michael, "'The Dullest Man That Ever Travelled?' – A re-Assessment of Richard Pococke and of His Portrait by J.-E. Liotard", in *Apollo*, vol. 143, 1996, pp. 25-29

McCarthy, M., "Extending the Grand Tour: Dr Pococke's Travels, 1735-41", in Ní Chatháin, P. and Fitzpatrick, S. (eds), *Pathfinders to the Past: The Antiquarian Road to Irish Historical Writing, 1640-1960,* Four Courts Press, Dublin, 2012, pp. 30-34

McVeagh, John, *Richard Pococke's Irish Tours*, Irish Academic Press, Dublin, 1995

Messenger, A.W.B., "An Eighteenth Century Dean of Exeter and his Family", *Report and Transactions of the Devonshire Association for the Advancement of Science,* 83, 1951, pp. 29-30

Mézin, A., *Les Consuls de France, au Siècle des Lumières* (*1715-1792*), Paris, 1998

Milles, T., *The Happiness of Those That Suffer for Righteousness Sake: in a Sermon Preached at St. Maries in Oxford, on the xxxth. of January 1700/01*, Oxford, 1701

Milles, T. (ed.), *S. Patris Nostri Cyrilli Hierosolymorum Archiepiscopi Opera Quae Supersunt, Omnia; Quaedam Nunc Primum ex Codd. MSS. Edidit, Reliqua cum Codd. MSS. Contulit, Plurimis in Locis Emendavit, Notisque Illustravit*, Oxford, 1703

Milles, T., *The Natural Immortality of the Soul Asserted: and Proved from the Scriptures, and First Fathers: in Answer to Mr Dodwell's Epistolary Discourse, in Which he Endeavours to Prove the Soul to be a Principle Naturally Mortal*, Oxford, 1707

Müller, L., *Consuls, Corsairs, and Commerce. The Swedish Consular Service and Long-distance Shipping, 1720-1815*, Uppsala, 2004

Mulvin, L. (ed), The New Griffon. 13, A Culture of Translation: British and Irish Scholarship in the Gennadius Library (1740-1840), Athens, 2012

Musson, A. E. and Robinson, E., *Science and Technology in the Industrial Revolution*, Manchester, 1969

Nanninga, J.G, *Bronnen tot de Geschiedenis van den Levantschen Handel Vierde Deel: 1765-1826 – Eerste Stuk*, The Hague, 1964

Nash, P.W., Savage, N., Beasley, G., Meriton, J. and Shell, A. (eds), *Early Printed Books, 1478-1840: Catalogue of the British Architectural Library Early Imprints Collection*, K.G. Saur, Munich, 1994-2003, vol. 5

Nelson, C. E., "The Dublin Florists' Club in the Mid Eighteenth Century", *Garden History*, vol. 10, no. 2, 1982, pp. 142-48

Ní Chatháin, P., and Fitzpatrick, S. (eds), *Pathfinders to the Past: The Antiquarian Road to Irish Historical Writing, 1640-1960*, Four Courts Press, Dublin, 2012

Nichols, J., *Illustrations of the Literary History of the Eighteenth Century, Consisting of Authentic Memoirs and Original Letters of Eminent Persons; and Intended as a Sequel to The Literary Anecdotes*, vol. 2, London, 1814

Nichols, J. and Bentley, S., *Literary Anecdotes of the Eighteenth Century*, vol. 3, London, 1812

Notes and Queries: A Medium of Intercommunication for Literary Men, General Readers, Etc., vol. 10, London, 1885

Ó Drisceoil, C., Excavation & Archaeological Building Survey at the Robing Room, Bishop's Palace, Church Lane, Kilkenny 2011 (11E157) Preliminary Report, Heritage Council, Kilkenny, 2011

Özkul, A.E., "The Consuls and their Activities in Cyprus under the Ottoman Administration (1571-1878)", in *Turkish Studies: International Periodical for the Languages, Literature and History of Turkish or Turkic*, vol. 8, no. 2, Winter. Ankara, 2013, pp. 239-83

O'Mahony, O., "Some Henrietta Street Residents, 1730-1849", *Irish Georgian Society Quarterly Bulletin* 2, no. 2, April-June 1959, pp. 9-19

Osmanlı Arşivi (Ottoman Archives), Ecnebi Defteri 35/1 [England]

Page, W. (ed.), *A History of the County of Northampton*, vol. 3, 1930. Available at: <http://www.british-history.ac.uk/source.aspx?pubid=538>

Payne, T., *A Catalogue of a Very Large and Curious Collection of Books, in All Branches of Learning...*, London, 1754

Payne, T., *A Catalogue of a Large Collection of the Best Books, in All Parts of Learning, etc.*, London, 1761

Payne, T., *A Catalogue of a Large and Curious Collection of Books...*, London, 1762

Payne, T., *A Catalogue of Twenty Thousand Volumes...*, London, 1764

Payne, T., *A Catalogue of the Library of the Late John Grey, Esq...*, London, 1770

Payne, T., *A Catalogue of a Large and Curious Collection of Books...*, London, 1781

Payne, T., *A Catalogue of Valuable Books, in Various Languages, and in Every Branch of Science and Literature*, London, 1796

Perdrizet, P., "Les dossiers de P. J. Mariette sur Ba'Albek et Palmyre", *Revue des Études Anciennes* 3/3 (1901), 225-264

Perry, C., *A View of the Levant: Particularly of Syria, Egypt, and Greece. In Which Their Antiquities, Government, Politics, Maxims, Manners, and Customs, (with Many Other*

Circumstances and Contingencies) are Attempted to be Described and Treated on, London, 1743

Pitton de Tournefort, J., *A Voyage into the Levant Perform'd by Command of the Late French King, Translated by J. Ozell*, London, 1718

Pococke, R., *A Description of the East and of Some Other Countries, Volume I, Observations on Egypt*, London, 1743

Pococke, R., *A Description of the East and of Some Other Countries, Volume II Part I: Observations on Palaestine or the Holy Land, Syria, Mesopotamia, Cyprus and Candia; Volume II. Part II: Observations on the Islands of the Archipelago, Asia Minor, Thrace, Greece, and Some Other Parts of Europe*, London, 1745

Pococke, R. and Milles, J., *Inscriptionum Antiquarum Græcae. et Latinae Liber. Accedit, Numismatum Ptolemæorum, Imperatorum, Augustarum, et Cæsarum, in Ægypto Cusorum, e Scriniis Britannicis, Catalogus...*, London, 1752

Pococke, R., "An Account of some Antiquities found in Ireland; communicated by the Right Reverend Richard Pococke, Late Lord Bishop of Meath", *Archaeologia: or Miscellaneous Tracts Relating to Antiquity*, vol. 2 Society of Antiquaries of London, 1773, pp. 32-41

Pococke, R., *The Happiness of Doing Good: A Sermon* [*on Hebrews xiii. 16*] *Preached before the Right Hon. the Earl of Hertford, President; the Vice-Presidents, Treasurer, and Governors of the Magdalen-House Charity, on Thursday the 12th of March, 1761, at the Parish Church of St. Brides, Fleet-Street...*, London, 1761

Pococke, R., *A Sermon* [*on Thess. ii. 19, 20*] *Preached at Christ-Church, Dublin, on the 27th of June, 1762, before the Incorporated Society, for Promoting English Protestant Schools in Ireland...*, Dublin, 1762

Pomet, P., *A Compleat History of Druggs*, London, 1712

Quane, M., "Pococke School, Kilkenny", *The Journal of the Royal Society of Antiquaries of Ireland*, 1950, pp. 36-72

Ragette, F., *Baalbek*, Chatto & Windus, London, 1980

Rear, M., *William Barker Member of the Right Worshipful Levant Company 1731-1825: A Life in Smyrna*, 2015. Available at: <http://www.levantineheritage.com/pdf/Biography-of-William-Barker-Levant-Company-Merchant-Marjorie-Rear.pdf>.

Rennison, W., *Succession List of the Bishops, Cathedral and Parochial Clergy of the Dioceses of Waterford and Lismore, from the Earliest Times, Together with Some Hitherto Unpublished Records*, Ardmore, 1921

Robinson, G., *Travels in Palestine and Syria*, vol. 1, London, 1837

Russell, A., *The Natural History of Aleppo*, London 1794

Russell, C. F., *A History of King Edward VI School Southampton*, Cambridge, 1940

Schutte, O., *Repertorium der Nederlandse Vertegenwoordigers, Residerende in het Buitenland, 1584-1810*, Leiden, 1976

Shaw, T., *Travels or Observations Relating to Several Parts of Barbary and the Levant*, Oxford, 1738

Shaw, T., *A Supplement to a Book Entituled Travels, or Observations, &c. Wherein Some Objections, Lately Made Against it, Are Fully Considered and Answered: with Several Additional Remarks and Dissertations*, Oxford, 1746

Sotheby, S. L., *Catalogue of a Valuable and Interesting Assmblage of Autograph Letters …*, London, 1843

Sotheby, S. L., *Catalogue of the Remaining Portion of the Books and Manuscripts of a Distinguised Collector Deceased*, London, 1843

Sotheby, S. L., *Catalogue of the Valuable Topographical, Theological, Classical and Historical Library: with Some Highly Interesting Manuscripts*, London, 1843

Spon, J. and Wheler, G., *Voyage d'Italie, de Dalmatie, de Grèce, et du Levant, Fait aux Anées 1675 & 1676*, The Hague, 1724

Stanford, W.B., *Ireland and the Classical Tradition*, Dublin, 1976

Stanford, W.B., "The Manuscripts of Lord Charlemont's Eastern Travels", in *Proceedings of the Royal Irish Academy*, Section C, vol. 80, no. 5, 1980, pp. 69-90

Stanford, W.B., and Finopoulos, E.J. (eds), *The Travels of Lord Charlemont in Greece & Turkey 1749*, Trigraph, London, 1984

Stokes, G.T. (ed.), *Pococke's Tour in Ireland in 1753*, Dublin, 1891

Stoney, S., *Some old Annals of the Stoney Family … Compiled by Major S. … Illustrated by C. Stoney of Portland*, London, 1879

Sullivan, M.G., "Wale, Samuel (1721?-1786)", in *Oxford Dictionary of National Biography*, Oxford University Press, 2004. Available at: <https://doi.org/10.1093/ref:odnb/28453>

Tappe, E.D., "The Greek College at Oxford, 1699-1705", in Oxoniensia 19, 1954, pp. 92-111

The Aldine Magazine of Biography, Criticism and the Arts, vol. 1, London, 1839.

The British Chronologist, vol. 2, London, 1775

The Clergy of the Church of England Database 1540-1835. Available at: <https://theclergy database.org.uk/>

The Integrated Conservation Group, "The Bishop's Palace, Kilkenny", *Old Kilkenny Review, Journal of Kilkenny Archaeological Society*, 2003, pp. 30-53

The Peerage. Available at: <http://www.thepeerage.com/p13998.htm>

The Scots Magazine: Containing a General View of the Religion, Politicks, Entertainment, &c. in Great Britain and a Succinct Account of publick Affairs Foreign and Domestick for the Year MDCCXLIII, vol. 5, Edinburgh, 1743

Thorpe, T., *Manuscripts, upon Papyrus, Vellum and Paper, in Various Languages. A Catalogue of a Most Valuable Collection. Recently Purchased from the Libraries of the Late Right Hon. William Berwick. The Late Thomas Milles, D.D. Bishop of Waterford and Lismore. The Late Very Reverend Jeremiah Milles, D.D. Dean of Exeter, and President of the Antiquarian Society. The Late Learned Traveller, Richard Pococke, L.L.D. Bishop of Meath and Ossory, Author of the Description of the East, etc…*, London, 1843

Urban, S., *The Gentleman's Magazine: and Historical Chronicle. Volume* XIII. *For the Year* M.DCC.XLIII, London, 1743

Urban, S., *The Gentleman's Magazine, Volume* XX, *New Series, July to December Inclusive*, London, 1843

Vlami, D., "Entrepreneurship and Relational Capital in a Levantine Context: Bartholomew Edward Abott, the 'Father of the Levant Company' in Thessaloniki (eighteenth – nineteenth Centuries)", *The Historical Review/La Revue Historique*, Institute for Neohellenic Research, vol. 6, 2009, pp. 129-64

Waddell, J., *Foundation Myths: The Beginnings of Irish Archaeology*, Wordwell Ltd., Dublin, 2005

Walpole, H. and Montagu, G., *The Correspondence of Horace Walpole, with George Montagu, esq.,* [*and others*], New Edition, London, 1837

Warburton, W., *Remarks on Several Occasional Reflections in Answer to The Rev. Dr Middleton, Dr Pococke... and Others... Serving to Explain and Justify Divers Passages, in the Divine Legation, Objected to by Those Learned Writers...*, London, 1744

Ware, J., *The Whole Works of Sir James Ware Concerning Ireland, Revised and Improved by Walter Harris*, Dublin, 1764

Warren, P., "Early Travellers from Britain and Ireland", in D. Huxley (ed.), *Cretan Quests: British Explorers, Excavators and Historians,* British School at Athens, 2000, pp. 1-8

Wells, J., *Wadham College*, London, 1898

White, B., *A Catalogue of the Entire Libraries of the Rev. Mr. Joseph Spence...*, London, 1769

Wilson, D., *List of British Consular Officials in the Ottoman Empire and Its former Territories, from the Sixteenth Century to about 1860*, 2011 Available at: <http://www.levantineheritage.com/pdf/List_of_British_Consular_Officials_Turkey(1581-1860)-D_Wilson.pdf>

Wilson, D., "More Notes on Richard Pococke's Letters from Abroad, vol. 3", unpublished document, 2014 (document supplied by author)

Wilson, D., *Levant Company: Admissions of Freemen and Grants of Liberty of Trade, 1695-1824*, 2017. Available at: <http://www.levantineheritage.com/pdf/List_of_British_Consular_Officials_Turkey(1581-1860)-D_Wilson.pdf>

Windham, W., *An Account of the Glaciers or Ice Alps in Savoy, in Two Letters, One from an English Gentleman to His Friend at Geneva; The Other from Peter Martel ... Illustrated with a Map, and Two Views of the Place, &c as Laid before the Royal Society*, London, 1744

Wood, A.C., *A History of the Levant Company*, Oxford, 1935

Wood, R., *An Essay on the Original Genius and Writings of Homer,* London, 1775

Wood, R., *The Ruins of Balbec, Otherwise Heliopolis in Coelosyria*, London, 1757

Index of Selected People and Places

Explanation of the Index

People: this index gives the correct/official names and titles of individuals encountered by Pococke during his travels, with his own idiosyncratic spelling in brackets. It also states the nationality of non-British people, if known, and either the place(s) at which Pococke met them or the centre at which they were normally based.

The index also lists the names of other people mentioned in the correspondence, such as Pococke's friends and acquaintances in England and Ireland. These entries give their country (or occasionally their city) of residence and their occupation, if relevant.

Places: the index lists a selection of the places visited by Pococke during his travels. Each place is listed by its modern name, if known, with any contemporary or ancient versions of the name, or any misspellings, appearing in brackets. Most entries relate to his experiences in these centres; for any general accounts of the main centres visited, the reader is directed to the chapters containing his "Accounts".

www.ingramcontent.com/pod-product-compliance
Lightning Source LLC
LaVergne TN
LVHW010638300926
849161LV00005B/970